THE PHARMACOLOGY OF

Chinese Herbs

THE PHARMACOLOGY OF

Chinese Herbs

KEE CHANG HUANG
Emeritus Professor
Pharmacology and Toxicology
University of Louisville
Louisville, Kentucky

CRC Press
Boca Raton Ann Arbor London Tokyo

Library of Congress Cataloging-in-Publication Data

Huang, K. C. (Kee Chang), 1917–
The pharmacology of Chinese herbs/Kee Chang Huang.
p. cm.
Includes bibliographical references and index.
ISBN 0-8493-4915-X
1. Herbs—Therapeutic use. 2. Medicine, Chinese. I. Title.
[DNLM: 1. Drugs, Chinese Herbal—pharmacology. 2. Drugs, Chinese Herbal—therapeutic use. 3. Medicine, Chinese Traditional. 4. Plants, Medicinal—China. QV 770 JC6 H835p]
RM666.H33H83 1993
615'.321—dc20
DNLM/DLC
For Library of Congress 92-11376
CIP

Direct all inquiries to CRC Press, Inc., 2000 Corporate Blvd., N.W., Boca Raton, Florida, 33431.

International Standard Book Number 0-8493-4915-X

Library of Congress Card Number 92-11376

Printed in the United States of America 2 3 4 5 6 7 8 9 0

Printed on acid-free paper

Chang-Shou Jang, M.D., Ph.D. (1906–1967)

Professor of Pharmacology
Shanghai Medical College, China

*This book is dedicated to my mentor, Professor Chang-Shou Jang,
who devoted his life to research on Chinese herbs
and was wrongly persecuted during the Cultural Revolution.*

FOREWORD

This encyclopedic compendium could not have been written by a more appropriate scholar than K.C. Huang. It has been my pleasure to be his colleague and friend for many years and to follow the evolution of this book from its inception more than five years ago.

K.C. is recognized internationally as a teacher and researcher. He has been invited to lecture at many schools and spent extended periods of time teaching in locations such as China, Germany, Sudan, Kuwait, and Egypt. His teaching is inspirational; for the first three years in which his lectures were presented, he was the unanimous choice for the University of Louisville medical students' Golden Apple Teaching award.

He is the author of many research publications on a wide variety of topics in pharmacology. His special interest was in renal and intestinal transport of amino acids and sugars, as well as electrolyte and fluid balance. His book on pharmacology for medical students was well received and used by many students both in the U.S. and other countries.

There are at present only a handful of examples of Western drugs which have been developed from Chinese herbs. One classic example is the 1923 isolation of ephedrine from *ma huang,* after K.K. Chen had shown this herb's potency as a drug. The problem, of course, has been in identifying the truly useful and potent herbs in the armamentarium of traditional Chinese herbs. The list of supposedly useful herbs is endless, but few have been verified by objective evaluation. In many cases, scientific trials reveal the herb to be nothing more than a placebo.

This treatise on the pharmacology of Chinese herbs has been in K.C.'s mind for a long time. He is uniquely qualified to evaluate these herbs from both the traditional Chinese and Western points of view. His early medical education gave him familiarity with the herbs and their uses, while throughout his career he has had occasion to ponder the possible correlations of Western pharmacology with the teachings of traditional Chinese medicine. He has carefully evaluated a long list of Chinese herbs and included in this book only those he considers to have potential for further investigation by objective research. The result is a definitive treatise on the subject which will be useful to medical researchers as well as physicians, historians, pharmacologists, and students.

William J. Waddell, M.D.
Professor and Chairman
Department of Pharmacology and Toxicology
University of Louisville

PREFACE

As a pharmacologist trained primarily in Western traditions, I am constantly astounded by the abundance and richness of the herbal lore available in Chinese medicine. Records cataloguing the healing powers of natural substances — plants, chemicals, and animal byproducts — date back more than four thousand years. Even more remarkable is the fact that much of this knowledge is still in wide usage, neither obsolete nor forgotten.

Over half of the Chinese population still relies on herbal prescriptions rather than Western medicine. After the communist revolution of 1949, the Chinese government reversed the Kuo Min Tong government's ban on traditional Chinese medicine, establishing more than ten traditional medical colleges and institutes whose role was to train physicians and further investigate the uses of herbs. Today, even Western hospitals in China are equipped with apothecaries which dispense Chinese herbs upon request.

While there is no denying the effectiveness of traditional Chinese medicine, the roots of this knowledge are largely lost in superstition and folklore. Even recent research in herbal medicine is somewhat flawed, lacking any critical review. This contrasts sharply with Western medicine, which emphasizes a healing system based on defined scientific rules and technology.

The purpose of this book is not to debate the value of Eastern (traditional) or Western medicine. Instead, I hope to bring together ancient Chinese herbal lore and modern Western scientific methods. Within this manuscript, herbs are classified based on their therapeutic value, rather than pharmacognostically. Each herbal medicine is described in terms of its chemical composition, pharmacological action, toxicity, and therapeutic uses. It is my hope that this work will be of benefit to colleagues who may one day apply the successes of Western research, particularly in emerging gene-modification technology, to develop more effective treatments from Chinese herbs. Of special interest may be the employment of modern medical chemical techniques to modify the structures of certain purified herbal ingredients into better agents with higher efficacy and activity.

I would like to acknowledge my gratitude for the contributions of the following:

W. Michael Williams, M.D., Ph.D., Professor of Pharmacology and Toxicology, University of Louisville, Health Science Center, for his inspiring discussions throughout the preparation of this manuscript;

The librarians of the University of Louisville Medical Science Library, the Tianjin Medical College Library (People's Republic of China), the Tianjin Traditional Chinese Medical College Library (PRC), and the Beijing Traditional Chinese Medical College Library (PRC) for their unstinting help in the search for literature;

Professor Ying Li, Department of Pharmacognosy, Shanghai Medical University, School of Pharmacy (PRC), for her help in obtaining pictures of certain herbs;

The authors who kindly allowed reproduction of figures and tables from their texts;

And my wife and children, who encouraged and supported me throughout the process of preparing this work. Special thanks go to my youngest daughter, Karen Tsin Huang Soo, who patiently edited the manuscript prior to publication.

KCH (1992)

THE AUTHOR

Kee Chang Huang, M.D., Ph.D., was born in Canton, China, in 1917. He attended Sun Yat-sen University, receiving his medical degree in 1940. He was a research fellow in pharmacology at the Chinese National Institute of Health for six years before joining the faculty of the National Shanghai Medical School in 1946.

In 1949, he left China to enter graduate school at the Columbia University College of Physicians and Surgeons. After receiving a Ph.D. in Physiology in 1953, he was recruited by the University of Louisville as a Research Associate in Pharmacology, rising through the ranks to become Professor of Pharmacology in 1963. During his tenure he was awarded the Fulbright Professorship twice and the Distinguished Professorship by the University of Louisville and the Tianjin Medical College of China. He is the author of two books, *Absorption, Distribution, Transformation and Excretion of Drugs* and *Outline of Pharmacology,* and the author of many research papers. Although he retired from the University in 1989, he remains active in research.

TABLE OF CONTENTS

Section I
Introduction

INTRODUCTION

The Chinese are proud of their heritage, particularly the ingenuity shown in their early scientific and cultural achievements. One of their most remarkable contributions to civilization undoubtedly is the abundance of information documenting the uses of natural substances, plants, chemicals, and animal products, in treating illnesses. Medical and pharmaceutical literature, either commissioned by the government or written privately as scholarly exercises, has existed in China for more than 4000 years.

Today this knowledge is still in use, neither obsolete nor forgotten. Western medicine and drugs, which were brought to China over 2 centuries ago, have replaced some but not all Chinese herbs in medical treatment. Over half of the Chinese population still uses traditional herbal prescriptions, particularly when Western medicines do not produce the desired results. It is no surprise that most hospitals in China, whether Western or traditional, are equipped with traditional herbal apothecaries to distribute Chinese herbs upon request.

However, the Chinese can still learn much from occidental methods. Foremost is the fact that a healing system based on scientific method and technology is superior to the superstition-bound medicines of antiquity. Secondly, Western successes in medicinal chemistry, structure-activity studies, and genetic modification have opened a promising route to the development of better and more effective drugs from Chinese herbs.

This book is not intended to glorify China's past, nor to examine myths of herbal rejuvenation or resurrection of the dying (see Unschuld,[14] pp. 230–231). Instead, we will examine scientific materials on herbal medicines which have been published in China during the past 4 decades. Since the Communist revolution in 1949 and the subsequent establishment of more than 10 traditional medical colleges and institutes in China researchers have worked to isolate and identify the ingredients of many important herbs, testing their effectiveness on hospitalized patients. Volumes of material on these cases have been published in the Chinese language, but have seldom been critically reviewed.

This book will introduce the pharmacology and uses of certain Chinese herbs whose values have been analyzed convincingly by modern scientific methods. To unify Western and traditional Chinese views on these drugs, we have classified the herbs based on their therapeutic value, rather than pharmacognostically. We describe each herbal medicine in terms of its chemical composition, pharmacological action, toxicity, and therapeutic uses.

Some Chinese Medical Terminology

Perhaps the most difficult thing for non-Chinese to understand about Chinese herbal medicine is the terminology used to describe symptoms for which various herbs were effective. Nearly all symptomatic diagnoses are based in the philosophy of *yin* and *yang*, the two forces which the Chinese believe control the workings of the universe.

Yin represents the feminine side of nature, encompassing darkness, tranquility, depth, cold, and wetness; the earth, the moon, and water are all *yin* elements. *Yang* represents a masculine principle, encompassing light, activity, height, heat, and dryness; heaven, the sun, and fire are *yang* elements.

Although *yin* is commonly interpreted to be a negative force, while *yang* represents a positive force, the philosophy of *yin* and *yang* does not parallel Western philosophy's dualism between good and evil. Instead, the two are complementary, and neither can exist without the other. Thus, Chinese medicine attempts to achieve a balance between *yin* and *yang*.

Associated with *yin* and *yang* are the terms "coolness" and "heat". Both are conditions which are generally believed to stem from types of foods which affect the body in different ways. ''Heat'' refers to a condition in which the body consumes large amounts of energy, thus generating heat. ''Heating'' foods include nearly all fried foods, red meats, and vegetables and fruits with high fat content, such as avocado and the durian fruit of southeast Asia. Surprisingly, fruits such as lychee and mangoes are also considered to be ''heating'', despite their refreshing nature. ''Cooling'' foods include most fruits and vegetables, and many herbs. Excessive ''heat'' requires treatment with ''cooling'' herbs, while excessive ''coolness'' requires treatment with ''heating'' herbs.

Another term used extensively in Chinese medical texts is the word *qi* (气). This refers to the total energy of the body. Again, herbs are used to achieve an optimum balance of *qi*; that balance is believed to manifest itself in the overall health and vigor of the patient.

Finally, traditional Chinese medicine made reference to a condition called ''wind''. To some extent, ''wind'' could be described as a gassy condition developing from acid indigestion, but the Chinese imbued the term with additional meaning. Western folklore commonly speaks of people who could foretell bad weather by ''feeling it in their bones''. That subjective feeling is also part of that which the Chinese call ''wind''. It was believed that excessive ''wind'' would travel through the body, causing indigestion and a rheumatic ache of the joints; the latter symptom is also referred to as ''dampness''.

Scope and Structure

The earliest known work on Chinese herbs is *Sheng Nung Ben Cao Chien*, or *The Herbal Classic of the Divine Plowman*. This pharmacopoeia was written anonymously in approximately 100 B.C., predating Dioscorides' *De Materia Medica*. According to this work, *Sheng Nung*, the Divine Plowman, tested and recommended a total of 365 herbs — one for each day of the year.

The herbs were categorized into three classes, based on ''a macrocosmic concept of heaven, man, and earth.'' These included the upper, middle, and lower classes, which were then subdivided according to whether the ''herb'' originated from living material (animal and plant) or from the earth (mineral). The upper class of herbs was supposed to sustain life (*mien*, 命); the middle one to smooth the mental state (*shien*, 性); and the lower class to cure illnesses (*pien*, 病). For example, ginseng, licorice root, chrysanthemum, and *Wu wei zi* (五味子, an herb considered to have a potent sperm-stimulating effect) belonged to the upper class. *Ma Huang* (the ephedra), *Dong Gui* (angelic root), *Wu Jia Pi* (*Acanthopanax* bark), *Fang Gi* (tetrandra root), and *Xi Jiao* (rhinoceros horn, an aphrodisiac) were listed in the middle class. Iron, lead, lime, rheum root, croton oil, and *Chang Shan* (*Dichroa* root) belonged to the lower class.

As knowledge of medicinal herbs increased through the years, this system of classification became meaningless. Instead, most of the written works on herbs created during the last 4 centuries were based on the growth, nature, and characteristics of herbs, namely, their pharmacognostic properties. Such written works were typified by the monumental work of Li Shih-Chen. After more than 30 years of study, collection, and field investigation, he finally published the famous pharmacopoeia *Ben Cao Kong Mu* (本草纲目). Li's work consists of a total of 52 volumes, comprising 1,898 herbs or agents and 11,096 recorded prescriptions.

Li classified these agents on the basis of their nature, shade, and growth; they were further subdivided based on their mineral, botanical, or animal origins. ''Herbs'' originating from plants were subdivided still more according to their habitats, shades, and the part of the plant that was used. This work is held in high esteem among Chinese medical practitioners, and was the first Chinese medical work to reach Europe, where it was well received.

The prescription format used by traditional Chinese medical practitioners (Tai Fu, 大夫, or *Long Chong*, 郎中) has remained unchanged for thousands of years. They tend to prescribe four or more herbs together, believing that drug interactions would synergize the major action of the principal herb or smooth its possible side effects. A rhyme about this prescription of multiple herbs has become a well accepted motto in the practice of herbal medicine: *Yi Jun, Er Chen, San Zuo, Si Shi* (一君,二臣,三佐,四使), or ''One ruler, two ministers, three aides, and four guides''. The ruler drug is the active principle, while the other nine are helpers of varying degrees of strength.

Such combinations are prescribed based on the assumption that supporting herbs can help or aid the effectiveness of the principal one. However, the possible incompatibility of these herbs is never considered. Frequently, the reasons for adding a vehicle (one of the ''four guides'' or a *Shi*) to a preparation is far beyond scientific comprehension. For example, many prescriptions included the use of one or two teaspoons of a child's urine, which was considered to be helpful in strengthening of effect of the principal herb. Recently, Lang et al.[11] published a book on Chinese basic herbs and prescriptions, describing 700 different preparations, each made up of three or more herbs. Examples of such combination prescriptions can also be found in other references.[10,12,15] It is beyond the scope of this book to assess the value of these combination prescriptions.

Another problem in studying Chinese herbs is that numerous herbs bearing the same name are actually different. Conversely, one herb may frequently be known by several names. Additional complexity is added because of the lack of an alphabet in the Chinese language: dictionaries and literature index their contents on the basis of the number of strokes contained in each word, while the phonetic translation of a Chinese word also varies. In order to eliminate such confusion, we have used the recently published Chinese-English Medical Dictionary (*Han Yin Yi Xue Tai Zi Dian* 汉英医学大詞典, Peoples Health Publisher, Beijing, 1987) as a guideline for translation. For indexing, we have opted to use the botanical name of each herb, in alphabetical order. The Appendix, pages 365 through 369, provides a cross-reference by the phonetic Chinese names.

One of the major dilemmas faced while preparing this book was the question of which herbs to include, and which should be considered therapeutically valuable. Many of the herbs listed in the *Sheng Nun Ben Cao* and the *Ben Cao Kong Mu* belong more appropriately in biology texts, such as pine wood, mulberry, elephant tusks, and camel. Others are more suitable for a cookbook or food manual, such as vegetables, fish, fruit, rice, chicken, duck, crabs, beef, pork, and lamb. Some medicinal herbs are of dubious value and are not discussed in this work: swallow's nest (*Yen Ju Cao*), rafter dust (*Liang Shang Ch'en*), child's urine, and human feces are among these. Certain plants imported from foreign lands, such as the opium poppy, are optionally omitted here because their pharmacological actions and therapeutical uses are well described in all Western pharmacological texts. Furthermore, *Cannabis* (or hemp, marihuana), grown in Southwest China and worldwide, was never mentioned in the Chinese pharmacopoeia as an hallucinogenic or psychotomimetic agent, but its semen was used in Chinese medicine as purgative herb.

The second issue requiring resolution was the question of how to handle combination formulas. Several recent Chinese publications have discussed hundreds of combination formulas, some reputed to be secrete formulas from ancient times and others from ''renowned practitioners''. Lang listed more than 700 such prescriptions in his book.[11] And recently, an international bulletin published a detailed list of 711 Chinese herbal combinations, describing their ingredients and therapeutic uses very explicitly.[12] Yet, with all of these combinations, how do we decide whether the components of combination formulas will produce a synergistic effect or an antagonistic effect? Drug interaction is still the primary issue for any physician when prescribing drugs to patients. In principle, we should avoid placing two or more drugs together if we don't know what kind of interaction a combination would produce. Because of this, the author decided not to include these combination prescriptions in this book.

To simplify the task of selecting material, the author decided to use Chinese pharmacological works published since 1950 — after the proletarian revolution — as a guideline, and to categorize the herbs according to their therapeutic uses. Sometimes, this grouping is imperfect and can be confusing. For example, several herbs were claimed to be antipyretic but are also antibacterial agents: in ancient times, physicians were not aware that bacterial or viral infections cause fever, and that herbs which kill the infecting microorganism would produce, indirectly, an antipyretic effect.

This author spent his early professional life researching Chinese herbs under Jang Chang-Shou.[2] He specifically investigated the pharmacological action of *Jai Zhu Tao* (a cardiac glucoside), *Ya Dan Zi* (an antiamebial agent), and *Chang Shan* (an antimalarial agent). He soon learned that most early studies on Chinese herbs were based on the use of their extracts or other very crude and impure forms. When injected into animals, these preparations commonly produced a hypotensive effect. Yet no follow-up studies were made to elucidate whether the hypotensive effect was due to the active principle or to impurities found in the preparations. Here again, this book does not want to prejudge itself, but must quote the results of early Chinese studies based on the use of crude extracts.

As mentioned earlier, the materials used for this book were abstracted mainly from several Chinese works which were published after 1950. Since they are used frequently and will be quoted in nearly every chapter, these references are listed below, rather than as general references in each chapter. Additional specific literature is referenced in the corresponding chapter.

Readers interested in the historical backgrounds of Chinese herbs are encouraged to see Unschuld's work, *Medicine in China*.[14] In his book, he introduces much old Chinese medical literature which has been preserved either in archives or museums. Although he seldom discusses individual herbs and does not validate the medical value of the formulas, his work is a valuable reference on the history of these preparations.

GENERAL REFERENCES

Reference Books (in Chinese) Used Repeatedly in the Text

1. **Hsu, Hong-yen,** *Zhong Yao Chien Fan Jui Tsin Yian Jiu* [*The Progress on Chinese Herbal Study*], National Research Institute of Chinese Medicine, Taiwan, 1968.
2. **Jang, Chang-shou,** *Xian Dai Zhong Yao Yian Jiu* [*Recent Research on Chinese Herbs*], China Science Library and Equipment Co., Shanghai, 1954.
3. **Koo, Wen-wah,** *Zhong Yao Gi Yian Jiu* [*A Study of Chinese Drugs,*] Taiwan University Medical College Publisher, Taiwan, 1967.
4. **Ling, I-qi,** *Zhong Cao Yao Yian Jiu* [*Chinese Herbal Studies*], Shanghai Science Technology Publisher, Shanghai, 1984.
5. **Nanjing Pharmacy College, Ed.,** *Zhong Cao Yao Xue* [*Chinese Herbs Studies,* 3 vol.] Jiangsu Science and Technology Publisher, Jiangsu, 1980.
6. **Szechuan Medical College, Ed.,** *Zhong Cao Yao Zhi Yao Mai Xue* [*Pharmacy of Chinese Herbs*], People's Public Health Publisher, Beijing, 1978.
7. **Wang, Yue-seng et al., Eds.,** *Zhong Yao Yao Lee Yue Ying Wan* [*The Pharmacology of Chinese Herbs and Their Uses*], People's Public Health Publisher, Beijing, 1983.

8. **Wu, Po-jie,** *Zhong Yao Zhi Yao Lee Xue [Pharmacology of Chinese Herbs],* People's Public Health Publisher, Beijing, 1982.
9. **Zhou, Jiang-huang et al., Eds.,** *Zhong Yao Yao Lee Xue* [*Chinese Herbs Pharmacology*], Shanghai Science Technology Publisher, Shanghai, 1986.

Other Reference Books

10. **Beijing Traditional Chinese Medical College and Hospital, Ed.,** *Xue Shu Luen Wen Huei Bain* [*A Collection of Herbal Prescriptions*], Beijing Traditional Chinese Medical College Publisher, Beijing, 1978–1981.
11. **Lang, Fon-nan et al., Eds.,** *Zhong Kou Ji Ben Zhong Cheng Yao* [*Chinese Basic Herbs and Prescriptions*], People's Health Publisher, Beijing, 1988.
12. *Oriental Healing Arts International Bulletin,* 11(4), 215–345, 1986.
13. *Oriental Materia Medica, A Concise Guide,* Oriental Healing Arts Institute, Long Beach, CA, 1986.
14. **Unschuld, Paul U.,** *Medicine in China. A History of Pharmaceutics,* University of California Press, Berkeley, CA, 1986.
15. **Wu, Yeo-chin:** *Yi Fon Chin Ji* [*New Explanation of Medical Formulae*], Shanghai Science Technology Publisher, Shanghai, 1980.

Section II
History of Chinese Medicine

A BRIEF HISTORY OF CHINESE MEDICINE

PREHISTORIC ERA (BEFORE 22ND CENTURY B.C.)

The archeological excavation of the "Peking Man" in northern China has suggested that the origin of the Chinese occurred about half a million years ago. Primitive man lived in caves and showed signs of making fire. And, where there is life, there will be death and illness. Man not only lived and worked hard, but had to struggle against natural enemies and the enemy within — disease. The "Peking Man" excavation revealed that over one third of the people died in prepubescence and only 2.6% reached their fifth or sixth decade.

In another location of the excavation, many bone fractures, regenerated broken bones, and palate abnormalities were found among the fossil bodies. To combat sickness, primitive man learned many things. He made stone needles for treatment and for use in minor operations. He skillfully selected plants and minerals to ease suffering, and began to avoid things found to be toxic after ingestion. In some cases, man may have learned these skills from the wild animals in his surroundings.[2]

In this early age, man eventually became tired of hunting for his food and started to identify edible plants which could be replanted near his dwellings. He developed an agrarian society and enjoyed a more comfortable life. Accidentally, he found that some of the plants he ate or chewed could help in curing illnesses. The Chinese called this era the *Sheng Nung* Period (神农時代, ca. 2800 B.C.). In that time, man had already learned to make pottery and had developed sign language.

It was recorded in later documents that *Sheng Nung,* the Divine Plowman, plowed the fields and established agriculture for his people. He tasted hundreds of plants and identified those helpful in curing certain diseases. From these experiments, he taught his people how to use these herbs and to avoid poisonous substances. Some people reportedly followed his advice and lived beyond 70 years of age. His experiences and works on drugs were noted in the book *Sheng Nung Ben Cao Chien* (神农本草经), or *The Herbal Classic of the Divine Plowman*, which was recorded for posterity by an anonymous scribe in approximately 101 B.C.

China's first emperor, *Huang Ti* (黄帝) or the Yellow Emperor, ascended the throne in 2697 B.C., approximately 100 years after Sheng Nung. He and his cabinet members, Gi Po (岐伯) and Lei Kun (雷公), developed techniques to diagnose diseases and to use herbs in their treatment. A well-reputed medical writing called *Huang Ti Nei Chien* (黄帝内经) documenting these techniques was recorded in approximately 100 B.C.

The introduction of fire and wine or alcohol advanced man to an even better life. Fire was not only used to cook and make food more tasty and hygienic, but was used to modify various herbs' characteristics and properties; man also learned to use fire to dry plants, instead of relying solely on the sun. Wine or *jiu* (酉), as it is called in Chinese, was used not only as a beverage, but as a vehicle to preserve drug activity. The Chinese also learned to fortify drug action by extracting drugs with wine and then condensing the extract. The word "medicine" in Chinese, *yi* (醫), has as one of its components the term for wine, suggesting that in early ages, wine was one of the important components of medicines.

SHANG-ZHOU DYNASTIES (2000 to 220 B.C.)

In this period, the knowledge of herbs and their uses in treating illnesses were far more advanced than in the earlier age. In the 1973 excavation of a ruined Shang Dynasty village in Kuo Chien County, Hepei Province, more than 30 kinds of plant seeds were found.

Among them were the seeds of almonds, peaches, and *Yonk Li Yen* (郁李仁), which were commonly used as medical herbs according to the pharmaceutical records.

Yi Yuen (伊尹) was the talented and intelligent chef of Emperor Sheng Tong Wong (1600 B.C.). He helped the Emperor to conquer and unify the country and was later appointed prime minister. He shared his interest in cooking with the common people and taught them how to select plants which are edible and useful for medicine. He was the first to introduce the technique of decoction in preparing herbs, and adapted the teachings of *The Herbal Classic of the Divine Plowman* in making certain formulas, such as herbal soups and extracts.

One of the earliest recorded Chinese history texts, *Si Gi* (史記), which was written in the early Han Dynasty by See Ma Chin, mentions Pin Choi (扁鵲, 407 to 310 B.C.) as the pioneer of sphygmology. He was the most famous medical practitioner of the Qin Period, and traveled around the country helping the sick. He diagnosed illnesses by detecting abnormalities of the pulse. In addition to his medical practice, he wrote several books on diagnosis and sphygmology, such as the 9 volume *Pin Choi Nai Chien* (扁鵲内经) and the 12-volume *Wal Chien* (外经). In his books, he denounced the beliefs and practices of witch doctors.

Some of the first real records of Chinese history were preserved from the Zhou Dynasty (1100 to 207 B.C.). In one record, Zhou Li (周礼) mentions some medical practices, noting that " . . . with the help of five tastes, five seeds, and five drugs, we can ease the ill . . . ." Zhou Li is also credited with formally grouping living materials into two categories — plants and animals.

The *San Hai Chien* (山海经) was written in the period of the East Zhou Dynasty (770 to 256 B.C.). This work recorded the medical uses of plants, animals, and minerals, and the techniques for using them. It also stated the purposes behind the use of those herbs, noting that they would maintain health and reproduction, and serve as cosmetics and insecticides, for internal or external application.

The *Si Chien* (詩经) was written in the "Spring-Autumn" and the "Warring States" Periods (475 to 221 B.C.). It mentioned more than 100 herbs and their uses.

QIN-HAN DYNASTIES (206 B.C. to 265 A.D.)

It is believed that most pharmaceutical and medical writings recorded earlier than this period were not authentic, but were actually written anonymously during the Qin and Han dynasties. This is understandable, given that paper was invented in the early Han era, and that printing was developed much later. Thus, any writings prior to the Qin and Han dynasties were not maintained as permanent records and must have been copied to paper only by later generations. Discovering who the actual authors were is also made difficult because of the common practice of attributing works to famous authorities and heroes, rather than publishing them under the name of the real author. Therefore, it is not known exactly who wrote *The Herbal Classic of the Divine Plowman,* nor the *Huang Ti Nai Chien.*

The *Herbal Classic* contains three volumes describing 365 drugs — one for each day of the year. Among them, 252 are derived from plants, 67 from animals, and 46 from minerals. The main contents include the basic theory of materia medica, the so called *Yi Jun, Er Chen, San Zuo, Si Shi* (一君, 二臣, 三佐, 四使), which stated that a prescription should contain one major active drug and several supporting ones, to smooth its adverse effects or synergize its pharmacodynamic actions.

The book describes the properties of each drug in detail, including its taste, source of collection, pharmacological action, and therapeutic uses. The book also outlines approximately 170 kinds of diseases which are effectively treated with these 365 herbs. Some of the descriptions have been proven to be correct and accurate by recent scientific investigations. For example, ginseng (*Panax ginseng*) was said to "strengthen mentality, relax stress, and calm the nerves . . . ." It is now known that the active principles of ginseng, panaxosides, do exert effects on the central nervous system and the cardiovascular system, and stimulate the adenocorticotropic hormone-cortisol axis. Likewise, *Ma Huang* (the *Ephedra*, 麻黄) was said to increase sweating and to stop asthmatic cough. *Chang San* (*Dichroa*, 常山) was described as an antimalarial agent. Algae were used as antigoiter agents.

The *Herbal Classic* categorizes the 365 drugs into three classes: upper, middle, and lower class. The upper class, totaling 120, strengthen the body, are harmless, and support life. The middle class, also 120 in number, do not produce harm if the dose is measured carefully. The lower class, 125 in all, cure illnesses but would be harmful if taken excessively or too often. Generally, each class of herb was prescribed according to the corresponding ranking of patients within the social hierarchy.

Zhang Chong Jin (張仲景), born in the late Han Dynasty (ca. 2nd to 3rd Century A.D.), was China's most popular physician, standing in great repute even today. In his time, the late Han period, the country suffered from continual war and epidemics. Zhang devoted his time to medicine and offered his services to the poor. He wrote several books about herbal therapy and typhoid, calling the latter "harmful cold" (傷寒). Mandatory reading for all Chinese traditional medical practitioners is his 16 volumes on *Diagnosis and Treatment*, which includes 397 treatment methods, rules, and formulas.

Hua Tau (華陀 , 145 to 203 A.D.), the most famous internist and surgeon in recorded Chinese history, left many formulas and manuscripts on healing. One of the most famous, which remained in use for a thousand years, described the technique of general anesthesia prior to operations. He used these techniques when operating on his patients, performing such procedures as opening the abdomen, amputation of extremities, and cutting out abscesses. He also made great contributions to acupuncture and sphygmology. In China, many clinics today still follow his teachings, hanging mottos on their walls which proclaim: "The Reborn Hua Tau" (華陀再生).

Taoism, still one of the primary religions in China, flourished in the Han Dynasty. Priests developed the science of alchemy to prepare pills and drugs for sustaining life. The alchemist Yi Po Yang (魏伯阳) wrote a book on the methods of alchemy, instructing how to purify metals and to make pills. He described the characteristics of mercury, the making of mercurial products, and the use of mercury products in the treatment of skin diseases. This book is considered to be the oldest alchemy publication in the world.

JIAN DYNASTY AND THE SOUTH-NORTH DIVISION (265 to 560 A.D.)

Tou Yu Gin (陶弘景, 452 to 536 A.D.) was the most famous medical herbalist of this time. He authored the *Ben Cao Chien Collection* (本草经集註), a collection of herbal classics, and the *Yung San Chian* (養生经), a work about how to sustain longevity. The first of these is a seven-volume collection of formulas from prominent physicians of the early Han and Jian periods. It describes the uses of 730 herbs, substantially improving the materials and contents of the *The Herbal Classic of the Divine Plowman*. It not only grouped the herbs into three categories (the upper, middle, and lower classes), but also subgrouped them according to their habitat and nature. It describes the characteristics of each herb, its origin, and the method of collection. This book also contains contraindications and antidotes for certain toxic substances.

Guo Hong (葛洪, 283 to 363 A.D.) was an alchemist who wrote the book *Bou Po Shi* (抱樸子), an alchemical introduction to medical chemistry. He explained the techniques for preparing some medical agents from the interactions of two or more compounds. He separated mercury and sulfur from mercury sulfite, HgS, and prepared lead carbonate. Guo Hong also contributed greatly to medical writing. He wrote the three-volume *Yue Han Fan* (玉函方), a collection of many classical formulas used in folk medicine. For example, he described the use of areca seed as an insecticide, the treatment of goiter with algae tincture, and the use of a rabid dog's brain to treat rabies-bite wound infections.

TSUAI-TANG DYNASTIES (580 to 900 A.D.)

It was claimed that the world's first pharmacopoeia was written during the Tang Dynasty. The Emperor commissioned Li Chi (李勣, 594 to 669 A.D.), Su Chiang (苏敬), and 20 other scholars to edit a complete pharmacopoeia of Chinese herbs. This work was named *Tang Hsin Hsu Ben Cao* (唐新修本草, 659 A.D.), or "Tang's Newly Revised Materia Medica". It contains 54 chapters and 850 drug descriptions. Among them were 20 imported from foreign countries. It details the taste, character, habitat, time of collection, effect, and therapeutic uses of each drug, and is illustrated with pictures. The book specifically points out several errors of early publications and corrects them. It not only had great influence on medical practice in China, but in other countries as well, including Japan and Korea. Traditional medical schools consider this book mandatory reading for their trainees.

Writings on nutrition also flourished in the Tang period. Meng Seng (孟詵, 621 to 713 A.D.), a prominent physician, wrote the three volume *Diet Ben Cao* (食療本草), which was preserved in the Dung Huang caves in the western part of China. It covers many important methods of identifying varieties and characteristics of plants which are nutritious and able to improve health and life. This book also describes in detail the preparation of food, and diseases due to malnutrition or deficiency of certain food ingredients.

There were many writings on the uses of drugs and methods of herb collection, preparation, and storage. These had a significant influence on the scholars of later generations in their writings on drugs. Also, cultural exchange between China, India, Korea, and Japan was well developed in this period. Many herbs not present in previous Chinese writings appeared in this period; these were grown and imported from outside the country. The Chinese pharmacopoeia, *Tang Hsin Hsu Ben Cao*, was brought to the attention of China's trading partners and was translated into their languages. In 733 A.D., Japan sent two monks, Ying Do and Po Chow, to China to study under the High Monk, Chan Jian (鑑真, 688 to 763 A.D.), who was a specialist in the identification and preparation of Chinese herbs. He practiced medicine to serve the people and was later invited to go to Japan six times to teach medicine and pharmaceuticals.

SUNG-JIANG-YUAN DYNASTIES (960 to 1368 A.D.)

The *Tang Hsin Hsu Ben Cao* was revised under the sponsorship of the Sung government (974 A.D.). In the revised version, 137 new drugs were added to make a total of 983; certain corrections to the classification of drugs were also made. A revised *Sheng Nung Ben Cao Chien,* which included 1082 drugs and the addition of some illustrations of plants and animals, was also published. It was also a government-commissioned writing.

Besides these two works, several private writings were published. One was the 33-volume *Jian Lui Ben Cao* (證類本草), written by Tang Chian Mei (唐慎微, 11th Century A.D.). This book contains descriptions of 1558 drugs and details of their method of preparation, most with illustrations. The material described in this book broadly covered the knowledge in this field and was the model for later herbal studies in China for the next 500 years, when it was replaced by the highly respected *Ben Cao Kong Mu* of Li Shih Chen.

Another writer in this period worthy of mention is Zhang Yu Su (張元素, 12th Century A.D.). He proposed a theory about the relationship between the *qi* and circulation, and its correlation to health and illness. He wrote several books on medicine and drugs, including *An Introduction to Medicine, Difficulties in Pharmaceuticals,* and *Herbal Gems* (*Chian Chu Nong*, 珍珠囊).

MING-CHIN DYNASTIES (1368 to 1911 A.D.)

The most important work on Chinese herbs written in this period was the *Ben Cao Kong Mu* (本草纲目), recognized worldwide as an outstanding work. Its author was Li Shih Chen (李時珍, 1518 to 1593 A.D.), the scion of a family of medical practitioners. Through hard study and by reading every book he could obtain, Li became a well-known physician whose exact diagnoses and treatments were widely esteemed.

Li's sense of observation was keen. He spent 30 years studying and surveying all types of plants and drugs, and traveled to rural areas to collect special specimens. He drew pictures of these specimens and questioned the locals to substantiate the truth of their effectiveness in treatment. Reportedly, he read more than 800 references. Finally, he compiled all of his data into a single book, mastering source material that seemed nearly boundless. As Unschuld has stated, ''He expanded the *Ben Cao* from a work on drugs to a comprehensive and detailed encyclopedia of medicine, pharmaceutics, mineralogy, metallurgy, botany, and zoology.'' (see Unschuld, *Medicine in China*, p. 147 [General Reference 14]).

The *Ben Cao Kong Mu* contains 52 volumes and includes 1892 drugs. Among these, 350 are derived from minerals, 1099 from plants, and 443 from animals. An appendix of 11,096 prescription formulas is included. Some herbs, 1160 in total, are illustrated with pictures drawn by his son. This book is a complete, detailed pharmacopoeia. It describes the shape, character, taste, thermal influence, source, method of collection (when and how), preparation, and pharmacological action of each drug.

Li made a significant improvement in the classification of the drugs listed in his book. He categorized them according to biological terminology, from the simplest to the most complex. He divided the total number of drugs into 16 groups, then into subgroups. The groups are: water, fire, earth, gold-mineral, plant, cereal, vegetables, fruit, wood, utensils, worm, scaly animals, crustacean, cattle, domestic fowls, and human. Subgroups would follow logically. For example, subgroups of water would include well water, rain, snow, dew, and hail. Prescription formulas were classified in four groups, depending upon how many drugs were used, or whether they were popular at that time. Some preparations contained only a single drug, and some originated from the experience of individual physicians or a traditional family recipe.

Li's book also corrected errors in previous pharmaceutical writings. After his death, the book was finally printed and published in 1596. It was brought to Europe and Japan and translated into Latin by Michael Boyn (1612 to 1650 A.D.). Later, it was also translated into English, French, German, Russian, and Japanese.

More books on medicine and herbs were written during the Ming Dynasty. They were more specific and theorized the pharmaceutics necessary to accommodate readers' demands and basic uses.

Ben Cao Fa Hsu (本草發揮), written by Hsu Yung Dung (徐彥纯, 14th Century A.D.), collected the theories and practices of herbs used by some well-known physicians in the Jiang Dynasty.

Gu Fon Ben Cao (救荒本草), by Chu Shou (朱橚 14th Century A.D.), described the drugs used to combat famine and emergency. Chu was the fifth son of the first Ming Emperor. There were originally 138 herbs in this book, but in a later revision, some other herbs were added, bringing the total to 444.

Ben Cao Chi Yao (本草集要, 1492 A.D.), by Wang Lun (王纶, 15 to 16th Century A.D.), is a collection of certain important elements in pharmaceutics. It contains eight chapters and 545 drug descriptions. A revision of this book was written by Chen Ja Mu (陈加謨), and an introduction was added to advise beginners on how to use the herbs.

Tin Nan Ben Cao (滇南本草), written in the early Ming Dynasty, specialized in herbs grown locally in the southwestern province of Yunnan. It includes 450 plants and herbs grown in that area, which were used primarily by local tribes. One of the herbs, *Tou Fou Lin* (土茯苓) was used as an antisyphilitic agent.

In the Chin Dynasty (1644 to 1911 A.D.), scholars continued the works carried over from the Ming Dynasty. Zhou Xue Meng (趙学敏, 1719 to 1805 A.D.) was a scholar who specialized in materia medica. He wrote the 10-volume *Ben Cao Kong Mu Shi Hui* (本草綱目拾遺, 1765 A.D.), which described 921 drugs. Among them were 716 folk medicines which were not listed in the original *Ben Cao Kong Mu* by Li. Zhou also wrote the 11-volume *Chui Ya Nai Pin* (串雅内编) and the *Chui Ya Wal Pin* (串雅外编), also 11 volumes. These were collections of special and rare prescriptions and medical techniques used in folk medicine.

Wu Kee Tsin (吳其濬, 1789 to 1847 A.D.), the Viceroy of the southwestern provinces, was interested in botany and materia medica. He edited a 38-volume botanical work, which was a collection of 1714 plants and pictures. It was the first botanical literature written in China and was later brought to Japan and Korea, where it was received with great esteem.

POSTREVOLUTION PERIOD (1911–PRESENT)

The Chinese Republic was born after the 1911 Revolution, which was led by a Western-educated physician, Dr. Sun Yet Sen. The details of medical progress in this period has been thoroughly reviewed by Croizier.[3] As he states, the record of Chinese medical developments prior to 1949 "was not impressive, and before 1927 it was dismal."

The Ministry of Health of the Nationalist government sought to curtail or eliminate traditional Chinese medicine. After the 1949 Communist Revolution, the People's Republic government reemphasized the importance of Chinese herbs and medicine. A campaign to rehabilitate Chinese medicine commenced, intended to break China's dependence on the West. More than ten traditional medical colleges and institutions were established in the coastal provinces and Szechuan Province. Today, even students within China's Western-style medical colleges are taught to use herbal therapy and acupuncture in treating their patients.

The third revolution, the Cultural Revolution, which started in 1966 and quieted down by 1976, has brought Chinese culture as a whole to a standstill. Barefoot doctors with no more than 6 months training and no high school education were dispersed to rural areas to replace the denounced "capitalist-trained" physicians and intellectual herbal practitioners. There is not much we can say about medical progress in this period.

REFERENCES

1. **Chen, Sui-lou,** *Zhong Kou Li Dai Ming Yi Tu Zhuan* [*The Renown Medical Physicians and Scholars in China*], Jiangsu Science Technology Publisher, Jiangsu, 1987.
2. **Cowen, R.,** Medicine on the wild side, *Sci. News,* 138, 280–282, 1990.
3. **Croizier, R.,** in Medicine in Chinese Culture, Kleinman, A. et al., Eds., U.S. Dept. Health, Education, and Welfare Publ. 1974, 26–36.
4. **Kuo, Lou-chun,** *Zhong Kou Yi Shi Nin Bil* [*Chinese History Calender*], Hai-Lun-Jiang People's Publisher, Hai-lun-jiang, 1984.
5. **Quinn, J. R., Ed.,** Medicine and Public Health in the People's Republic of China, U.S. Dept. Health, Education, and Welfare Publ. 1972.
6. **She, Yue et al.,** *Zhong Kou Yi Xue Shi Liu* [*Chinese Pharmaceutical History*], People's Health Publisher, Beijing, 1984.
7. **Yin, Gi-ye,** *Zhong Kou Yi Xue Shi* [*Chinese Medical History*], Jiangsi Science and Technology Publisher, Jiangsi, 1987.

Section III
Cardiovascular System

THE CARDIOVASCULAR HERBS

For thousands of years, Chinese traditional medicine has made use of a myriad of herbs with great success, though lacking a thorough understanding of their mechanisms of action. Correspondingly, diagnostic techniques were less than scientific. For example, physicians — *lang-chong* (郎中) or *tai-fu* (大夫), as the Chinese called them — diagnosed cardiac disease by merely feeling the pulsation of the patient's radial artery or by examining the color of the patient's tongue and the rapidity of his breathing — with surprisingly frequent precision. A common term, *yin shue-tee yurk* (阴虚体弱), meaning ''emptiness in negativity and weakness of the body'', translated into a diagnosis that the heart was failing, indicating the need for a cardiac drug. No proof existed that the herbs so prescribed were good for the heart: all that was said was that the herbs would ''fill the emptiness and support the weakness''.

Nor is the use of such herbs limited to a single function. Just as foxglove or digitalis was used early on as a remedy for dropsy, many Chinese cardiac herbs were first prescribed as diuretics. Even now, Chinese traditional medicine has not stopped to consider what that relationship is between diuretic effects and heart function improvement. In many cases, a Chinese herb may contain more than ten constituents. Thus, their actions are so diverse and multitudinous that we may opt to classify them in one category though they can arguably belong to many other chapters.

This chapter deals with herbs which have been proven scientifically to be effective in the treatment of cardiovascular diseases, via their extracts or constituents. Other herbs which are used by the Chinese *lang-chong*, but with little-known reasons or values, will not be described here.

Chapter 1

HERBS WITH MULTIPLE ACTIONS

GINSENG — *Panax ginseng* C. A. Meyer, "A Dose of Immortality"

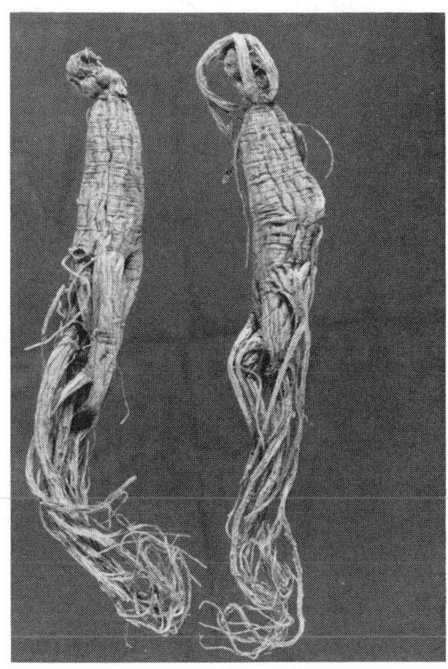

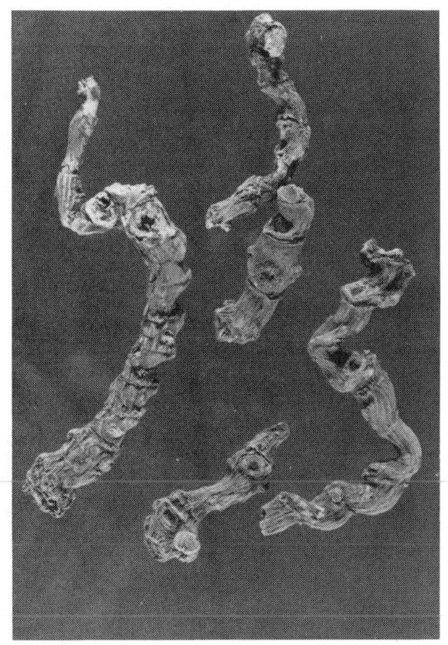

A B

FIGURE 1. (A) *Panax ginseng* C. A. Meyer from Kilin Mountain; (B) *P. japonicus* C. A. Meyer.

In Chinese, the word ginseng (人参) is directly translated as "the essence of man". It is the most valued herb used in China, and is also widely used in other Asian countries such as Japan and Korea, as well as among the overseas Chinese in the U.S. For thousands of years, ginseng has been used by the common people as a tonic and emergency medicine to rescue dying patients and used by the rich as a rejuvenating and revitalizing agent.

History

According to the *Herbal Classic of the Divine Plowman*,[1] published around 100 B.C. during the West Han Dynasty, ginseng is able to "support the five visceral organs, calm the nerves, tranquilize the mind, stop convulsions, expunge evil spirits, clear the eyes, and improve the memory". Daily administration was supposed to restore the yin and the yang, increasing the longevity of men or women. The Chinese called it a "Dose of Immortality".

Because of such claims, many names were coined glorifying the herb, including "the spirit of earth" and "the spirit of man". But the name "ginseng", or the essence of man, is considered especially apt. There are many scholarly writings about ginseng, particularly on how its use can bring happiness. Some books described the shape, the characteristic methods of cultivation, and the time of harvest of the root. As described by the Herbal Classics and other old Chinese traditional medicine textbooks, its effectiveness values reached almost mythical proportions. Even in the U.S. prior to World War I, many Americans cultivated American ginseng in their backyards or in the Appalachian mountain range, describing it as "nuggets of gold" (see Hardacre[2]).

TABLE 1
Sources and Species of Ginseng (Wild.)

Region	Species
Wild growth	
Kilin Mountain, Manchuria	*Panax ginseng* C. A. Meyer
Other areas in China	*Panax notoginseng* (Burk.) F. H. Chen
	Panax japonica var. *augustifolius* Cheng et Chu (*Chu-j-ginseng*) (竹节人参)
	Panax japonica var. *major* (*Chu-shi-chek*)
	Panax japonica var. *bipinnatifidus*
	Panax stipulenentes (*Pin-bin-san-chek*)
	Panax pseudoginseng Wall
	Panax zingiberensis, *San-Qi* (三七)
Korea	*Panax ginseng* C. A. Meyer
North America	*Panax quinquefolius* Linn.
Japan	*Panax trifolius* C. A. Meyer (*Chikusetsu-ninjin* and *Satsuma-ninjin*)
Cultivated (3 to 6 years)	

Ginseng occupies an important role in the folk medicine of not only China, but also Korea and Japan. Its therapeutic effects and doses were well described in the famous Chinese pharmacopoeia, *Ben Cao Kong Mu* (本草纲目) (The Categoric Lists of Herbs and Plants) by Li Shih Chen.[3]

Ginseng did not appear in Europe until the 12th Century, when Marco Polo brought the herb to the west. Because of its rarity and the prevailing myth of its resurrecting and revitalizing effects, the herb never appeared in common use in Europe. In 1697, Baldelin read a paper to the French Science Academia, describing the medical uses of ginseng. Gradually, the herb was brought to the attention of scientists and the aristocracy in Russia, Europe, America, and Canada.

Because the Chinese originally only identified ginseng by its shape, leaves, and flower, its actual botanic species was varied and confusing. The root collected from the hills of Korea and Manchuria was named *Panax ginseng* C. A. Meyer; when collected from the North America's Appalachian mountains it was *Panax quinquefolius* — *Shi-yang Seng* (西洋参) or American ginseng. Frequently, Chinese drug stores and pharmacopoeia erroneously classified ginseng with other plants belonging to the Araliaceae family, such as *Adenophora tetraphylla* (Thun.) Fisch. — *Sha-seng* (沙参).

Genuine Chinese ginseng was collected from the Kilin Mountains of Manchuria and northeastern China. This particular species was described in detail in the *Herbal Classic of the Divine Plowman.*[1] The *Ben Cao Kong Mu* includes a drawing of the plant, and describes the method and season for collecting the herb.[3] Table 1 lists different commercially used ginsengs and their sources.

For thousands of years, the cultivation and harvesting of ginseng was considered to be a royal prerogative. Areas in which ginseng was collected were not open for public use. Many local residents would risk their lives to steal the herb for their own use or to sell the root to the wealthy. During the Chin dynasty, a law was passed that anyone caught stealing ginseng would be beheaded, his helpers would receive 40 strokes with a wooden plank, and their families and animals sent to government storage areas for disposition.

Because of the rarity of the actual herb and the lack of a scientific identification of the true ginseng species, many counterfeit ginseng-like plants were made available on the market. Some of the most common were *Sha-seng* (*Adenophora tetraphylla*) and *Dong-seng* (党参) (*Radix Codonopsis Philosulae*).

Cultivation of ginseng began during the Ming Dynasty (14th Century). In *Ben Cao Kong Mu*,[3] Li Shih Chen wrote a chapter on methods of cultivation, instructing readers to seed the ginseng in October for sprouting the next spring. The procedure described was very similar to common vegetable planting methods. Hardacre has also described cultivation methods for American ginseng.[2] The technique was further supplemented by others.[7]

The scientific study of ginseng began in the early 19th Century. Chemical structures were identified and their principals studied. Gradually, attention was paid to the pharmacological action of the herb, together with its clinical applications. But ginseng contains multiple active elements, not just a single agent. Some of these are well studied and recognized, but many are not yet fully understood. The exact value of ginseng as a whole is still a mystery to many pharmacologists.

Pharmacognostic Characteristics

The optimum climate for the Kilin ginseng is between −10 to +10°C with an annual rainfall of 50 to 100 cm. Figure 1A illustrates a sample of Kilin ginseng which is slightly different than the shape of *Panax japonicus* C. A. Meyer (see B) or *San-Chie* (see Chapter 5, Figure 2).

Although the seeds, flowers, leaves, and stems of ginseng are available for use, common references to the herb allude to its root or rhizome. In the apothecary or the market, only the root or rhizome is generally seen.

Classification

1. Ginseng root and rhizome
   A. Round Seng or cultivated ginseng. There are two common preparations of the cultivated root: sun-dried or roasted ginseng, vs. sweet red ginseng (prepared by boiling in water with sugar, then steamed and dried).
   B. Wild Seng, collected in certain regions such as Kilin, Korea, or the Appalachian mountains. Generally, the body of the wild root is stockier and shorter. It is cylindrical in shape, approximately 2 to 10 cm in length and 1 to 2 cm in diameter. Detailed descriptions of American ginseng, including its historical uses and habitat, are available in two early publications, *The Herbalist*[5] and *American Medicinal Plants*.[6]
2. The leaf is elongated or egg-shaped, with some fragile ciliar hairs.
3. The flower generally grows in clusters of 40 to 70, arranged in a circle around a stem 3 to 7 cm in length.
4. The seed consists of a kernel enclosed by a thin skin. It is usually 4.8 to 7.2 mm in length and 4 to 5 mm in width.

The Chemistry of Ginseng

Besides the proteins and carbohydrates which usually exist in most herbs and plants, ginseng contains many valuable ingredients, including:

1. Volatile oil, about 0.05%
2. Saponins, known as panaxosides or ginsenosides
3. Antioxidants
4. Peptides
5. Polysaccharides, fatty acids, alcohols, and vitamins

Scientists have attempted to isolate the chemical elements of ginseng for over a century. In 1854, Garrigues reported success in isolating a glycoside from the *Panax quinquefolius*

TABLE 2
Properties and Structures of Some Ginsenosides

Ginsenoside	Form	Melting point (C°)	Empirical formula
Ro	Colorless needle	239–241	$C_{48}H_{76}O_{10}$
Ra_1	White powder	202–206	$C_{58}H_{98}O_{26}$
Ra_2	White powder		$C_{58}H_{98}O_{26}$
Rb_1	White powder	197–198	$C_{54}H_{92}O_{23}$
Rb_2	White powder	200–203	$C_{52}H_{90}O_{22}$
Rb_3	White powder	193–195	$C_{53}H_{90}O_{22}$
Rc	White powder	199–201	$C_{53}H_{90}O_{22}$
Rd	White powder	206–209	$C_{48}H_{82}O_{18}$
R-20-glc-Rf	White powder	182–184	$C_{48}H_{82}O_{19}$
Re	Colorless needle	201–203	$C_{48}H_{82}O_{18}$
Rf	White powder	197–198	$C_{42}H_{72}O_{14}$
Rg_1	White powder	194–196.5	$C_{42}H_{72}O_{14}$
Rg_2	Colorless needle	187–189	$C_{42}H_{72}O_{13}$
20(R)Rh_1	White powder		$C_{36}H_{62}O_9$
20(R)Rg_2	Colorless needle	214–216	$C_{42}H_{72}O_{13}\cdot H_2O$
Rh_2	Colorless needle	218–220	$C_{36}H_{62}O_8\cdot H_2O$
Rh_1	White powder		$C_{36}H_{62}O_9\cdot 1^1/_2H_2O$

L., calling it panquilon ($C_{12}H_{25}O_9$). It was observed that this glycoside tastes much like glycyrrhizin. It breaks down under the action of sulfuric acid to form panacon:

$$C_{12}H_{25}O_9 = CO_2 + C_{11}H_{10}O_4 + 3\ H_2O$$

Panaquilon Panacon

In 1906, Yasudiko isolated a pure saponin from Korean ginseng, noting a melting point of 270°C. A saponin called panaxin was also isolated from the plant. Subsequently, several investigators expanded the chemical separation of ginseng and obtained many components (see Wang[8], p. 37). The structures of these components were then determined and named. In total, 77 saponins have been listed by Tanaka et al.[31] Their chemical structures were grouped separately under five tables, depending upon the hydrolyzed product, genin, of the saponin. (For details, see Tanaka's article.) Tables 2 and 2A present 17 known ginseng saponins, together with their melting points and structures.

Hydrolysis of each saponin will produce sapogenin and several sugar molecules. For example, Ro gives three molecules of sugar, Rb_1 produces four molecules of glucose, and Rb_2 has three glucose and one arabinose. Ra_1 has three sugars at the C-20 position and a fourth sugar at the C-3 position. With the exception of Ro, the sapogenins of ginseng are either a dihydroxyl or a trihydroxyl sterol, named as panaxadiol or panaxatriol.

In their article, Tanaka et al.[31] pointed out the sensitive reactivity of these compounds, such as the hydration of the side-chain double bond. This suggests that structure of saponins may be modified during processing or decoction of the ginseng root, or possibly during metabolism within the body.

It has been shown that the leaf and stem of ginseng contain saponins identical to those contained in the root. On a per-weight basis, the leaf contains a higher percentage of those saponins, but Chinese apothecaries seldom dispense the leaf for medicinal uses; instead, its use is primarily reserved for the brewing of teas — nor are ginseng flowers or buds used medicinally in China.

TABLE 2A
Structures of Ginsenosides

Structure of Some Ginseng Saponins

Ginseng saponin	Structure	
Ro	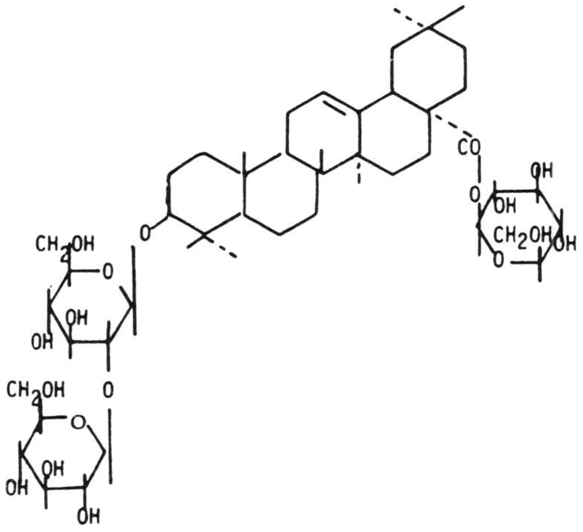	
Rb_1 Rb_2 Rc Rd	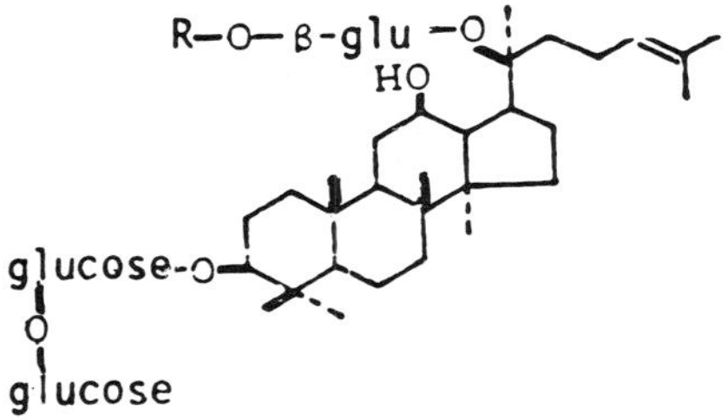	R = β-D-glucose Arabinose (pyr.) Arabinose (fur.) H

TABLE 2A (continued)
Structures of Ginsenosides

Structure of Some Ginseng Saponins

Ginseng saponin	Structure		
	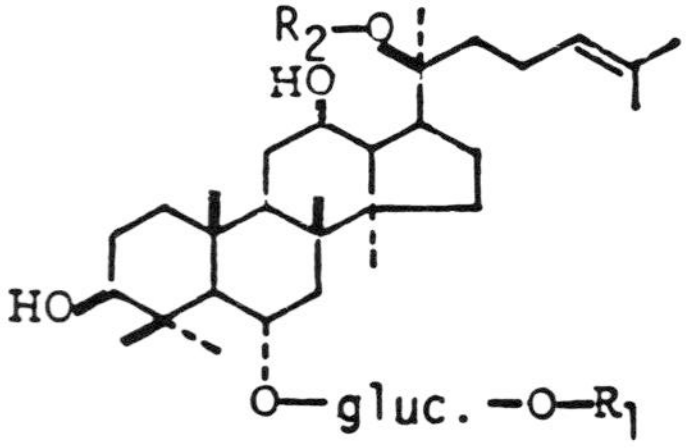	R_1 =	R_2 =
Re		Rhamnose	β-Glucose
Rf		Glucose	H
Rg_1		H	β-Glucose
Rg_2		Rhamnose	H
Ra_1	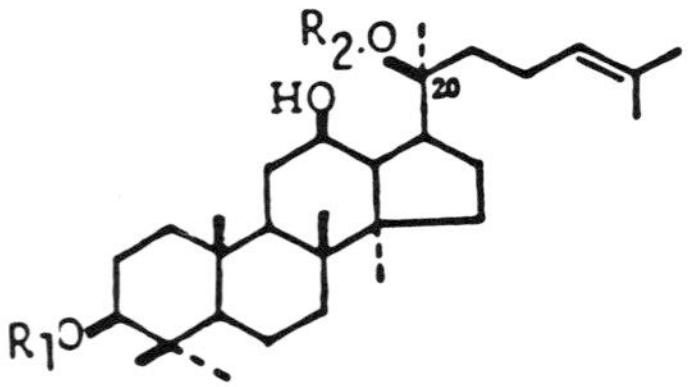	H	β-Rham–β-glucose
I		β-Glucose	β-Glucose
II		H	β-Glucose
III		β-Glucose	β-Zylose-α-rham–β-glucose
IV			

Structure of Some Ginseng Saponins Isolated from *Zhu Ji Ginseng* (*Panax japonicum*)

Ia	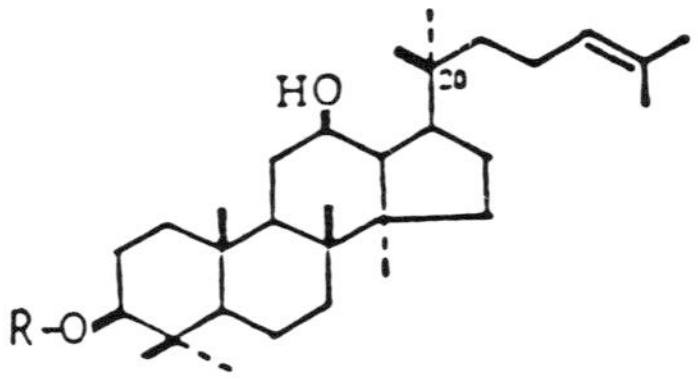	Glu–glu–
III		Glu–glu– | glucose

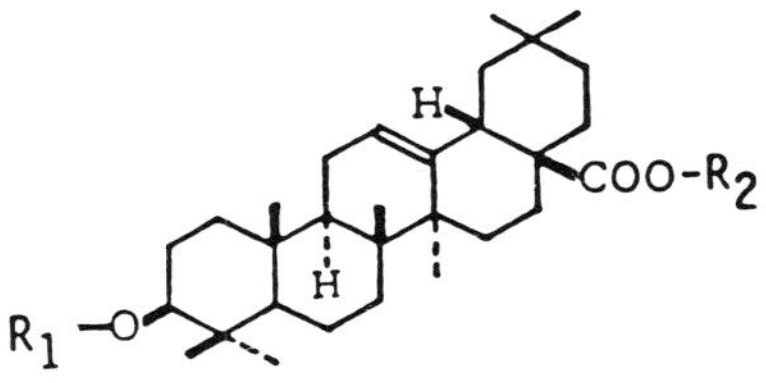

	$R_1 =$	$R_2 =$
Ib	Glu–glu– | glucose	H
IV	Glu–glu–	–glucose
IVa	Glucose–	–glucose
V	Glu– | glu	–glucose

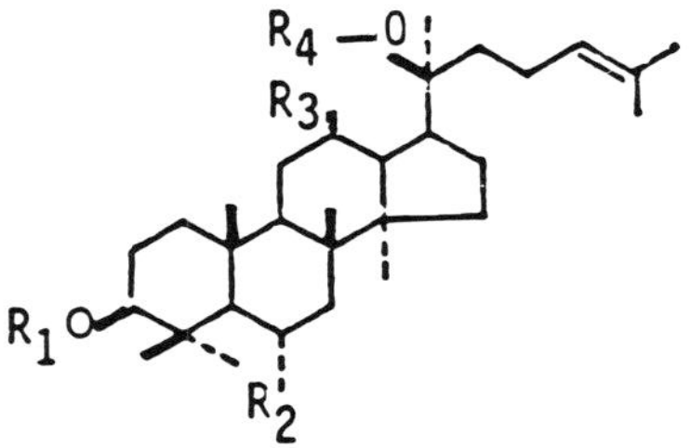

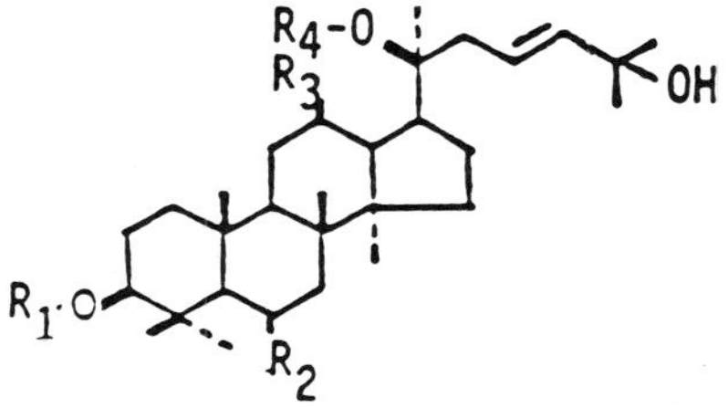

	$R_1 =$	$R_2 =$	$R_3 =$	$R_4 =$
L_5	H	HO	HO	–glu–ara–xyl
L_{10}	H	HO	–O–glu	H
LT_5	–glu	H	=O	–glu–glucose
LT_8	–glu	H	=O	–glucose
LN_4	–glu–xyl	H	=O	–glu–ara
L_{9a}	H	HO	–O–glu	H

TABLE 2A (continued)
Structures of Ginsenosides

Structure of Two Other Ginseng Saponins Isolated from *Panax pseudoginseng*

Ginseng saponin	Structure	R_1 =	R_2 =	R_3 =	R_4 =
F_8	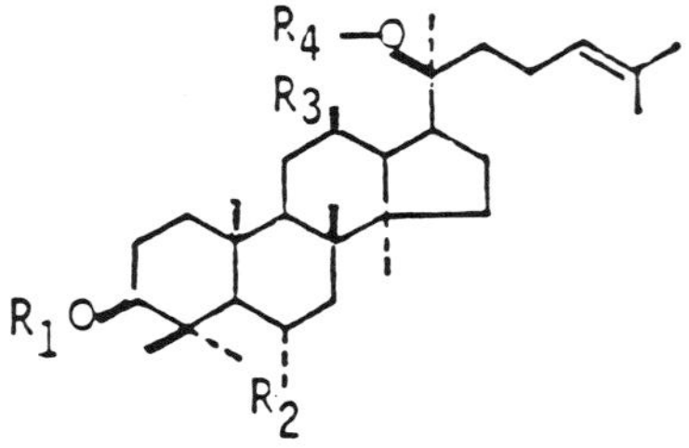	–glu–glu	OH	OH	–glu–xyl
F_{11}	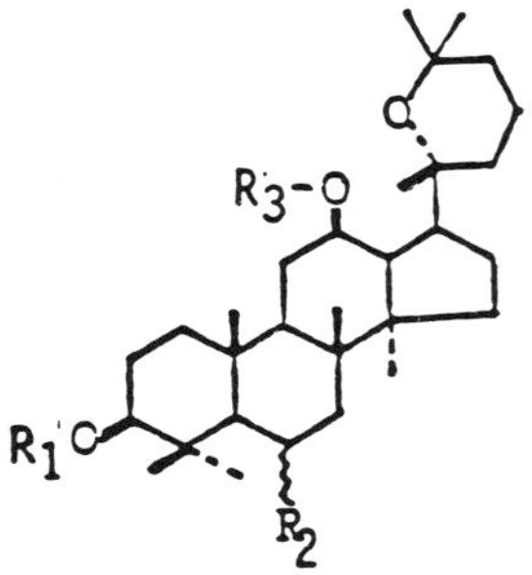	H	–glu–rham	H	

TABLE 3
Saponin Content of Different Ginseng Herbs

Species	Total saponins (% by weight)
Panax ginseng C. A. Meyer	2.1–4.4%
Panax quiquefolius (American ginseng)	4.9%
Panax notoginseng and *Panax japonica (Chu-j-ginseng)*	13.6–20.6%
Panax japonica var. major (*Chu-chi-seng*)	9.34%

TABLE 4
Saponin Content in Different Parts of the Ginseng Herb

Portion	Total saponins (% by weight)
Rhizome	6.23
Root	
Main	2.75
Branch	8.28
Leaf	12.2
Skin (outer part)	8.0
Bud and flower	15.0
Seed	0.7

Recent analysis by Chinese investigators showed that the ginseng bud contains at least seven different saponins: Rb_2Rc, Rb_3, Re, Rd, Rg_2, and 20-glucoginsenosaponin-Rf.[8] One saponin not found in the root was also isolated within the flower: RM7cd.

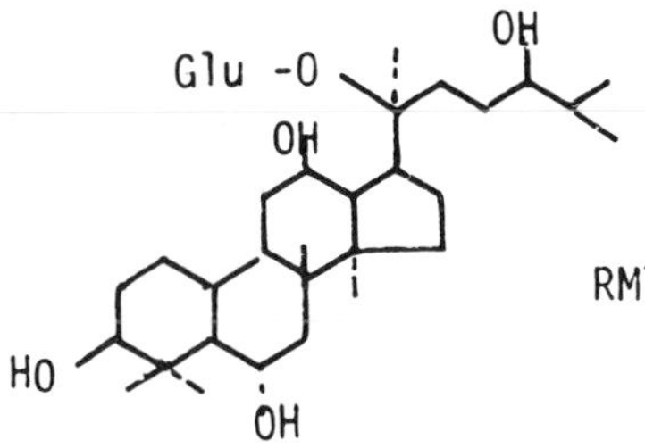

Glu -O OH OH HO OH RM7cd

The saponins identified in Table 2 are primarily isolated from *Panax ginseng* C. A. Meyer, grown either in Manchuria or in Korea. However, as described in Table 1, there are other species of ginseng commercially used all over the world. The saponins contained in each species varies. *Panax notoginseng* (Burk.) F. H. Chu, which grows in the wild in China's southwestern Yunnan and Kwangsi provinces, contains a much higher percentage of saponins than *Panax ginseng* — about 4.5 to 5 times higher (see Table 3). It has been widely used in traditional Chinese medicine as an anti-inflammatory, an analgesic, an antibleeding drug, and also as a tonic to promote rapid recovery from chronic illnesses.

The work of Tanaka et al.[31] quantified the saponin content of ginseng rhizomes and roots, as well as that of the leaves or aerial parts of *Panax pseudoginseng* and *Panax japonica*. These data would be of interest to investigators wanting to isolate certain saponins from the herb in greater yield (see Table 4.)

Other Constituents of Ginseng

In Korea, ginseng is known as the ''elixir of life''. Its rejuvenating effects are not due solely to its saponins, but also to other components. The most important nonsaponin component of ginseng is the antioxidant which has been shown to be the principle in improving the functions and activity of the central nervous system (CNS), that is, ginseng's observed antiaging effects. The antioxidant is also the major factor contributing to the herb's anti-irradiation effects.

The ether-soluble acidic fraction of ginseng contains three components which can be purified to have antioxidant effects:

1. Maltol
2. Salicylic acid
3. Vanillic acid

The structure of maltol shown below is very close to that of vitamin E, an effective protective agent for cardiac and brain tissue. Tanaka et al.[31] suggested that maltol is formed from maltose and an amino acid, as the result of a Maillard reaction during the steaming process in ginseng preparation (red ginseng).

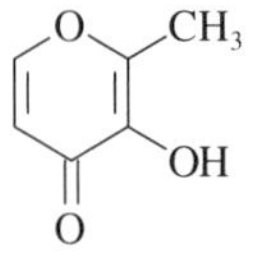

O CH$_3$

OH

O

Structure of maltol.

The strong antioxidant activity of these components most likely stems from their chelation with ferric ions. In an experiment with mice, these three antioxidants have successfully increased the endurance of animals while swimming, unlike similar tests involving the ginsenosides Rg, Re, Rb_1. These antioxidant components have also been proven to exert a protective effect against irradiation, liver damage, thrombosis, and atherosclerosis.

Ginseng contains fatty acids, several essential oils, amino acids, peptides, and polyacetyleinic alcohols such as panaxynol ($C_{17}H_{25}O_7$), which is identical to the carotatoxin produced from the carrot plant. In addition, vitamins C, B_1, B_2, B_{12}, and nicotinic acid are present, plus a small quantity of minerals (Mn, Cu, Co, As). Five polysaccharides, panaxans A, B, C, D, and E, have also been isolated from the ginseng root. They are the principal constituents in the herb's hypoglycemic effects. Other investigators have reported that a glycogen-like polysaccharide was isolated from the ginseng root, exhibiting a reticuloendothelial system activating effect.[31]

Gotirner et al.[15] isolated low molecular weight peptides from Korean white ginseng, plus uracil, uridine, guanidine, alanine, and β-sitosterol from the ginseng root.

Pharmacodynamic Actions of Ginseng

Cardiovascular System

In an experiment, healthy male volunteers were given ginseng orally for 9 weeks. At the 10th and 12th weeks, after medication had been terminated, subjects showed substantial decreases in heart rate. The difference between these results and those of the placebo control group were statistically very significant (Figure 2).

In experiments with dogs, the intravenous administration of ginseng extracts caused an immediate fall of blood pressure which could be blocked by β-adrenergic agents. Certain ginseng preparations, such as red ginseng, cause a release of plasma histamine, but the fall of blood pressure was maintained for a much longer period even when the plasma concentration of histamine was reduced to a normal level. Such prolonged hypotensive effect is probably due to a blocking effect of Ca^{2+} mobilization in the vascular smooth muscles.

Administration of ginseng to rats with hypertension induced by partial renal arterial constriction (Goldblatt's hypertension) also produced a hypotensive effect. Zhang and Chen[40] reported that ginsenoside is an α_2-adrenergic agonist which can be blocked by yohimbin.

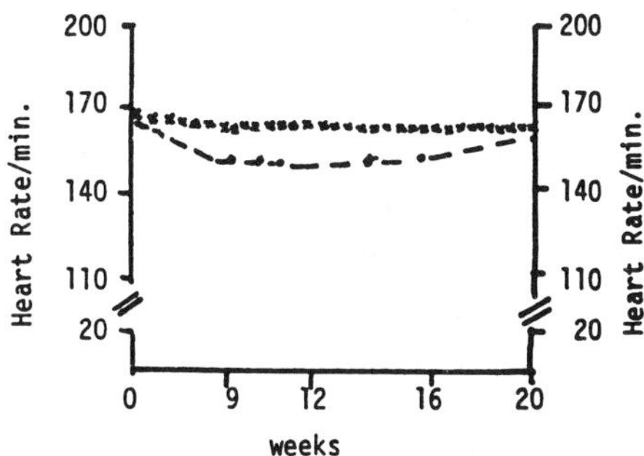

FIGURE 2. Effect of ginseng on the heart rates of young males. (xxx) Placebo; (—-) received G-115 (contains 4% total ginsenosides). (From Kirchorfer, A. M., in *Advances in Chinese Medicinal Materials Research,* Chang, H. M. et al., Eds., World Publishing Co., Singapore, 1985, 536. With permission.)

TABLE 5
Survival Time of O_2-Deprived White Rats

Experiment	Dose (mg/kg)	Survival Time (min)
I. Control	—	45.0 ± 4.3 (10)
Total ginseng saponins	270	84.7 ± 9.7 (10)
II. Control	—	47.8 ± 2.5 (9)
Isoproterenol	2.5	27.5 ± 2.0 (9)
Ginseng saponin	270	
+ Isopren.	2.5	40.7 ± 2.8 (9)
Propranolol	5.0	
+ Isopren.	2.5	40.4 ± 2 (9)

It has also been found that Rb, not Rg, can offer a protective action on experimental acute myocardial infarction in rabbits and myocardial necrosis induced by isoproterenol in rats.

Other significant experiments have been performed on oxygen-depleted animals. Rats were placed in a closed glass chamber for 30 min and were taken out to see how long they could survive after oxygen depletion (at the end of 30 min, the O_2 content of the chamber was 7.6 ± 0.2%). As shown in Table 5, ginseng saponins can prolong survival times and block the aggravated oxygen lack by isoproterenol. Such blocking effects are similar to those observed with β-adrenergic blockers.

In an isolated perfused heart preparation, ginseng saponins, at a concentration of 0.27 mg/ml, prolonged heart contraction; this has been seen with ATP at a concentration of 0.01 mg/ml.

Ginseng extract causes vasoconstriction in small doses and vasodilatation in large doses.[19] But in cerebral and coronary vessels, ginseng extract exhibits only a vasodilating effect, resulting in an improvement in brain and coronary blood flow. Among the ginsenosides tested, Rc and Rg_1 exhibit the vasodilatation effect, but Rb_1 does not.

The vasodilatation effect of ginsenosides is due to the relaxation of vascular smooth muscles. With isolated aorta strip preparation, ginsenosides have blocked the constricting effects induced by norepinephrine. Experiments with membrane and sarcolemma of rabbit heart tissue show that ginsenosides inhibit the $^{45}Ca^{2+}$ uptake (under normal ATP concentration). Among the ginsenosides tested, panaxidiols were much more potent in such inhibiting effects than panaxitriols. It is believed that such inhibition of Ca^{2+} uptake by the muscle membrane contributes to the mechanism of vasodilatation.

Ginseng extract can cause hemolysis of red blood cells. The degree of hemolysis varies depending on which portion or preparation of the herb is used to make the extract. For example, extracts of the ginseng leaf exert the greatest hemolytic effect, while root extracts have the least hemolytic effect.

Ginseng stimulates the hemopoietic tissue of the bone marrow. The herb is not only effective in the treatment of hemorrhagic anemia, but also has a protective effect on leukopenia caused by bacterial infections. Administration of ginseng to animals has increased ^{59}Fe incorporation into erythrocytes from a control value of 8.06 to 14.62%. Such hematopoietic-stimulating effects can be observed even in nephroectomized animals, implying a direct effect on bone marrow rather than an indirect effect through the erythropoietin stimulation of the kidney.

Extracts of red ginseng have been shown to definitely lower serum cholesterol levels, producing a protective effect against atherosclerosis. This may also be a factor contributing to lower blood pressure.

Recently Chen and his associates[35] used microelectrode techniques to study the effects of *Panax notoginseng* total saponins and its pure substances, Rb_1, Re, and Rg_1, on the cardiac muscles. They found that these saponins have a selective vasodilating action. Rb_1 can block Ca^{2+} influx into the muscle fibers, while Re and Rg_1 can inhibit intracellular Ca^{2+} release.[35,36] This result was confirmed by other researchers, in the ^{45}Ca uptake studies on the right ventricular papillary muscles of guinea pigs.[22]

Chen et al.[35] also showed that ginseng saponins can protect against cardiac ischemia in animals by decreasing cardiac CPK release, thereby attenuating myocardial Ca^{2+} accumulation and preventing the reduction of superoxide dismutase activity. In their early report, they claimed that *Panax notoginseng* saponins can produce a negative chronotropic and inotropic effect very similar to that of verapamil.[35] They did not, however, correlate such effects as a factor in ginseng's protective effect on cardiac ischemia. Furthermore, Shi et al.[29] demonstrated that ginseng saponins do have an antiatherosclerotic effect from a combined action of increasing prostacyclin in the carotid artery and decreasing thromboxane A-2 in blood platelets.

Central Nervous System

Sakai[38] reported several effects of ginseng extract, including a tranquilizing effect on the CNS, stimulation of medullar vomiting and the respiratory center in small doses, and inhibition of medullar vomiting and the respirator center in large doses. This type of stimulation can antagonize barbiturate or alcohol inhibition on the medullar center.

Russian investigators reported that oral administration of ginseng would raise mental activity, improve concentration and intelligence, and increase night adaptation of the eyes. It is believed that ginseng can strengthen conditional reflexes by improving the association center between neurons and increasing the sensitivity of the cortex to stimulants (external and internal).

The tranquilizing effects of ginseng are observed when the herb is given in large doses. Intraperitoneal injection of water extract of ginseng would exert an anticonvulsive effect on strychnine poisoning in mice and antagonize the shock syndrome produced by cocaine intoxication. Such effects are predominantly exerted by the ginsenoside Rb_1. Ginsenosides or Rb_1 can also lower the body temperature of mice.

TABLE 6
Effect of Ginsenosides on Corticosterone Secretion

Compound	Dose (mg/100 g)	Serum corticosterone (μg/dl)
Control (saline)		3.9 ± 0.6[a] (8)
Ginsenoside		
Rb_1	3.5	19.0 ± 5 (10)
	7.0	41.0 ± 4 (5)
Rb_2	3.5	31.0 ± 9 (5)
	7.0	31.0 ± 5 (5)
Rc	3.5	50.0 ± 3 (4)
Rd	3.5	38.0 ± 12 (4)
Re	3.5	16.0 ± 6 (5)

[a] Mean ± S.D. (number of animals).

Peripheral Nervous System

Ginseng extract can contract a frog's rectus muscle which is blocked by curare. Small doses of ginseng extract can stimulate rabbit intestine, while large doses will inhibit it.

Endocrine System

Among the multiple actions of ginseng reported by many investigators, the most clear cut is its effect on the endocrine system. Most notable is its stimulating effect on the hypothalamopituitary suprarenal cortical system.

Adrenal Cortex

In their early experiments with rats, Liu and Song[26] reported that ginsenosides had no direct effect on adrenal cortex secretion. However, several researchers from other Chinese institutions reported quite different results. Oral administration of ginseng rhizome or leaf to rats can decrease the vitamin C content of the adrenal cortex, indicating increased synthesis activity of the tissue. Such a decrease of vitamin C content was not observed in hypophysectomized animals, suggesting that this is produced via the stimulating effect of adrenocorticotropic hormone (ACTH) in the pituitary gland.

Analysis of the plasma concentration of corticosterone in those animals also showed a remarkable increase after administration of the ginsenosides which coincided with the decrease of vitamin C content. When the group of animals was given cortisone simultaneously, the vitamin C content of the adrenal cortex did not decrease, suggesting that cortisone exerts a feedback mechanism needed to block the stimulating effect of ACTH by ginsenosides.

Among the ginsenosides being studied, Rc has a potent effect on stimulation of corticosterone secretion as shown in Table 6.

Rats receiving ginsenosides at a dose of 7 mg/100 g of body weight by intraperitoneal injection showed an increase in plasma concentration of ACTH and cortisol. Table 7 summarizes the results. The increase of plasma cortisol level which was related to the dose of ginsenosides administered is shown in Figure 3.

It has been shown that cAMP in the adrenal cortex plays an important role in the biosynthesis of corticosterone. Rats receiving ginsenosides also showed increased levels of cAMP in the adrenal cortex. A dose response curve between cAMP content and ginsenosides dose is presented in Figure 4.

An increase in cAMP by ginsenosides did not, however, occur in hypophysectomized rats, as shown in Table 8. This is additional evidence indicating that ginsenosides act on the hypothalamopituitary axis.

TABLE 7
Effect of Ginsenosides on ACTH and Cortisol Secretion in Rats

Time (min)	Plasma ACTH (ng/ml)		Plasma cortisol (μg/dl)	
	Control (saline)	Ginsenosides	Control	Ginsenosides
0	93 ± 21	—	4.6 ± 2.1	
30	96 ± 24	1250 ± 50	12 ± 8	48 ± 7
60	48 ± 6	1430 ± 280	2.9 ± 1.6	52 ± 5
90	133 ± 27	750 ± 110	6.8 ± 3.2	57 ± 5

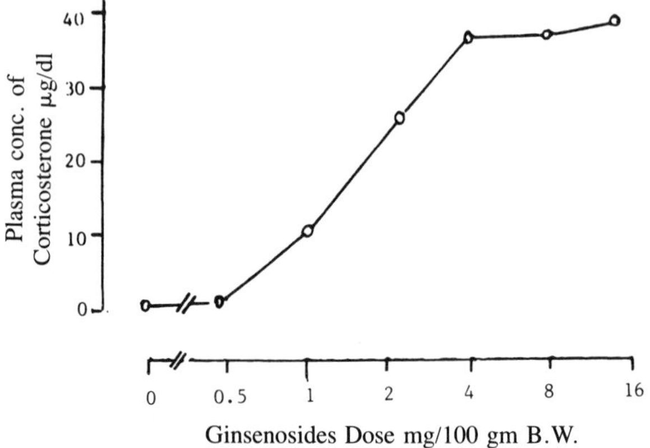

FIGURE 3. Dose response curve between ginsenosides and cortisol plasma concentration. (From Wang, B. C., *The Ginseng Research,* Tianjin Scientific Publisher, 1984, 135.)

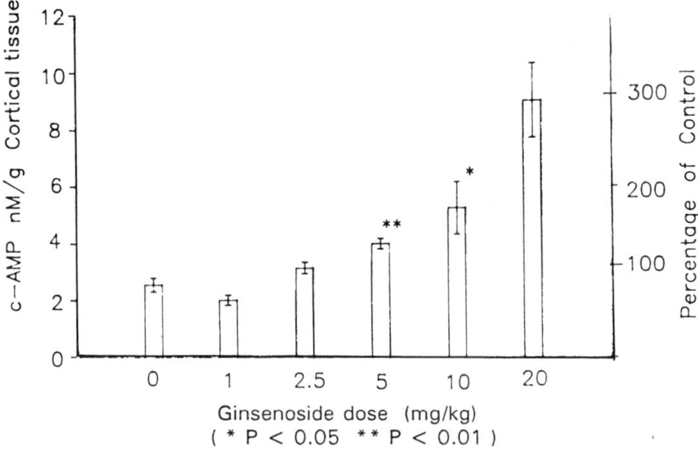

FIGURE 4. Dose response curve between ginsenosides and cAMP content in adrenal cortex of rats. (From Wang, B. C., *The Ginseng Research*, Tianjin Scientific Publisher, 1984, 135.)

TABLE 8
cAMP Content in Adrenal Cortex after Ginsenosides

Group	cAMP Content (ng/g adrenal cortex tissue)	
	Normal rats	Hypophysectomized rats
Control	4.5 ± 0.3 (6)[a]	3.5 ± 0.2 (4)
Ginsenosides (10 mg/kg i.p.)	17.0 ± 4 (6)	3.7 ± 0.4 (6)
ACTH (200 Units i.p.)		102 ± 20 (4)

[a] Mean ± S.D. (number of animals).

Histochemical analysis of animals which received ginsenosides showed an increase in the number of eosinophilic cells of the anterior pituitary. This suggests that ginsenosides act on the release of corticotropin releasing factor (CRF) from the hypothalamus, rather than directly on the pituitary gland.

The ginseng saponin effect has also been shown to follow a circadian pattern. Li and Shi[22] gave the ginsenosides to a group of mice and found the serum corticosterone level to increase sevenfold during the daytime and increase less in the darkness period. The increase of 5 hydroxy tryptamine (5HT) in the plasma level followed the same pattern. When ginseng saponin was given to female rats at a dose of 5 to 20 mg/kg, it increased the plasma prolactin level markedly. Du et al.[13] claimed this is due to a direct effect on the pituitary or hypothalamus. Ginsenosides also have a stimulating effect on thyroid glands via the stimulation of the anterior pituitary.

Sexual Development

For thousands of years, folk medicine has touted ginseng as a rejuvenating drug, with particular popularity among the Chinese middle and upper classes. Traditional Chinese medical texts claimed that ginseng is a drug which ''enriches the juices and blood of the body'' and ''strengthens exhausted sperm and impotent genitals.''

To test these claims, several Russian investigators administered ginseng extract to castrated mice and reported that even a large dose of ginseng extract would neither prevent atrophy of the male prostate and seminal vesicles, nor promote puberty development in female mice. These findings were confirmed by several groups of Chinese researchers, who gave castrated male rats either 20% ginseng alcoholic extract or pure ginsenosides continuously for 18 d.

However, in experiments with young rodents, ginseng extract showed a definite stimulation of sexual functions. Rabbits receiving ginseng extract showed an increased sperm count in the epididymis and an increase in motility in the seminal vesicles. Female mice aged 6 to 7 weeks which received ginseng extracts developed into puberty much earlier than mice receiving saline; the size and weight of their uterus and ovaries were greater than the control group, and the concentration of luteotropic hormone in the plasma increased.

Such stimulating effects were not observed in hypophysectomized rats, nor in animals which received morphine and ginsenosides in combination. It is known that morphine possesses an inhibiting effect on gonadotrophin secretion from the hypothalamus. Thus, these results indicate that ginsenosides act on the sex organs through stimulation of the hypothalamopituitary axis by increasing the secretion of gonadotropins. Table 9 presents a summary of the study.

In cell culture experiments with anterior pituitary tissue, it was found that ginsenosides increase the concentration of lactogenic hormone in the medium.

TABLE 9
Effect of Ginseng Preparations on Puberty Development in Female Rats

Experimental group	Animal no.	Puberty development (%)	Weight of Ovary (mg/100 body wt.)	Weight of Uterus (mg/100 body wt.)
Normal rats				
Control	12	33	46.1 ± 2.99	164.05 ± 19.1
Ginseng root (2 g/kg)	11	63	48.7 ± 3.42	221.12 ± 30.3
Ginsenosides (100 mg/kg)	11	100	49.9 ± 2.36	225.61 ± 21.3
Ginsenosides (100 mg/kg) + morphine (2 mg/kg)	10	10	44.5 ± 5.36	171.4 ± 30.1
Hypophysectomized rats				
Ginsenosides (100 mg/kg)	10	0	21.21 ± 2.28	129.41 ± 18.9

TABLE 10
Antidiuretic Effect of Ginseng

Experiment group	Dose	Urine volume (ml/kg, in 5 h)	Electrolytes in the urine (mEq/l) U_{Na^+}	U_{K^+}
I.	Control (saline)	49 ± 3	1.7	1.4
	Ginseng leaf			
	5 g/kg	26 ± 4	1.67	2.5
	12 g/kg	10 ± 2	1.0	2.2
II.	Control (saline)	43 ± 2	0.82	0.66
	Ginsenosides			
	63 mg/kg	35 ± 4	0.66	1.53
	126 mg/kg	31 ± 1	0.74	1.48
III.	Control (sunflower oil + H_2O)	59 ± 9	1.5	0.33
	DOCA	26 ± 4	0.56	0.68
	Ginsenosides + DOCA (126 mg/kg)	22 ± 4	0.5	0.5

Experiment group	Dose	Urine volume (ml/kg) 1 h	2 h	3 h
IV.	Control	15.8 ± 1.0	19.0 ± 4	20.4 ± 4.0
	Ginsenoside 126 mg/kg	0	2.6 ± 0.7	8.8 ± 0.5
	Pitressin 50 U/kg	1.0 ± 0.7	7.8 ± 1.2	18.4 ± 4.0
	Ginsenoside 126 mg/kg + pitressin 50 U/kg	0	1.4 ± 0.1	13.0 ± 1.4

Data from References 8 and 33.

Antidiuretic Effect

Since ginseng stimulates the anterior pituitary, resulting in increased secretion of ACTH, thyroid stimulating hormone, and gonadotropic hormones, it is not surprising to observe that ginseng also affects the posterior pituitary in exerting an antidiuretic action. Rats were anesthetized with ether and given a 0.45% saline injection by stomach tube at a dose of 50 ml/kg. Their bladders were emptied and urine was collected in a metabolic cage for 5 h. In a group of animals which received either ginseng extract or ginsenosides, urine excretion was markedly reduced and the urinary Na^+ concentration decreased, but urinary K^+ increased. The results are summarized in Table 10.

TABLE 11
Effect of Cortisone and Ginsenosides on Blood Sugars and Hepatic Glycogen Content in Mice

	Group and drug(s)	Blood sugar (mg/dl)	Hepatic glycogen (mg/100 g liver tissue)
I.	Control	96 ± 7 (9)[a]	0.401 ± 0.059
	Cortisone	205 ± 8 (9)	0.312 ± 0.028
	Cortisone + ginsenoside (50 mg/kg)	127 ± 7 (9)	0.308 ± 0.232
II.	Epinephrine		
	(0.1%) 0.02 ml/Mouse	215 ± 7 (8)	
	+ Ginsenoside (25 mg/kg)	164 ± 5 (8)	
	+ Ginsenoside (50 mg/kg)	156 ± 10 (9)	

[a] Mean ± S.D. (the number of animals).

Data modified from Reference 32.

Metabolism

Ginseng is described in Chinese medicine as a tonic or a drug to replenish vital energies, used to restore body fluids. Contributing to such effects are its actions on body metabolism.

Carbohydrates

The effects of ginseng on carbohydrate metabolism are complicated, exhibiting a combination of direct and indirect effects.

Ginseng was used in Chinese medicine to reduce blood sugar levels in diabetic patients. But ginseng itself is a powerful corticosterone-stimulating agent, as described in the section on the effects of the herb on the endocrine system. In glucose tolerance tests, blood sugar levels fell faster in the ginseng group than in the control group. The plasma concentration of injected radioactive insulin also fell faster in the ginseng group than in the control.

Table 11 presents the experimental results on blood sugar and hepatic glycogen content in rats receiving cortisone, or cortisone and ginsenosides in combination.

Yang et al.[37] injected ginseng polysaccharides in mice and found they reduce the blood sugar level and liver glycogen by promoting the oxidative phosphorylation of carbohydrate, secondary to an increase of insulin release.

Gong et al.[14] reported that ginsenoside Rg_1 isolated from *P. notoginseng* (Burk.) produced a significant hypoglycemic effect on alloxan-induced diabetic mice. Such an effect lasted more than 4 h and was not synergistic to insulin action. In alloxan-induced diabetic rats, ginsenosides also can reduce blood sugar level as shown in Table 12.

Experimental data suggest that ginsenosides have a hypoglycemic effect, though probably directly on glucose metabolism rather than via insulin regulation. The mechanism of this action is still unclear.

RNA and DNA Metabolism

Ginseng extract can increase the biosynthesis of RNA and incorporation of amino acids in the nuclei of hepatic and renal cells, resulting in an increase of serum protein. Among the ginsenosides tested, Rd exhibits the greatest effect on protein synthesis. Experiments also show that ginsenosides increase the synthesis of high density lipids (HDL) and

TABLE 12
Effect of Ginsenosides on Alloxan-Induced Diabetic Rats[8]

Alloxan rats	Blood sugar level (mg/dl) 1 d	4 d (after alloxan administration)	7 d
Control	290 ± 33	384 ± 23	426 ± 22
Ginsenosides			
30 mg/kg i.p.	144 ± 16	341 ± 23	358 ± 25
100 mg/kg i.p.	160 ± 18	130 ± 4	289 ± 27

apoprotein in serum. After administration of ginseng extract to rats, the activity of DNA-dependent RNA polymerase in the hepatic cellular nuclei increased markedly. Subsequently, the rate of RNA synthesis in the nuclei increased and was followed by cytoplasm. An increase of serum protein was observed later — 8 to 12 hours after drug administration.

Administration of ginseng also greatly enhances the biosynthesis of DNA and protein in the bone marrow. It is thus seen to be beneficial to use ginseng in the treatment of hemorrhagic anemia. In animal experiments, ginseng extract has increased ^{59}Fe incorporation into red blood cells. Furthermore, the hemopoietic stimulating effects of ginseng can be blocked by administration of cycloheximide, a bone marrow depression agent.

Ginseng contains some vitamin B_{12}. It seems that the hemopoietic action of ginseng is its direct effect on DNA synthesis of the hemopoietic tissue. *In vivo* and *in vitro* experiments show that ginseng extract also exhibits a stimulating effect on DNA and protein synthesis of the testes, thymus, and adrenal cortex.

Fat Metabolism

Ginseng extract can increase the total fat content of tissue. Rats which received ginsenoside Rc at 5 mg/100 g body weight intraperitoneally showed a threefold increase of epididyms fat synthesis from injected ^{14}C-acetic acid relative to the control group, and a sixfold increase in intestinal fat synthesis.

It is known that ACTH or insulin possesses lipolytic effect. This can be antagonized by the administration of most ginsenosides, such as Rb_1, Rb_2, Rc, Rd, Re, Rg_1, Rg_2, or Rh_1. Several investigators also reported that oral administration of ginsenosides or the essential oils from ginseng root to rats for more than 4 weeks markedly reduces cholesterol levels in the plasma and liver. Among the ginsenosides being tested, Rb_1 was seen to have the most potent stimulating effect on cholesterol synthesis. This ability to lower plasma cholesterol levels is believed to stem from increased biliary excretion of cholesterol in the bile. Aburirmeileh[10] gave red ginseng to rats for 90 d and found cholesterol levels and triglycerol concentrations in plasma were much lower than that of the control group. Moreover, the HDL level increased. Results of his experiment are presented in Table 13.

Zheng and Yan[41] fed ginsenosides to high-lipid-diet rabbits for 24 d and found that the total cholesterol level in the plasma was down from 33.75 ± 7.75 to 5.98 ± 1.46 mmol/l, and the HDL was up from 0.61 to 0.89 mmol/l.

In another animal experiment, rabbits were fed with total ginseng saponins isolated from *P. notoginseng* at a dose of 100 mg/kg/d for 8 weeks by a stomach tube. When the animals were sacrificed a remarkable reduction in their aortic endothelial damage and atherosclerosis formation was shown as compared with the control animals.[25] The authors also reported that the ginseng saponins increased the 6-keto-$PGF_{1\alpha}$ level in the carotid artery and decreased the thromboxane A_2 concentration of the platelet.

TABLE 13
Effect of Ginseng on Cholesterol, HDL, and Triglycerol Plasma Level in Rats[8,10]

	Serum concentration (mg/dl)		Hepatic content (mg/100 g tissue)	
	Control	Red ginseng (100 mg/100 g b.w. oral × 90 d)	Control	Red ginseng
Total cholesterol (TC)	93.6 ± 14.6	64.7 ± 3.3	72 ± 6	50 ± 4
HDL	24.9 ± 2.8	31.7 ± 26		
Ratio TC/HDL	3.8	2.0		
Triglycerol	230 ± 35	161 ± 15	124 ± 12	72 ± 7
Phospholipid	135 ± 6	148 ± 5	105 ± 5	128 ± 4

Body Growth

In yet another experiment, young pigs weighing 20 to 40 kg were given the ginsenosides of ginseng leaf or rhizome every other day for 14 d. As a result, their body weight increased significantly more than the control group. Analysis of the treated animals showed an increase of protein content as well as RNA and DNA concentration in the liver cells and muscle cells, but fat content did not change.

Immune System

Small doses of ginseng saponins can increase the level of serum antibody. Large doses lower antibody concentration. Tumor-carrying mice which are given ginseng saponins extracted from either the root or the leaf have seen an increase in the phagocytosis of the reticuloendothelial system and a reduction in tumor weight.

Li et al.[21] reported that ginseng can stimulate the production of specific antibodies in animals, resulting in an increase of immune function. Ginsenosides were found to be effective in the treatment of cancer patients, improving their appetites and aiding in sleep. Objective signs such as a decrease in tumor size and increase of serum immunoglobulins have also been observed in those patients treated with ginseng.

Ginseng extract, especially *Panax zingiberensis*, has a potent antiirradiation effect. Some investigators have recommended a combination of ginseng and X-ray irradiation therapy in cancer treatment. Ginseng not only would reduce the adverse effects of X-ray irradiation, but can also accelerate the recovery of immune system and stimulate the hemopoietic system of the bone marrow.

Anticancer Activity

The alcoholic extract, though not the water extract, of ginseng can inhibit several types of cancer cell growth *in vitro*. Experimental data also indicate that giving ginsenosides to mice inoculated with S-180 cancer cells can prolong animal survival. The mechanism of the anticancer activity of ginsenosides is not well understood, however. When ginsenosides were added to the culture medium of cancer cells, a morphological change of the cancer cells was observed. The number of cytosol mitochondria increased significantly and they were arranged in a more orderly fashion. At a concentration of 2.5 to 5 μg/ml, ginseng alcoholic extract can inhibit the growth of S-180 or U-14 cancer cells in the culture media.

Clinical trials giving ginseng to patients suffering from stomach cancer showed a shrinking of the cancer size and an increase in patients' appetites. Body weight increased, and survival times were prolonged.

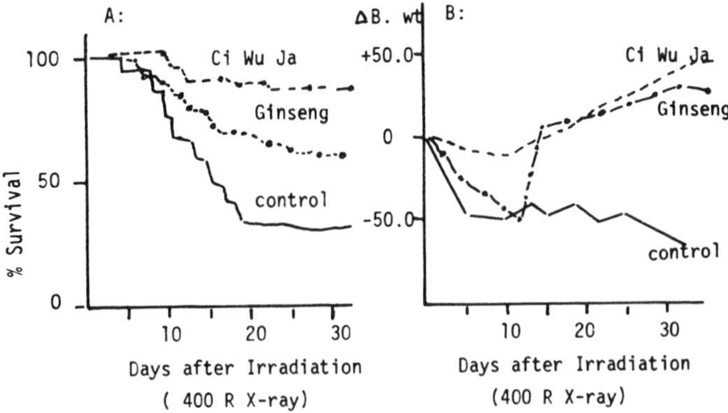

FIGURE 5. The antiirradiation effect of ginseng and *Ci Wu Jia*. (A) Survival rate; (B) change of body weight; (control) saline (i.p.) daily for 1 month; (ginseng) extract 10%: 0.1 ml/20 g mice daily for 1 month; (*Ci Wu Jia*) extract 10%: 0.1 ml/20 g mice daily for 1 month.

Antiirradiation Activity

An experiment testing the antiirradiation effects of ginseng was performed on male mice exposed to a dose of 400 Roentgen total body X-ray irradiation. These were then treated with either 10% ginseng extract at 0.1 ml/20 g body weight intraperitoneally daily or 10% extract from *Ci Wu Jia* (刺五加), *Acanthopanax senticosus* (Rupr. et Maxim.) Harns. In comparison with the control group which received an injection of saline only, both extracts were shown to have a significant effect in protecting the animals, especially the latter. The mortality rate was reduced, the damage of bone marrow as measured by red blood-cell count was much milder, body weight recovered rapidly, and the decreased ratio of serum albumin to globulin was smaller. Figure 5 illustrates the antiirradiation effect of ginseng and *Acanthopanax*.

The difference between these two herbs is the content of antioxidant maltol. *Acanthopanax* contains a high amount of maltol, which may be the principal component in the antiirradiation effect. Table 14 illustrates the dose response relationship of ginseng extract on anti-X-irradiation effect. Generally, the ginseng-treated group showed a rapid recovery of red blood cell and platelet count.

Antiaging Effects

Chinese folklore acclaims ginseng as a rejuvenizing agent. Because of the multiple functions exerted by ginseng, antiaging effects may be indirectly contributed to by many factors. As previously described, ginseng extract can increase the biosynthesis of DNA, RNA, and protein, as well as stimulating the secretion of gonadotropin and ACTH from the hypothalamopituitary axis, resulting in an increase of sex hormones and plasma corticosterone concentration. Such effects can exert some indirect action in prolonging cell life. Bittles et al.[11] reported that ginseng can prolong the life span of young female mice, though not older animals.

Anti-Inflammation Activity

Inflammation and swelling were induced in both normal and adrenalectomized rats by inoculation of 0.05 ml of 6% sugar solution at the ankle area of the animals. In $^1/_2$ to 2 $^1/_2$ h before the inflammatory substance was applied, the animals received a subcutaneous

TABLE 14
Survival Percentage of X-Irradiated Mice Receiving Ginseng Extract[8]

X-irradiation (roentgen)	Dose of ginseng extract (mg)	% Survival in 30-d period
650	0	33 (15)[a]
	1.0	78 (15)
675	0	28.5 (55)
	4.0	80.5 (55)
720	0	5 (40)
	1.8	45 (40)
	3.4	75 (40)
	6.8	82.5 (40)

[a] Number in parenthesis is the number of animals.

injection of 50% alcoholic extract of ginseng at a dose of 0.2 to 0.4 ml (equivalent to 1 to 2 g of crude herb). The inflammation and swelling of these animals were greatly reduced compared with the control animals which did not receive ginseng treatment. Such anti-inflammation effects were equal to those recorded in a group receiving cortisone injections at a dose of 15 mg/kg. The adrenalectomized group receiving ginseng extract also showed a reduction of inflammation and swelling, but not as remarkable as in the normal group.

Ginseng extract was also effective in reducing inflammation and irritation due to burns. Local application of 1% ginseng tincture to the eye can reduce eye irritation and improve the healing of wounds.

The mechanism of the anti-inflammation effect is not clear, partially due to the indirect effects of ACTH-cortisone stimulation, but partially due to a direct effect. Some investigators reported that they have isolated two anti-inflammatory components from ginseng root which were panaxatriol derivatives. Some Russian investigators also believe that ginseng extract can inhibit increased capillary permeability induced by injection of egg white. Hao and Yang[16] reported that the ginseng saponins can decrease the capillary permeability induced by histamine injection and the inflammatory edema of the ear induced by injection of dimethyl benzene.

ATPase

The structures of most ginseng sapogenins, either panaxadiol or panaxatriol, are very similar to those of the digitalis glucoside genin. However, they lack the hydroxyl (OH^-) group at the C-14 position. Therefore, one would expect that the ginsenosides act less effectively on Na-K ATPase activity than that of digitalis.

Early investigation by a Chinese team showed that ginseng inhibited ATPase activity and resulted in increased myocardial contractility.[27] Wo[34] reported that ginseng saponins exerted an inhibitory effect on mouse heart muscle ATPase, Mg^{2+} ATPase and on O_2 consumption. Hu et al.[17] also reported that the total ginseng saponins caused a dramatic inhibition of the activity of Na-K ATPase, Mg^{2+} ATPase, and Ca^{2+} ATPase in rabbit cerebral microsomes. Cao et al.[12] showed that Rb_1 exerted a noncompetive inhibition on Na-K ATPase activity of rat brain microsomes. The results are summarized in Table 15.

TABLE 15
Effect of Ginseng on ATPase Activity

Tissue	Total ATPase	Ca^{2+} ATPase	Mg^{2+} ATPase	O_2 consumption	Ref.
		(μM/Pi/mg protein/h)		(nM O_2/mg tissue)	
Heart muscle					
Control	10.45 ± 0.11		10.3 ± 0.15	13.89 ± 0.17 (10)	34
Ginseng No. 6 (500 μg/tube)	3.45 ± 0.06		2.68 ± 0.09	10.08 ± 1.48 (10)	
Brain microsomes					
Control	42.5 ± 0.7	19.7 ± 1.0	17.9 ± 1.2		17
Ginsenoside					
0.1 mg/ml	33.7 ± 0.9	14.2 ± 1.3	15.9 ± 0.9		
1.0 mg/ml	8.27 ± 1.18	9.6 ± 1.1	13.2 ± 1.3		

TABLE 16
A Summary of Pharmacokinetic Data Obtained from Different Animal Species

		Parameter					
Animal	Ginsenoside	$t_{1/2/\alpha}$ (min)	$t_{1/2\beta}$ (h)	Vd (l/kg)	Cl (l/kg/h)	K_{el} (min^{-1})	Ref.
Mini-pig (i.v.)	Rb_1	20	16	0.16	0.18	0.13×10^{-3}	19
	Rg_1	27		0.36	13.6	0.025	
Rat (i.v.)	Rb_1		14.5				
	Rg_1	6.3					
Mice (i.v.)	Rg	4.2	25	44	1.2		18

Pharmacokinetics

Absorption and Distribution

Early on, Rg_1 had been found to be rapidly absorbed from the intestine and distributed into the liver, spleen, kidney, lung, and brain. The concentration of saponin in liver and kidney was thought to be about twice as high as in other organs. Recent reports by Takuno et al.[30] showed a contrary result. They found that after oral administration of Rg_1 to rats, only 23% of the dose was absorbed after 2.5 h; the rest remained in the small intestine.

Rb_1 was not completely absorbed from the intestine. If it was administered by intravenous injection, a two-compartmental curve was indicated. Data obtained from different investigators varied greatly, depending upon the kind of animal used (Table 16).

Excretion

Rb_1 is excreted in the urine mainly in an unmetabolized form. Twenty hours after intravenous injection, the amount appearing in the urine is 44.4 ± 2.6% of the dose given. A very small amount (0.8 ± 0.6%) is excreted in the bile and appears in the feces.

Adverse Effects

The *Herbal Classic of the Divine Plowman*[1] states that ginseng is nontoxic. Recent investigations, however, report that LD_{50} in mice was 5 g/kg for ginseng powder (oral) or 16.5 ml of ginseng extract administered subcutaneously. Clinical trials on patients using 3% ginseng tincture up to 100 ml showed slight irritation and excitation. If the dose was increased to 200 ml, urticaria, itching, headache, dizziness, hemorrhaging, and insomnia were reported. But because of the expense and rarity of the herb, no report has ever been recorded of death or poisoning from overdose.

Therapeutic Uses

Revival of Dying Patients

Ginseng is reputed to be effective in treatment of shock, collapse of the cardiovascular system, hemorrhaging, and heart failure. Clinical trials on volunteers show that ginseng extract can slow down the heart rate and reduce O_2 demand but the mechanism of action is very debatable.

In Chinese medical journals, there have been some reports on the effectiveness of ginseng decoction in combination with other herbs in the treatment of cardiac shock and acute myocarditis. Shen et al.[28] used ginsenosides in the treatment of *Escherichia coli* endotoxin-induced septic shock in dogs, and reported that ginsenosides can prevent the inhibitory effect of endotoxin on myocardial function, as well as restore failing blood pressure.

Antihypertension, Antianemia, Antiangina, and Coronary Atherosclerosis

The effectiveness of ginseng tincture in hypertension is inconsistant and controversial. Sometimes it lowers blood pressure, at others, it raises blood pressure. In traditional Chinese medicine, chewing ginseng was recommended as a means of revitalizing the *yin* and the *yang* by raising blood pressure.

Because of its definite effect in lowering serum cholesterol levels and raising HDL levels, continued administration of ginseng has been widely applied as an antihypertensive remedy. It was recommended primarily in the form of a tincture.

Diabetes Mellitus

Ginseng is effective in the treatment of mild hyperglycemia. Usually, blood sugar levels can be lowered 40 to 50 mg/dl after continuous administration for 2 weeks. In severe diabetic cases, ginseng alone is less effective. However, it was reported that ginseng in combination with insulin treatment will reduce the required dose of insulin and prolong the hypoglycemic effect of insulin.

Rejuvenization and Revitalization

Middle-aged Chinese males still use ginseng as a remedy for impotence and to reinforce sexual desire.

Neurological Symptoms and Mental Fatigue

Ginseng can reduce the effects of O_2 deprivation of brain tissue, and improve mental and neurological performance. Ginseng has been effective in the treatment of persons suffering from neurasthenia, bad moods, headaches, and insomnia. Reports from Shenyang Military Hospital showed that in 50 cases of polio, ginseng extract greatly improved patients' muscle strength and skin circulation.

Coughing and Headaches

American Indians have used ginseng (*Panax quinquefolius*) in treating coughs, headaches, and fevers. They also used the herbs to "strengthen mental processes".[4]

Gastrointestinal Malfunction

Ginseng is listed in most Chinese pharmacopoeias as a tonic to stimulate gastric secretion and accelerate recovery in chronic illness. Elderly persons take ginseng tincture to stimulate their appetite and reduce weakness.

Anticancer Activity and Stimulation of the Immunologic System

Clinical trials using ginseng in the treatment of cancer of the stomach, liver, pancreas, or intestine showed a progressive improvement. Patients reported an increase of appetite, body weight, and reticuloendothelial cell counts.

Hematopoietic Stimulation

Ginseng can stimulate reticulocyte formation from the bone marrow and increase the ^{59}Fe uptake. Folk medicine used it in the treatment of anemia.

Preparations

The following are examples of ginseng decoctions used in Chinese folk medicine:

1. *Du-seng* soup (独参湯)
   Ginseng 25 to 50 g
   Water decoction
2. *Seng-fu* soup (参附湯)
   Ginseng 25 g
   Fu-zi 10 to 20 g
   Water decoction
3. *Sen-mag* mixture (生脉散)
   Ginseng 5 g
   Mag-don 15 g
   Wu-mei-zi 5 g
   Water decoction
4. *Qui-pi* soup (归脾湯)
   Ginseng 15 g
   Acid date 9 g
   Pei-sue 15 g
   Moi-shan 6 g
   Foo-ling 15 g
   Licorice 6 g
   Huang-gi 15 g
   Long-yen-yu 6 g
   Water decoction
5. Ginseng *Fou yi* soup (四逆湯)
   Ginseng 15 g
   Choi-fu 9 g
   Pei-chai 9 g
   Chi-si 9 g
   Licorice (roasted) 9 g
   Water decoction
6. Ginseng tincture 3 to 10% in alcohol, 10 ml t.i.d.

7. Ginseng injection 10% 2 ml/ampule for i.m. injection
8. Ginseng tablet 0.15 g/tablet

References

Books

1. *The Herbal Classic of the Divine Plowman* (Shen-nong Ben-cao Chien), written about 100 BC, author(s) unknown.
2. **Hardacre, V.,** *Woodland Nuggets of Gold,* Vantage Press, 1968.
3. **Li, Shih-Chen,** *Ben Cao Kong Mu* (本草纲目), 1580.
4. **Martin, L. C.,** *Wildflower Folklore,* Globe Pequot Press, Chester, CT, 1984, 107.
5. **Meyer, J. E.,** *The Herbalist,* Hammond Book, Hammond, IN, 1934, 106.
6. **Millspaugh, C. F.,** *American Medicinal Plants,* Dover, New York, 1974, 276.
7. **Roberts, C. R. and English, J., Eds.,** *Proc. 4th National Ginseng Conf.,* 1982.
8. **Wang, B. C.,** *The Ginseng Research,* Tianjin Scientific Publisher, 1984.
9. **Wu, S. J.,** [*The Pharmacology of Chinese Herbs*], People's Public Health Press, China, 1982.

Journals

10. **Aburirmeileh, N., et al.,** *5th Asian Symp. Medical Plants (Korea),* 1984, 79.
11. **Bittles, et al.,** *Proc. Int. Geront. Symp.,* Singapore Press, Singapore, 1977, 167.
12. **Cao, J. et al.,** *Acta Pharmacol. Sin.,* 11, 10, 1990.
13. **Du, W. et al.,** *Chin. J. Appl. Physiol.,* 7, 125, 1991.
14. **Gong, Y. H. et al.,** *Acta Pharmaceutica Sin.,* 26, 81, 1991.
15. **Gortirner, F. and Vogt, J.,** *Arab. Pharmaceutical,* 299, 916, 1966.
16. **Hao, C. O. and Yang, F.,** *Acta Pharmacol. Sin.,* 7, 252, 1986.
17. **Hu, G. et al.,** *Acta Pharmacol. Sin.,* 9, 486, 1988.
18. **Huo, Y. S. et al.,** *Acta Pharmacol. Sin.,* 7, 519, 1986.
19. **Jenny, E. and Soldati, F.,** in *Advances in Chinese Medicinal Materials Research,* Chang, H. M. et al., Eds., World Scientific Publishing Co., Singapore, 1985, 499.
20. **Kirchdorfer, A. M.,** in *Advances in Chinese Medicinal Materials Research,* Chang, H. M. et al., Eds., World Scientific Publishing Co., Singapore, 1985, 529.
21. **Li, S. et al.,** *Jilin Chin. Med. Mater. Med.,* 1:67–71, 1980.
22. **Li, H. T. and Shi, L.,** *Acta Pharmacol. Sin.,* 11, 213, 1990.
23. **Li, J. C. et al.,** *Acta Pharmacol. Sin.,* 9, 22, 1988.
24. **Li, X., Chen, J. X., et al.,** *Acta Pharmacol. Sin.,* 11, 26, 1990.
25. **Lin, S. et al.,** *Acta Pharmacol. Sin.,* 11, 29, 1990.
26. **Liu, G. T. and Song, C. Y.,** *Physiol. J.,* (Chin.) 25, 129, 1962.
27. Research Team of the Traditional Medicine Institute, *New Med. Pharmacol. J.,* 10, 27, 1973.
28. **Shen, Y. T. et al.,** *Acta Pharmacol. Sin.,* 7, 439, 1986.
29. **Shi, L. et al.,** *Acta Pharmacol. Sin.,* 11, 29, 1990.
30. **Takuno, Y. et al.,** *Chem. Pharm. Bull.,* 30, 2196, 1982.
31. **Tanaka, O. et al.,** *Abstr. Chin. Med.,* 1, 30, 1986.
32. **Wang, B. X. et al.,** in *Advances in Chinese Medicinal Materials Research,* Chang, H. M., Ed., World Scientific Publishing Co., Singapore, 1985, 519.
33. **Wang, B. X. et al.,** *Acta Pharmacol. Sin.,* 1, 126, 1980.
34. **Wo, X. D.,** *Pharmacol. Clin. Chin. Herbs,* 3, 27, 1987.
35. **Wu, J. X., Chen, J. X., et al.,** *Acta Pharmacol. Sin.,* 9, 147, 409, 1988.
36. **Xiong, Z. G., Chen, J. X., et al.,** *Acta Pharmacol. Sin.,* 10, 122, 1989.
37. **Yang, M., et al.,** *Acta Pharmacol. Sin.,* 11, 520, 1990.
38. **Sakei, Y.,** *Tokyo Med. J.,* 28, 8, 1915.

38a. **Sakei, Y.,** *Tokyo Med. J.,* 29, 6, 1916.

39. **Zhang, J. S., Sigdestad, C. P., et al.,** *Radiat. Res.,* 112, 156, 1987.
40. **Zhang, F. L. and Chen, X.,** *Acta Pharmacol. Sin.,* 8, 217, 1987.
41. **Zheng, X. L. and Yan, Y. F.,** *Chin. Pharm. Bull.,* 7, 110, 1991.

GINSENG-LIKE HERBS

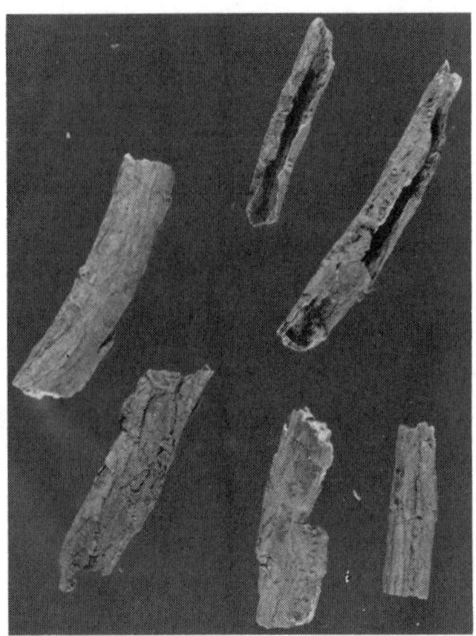

FIGURE 6. *Ci Wu Jia* or *North Wu Pie Pi*.

CI WU JIA (刺五加) — the dried root and rhizome of *Acanthopanax senticosus* (Rupr. et Maxim.) Harms; also known as *Eleutherococcus senticosus*

This herb was commonly sold on the market as Siberian ginseng (Figure 6). The Chinese also called it North *Wu Jia Pia* (北五加皮). It has been frequently confused with two similarly named herbs: (1) *Wu Jia Pi* (五加皮, or South *Wu Jia Pi*), the root of *Acanthopanax gracilistylus* (W. W. Smith) (see Chapter 11); and (2) *Xiang Jia Pi* (香加皮, the fragrant *Jia Pi*), the bark of *Periploca* (see Chapter 2 and Figure 6).

In his article, Hu[1] described the differences between these herbs, giving a detailed picture of each plant. Through a miscroscope, it can be seen that the parenchyma cells of these plants are different. For example, the parenchyma cells of the periploca have prismatic crystals; the herb has a distinct bitter taste. *Eleutherococcus*, on the other hand, has druse-type crystals in the parenchyma cells and the bark has no taste.

Chemistry

Several substances have been isolated from the *Ci Wu Jia*

1. The glucosides
   β-Sitosterol glucoside, $C_{35}H_{60}O_6$
   Eleutheroside B_1, $C_{17}H_{20}O_{10}$
   Eleutheroside C, $C_8H_{16}O_6$
   Eleutheroside D, E, F, and G
2. The nonglucosides
   l-Sesamen, $C_{20}H_{18}O_6$
   Syringaresinol, $C_{22}H_{26}O_6$

Other glucosides have been isolated from the leaf. They are eleutherosides I, K, L, and M; their structures differ significantly from those isolated from the bark.

Polysaccharides — according to Wagner there are four polysaccharides isolated from *Acanthopanax* with a molecular weight ranging from 15,000 to 200,000. They are immunostimulating agents.

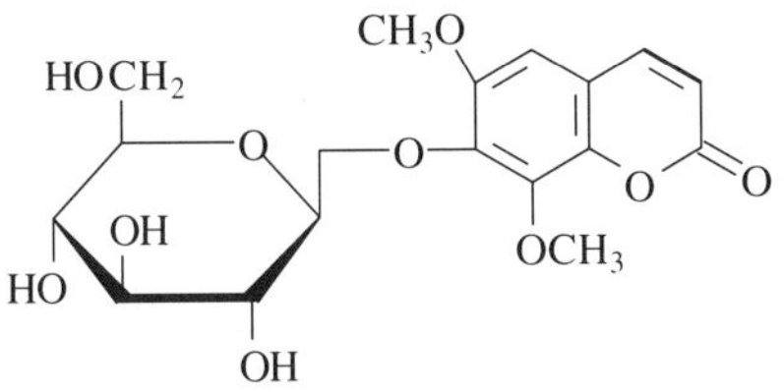

Eleutheroside B_1

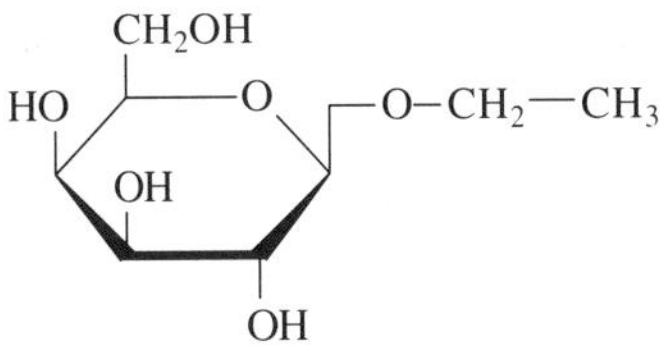

Eleutheroside C

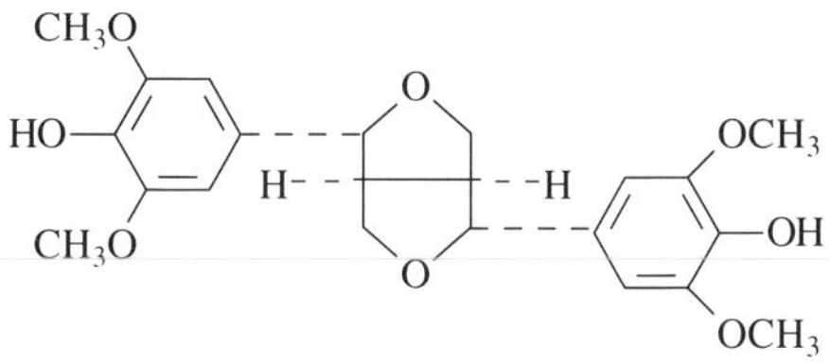

Syringaresinol

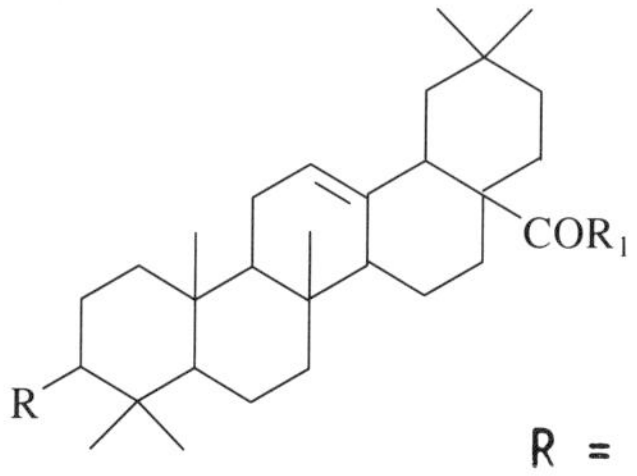

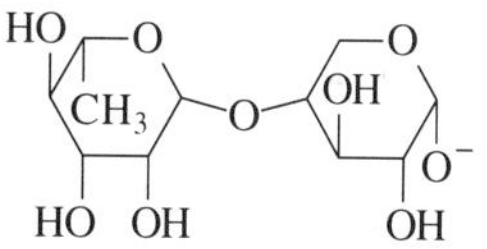

	R =	R_1=
Eleutheroside I	a	OH
Eleutheroside K	b	OH
Eleutheroside L	a	c
Eleutheroside M	b	c

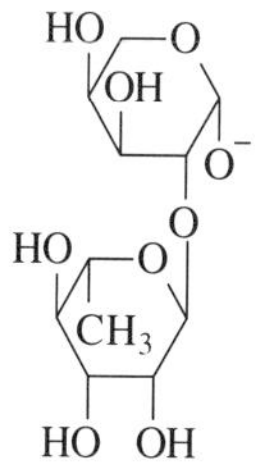

a

b

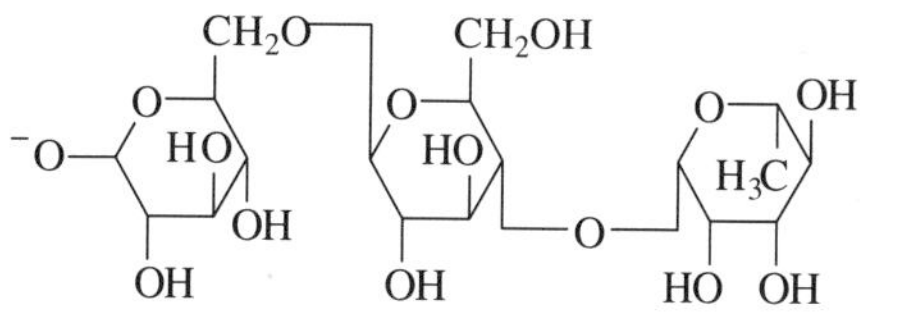

c

A related species from the genus *Acanthopanax* is Short Stem *Wu Jia* (短干五加), or the root of *A. sessiliflorus* Seem. It contains the following substances

- Acanthoside A, B, C, and D
- 1-Sesamen
- 1-Savinin, $C_{20}H_{16}O_6$

Both *Ci Wu Jia* and Short Stem *Wu Jia* are used as ginseng substitutes, thanks to their wide availability within China and the low cost.

Actions

1. The *Ci Wu Jia* glucosides have ginseng-like effects, including stimulation of the ACTH-cortisol system and lowering of blood pressure and blood sugar levels.
2. On the CNS, the herb exerts a tranquilizing effect.

Therapeutic Uses

Chinese traditional medicine recommends this herb for the replenishment of body functions and to promote digestion.

It is also said to enhance male sexual functions and to relieve mental strain.

References

1. **Hu, S. Y.,** in *Advances in Chinese Medicinal Materials Research,* Chang, H. M., et al., Eds., World Scientific Publishing Co., Singapore, 1985, 17–33.
2. **Wagner, H.,** ibid., 159–170.

ZHU JE GINSENG (竹節人參) — the root of *Panax japonicum* C. A. Meyer

Zhu Je Ginseng contains the following active ingredients

- Chikusetsa saponin II, $C_{47}H_{80}O_{17} \cdot 2H_2O$
- Chikusetsa saponin IV, $C_{47}H_{74}O_{18} \cdot 4H_2O$
- Ginsenoside Ro

This herb has been widely studied by Japanese investigators. It is commonly used commercially to substitute for the Manchurian or Korean ginseng.

Chapter 2

CARDIAC HERBS

For centuries, European folk medicine recommended foxglove or strophanthus for the treatment of congestive heart failure. Many Chinese herbs display pharmacological actions similar to foxglove, but were largely neglected due to the dominance of ginseng as the preferred cardiac treatment.

Only in the last century were digitalis preparations introduced by missionary medical personnel and subsequently adopted more widely by Chinese physicians. After the 1949 Communist revolution, the Chinese government encouraged medical professionals and the pharmaceutical industry to prepare and to grow Chinese cardiac herbs as a substitute for digitalis. However, the cost of isolating and purifying pure glucosides from herbs is higher than that of imported drugs. Additionally, there is little standardization in the decoction or tincture of herbs and even less scientific assessment of their therapeutic value. Thus, there is a long way to go before the use of these herbs can become practical.

JIA ZHU TAO (夾竹桃) — the dried leaf, root, and flower of *Nerium indicum* Mill.

FIGURE 1. *Jia Zhu Tao.*

Chemistry

The active principle of the herb is the glucoside oleandrin, $C_{22}H_{48}O_9$. After hydrolysis, it produces oleandrigenin.

TABLE 1
Cardiotonic Effect of *Jia Zhu Tao*

	Cardiotonic efficiency[b]				
		Jia Zhu Tao			
Dose[a]	**Digitalis leaf**	**Leaf**	**Stem bark**	**Wood**	**Flower**
0.08	44	47	80	33	0
0.10	75	64	100	100	66
0.12	95	90	90	100	100

[a] g/100 g frog's weight.
[b] Percentage of frog's heart arrested in systole.

$$\text{Oleandrin} \xrightarrow{H_2O} \text{oleandrigenin} + \text{oleandrose}$$

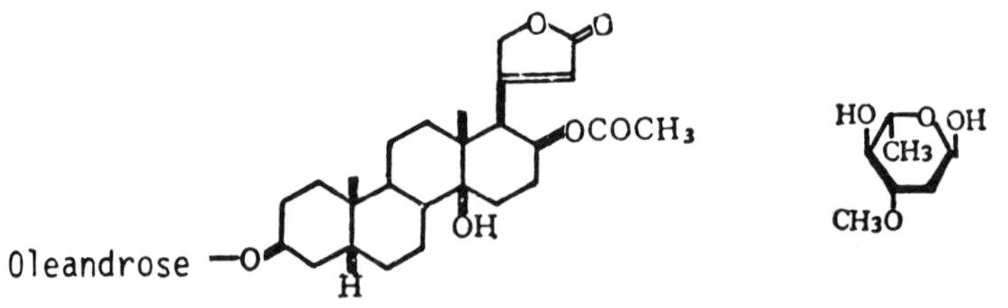

Oleandrin (folinerin)

Oleandrose

Actions

Similar to digitalis, oleandrin can increase myocardial contractility, slow the heart rate, and increase stroke volume. In their early studies, Chou and Jang and Huang and Jang reported that *Jia Zhu Tao* can accumulate in the heart muscle and produce a potent cardiotonic effect. Table 1 compares the cardiotonic effect of various parts of *Jai Zhu Tao* to the digitalis leaf.

Toxicity

Oleandrin is toxic. The minimum lethal dosage (MLD) for a pigeon is 0.44 mg/kg, compared to an MLD of 0.77 mg/kg for digitoxin. Common side effects include nausea, vomiting, and diarrhea. Occasionally, it may cause sleepiness and cardiac arrhythmia. The oral absorption rate is approximately 53%.

Therapeutic Uses

Once used by the Chinese to treat psychosis, *Jai Zhu Tao* is now primarily used to treat congestive heart failure. Generally, it takes about 1 to 3 d after administration to show an effect. The heart rate gradually decreases, urinary output increases, and the edema of lower extremities will disappear. Case reports, however, do not indicate that the benefits of oleandrin are equal to or better than those of digitoxin.

The herb is prepared as a tablet or capsule containing 0.05 to 0.1 g of crude powder. The total daily dose is 0.2 to 0.4 g, divided into two to three administrations per day.

References

1. **Chou, T. C. and Jang, C. S.,** *Chin. Med. J.*, 63A, 212, 1945.
2. **Huang, K. C. and Jang, C. S.,** *Technol. China,* 1, 25–36, 1948.

HUANG HUA JIA ZHU TAO (黄花夹竹桃) — the seed, flower, and leaf of *Thevetia peruviana*

Chemistry

The seed of this herb contains several cardiac glucosides. They are thevetin A and B, peruvoside A, B, C, and D, theveside, theviridoside, and cerberin. There is also a non-glucoside flavone, vertiaflavone.

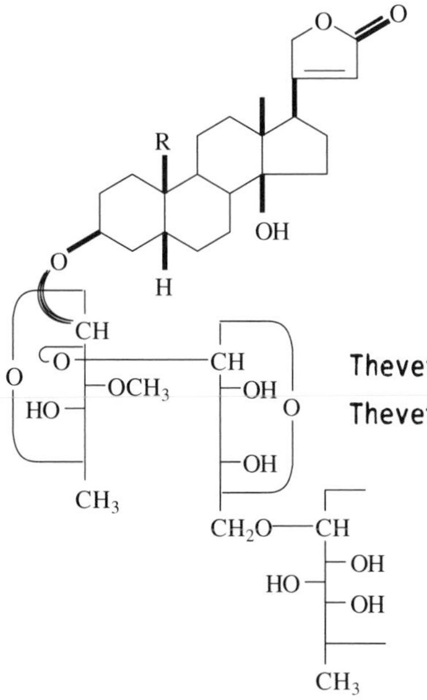

	R=
Thevetin A	CHO
Thevetin B	CH_3

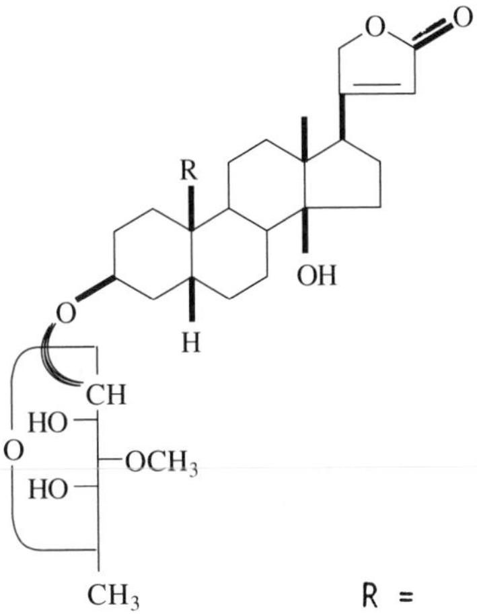

	R =
Peruvoside A	CHO
Peruvoside B	CH_3
Peruvoside C	CH_2OH
Peruvoside D	COOH

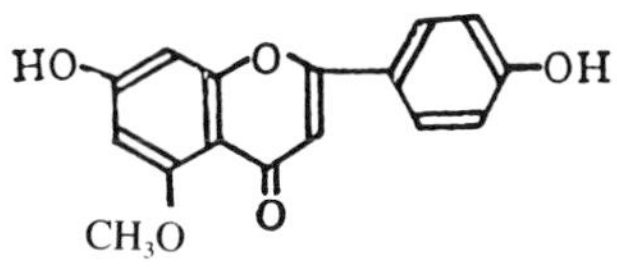

Vertiaflavone

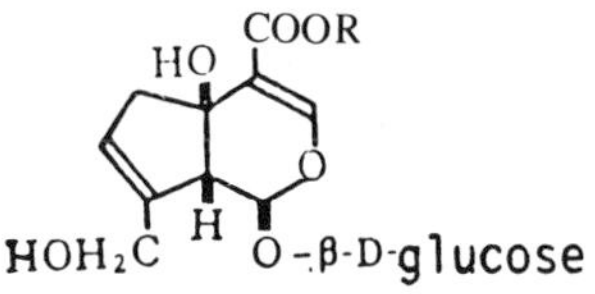

Theveside R= H
Theviridoside R = CH_3

The leaf also contains some glucosides, L-(*d*)-bornesitol, $C_7H_{14}O_6$, and lupeol acetate.

Actions

The pharmacological and toxic effects of this herb are similar to those of digitalis. Experiments on animals have shown that the potency of thevetin A and B on the heart is approximately 1/8 of that of strophanthin G. Table 2 presents a comparison of the effects of thevetin and strophanthin G.

It has been reported that thevetin A and B, plus the other glucosides of the herb, possess a tranquilizing effect and can prolong the sleeping time induced by barbiturates. Ye and Yang reported that peruvoside B showed a competitive inhibition on ^{3}H-ouabain binding to the Na-K ATPase isolated from guinea pig heart muscle. Clinical trials on 357 cases of congestive heart failure showed a 78.4% effective rate.

TABLE 2
Potency and Toxicity of Thevetin

Drug	Dose causing heart arrest in systole (mg/kg of body weight)	
	Frog	Cat
Thevetin	4.0–5.0	0.85
Strophanthin G	0.5–0.6	0.12

The herb is administered either as a thevetin tablet of 0.25 mg/tab, or a digitalized dose of 1.5 to 2.5 mg. A maintenance dose consists of 0.25 mg/d. For intravenous administration, an injection ampule is prepared from 0.25 mg/ml diluted in a 5% glucose solution.

Reference

1. **Ye, Y. X. and Yang, X. R.,** *Acta Pharmacol. Sin.*, 11, 491–494, 1990.

YANG GUO NAU (羊角扭) — the seed of *Strophanthus divaricatus* (Lour.) Hook. et Arn.

Chemistry

Cardiac glucosides constitute approximately 9 to 10% of the herb. These include

- Divaricoside, $C_{30}H_{46}O_8$
- Divostroside
- Sinoside
- Sinostroside
- Caudoside
- Caudostroside
- Sarmutoside

Among these glucosides, divaricoside occurs in greatest amounts within the seed and has the most potent cardiac effect.

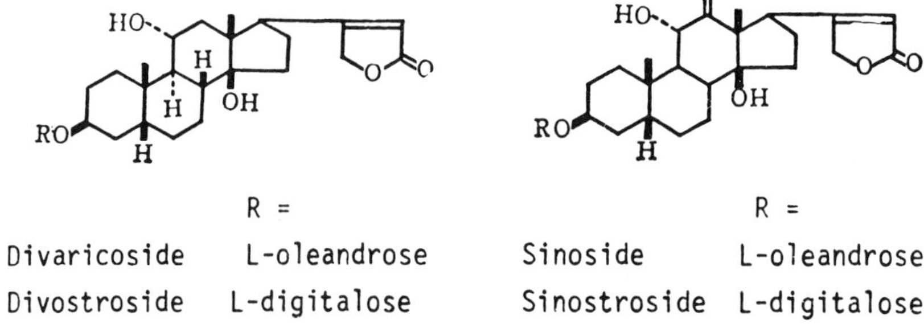

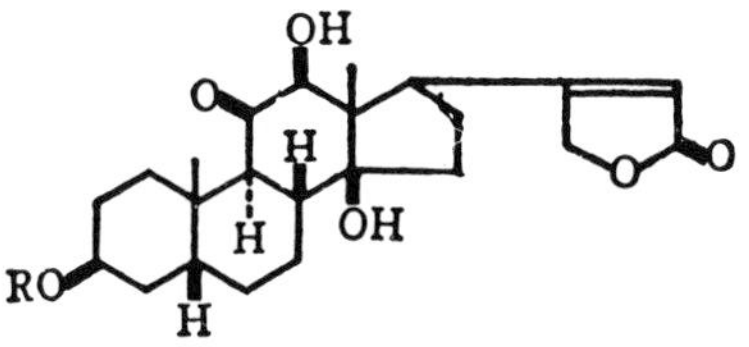

	R =
Sarmutoside	D-sarmentose
ψ-Caudoside	L-oleándrose
ψ Caudostroside	L-digitalose

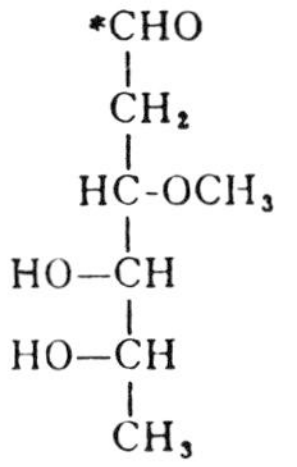

L-oleandrose	L-digitalose	D-sarmentose
*CHO	*CHO	*CH$_3$
CH$_2$	CH$_2$	CH$_2$
HC-OCH$_3$	H—C—OCH$_3$	H—C—OCH$_3$
HO—CH	H—C—OH	HO—C—H
HO—CH	HO—C—H	H—C—OH
CH$_3$	CH$_3$	CH$_3$

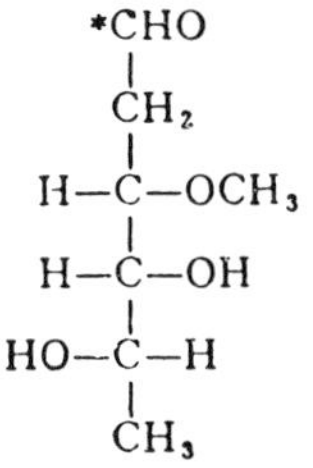

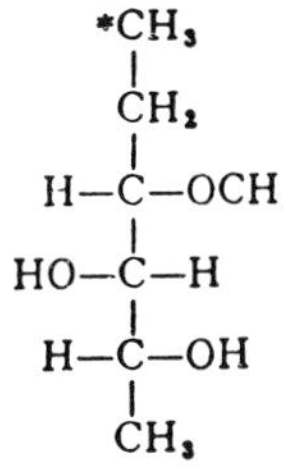

Actions

The cardiac glucosides of *Yang Guo Nau* have a cardiac stimulating action similar to strophanthin. They cause an increase of myocardiac contractility, slow the heart beat, and increase cardiac output. The onset of action is relatively fast and its cumulative effect is small.

The combined potency of *Yang Guo Nau* glucosides is about $^1/_4$ to $^1/_2$ that of strophanthin G, but the LD_{50} is almost equal. *Yang Guo Nau* glucosides exert a significant diuretic effect, probably due to the improvement of cardiac function. They also have a tranquilizing effect. In small doses, the glucosides can stimulate uterine contraction.

Toxicity

The intestinal absorption of these glucosides is slow and incomplete. Therefore, oral administration is not recommended, nor are subcutaneous or intramuscular administration advised, due to local irritating effects.

The glucosides are metabolized in the liver at a rate slower than that of strophanthin. Complete elimination of the drug from the body requires approximately 5 d.

The adverse effects are similar to those of strophanthin. These include gastrointestinal disturbances; nausea, vomiting, and constipation are common. Occasionally, cardiac arrhythmia may be observed.

Therapeutic Uses

Yang Guo Nau is used as a substitute for strophanthin K in treating heart failure. Contraindications include pericarditis, toxic myocarditis, acute and subacute endocarditis, shock and circulation collapse due to acute infections.

Divasidium is a preparation prepared from a combination of total *Yang Guo Nau* glucosides. Intravenous administration is recommended, diluting an ampule of 0.25 mg/ml with glucose solution.

LUO BU MA (罗布麻) — the dried leaf and root of *Apocynum venetum* L.

Chemistry

Luo Bu Ma contains several glucosides:

- Cymarin
- Strophantidin
- K-strophanthin-β
- Isoquercitrin
- Quercetin

The leaf also contains rutin and neoisorutin.

Actions

The glucosides of this herb can increase myocardial contractility; the action is similar to that of strophanthin, but is less potent. It lowers blood pressure, increases bronchial secretion, and produces a diuretic effect.

Toxicity

The toxicity of the herb is relatively low. Adverse effects generally include a parched mouth, nausea, vomiting, diarrhea, and loss of appetite.

Therapeutic Uses

Luo Bu Ma was used in traditional Chinese medicine to check hyperactivity, *yang* of the liver, to remove "heat", lower the blood pressure, and as a diuretic. Currently, it is primarily used in treating heart failure, hypertension, and edema. It can be used in the treatment of liver ascitis and pregnancy edema. On occasion, it is also used to treat chronic bronchitis and influenza.

The herb is prepared as an 8% decoction. This is taken in doses of 100 ml b.i.d. until the heart rate slows to 70 to 80 beats/min; subsequently, a daily maintenance dose of 50 ml is administered. The *Luo Bu Ma* leaf is also used as tea, with the dose titrated by the patient himself.

WON NIAN QING (万年青) — the dried rhizome and leaf of *Rhodea japonica* Roth

Chemistry

The plant contains four cardiac glucosides — rhodexin A, B, C, and D.

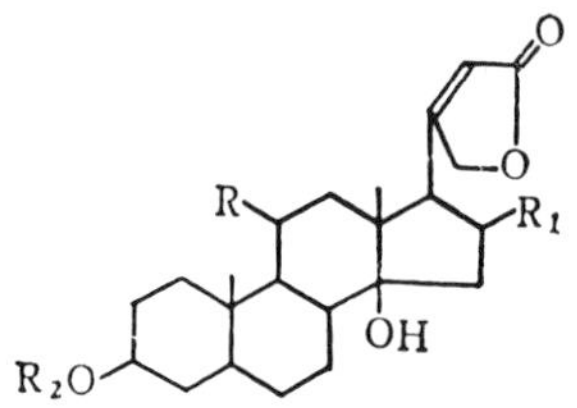

Rhodexin	R=	R_1=	R_2
A	OH	H	rhamnose
B	H	$OCOCH_3$	rhamnose
C	H	$OCOCH_3$	rham-O-glucose
D	H	$OCOCH_3$	glu-O-glucose

Actions

These glucosides have a digitalis-like effect on the heart muscle. Rhodexin A, B, and C are slightly more potent than digitoxin, although oral absorption of these glucosides is not complete. The oral dose required is approximately 10 times greater than an intravenous dose.

In low concentrations, the glucosides can cause vasodilatation of extremities and coronary vessels. In higher concentrations, however, they can cause vasoconstriction. Additionally, they increase gastrointestinal contraction, frequently resulting in severe abdominal pain and diarrhea. The glucosides also increase uterine contractions.

Won Nian Qing is also used as an emetic. It has an emetic effect via direct local irritation of gastric mucosa and stimulation of the chemoreceptor trigger zone (CTZ) of the medulla oblongata.

The plant is also reputed to have an antibacterial effect, especially against diphtheria bacteria. Toxicity is similar to that of digitalis, with nausea, vomiting, diarrhea, loss of appetite, dizziness, and sweating as common adverse effects. In severe cases, it may cause arrhythmia and heart blockage.

Therapeutic Uses

Folk medicine prescribed this herb as a cardiotonic, diuretic, antipyretic, detoxicant, and hemostatic. Primary uses of the herb today are in the treatment of heart diseases such as heart exhaustion or cardiac arrhythmia. Additionally, it has been administered in cases of diphtheria and dysentery. Reports show that diphtheria patients who received this herb daily for 5 to 6 d saw an improvement in symptoms and a removal of bacteria from the sputum; therapeutic effects in these cases were slightly below those of antitoxin treatments.

Preparation of the herb occurs in several forms

1. Decoction of the leaf or root, administered at 30 to 45 g/d
2. Injection ampule equivalent to 0.5 g crude herb per milliliter, diluted with glucose solution for intravenous injection
3. Rhodexin injection, containing rhodexin at 1.1 mg/ml
4. Vinegar extract, 40 g in 100 ml vinegar, extracted for 24 h. Remove the residue and dilute with water to 200 ml for a final concentration of 20%. This extract is used in diphtheria or dysentery treatments

FU SHOU CAO (福寿草) — or BING LIAN HUA (冰凉花) — the whole plant of *Adonis chrysocyathus* Hook F. et Thon., *A. brevistyla* Franch., and *A. vernalis* L.

Chemistry

The plant contains several cardiac glucosides: cymarol, $C_{30}H_{44}O_9$, corchoroside A, and convallatoxin, $C_{28}H_{42}O_{10}$. Other than these, there are many substances that have been isolated:

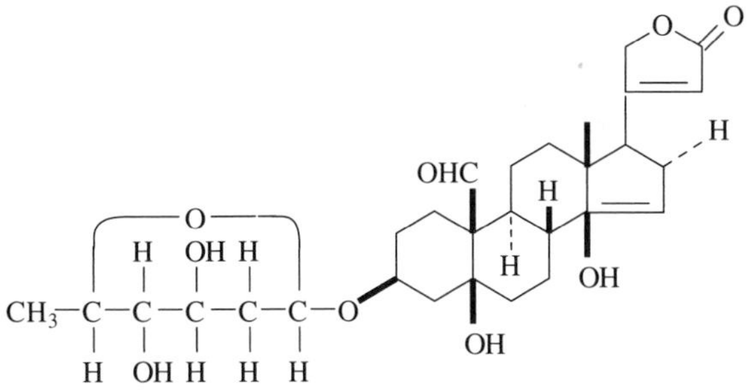

Corchoroside A

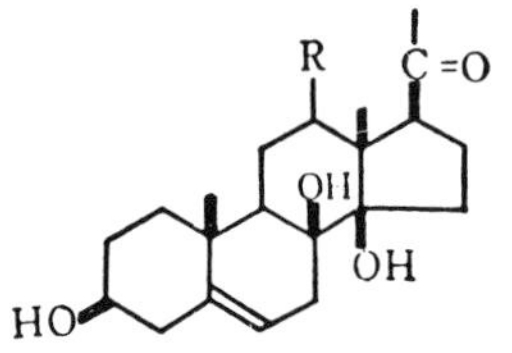

	R =
Fukujusone	H
12-O-Benzoylisolinelon	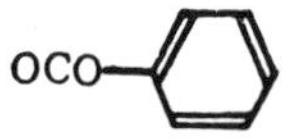 OCO–(phenyl)
12-O-Nicotinoylineolon	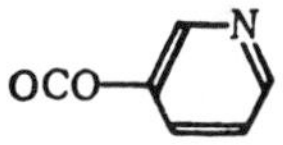 OCO–(pyridyl)

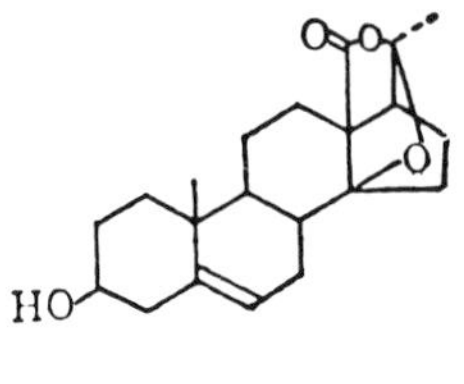

Adonilide

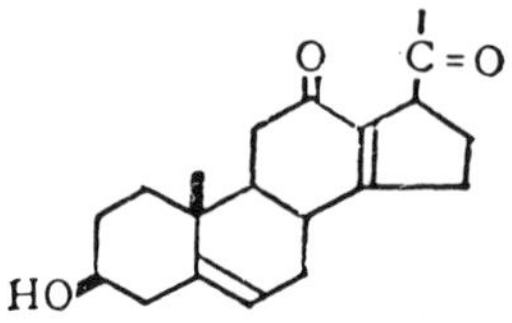

Fukujusonorone

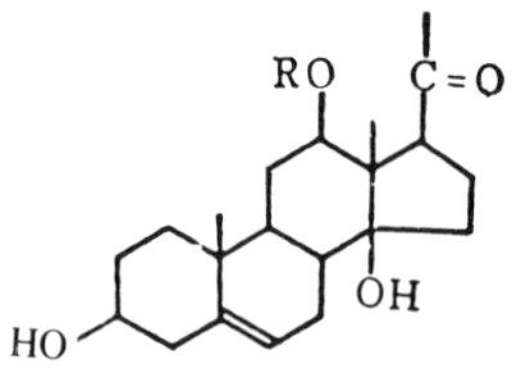

	R =
Isoramanone	H
Nicotinoylisoramanone	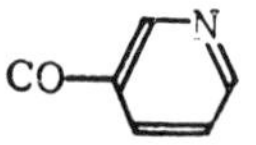 CO–(pyridyl)

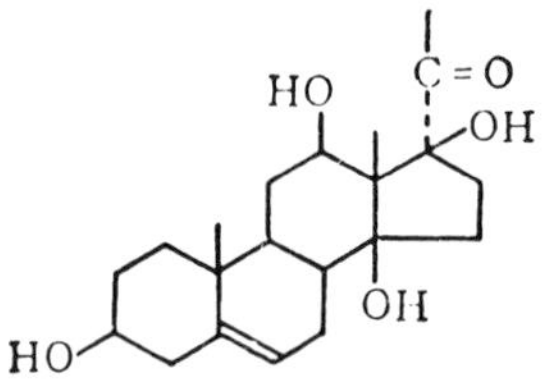

Pergularin

Actions

The herb exhibits a digitalis-like action on the heart muscle. Oral absorption is poor, requiring an effective dose approximately four times greater in quantity that an intravenous dose.

Toxicity

Toxicity is similar to strophanthin. The toxic dose for a cat is 0.75 mg/kg.

Therapeutic Uses

The herb is used in the treatment of heart disease and central depression. It can also be used as a diuretic, due to its direct inhibition effects on tubular reabsorption of Na, K, and Cl ions.

The herb is administered chiefly as a preparation called adonisidum, which contains all of the plant's glucosides at 0.5 mg/2 ml/ampule; this is diluted with a glucose solution for intravenous drip. For oral administration, the tablet form contains 0.5 mg glucosides per tablet; one tablet is taken t.i.d.

LING LAN (铃兰) — the whole plant or root of *Convallaria keiskei* Miq.

Chemistry

Ling Lan contains the following glucosides:

- Convallatoxin
- Convalloside
- Convallamarin
- Convallatoxol

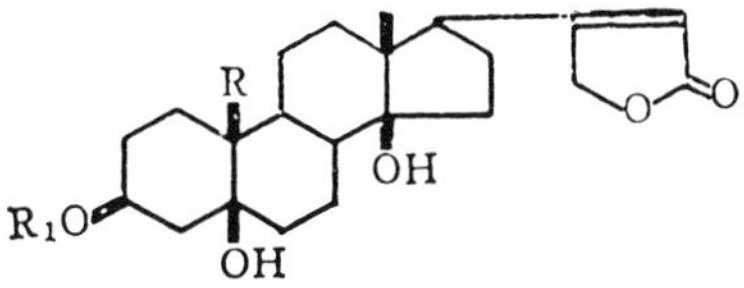

	R =	R_1 =
Convallatoxin	CHO	L-rhamnose
Convalloside	CHO	L-rham-O-glucose
Convallatoxol	CH_2OH	L-rhamnose

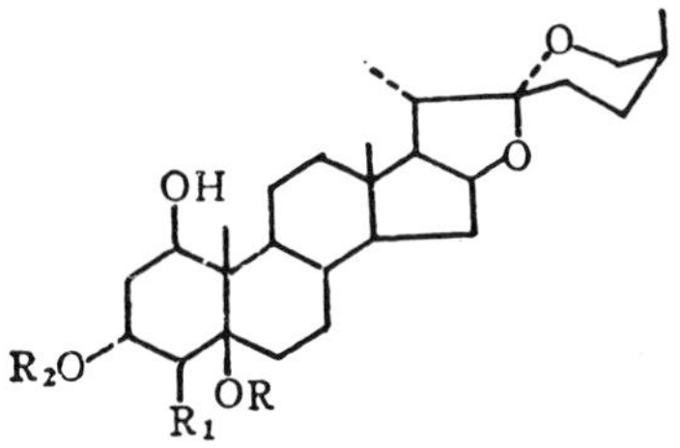

	R=	R_1=	R_2=
Convallasaponin A	H	H	α-L-arabinose
Convallasaponin B	α-L-arab.	OH	H
Glucoconvalla-saponin A	H	H	α-L-arab-O-β-D-glu.
Glucoconvalla-saponin B	α-L-arab	OH	β-D-glucose

Actions

This herb has a digitalis-like cardiotoxic action which is approximately 1.22 times more potent than that of strophanthin. The onset of action is fast, but the duration of action is short. It is mainly detoxificated in the liver.

Toxicity

The herb is toxic, with a lethal dose for mice of 1.61 ± 0.12 mg/kg (i.p.). Adverse effects include nausea, vomiting, and dizziness. Occasionally, arrhythmia may occur.

Therapeutic Uses

Ling Lan is used in the treatment of heart disease. There are two primary forms in which the herb is administered. It can be prepared as a 10% tincture in alcohol, with 1 ml given four times daily, or a total of 4 ml/d; the maintenance dose is 1 ml/d. It may also be injected, with injection ampules prepared at 0.1 mg/ml.

XIANG JIA PI (香加皮) — the dried root bark of Chinese silkvine, *Periploca sepium* Bye.

Chemistry

The primary active agent of the herb is periplocin ($C_{36}H_{60}O_{13}$), a cardiac glucoside. The root of the plant contains 0.02% periplocin. On hydrolysis, periplocin gives two sugars and a genin, periplogenin. *Xiang Jia Pi* also contains glycoside K, $C_{40}H_{66}O_{16}$, which is a glucoside with three sugars, and glycoside H, $C_{56}H_{92}O_{24}$, which is a glucoside with five sugars. Their structures are as follows:

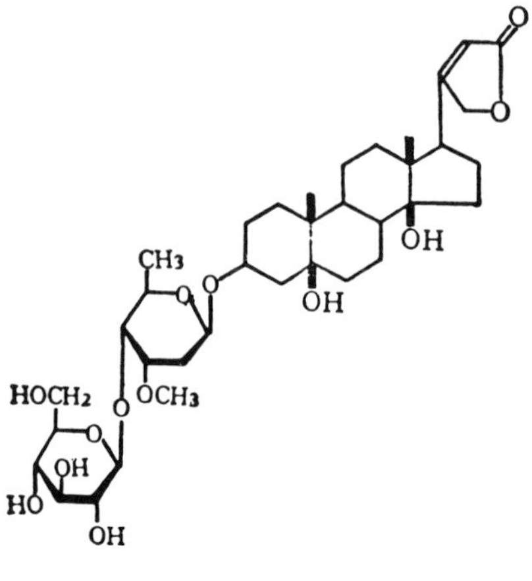

Periplocin

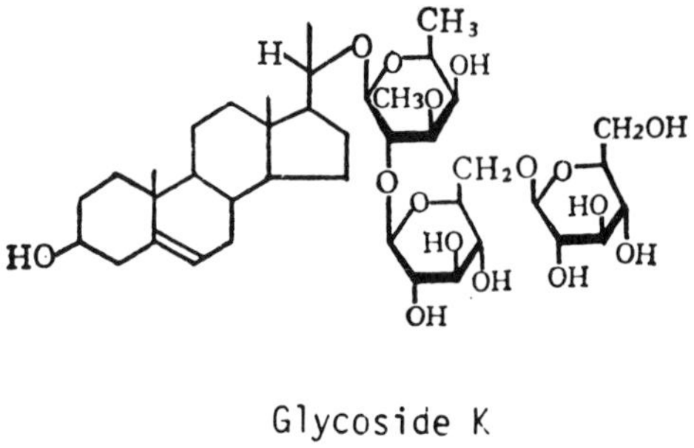

Glycoside K

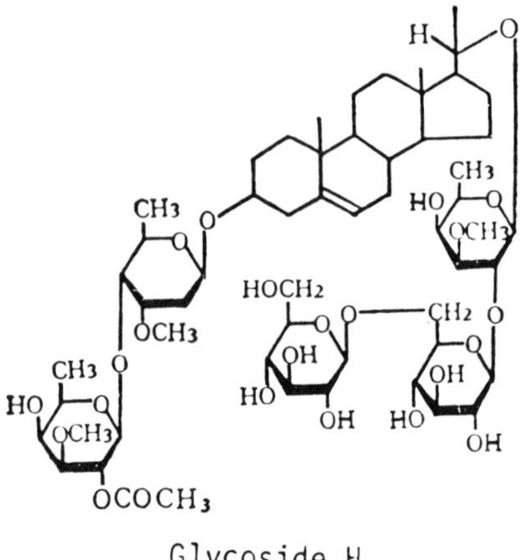

Glycoside H_1

Actions

Because of the similarity in names, this herb is frequently confused with *Wu Jia Pi* (五加皮), which is the active ingredient of a Chinese medicinal wine (*Wu Jia Pi* wine, 五加皮酒). *Wu Jia Pi* is the root of the *Acanthopanax sessiliflorus*, a ginseng-like herb popular in southwest China as a tonic and an aphrodisiac (see Chapter 11).

The chemistry and pharmacological action of *Ziang Jia Pi* are similar to that of strophanthin. Bioassay analysis reports that 1 cat unit of the glucoside, periplocin, is 0.121 mg.

Toxicity

The adverse effects and toxicity of the herb are similar to strophanthin. The lethal dose for a cat is 1 g/kg.

Therapeutic Uses

Ziang Jia Pi is used in folk medicine as both an antirheumatic and a cardiotoxic. It is also prescribed to reinforce bone and muscle tissue.

The herb is administered orally, with crude periplocin in tablet form at a concentration of 10 mg per tablet. The normal dose is two tablets t.i.d.

FU ZI (附子) or WU TAO (乌头) — the prepared aconite root, *Radix aconiti praeparata*

Chemistry

This plant contains several alkaloids:

- Aconitine, $C_{34}H_{47}O_{11}N$ (0.1%)
- Hypaconitine, $C_{33}H_{45}O_{10}N$ (0.1%)
- Mesaconitine, $C_{33}H_{45}O_{11}N$
- Talatisamine $C_{24}H_{39}O_5N$

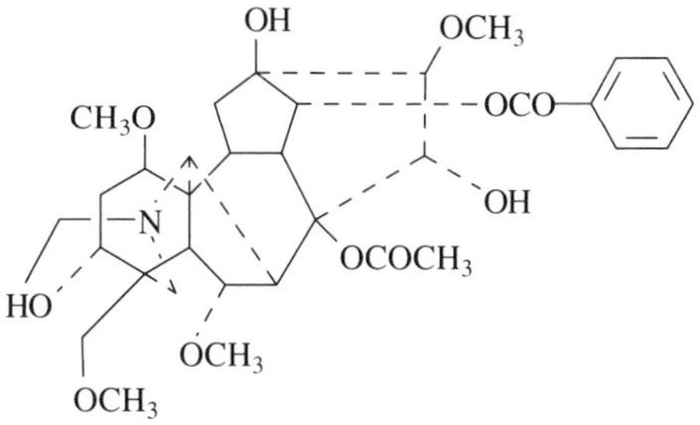

OH
OCH3
CH3O
OCO
N
OH
HO
OCOCH3
OCH3
OCH3

Aconitine

The raw root has a high content of these alkaloids. After preparation (oven-roasting, then decoction), the content of alkaloids and the potency are greatly reduced, as is toxicity.

Actions

Fu Zi is toxic to the heart muscle, and is used as a cardiotonic. In isolated heart experiments, it caused a temporary increase in heart contraction, followed by a decrease and then an irregular cardiac rhythm.

Fu Zi also acts as an anti-inflammatory agent, probably due to an indirect stimulation of the ACTH-corticosterone system. Urinary 17-ketosterone excretion increased after *Fu Zi* administration.

In addition, the herb can lower plasma cholesterol and reduce atheromata formation.

Toxicity

The herb and its alkaloids are very toxic. The LD_{50} for mice is 0.3 mg aconitine per kg (subcutaneous), or 17.42 g of prepared aconite root/kg (oral) or 3.5 g/kg (intravenous).

Toxic symptoms include bradycardia and irregular rhythm. The specific adverse effect depends on how the aconitine preparation is prepared and how long it is decocted. Patients usually lose sensation in the mouth and tongue after drinking *Fu Yi* decoction. Nausea and vomiting may occur, as well as extremity spasms and cardiac arrhythmia.

The intestinal absorption of the alkaloids is relatively quick. Because of this, immediate gastric lavage is recommended in cases of overdose.

Therapeutic Uses

Fu Zi is used in traditional medicine as a cardiotonic, to restore *yang* in the treatment of collapse and shock, to warm the kidneys, and to reinforce *yang* in the relief of pain.

Fu Zi is primarily prescribed in decoction. It is the major ingredient of several ancient Chinese formulas, including *Shi-I* soup (四逆湯), *Fu Zi* soup (附子湯), and *Sen Fu* soup (參附湯). Clinical trials using prepared *Fu Zi* injection solutions showed an effective result on angina and cardiac exhaustion.

JIAN SUI FUAN HOU (見血封喉) — the seed of *Antiaris toxicaris* (Pers.) Lesch.

Chemistry

This herb was used by the tribeman as arrow poison, which is how it earned its name. *Jian Sui Fuan Hou* means ''closing the larynx after reaching the blood''. It contains a number of cardiac glucosides, all of which are toxic

- α-Antiarin, $C_{20}H_{42}O_{11}$ (0.32%)
- α-Antioside, $C_{29}H_{44}O_{10}$ (0.15%)
- Convallatoxin, $C_{29}H_{42}O_{10}$
- Bogoroside, $C_{35}H_{52}O_{15}$
- Strophalloside, $C_{29}H_{42}O_{19}\cdot 2H_2O$ (0.12%)
- Peripalloside, $C_{29}H_{44}O_9\cdot 2H_2O$

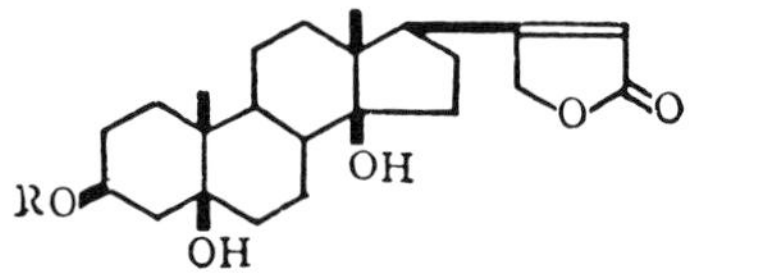

Peripalloside R= 6-deoxyallose

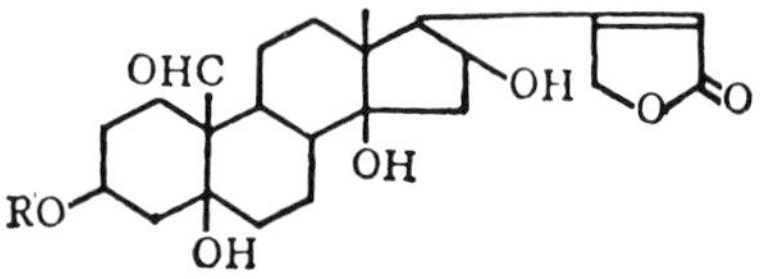

Strophalloside R = 6-deoxyallose

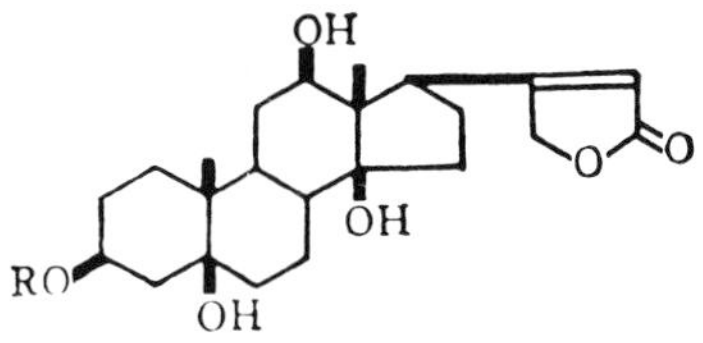

	R =
Antiogoside	6-deoxyallose
α-Antioside	6-deoxyglucose
Antioside	rhamnose

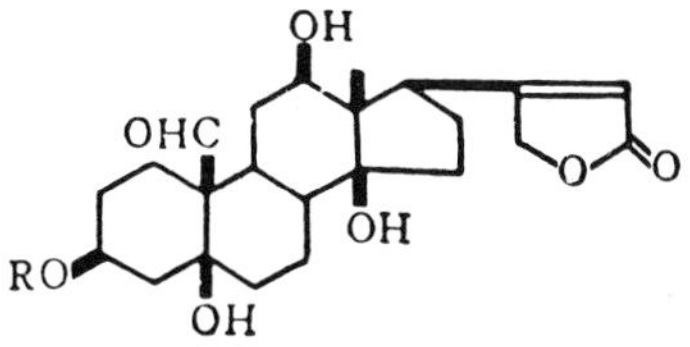

	R =
Antialloside	6-deoxyallose
α-Antiarin	6-deoxyglucose
β-Antiarin	rhamnose

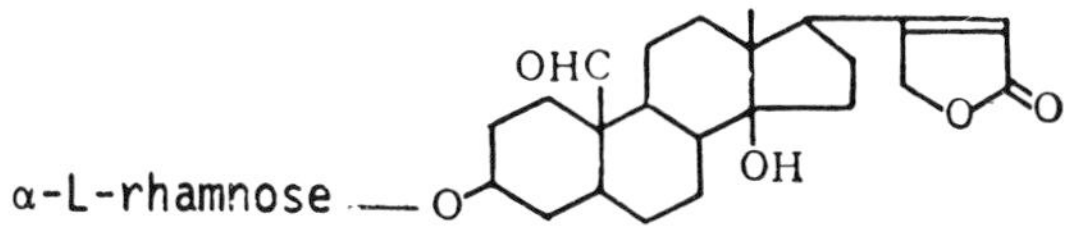

α-L-rhamnose — O

Malayoside

Action

The alcoholic extract of this herb exerts a cardiotonic effect and increases blood pressure and cardiac output. Feeding it to a dog would cause severe vomiting; large doses will cause cardiac arrhythmia and fibrillation. LD_{50} for the frog is 0.8 mg/kg, the MLD of α-antiarin for the cat is 0.116 mg/kg.

Therapeutic Use

The herb is a cardiotonic, emetic, and lactogenic agent. The seed was used as antipyretic and to treat dysentery.

References

1. **Chuan, T. K.,** *Chin. Med. J.*, 67, 261, 1949.
2. **Chen, K. K.,** *J. Pharm.*, 150, 53, 1965.

TANG JIE (糖芥) — the whole plant of *Erysimum cheiranthoides* L.

Chemistry

Tang Jie contains several glucosides and their hydrolized products. Some of these are similar to digitalis, but their pharmacological actions are not yet completely understood.

- Erysimoside, $C_{35}H_{52}O_{14}$
- Erysimosol, $C_{35}H_{53}O_{14}$
- Erychroside, $C_{35}H_{54}O_{14}$
- Helveticosol
- Erythriside
- Corchoroside A, $C_{20}H_{42}O_{9}$
- Erycodin glucoside L, $C_{35}H_{54}O_{14}{\cdot}2H_2O$
- Desglucorycordin glucoside I, $C_{35}H_{50}O_{14}{\cdot}2H_2O$

- Erysimotoxin, $C_{29}H_{42}O_9$
- Strophantidine
- Cannogenol
- Anhydrocanescegenin
- Canescein

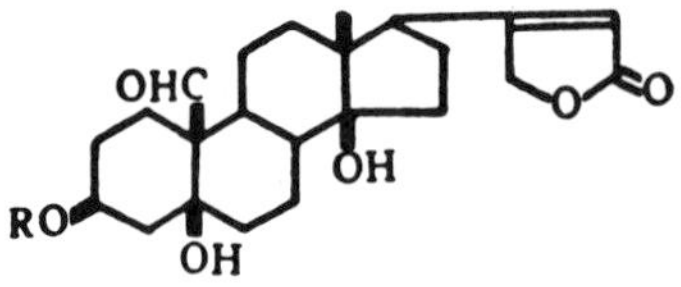

OHC RO OH OH

R =

Erysimotoxin -digitoxose

Erysimoside -digitoxose-O-glucose

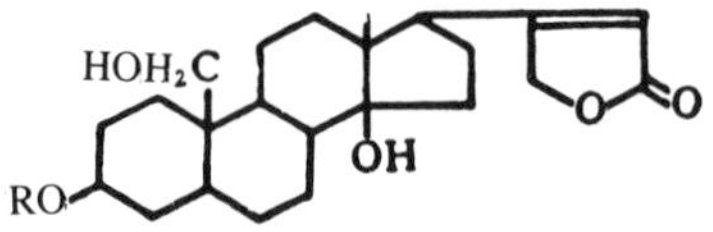

HOH_2C RO OH

R =

Cannogenol H

Erycordin -desoxygluc.-O-glucose

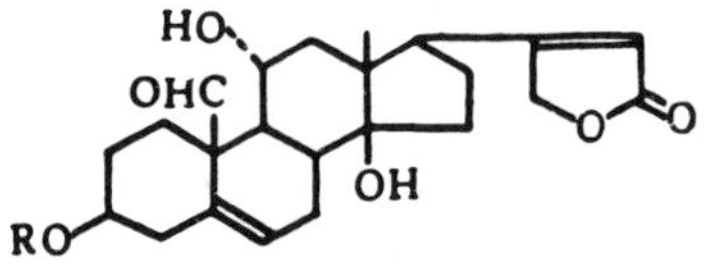

HO OHC RO OH

Canescein R = 6-Deoxy-D-glucose

Chapter 3

ANTIARRHYTHMIC HERBS

KU SENG (苦参) — the dried root of *Sophora flavescens* Ait. (Leguminosae)

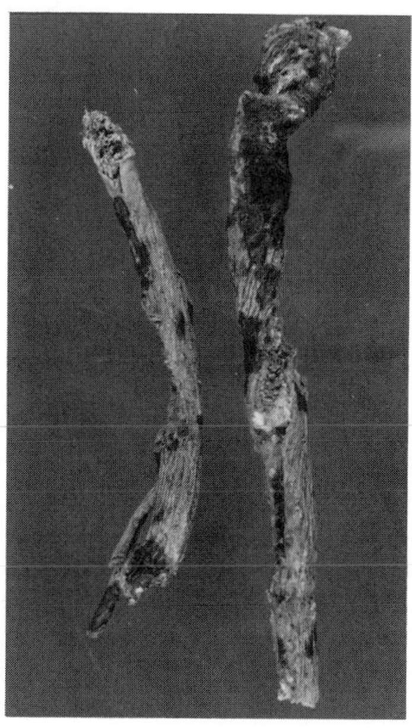

FIGURE 1. *Ku Seng* (the dry root).

Chemistry

This root contains several alkaloids, which are the major active principles responsible for its antiarrhythmic effect:

- *d*-Matrine
- *d*-Oxymatrine
- *d*-Sophoranol
- Cytisine
- *l*-Anagyrine
- *l*-Baptifoline
- *l*-Methylcytisine
- *l*-13-Ethylsophoramine
- Trifolirhizin
- Norkurarinone
- Kuraridin

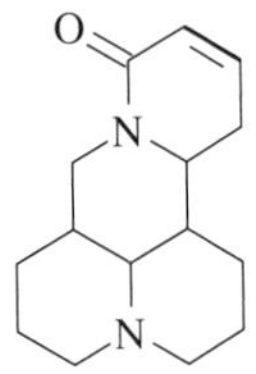

l-Sophocarpine

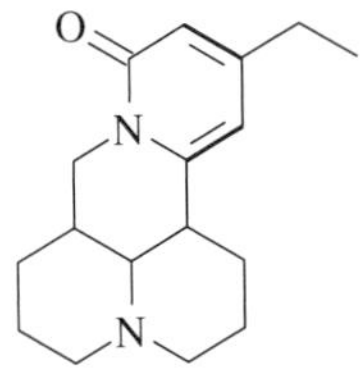

Ethylsophoramine

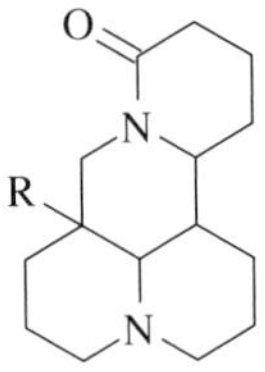

d-Matrine R = H
d-Sophoranol R = OH

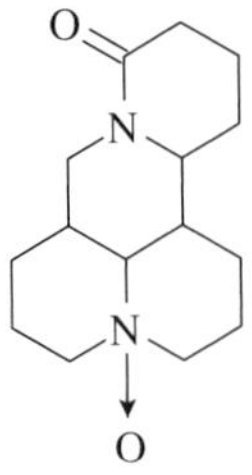

d-Oxymatrine

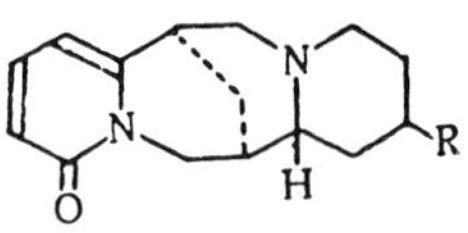

	R =
l-Anagyrine	H
l-Baptifoline	OH

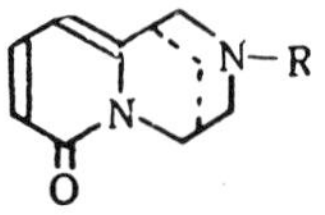

	R =
l-Methylcytisine	CH_3
Cytisine	H

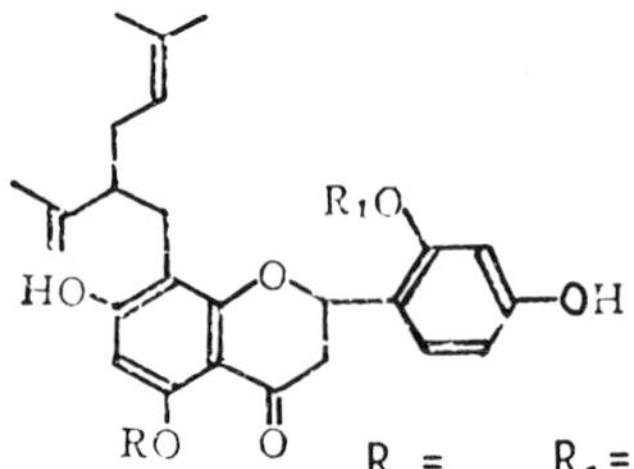

	R =	R_1 =
Kurarinone	CH_3	H
Norkurarinone	H	H
Isokurarinone	H	CH_3

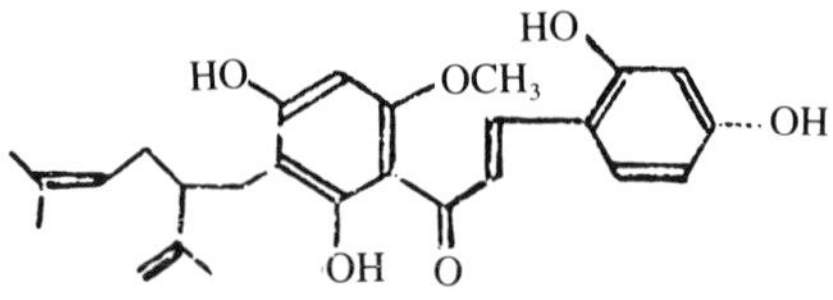

Kuraridin

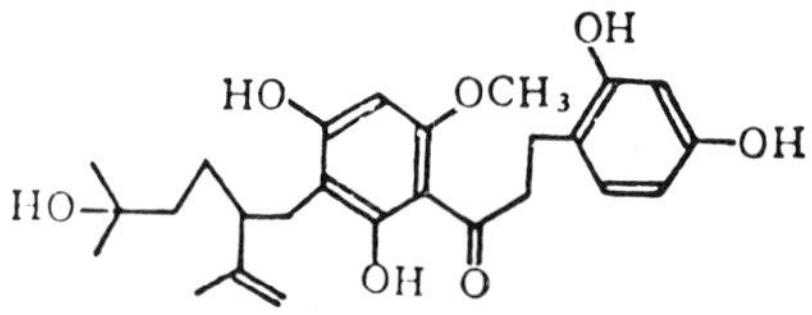

Kuraridinol

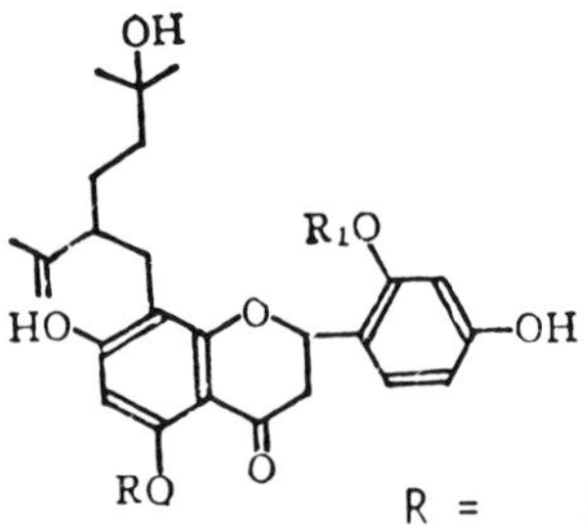

	R =	R_1 =
Kurarinol	CH_3	H
Neokurarinol	CH_3	CH_3

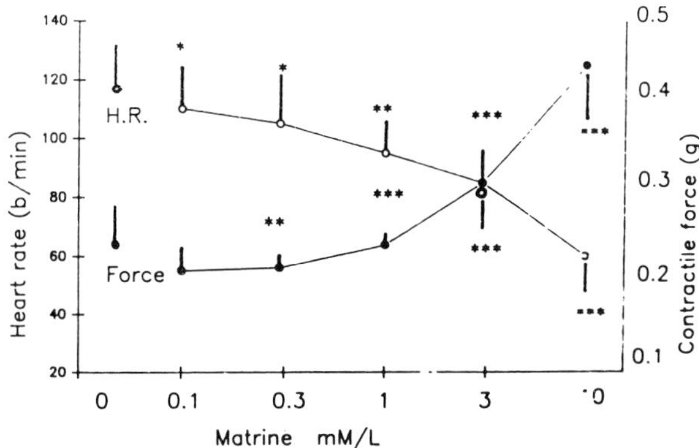

FIGURE 2. Effect of matrine on heart rate and myocardial contraction. (From Xin, H. B. and Lin, S. F., *Acta Pharmacol. Sin.*, 8, 501, 1987. With permission.)

Actions

Ku Seng's primary use is as an antiarrhythmic. It slows the heart rate, increases cardiac conduction time, and decreases myocardial excitability. Such effects are not influenced by atropine or β-adrenergic agents. Cats which received an intravenous injection of 100% *Ku Seng* solution at 1 ml/kg showed a decrease in heart rate and a simultaneous increase in coronary blood flow.

Xin and Lin[1] reported that *d*-matrine exerts an antiarrhythmic action via a direct inhibiting effect on atrial muscles. On guinea pig atrial muscle fibers, *d*-matrine produces a negative chronotropic and a positive inotropic effect. The agent also reduces maximal driving frequency and inhibits the automaticity of the left atrial muscles. Results are illustrated in Figure 2.

In other animal experiments, *d*-matrine was seen to exhibit a significant antiarrhythmic effect on arrthythmias induced by aconitine, $BaCl_2$, or coronary ligation.

Other properties of *Ku Seng* allow its use as an antiasthmatic, expectorant, diuretic, and natriuretic. It has been shown to increase leukocyte count, and has exhibited antibacterial and anticancer properties.

When given orally or by intramuscular injection, the herb is mainly transformed into matrine, which is excreted in the urine. Approximately 24% of the dose is eliminated within 24 h. It should also be noted that oxymatrine is less well absorbed from the intestine than matrine.

Toxicity

In mice, *Ku Seng* decoction has an LD_{50} of 43 ± 2.2 g/kg; for the crude alkaloid extract, it is 1.18 ± 0.1 g/kg; and for the crystalline alkaloids it is 297 ± 18 mg/kg (subcutaneous). The LD_{50} of matrine was reported to be 72.1 mg/kg in mice (intravenous).

Primary adverse effects are gastrointestinal disturbances, including gastric pain, nausea, vomiting, and constipation.

Therapeutic Uses

Ku Seng is used in Chinese medicine to remove ''heat'' and dampness from the body, and as anthelmintic and antipruritic. It is also used to treat irregular heart beat, eczema, acute dysentery, and trichomoniasis. Its protective effects against X-irradiation also allow its use against leukopenia, while its diuretic effects are useful for edema treatment.

Several preparations of the herb are used. In tablet form (2 g), general dosage is 3 to 10 tablets t.i.d.; the *Ku Seng* alkaloid may be administered in 50-mg tablet form, at a dose of 1 to 2 tablets t.i.d. In syrup form, 100 ml of syrup is equivalent to 50 g of raw herb.

For treatment of leukopenia, intramuscular injection is used at 100 mg/ml, 200 to 400 mg daily. Asthma patients may inhale the herb in aerosol form, at 400 to 600 mg of alkaloid per 12 ml vial.

References

1. **Xin, H. B. and Lin, S. F.,** *Acta Pharmacol. Sin.*, 8, 501–505, 1987.
2. **Zhang, B. H. et al.,** *Acta Pharmacol. Sin.*, 11, 253–257, 1990.

BA LI MA (八厘麻) or NAO YANG HUA (闹羊花) — the dried ripe Chinese azalea, or fruit of *Rhododendron molle* G. Don

Chemistry

The major active principles of *Ba Li Ma* are rhomotoxin, rhomedotoxin, and asebotoxin.

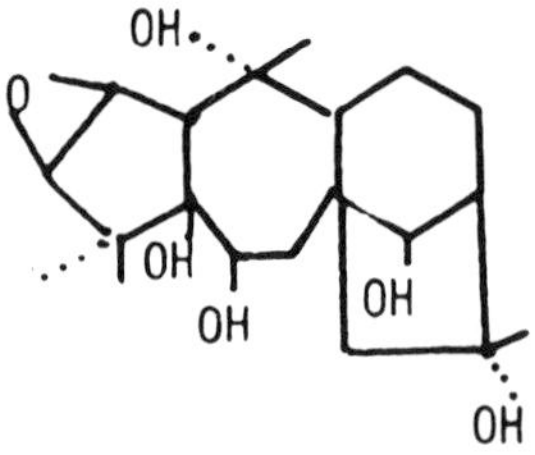

Rhomotoxin

Actions

This herb slows the heart rate. When rhomotoxin is given intravenously to anesthetized dogs in doses of 3.5 μg/kg, heart rates slow by 39%; at doses of 20 μg/kg, heart rates decrease by 70%. Atropine can block this bradycardiac effect.

Rhomotoxin also lowers blood pressure. Chen et al. gave rhomotoxin intravenously to rabbits at a dose of 8 μg/kg, reporting a marked drop in blood pressure. This can be blocked by either α_1- or α_2-adrenergic blocking agents. Chen et al. also reported that plasma renin activity is reduced after the injection.

Other properties of the herb include an analgesic effect and an ability to inhibit and kill many household insects.

Toxicity

In mice, the LD_{50} for rhomotoxin is 0.52 mg/kg (i.p.); in a decoction of the raw herb, it is 8.6 ± 0.9 g/kg (oral). The common toxic signs and symptoms are sleepiness, hypotension, nausea, vomiting, slowed respiration, and decreased heart rate.

Therapeutic Uses

Ba Li Ma is used in the treatment of tachycardia, palpitations, and hypertension. It can be administered as an adjuvant for anesthetic agents, and is also employed as an insecticide.

Preparations of the herb include a rhomotoxin tablet (0.25 mg), at doses of 2 to 3 tablets t.i.d., taken orally. Intramuscular injection can be applied with an injection ampule of 1 mg/ml.

Reference

1. **Chen, X. J. et al.,** *Acta Pharmacol. Sin.*, 8, 247–250, 1987.

CHANGROLIN (常咯啉) — a new antiarrhythmic agent isolated from *Radix dichroae*: a quinazoline derivative

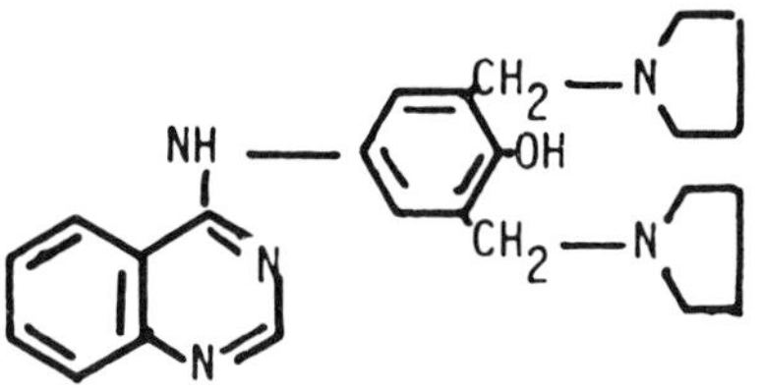

Changrolin

Actions

This agent causes reversible myocardial depression and inhibition of Na^{+}-K^{+}-ATPase activity, but no inhibition of Ca^{2+}-ATPase activity on myocardial sarcolemma. Electrocardiographic changes caused by this agent are similar to those found with quinidine and do not exceed the first degree block when blood concentration is below 8 μg/ml. It is more potent than quinidine. An additive effect is seen when it is given together with disopyramide; a more-than-additive effect is seen when given with procainamide. Nicotinamide or calcium chloride can potentiate its effect.[1]

Shen et al.[3] reported that, in experiments with dogs, changrolin administered intravenously in doses of 1 mg/kg/min for 6 min decreases the heart rate significantly and prolongs the PR interval of the electrocardiogram (ECG). Subsequently, they administered changrolin to 10 human patients at a dose of 3.3 mg/min for 60 min. In five of seven patients, ventricular ectopic beats (VEB) were completely suppressed when monitored approximately 20 min after beginning infusion. In two other patients, the number of VEB was markedly reduced. In one case, atrial fibrillation reverted to normal sinus rhythm after 45 min. There was no significant change in the cardiac index, total peripheral vascular resistance, or mean aterial pressure in those patients.[3]

Xu et al.[4] studied the antiarrhythmic effects of this agent in 37 patients with arrhythmias. Each received an intravenous infusion of changrolin, with a total dose of 500 mg administered at a rate of 3 to 3.5 mg/min. The plasma concentration of the drug was maintained at a level of 2.5 μg/ml. The herb prolonged both PQ interval and QRS duration, with no observable side effects.

Studies on guinea pig heart muscle have shown that changrolin prolongs the action potential duration significantly and prevents ventricular arrhythmia by decreasing the Na^{+} conductancy.[6]

Yu et al.[5] reported that changrolin can prolong the effective refractory period of atrial muscle. Li et al.[2] also showed that changrolin can protect chloroform-induced ventricular fibrillation in animals and nicotinamide can antagonize such antiarrhythmic effect of changrolin.

References

1. **Ding, G. S.,** in *Advanced Chinese Medicinal Materials Research,* Chang, H. M., et al., Eds., World Scientific Publishing Co., Singapore, 1985, 407–425.
2. **Li, R. S. et al.,** *Acta Pharmacol. Sin.,* 5, 26–29, 1984.
3. **Shen, Y. T. et al.,** *Acta Pharmacol. Sin.,* 2, 23–26, 1981; ibid., 4, 251–253, 1983.
4. **Xu, J. M. et al.,** *Acta Pharmacol. Sin.,* 8, 227–231, 1987.
5. **Yu, Y. et al.,** *Acta Pharmacol. Sin.,* 8, 421–425, 1987.
6. **Zhan, Y. M. et al.,** *J. Pharmacol. Toxicol.,* 5, 5–7, 1991.

TETRANDRINE (汉防己碱) — Alkaloid isolated from *Tetrandra* root, *Radix stephaniae tetrandrae*

Chemistry

Japanese investigators were the first to study this herb chemically and pharmacologically. They identified the structure of tetrandrine as follows (see Chapter 8).

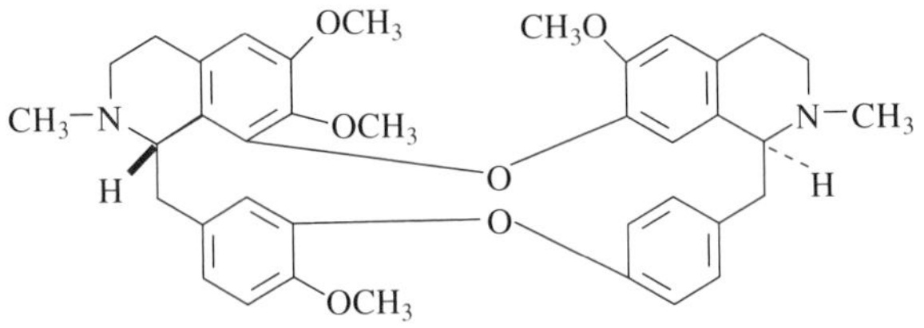

Tetrandrine

Actions

Tetrandrine can cause a hypotensive effect which may be related to peripheral vasodilatation or an impairment of vasoconstricter tone mediated by vascular postsynaptic adrenergic α_2 receptor. It can reduce the toxicity of ouabain, but has no effect on contraction amplitude increment. It antagonizes the positive chronotropic response to Ca^{2+} ion, but not the increase of cAMP or the force of contraction induced by isoproterenol. It decreases the contractility, automaticity, and prolonged functional refractory period.

Tetrandrine offers certain protection to myocardial infarct, by antagonizing the action of ouabain on the heart muscle. This action is abolished by the addition of extra calcium to the medium, suggesting an inhibitory effect on the potential-dependent channel, thereby preventing the influx of calcium ion. It prolongs the QT interval in ECG, depressing the automaticity of myocardial cells and displaying an antiarrhythmic effect. In addition, it is a calcium antagonist on the uterus.

Like verapamil, tetrandrine demonstrates a negative inotropic effect on atrial myocardium by inhibiting Ca^{2+} influx into the cells and by decreasing the intracellular Ca^{2+} release.

Therapeutic Uses

Chinese folk medicine used the root as a diuretic, antiphlogistic, and antirheumatic. Recent studies on tetrandrine demonstrate that it possesses an antiarrhythmic effect against either ouabain-induced or $BaCl_2$-induced arrhythmias. The antiarrhythmic effect is less potent than that of verapamil or quinidine, however.

Tetrandrine also exhibits anti-inflammatory, analgesic, and hypotensive properties. It has been used clinically in the treatment of hypertension, coronary disease, and silicosis.

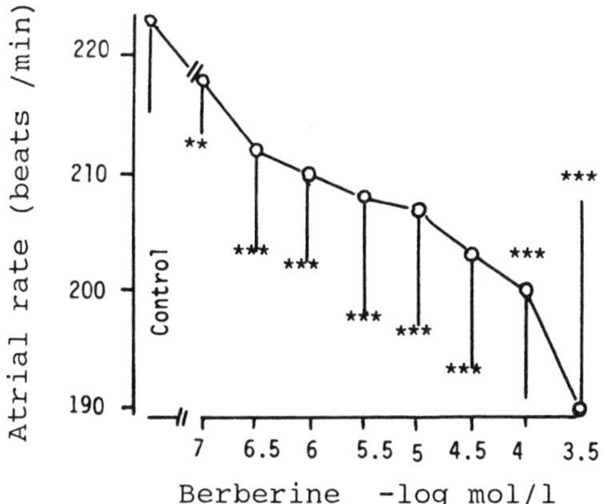

FIGURE 3. Effect of berberine on atrial rate. (From Wang, Y. H. et al., *Acta Pharmacol. Sin.*, 8, 220, 1987. With permission.)

References

1. **Cha, L. et al.,** *Acta Pharmacol. Sin.*, 2, 26–28, 1981.
2. **Ding, G. S.,** in *Advanced Chinese Medicinal Materials Research,* Chang, H. M., et al., Eds., World Scientific Publishing Co., Singapore, 1985, 307–425.
3. **Kondo, H. and Tomita, M.,** *Arch. Pharm.*, 274, 65–82, 1936.
4. **Shen, Y. R. et al.,** *Acta Pharmacol. Sin.*, 2, 23–25, 1981; *Acta Pharmacol. Sin.*, 4, 251–253, 1983.
5. **Wang, G. et al.,** *Acta Pharmacol. Sin.*, 8, 522–525, 1987.
6. **Yao, W. Y. et al.,** *Acta Pharmacol. Sin.*, 7, 128–130, 1986.
7. **Zhang, C. L. et al.,** *Int. Pharmacol. Congr. IX,* IUPHAR, Amsterdam, 1990.
8. **Zheng, X. F. and Bian, R. L.,** *Acta Pharmacol. Sin.*, 7, 40–43, 1986.

BERBERINE (小檗碱) — alkaloid isolated from many Chinese herbs; primary alkaloid found in *Huang Lian, Coptis chinesis* Franch. It is an isoquinoline derivative.

Berberine

Actions

Berberine has a negative inotropic effect. Wang et al.[3] reported that it can increase the functional refractory period of papillary fibers. As shown in Figure 3, this agent markedly reduces the atrial rate.

Fang et al.[1] induced ventricular fibrillation in anesthetized cats by electrical stimulation and subsequently administered berberine intravenously to the animals at a dose of 1 mg/kg. They reported that berberine increases the ventricular fibrillation threshold and increases the action potential duration (APD) and the refractory period of the isolated papillary fibers in guinea pigs.

Yao et al.[4] reported that berberine possesses a competitive α_1- and α_2-adrenergic blocking effect. Recently, Wang and his associates[2] studied isolated rabbit sinoatrial and atrioventricular node preparations and showed that berberine depresses the sinoatrial node by a prolongation of the action potential duration and effective refractory period. Berberine was shown to cause a decrease of atrial rate and prolongation of the refractory period of atrial muscle.[3] In studies on anesthetized cats, berberine was found to elevate the ventricular fibrillated threshold and to increase the duration of action potential. Such an effect was suggested to contribute its antiarrhythmic action.[1]

References

1. **Fang, D. C. et al.,** *Acta Pharmacol. Sin.*, 7, 321–324, 1986.
2. **Wang, Y. et al.,** *Acta Pharmacol. Sin.*, 11, 422–427, 1990.
3. **Wang, Y. X. et al.,** *Acta Pharmacol. Sin.*, 8, 220–223, 1987.
4. **Yao, W. K. et al.,** *Acta Pharmacol. Sin.*, 7, 511–515, 1986.

DAURICINE (蝙蝠葛碱) — one of the major active alkaloids isolated from *Menispermum dauricum* (see Chapter 8)

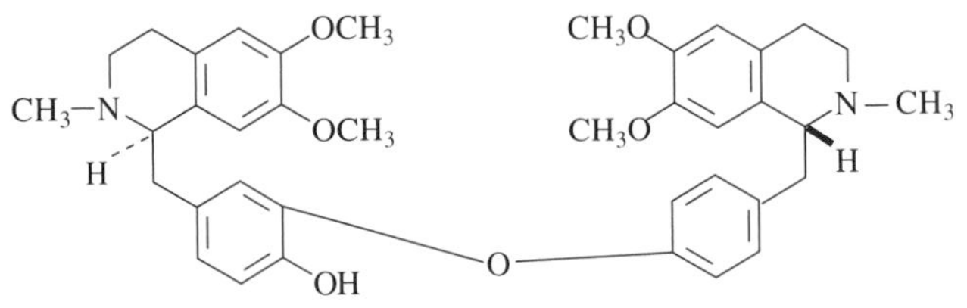

Dauricine

Actions

A group of investigators from Tonji Medical University, China, studied the antiarrhythmic effect of this alkaloid on the heart muscles of rabbit and dog. They reported that dauricine can increase the ventricular fibrillation threshold and exerts an additive synergistic effect with other antiarrhythmic agents such as amidarone, lidocaine, and propranolol. They also showed that this agent can prolong the action potential duration of rabbit atrial muscle, possibly due to a Ca^{2+}-antagonistic effect.

In anesthetized dogs, these researchers found that dauricine produced a dose-dependent depression effect on the action potential amplitude and prolongation of the refractory period of the Purkinji fibers and epicardial ventricular muscles from both infarcted and noninfarcted zone.

Arrhythmia was induced in animals by acute coronary artery ligation and was treated with dauricine, it was found that this agent can prolong the action potential duration, repolarization, and functional refractory period of the heart. It also significantly reduced the automaticity.

Hu and his colleagues treated 11 cases of arrhythmic patients with dauricine at a dose of 4 mg/kg intravenously by low infusion.[2] They reported that this agent does increase the effective refractory period in both atrium and AV node.

References

1. **Du, Z. H. et al.,** *Acta Pharmacol. Sin.*, 9, 33–36, 1988.
2. **Hu, C. J. et al.,** *Chin. J. Pharmacol. Toxicol.*, 6, 178–182, 1990.
3. **Zhu, J. Q. et al.,** *Acta Pharmacol. Sin.*, 11, 506–509, 1990.
4. **Zhu, J. Q. et al.,** *Chin. Pharmacol. Bull.*, 7, 23–26, 1991.
5. **Zong, X. G. et al.,** *Acta Pharmacol. Sin.*, 11, 241–245, 1990.

Chapter 4

ANTIHYPERTENSIVE HERBS

In Western medicine, patients whose systolic pressure exceeds 150 mmHg and diastolic pressure exceeds 90 mmHg are generally diagnosed as having hypertension. In traditional Chinese medicine, however, no mechanism exists for the scientific measurement of a patient's blood pressure. Thus, hypertension is a subjective diagnosis based solely on the patient's symptoms and pulse.

Because their described actions are based only on symptomatic improvement, the effectiveness of antihypertensive herbs discussed in Chinese pharmacopoeia or medical texts cannot be critically evaluated. In recent years, more scientific methods have been applied to investigations of these herbs. However, the number of recorded cases have been small and the statistical evaluations are incomplete.

Many Chinese herbs possess multiple pharmacological effects. For example, ginseng, the anticholesterolemic herbs, and the antiarrthymic herbs also have some antihypertensive effects, and they have been used in Chinese prescriptions as an aid or adjuvant. Another plant, *Lo Fu Mu* (萝芙木), the dried root and leaf of *Rauwolfia verticillata* (Lour.), was used in Chinese folk medicine for thousands of years as a tranquilizer, and not as a antihypertensive. After World War II, this plant was brought to America from India and its active principles, chiefly reserpine, were isolated. It is now used predominantly as an antihypertensive and tranquilizing agent. This chapter describes herbs which were used in Chinese folk medicine to lower blood pressure.

CHOU WU TONG (臭梧桐) — the dried leaf, stem, and root of *Clerodendrum trichotomum* Thunb.

Chemistry

The leaf contains the glycosides clerodendrin and acacetin-7-glucurono-(1,2)-glucuronide, as well as some alkaloids and some bitter substances. It also contains clerodendromin A and B, mesoinositol, and clerodolone.

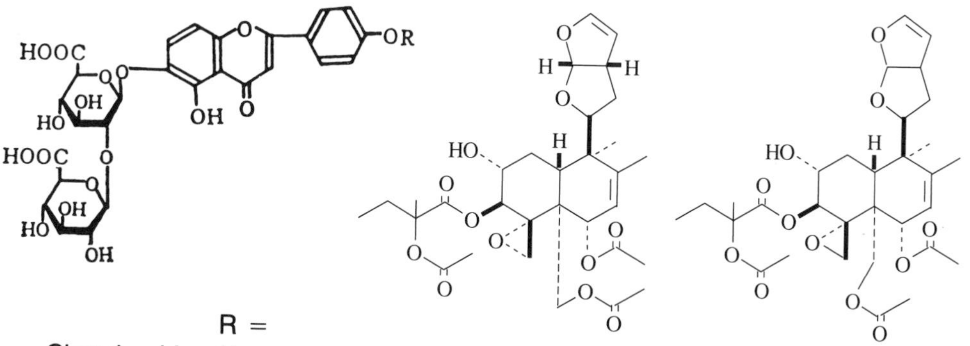

	R =
Clerodendrin	H
Acacetin-7-di-β-glucuronide	CH_3

Clerodendromin A

Clerodendromin B

Actions

When injected intravenously, the herb lowers the blood pressure in two stages. First, a sudden drop in blood pressure occurs, lasting 30 to 60 min. Then, a second drop in blood pressure occurs, for a duration of 2 to 3 h. Vasodilatation accounts for the first drop, while the second is probably due to a central inhibitory effect. This herb showed better results in lowering renal hypertension in dogs and rats: after 3 to 10 d of continuous administration, a 57.4% decrease in blood pressure was seen. The herb also has sedative, analgesic, and anti-inflammatory effects.

Toxicity is low. Dryness of mouth, loss of appetite, nausea, vomiting, and diarrhea are common side effects. In rare cases, cardiac arrhythmias, weakness of the extremities, edema, and skin eruptions have been reported; these symptoms disappear after administration is stopped.

Therapeutic Uses

Traditional Chinese medicine prescribes the herb to dispel ‘‘wind’’, remove ‘‘damp’’, and lower blood pressure. Today, it is primarily used to treat hypertension. After 4 to 5 weeks of daily administration of the herb, patients show a significant lowering of blood pressure. The daily dose is 9 to 16 g, given in divided doses three to four times per day.

Chou Wu Tong is also prescribed as an antimalaria treatment, though its effectiveness in this area has not been confirmed. The *Bei-Kou Wu-Tong* tablet (0.25 g) is administered in this treatment, with 14 tablets taken every 6 h for 2 d, followed by five tablets t.i.d. for an additional 5 d.

QIN MU XIANG (青木香) — the dried root of *Aristolochia debilis* Sieb. et Zucc.

Chemistry

The root contains aristolochic acid ($C_{17}H_{11}O_7N$) and debilic acid. It also contains several alkaloids: magnoflorine ($C_{20}H_{24}O_4N$), dibilone, and cyclanoline ($C_{20}H_{24}O_4N$). The essential oil aristolone ($C_{15}H_{22}O$) is also found.

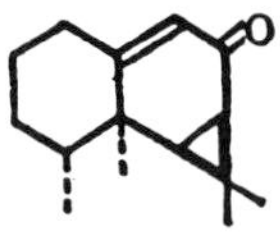

Aristolone

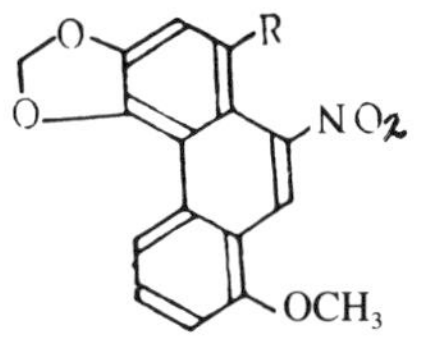

Aristolochic acid R = COOH

Debilic acid R = CH_2COOH

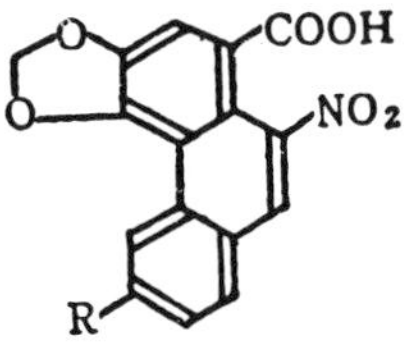

Aristolochic acid B R = H

Aristolochic acid C R = OH

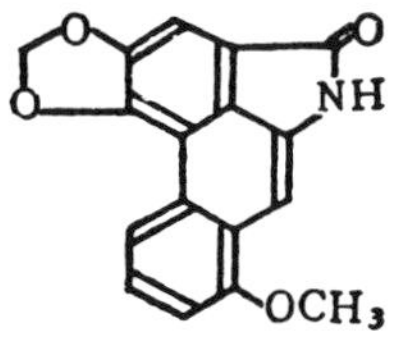

Aristolochialactam

Actions

The magnoflavine found in *Qin Mu Xiang* is the herb's primary antihypertensive agent, working through inhibition of central nervous system (CNS) centers. Other effects include vasodilatation, which disappears after sympathectomy; a lowering of the heart rate and myocardial contractility; and relaxation of smooth muscles, contributing to an antispasmodic effect.

The herb has also been observed to increase the immunological activity of the body's defense systems and to increase phagocytosis. Cases have been reported in which the herb has shown an anticancer effect.

Toxicity is low. Gastrointestinal symptoms such as nausea, vomiting, and constipation are the common side effects. Dryness of mouth and loss of appetite have also been reported.

The LD_{50} for magnoflorine in mice is 2 mg/kg (intravenous dose); for aristolochic acid it is 48.7 mg/kg (oral dose) or 22.4 mg/kg (intravenous dose).

Therapeutic Uses

Chinese folk medicine used the herb to promote the flow of *qi*, or ''bodily gas''. Traditionally, it is also used to relieve pain and to treat rheumatism and hypertension.

Modern day medicine prescribes the herb in the treatment of hypertension, gastric spasms, and gastric pain. Physicians have also used it in the treatment of infectious diseases, chronic bone marrow infections, chronic bronchitis, and chronic skin infections.

The herb is generally administered as an extract equivalent to 1 g/ml of the raw herb, in oral doses of 5 to 10 ml taken four times daily. A 6-g tablet is also available, while a 3% tincture is occasionally prescribed for gastric spasms.

For the treatment of infectious diseases, 0.15-mg aristolochic acid tablets are available. The standard oral dose in this form is one to two tablets t.i.d.

CHU MAO CAI (猪毛菜) — the whole plant of *Salsola collina* Pall. and *S. ruthenica*

Chemistry

The active principles are salsoline, salsolidine, and betaine. The herb also contains some organic acids and sugars.

Actions

The duration of the antihypertensive action of the herb is long, lasting from 7 to 30 d after administration is stopped. It has direct vasodilatatory and indirect central inhibitory effects. On the CNS, it acts as a sedative and anticonvulsive agent. Side effects can include skin eruptions, but these are rare.

Therapeutic Uses

The herb is commonly used in the treatment of early hypertension. The normal dose is a decoction or tea containing 30 to 60 g.

DU ZHONG (杜仲) — the dried bark of *Eucommia ulmoides* (Eucommiaceae)

Chemistry

The active principle of this herb is pinoresinol-di-β-D-glucoside. The herb also contains resin, organic acids, and some alkaloids; the actions of these components, however, have not been investigated.

The leaf contains approximately 3% of chlorogenic acid and some glucosides: aucubin ($C_{15}H_{22}O_9$), ajugoside, and reptoside. Also, harpagide acetate ($C_{26}H_{32}O_{12}$), and eucommiol ($C_9H_{16}O_4$), have been isolated.

Actions

The antihypertensive effect is mild but long in duration. The herb acts centrally, and its effect can be reduced by atropine or vagoectomy. In low doses, it can dilate the peripheral vessels, while in high doses it causes vasoconstriction. The herb also has a diuretic effect.

Therapeutic Uses

Du Zhong was used in folk medicine to improve the "tone" of the liver and kidney. It was also recommended as an agent to reinforce muscle and lung strength, to lower blood pressure, and, on occasion, to prevent miscarriages. It is prepared as *Du Zhong* tincture (5 or 10%), 1 to 5 ml orally t.i.d.

JIN GI ER (錦鸡儿) — the root of *Caragana sinica* Rehd., *C. microphylla* Lam., *C. intermedia* Kuang, and *C. franchetiana* Koma

Chemistry

The herb contains alkaloids and glucosides in small quantity.

Actions

The antihypertensive effect is due to a centrally mediated action on the sympathetic nervous system. It has also been reported that this herb has an anti-inflammatory effect.

There are few side effects. Occasionally, patients will experience dryness of mouth, nausea, vomiting, and hypersensitivity reactions such as urticaria and itching.

Therapeutic Uses

Chinese folk medicine prescribes this herb to lower blood pressure, and as an aid for supporting and strengthening general body functions. It has been used in the treatment of hypertension and chronic bronchitis, with a normal dose of 20 to 30 g, given in divided doses two to three times per day.

YAO JIU HUA (野菊花) — the dried flower and petal of *Chrysanthemum indicum* L., *C. boreale* Mak., and *C. lavandulaefolium* (Fisch.) Mak.

Chemistry

This herb contains several essential oils, including α-pinene, limonene, carvone, cineol, camphore, and borneol. It also contains three glucosides: chrysanthinin, chrysanthemaxanthin, and yejuhualactone.

Actions

The alcoholic extract of the herb, but not the water extract, has an antihypertensive effect which is slow in onset but long in duration. Its effect is not due to a central action, but rather to an antiadrenergic vasodilating action on peripheral vessels. It also displays antibacterial activity.

Therapeutic Uses

The traditional therapeutic use of this herb is to relieve certain hypertensive symptoms such as headache, insomnia, and dizziness. It is also widely used in treatment of the common cold, influenza, and meningitis.

The herb is prepared in tablet form at 2.4 g, as an injection ampule equivalent to 2 g/2 ml, as an extract of 20 g/ml, and as a syrup of 2 g/ml.

QIN CAI (芹菜) — celery, *Apium graveolens* L.

Chemistry

Several substances have been isolated from the plant, including apiin ($C_{26}H_{28}O_{14}\cdot H_2O$), graveobioside A ($C_{26}H_{28}O_{16}$), and graveobioside B ($C_{27}H_{30}O_{15}$). On hydrolysis, a genin and two sugars are produced.

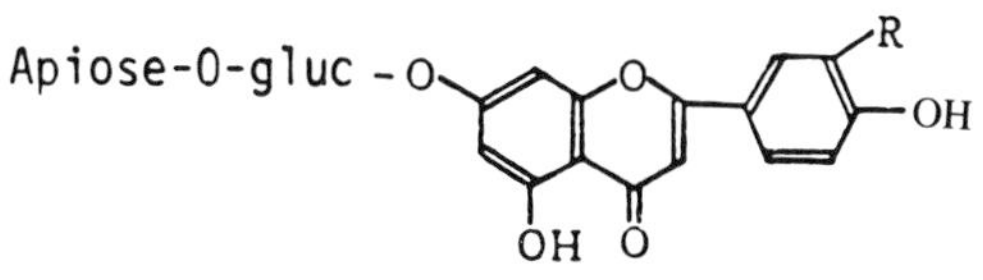

Apiose-O-gluc - O ... R ... OH ... OH O

	R =
Apiin	H
Graveobioside A	OH
Graveobioside B	OCH_3

Actions

Qin Cai can dampen the effects of epinephrine induction, reducing the increase in blood pressure generally seen. This antihypertensive effect can be antagonized by atropine or vagoectomy.

The herb lowers blood sugar levels and cholesterol concentrations and has a CNS inhibitory effect, working as an anticonvulsant agent.

Therapeutic Uses

This herb was prescribed by traditional Chinese physicians in the treatment of hypertension and hypercholesterolemia. The dose is 60 to 120 g as a water decoction.

WANG JIANG NAN (望江南) — the plant of *Cassia occidentalis* L.

Chemistry

Several glycosides have been isolated from this herb. They are anthraquinone derivatives, and include:

- *N*-Methylmorpholine
- Galactomannan
- Cassiollin, $C_{17}H_{12}O_6$
- Xanthorin, $C_{16}H_{23}O_6$
- Helminthosporin, $C_{15}H_{10}O_5$
- Apigenin
- Dianthronic heteroside

Actions

This herb can lower blood pressure and has antibacterial, antiasthmatic effect. It was also claimed to have antitoxic action, especially against snake bite. Dose: seeds or root are used in a 9 to 15 g dose as a water decoction, or roasted into powder form, served as tea.

SANG ZHI (桑枝) — the dried young twig of *Morus alba* L.

Chemistry

The herb contains several active substances, including

- Morin
- Dihydromorin

- Dihydrokaempterol, $C_{15}H_{12}O_6$
- 2,4,4′,6-Tetrahydroxybenzophenone
- Maclurin
- Mulberrin
- Mulberrochromene
- Cyclomulberrochromene

Morin

	R=
2,4,4',6-Tetrahydroxybenzophenone	H
Maclurin (penta-OH-benzophenone)	OH

Mulberrin

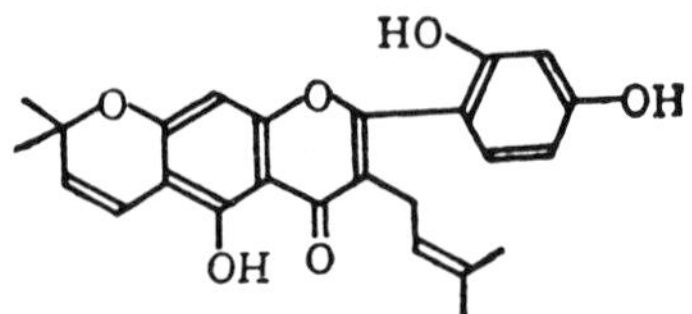

Mulberrochromene

Cyclomulberrin

Cyclomulberrochromene

Actions

The water extract of the bark was given to a rabbit intravenously and produced a significant hypotensive effect which can be blocked by injection of atropine. The extract can inhibit the frog heart and causes vasodilatation of rabbit ear. In mice it produces a tranquilizing effect.

Therapeutic Uses

Traditional Chinese medicine recommended this herb as an antirheumatic, antihypertensive, and diuretic. It was also used as an agent to remove obstructions of the intestinal tract.

LUO FU MU (萝芙木) — dried root of *Rauwolfia verticillata* (Lour.) Baill., and related species

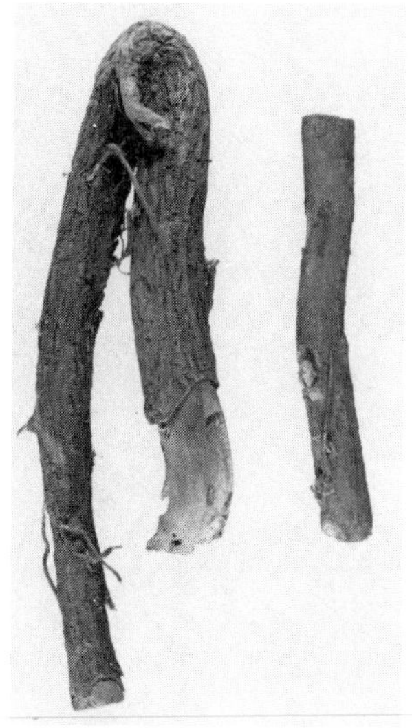

FIGURE 1. *Luo Fu Mu.*

Chemistry

There are a total of eight different species of *Rauwolfia* grown in China, chiefly in the provinces of Kwangtung, Kwangsi, Yunnan, Geizhou, and Hanan. There is reported to be a ninth variety of *Rauwolfia* found in Taiwan.

Approximately 1 to 2% of the root is made up of alkaloids. The major active principle is reserpine, which varies in quantity in different varieties of *Rauwolfia*. For example, the plant from Yunnan Province, *R. yunnanensus*, has an alkaloid content of 1.31 to 2.7%, with a reserpine content of 0.04 to 0.107%. The root also contains a small quantity of β-sitosterol.

The leaf of *Rauwolfia* contains aricine ($C_{22}H_{26}O_4N_2$), vellosimine ($C_{10}H_{20}ON_2$), peraksine ($C_{10}H_{22}O_2N_2$), serpentine, and robinin ($C_{33}H_{40}O_{10}\cdot{}^1/_2H_2O$).

The structure of *Rauwolfia* alkaloids follows.

The first group:

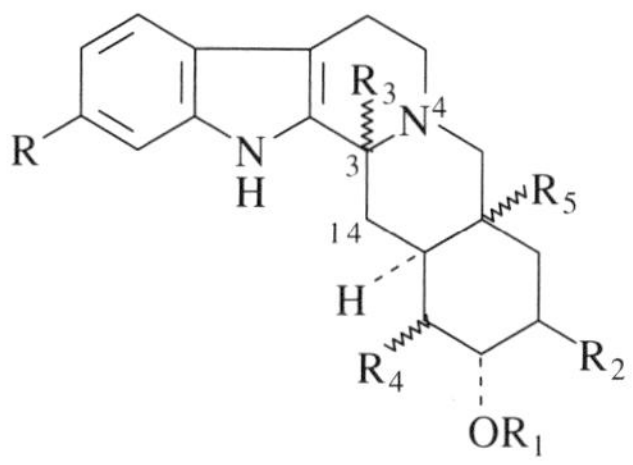

	R	R_1	R_2	R_3	R_4	R_5
Reserpine:	OCH_3	CH_3	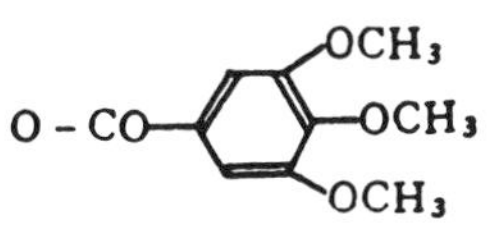 $O-CO$– (ring with OCH_3, OCH_3, OCH_3)	β-H	β-$COOCH_3$	α-H

	R	R_1	R_2	R_3	R_4	R_5
Rescinnamine	OCH_3	CH_3	O-CO-CH=CH-(3,4,5-trimethoxyphenyl: OCH_3, OCH_3, OCH_3) 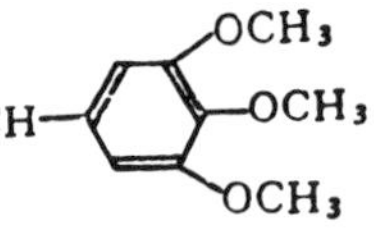	β-H	β-$COOCH_3$	α-H
Deserpidine	H	CH_3	O-CO-(3,4,5-trimethoxyphenyl: OCH_3, OCH_3, OCH_3) 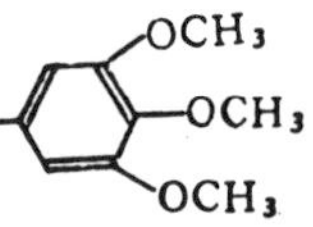	β-H	β-$COOCH_3$	α-H
Raunescine	H	H	O-CO-(3,4,5-trimethoxyphenyl: OCH_3, OCH_3, OCH_3) 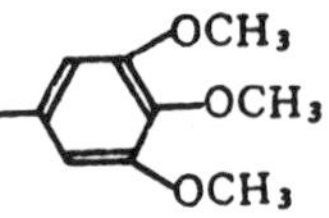	β-H	β-$COOCH_3$	α-H
Yohimbine	H	H	H	α-H	α-$COOCH_3$	β-H
α-Yohimbine	H	H	H	α-H	β-$COOCH_3$	α-H
Rauhimbine	H	H	H	α-H	β-$COOCH_3$	β-H
Isorauhimbine	H	H	H	β-H	β-$COOCH_3$	α-H

The second group:

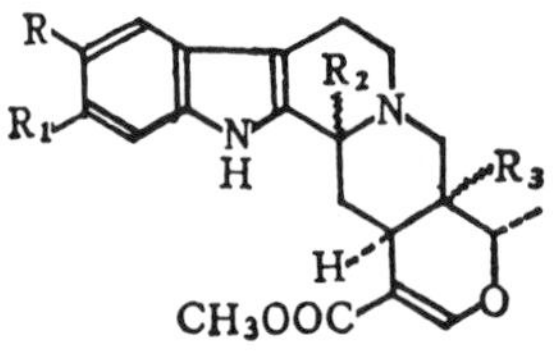

	R=	R_1=	R_2=	R_3=
Ajmalicine	H	H	α-H	β-H
Serpinine	H	OCH_3	α-H	β-H
Reserpiline	OCH_3	OCH_3	β-H	α-H
Reserpinine	H	OCH_3	α-H	α-H
Aricine	OCH_3	H	α-H	α-H

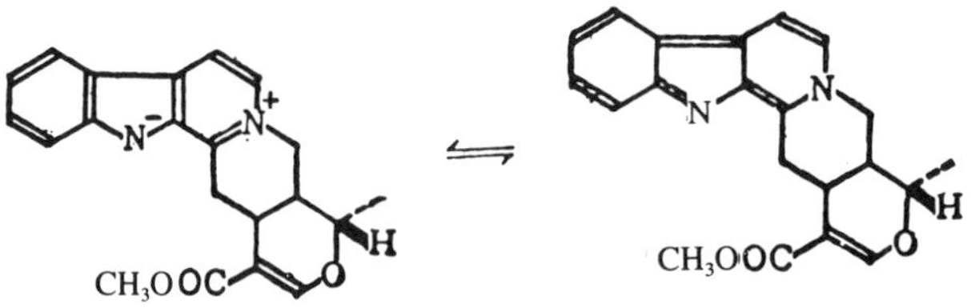

Serpentine

The third group:

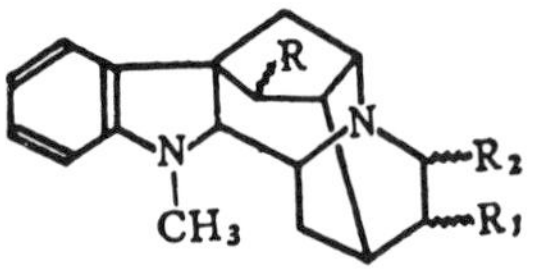

	R =	R_1 =	R_2=
Ajmaline	β-OH	β-C_2H_5	α-OH
Isoajmaline	α-OH	α-C_2H_5	β-OH

The fourth group:

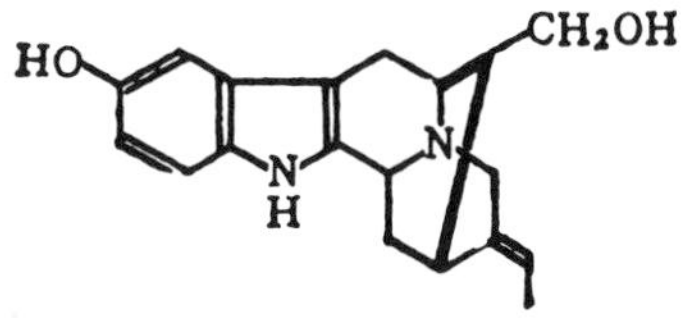

Sarpagine

Actions

Antihypertensive Effect

The alkaloids found in *Rauwolfia*, mainly reserpine, penetrate the adrenergic terminals and deplete norepinephrine (NE) from the granules, resulting in decreased α-adrenergic stimulation. The onset of action is slow and the duration is long.

Experiments in anesthetized animals showed that an intravenous dose of 1 mg/kg in dogs can lower blood pressure to 50% of the original level. In cats, the effective dose is 2 to 3 mg/kg, with an effect lasting 2 to 3 h.

Tranquilizing Effect

In the CNS, the alkaloids exert a tranquilizing effect by depleting NE in the brain stem, resulting in a depression of arousing action. The dose required for this CNS effect is relatively larger than that for antihypertensive effects. CNS effects include a calming and sedative action, a decrease in coordination, and a decrease in conditional reflexes.

Other Actions

Peripheral cardiovascular activity of the alkaloids includes depletion of NE in the myocardial terminals, resulting in a slowing of the heart rate. *Luo Fu Mu* has also been reported effective in delaying ovulation and in producing a curare-like effect.

Toxicity

The LD_{50} of *Rauwolfia* alkaloids was determined in mice by stomach administration. It ranged from 690 to 2000 mg/kg, depending upon the species of *Rauwolfia* used. For reserpine, the LD_{50} was 500 mg/kg (by stomach) and 16 mg/kg (i.p.).

The major side effects of *Rauwolfia* alkaloids in patients include a syndrome similar to Parkinson's disease and mental depression; some patients have developed suicidal tendencies.

Therapeutic Uses

In the treatment of hypertension, reserpine is generally administered in oral doses of 0.125 to 0.5 mg daily. In a recent study, alkaloids of the Yunnan *Rauwolfia* were administered to 200 patients with moderate hypertension, in doses of 6 to 15 mg daily for 3 weeks to 2 months. All 200 patients showed highly satisfactory results: blood pressure was reduced to 30 to 40% of the original level and minimal side effects were observed.

Reserpine has also been used to treat psychosis and schizophrenia. In these cases, it is given intramuscularly at an initial dose of 2 to 4 mg, and subsequently in oral doses of 2 to 6 mg/d.

The herb has been used in treating dermatitis, including neurogenic dermatitis, chronic eczema, contagious dermatitis, and chronic urticaria. The latter is especially responsive to treatment with *Rauwolfia* alkaloids, with an 89% effective rate reported at doses of 0.5 to 1 mg/d.

Luo Fu Mu is also used in some cases of malnutrition. Persons with malnutrition of unknown cause have usually responded favorably to *Rauwolfia* alkaloids when high energy or high protein diets are ineffective. Additionally, the herb is occasionally prescribed in the treatment of hyperthyroidism.

Chapter 5

ANTIANGINAL HERBS

TAN SENG (丹参) — the dried root or rhizome of *Salvia miltiorrhiza* Bge.

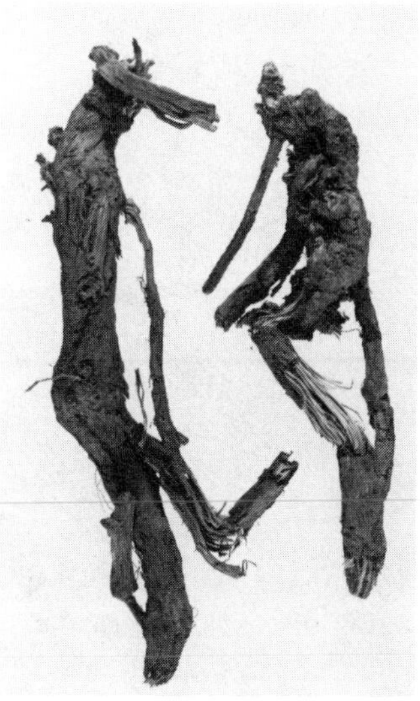

FIGURE 1. The root of *Tan Seng*.

Chemistry

The herb contains several ketone and alcoholic derivatives, including tanshinone (I, IIA, and IIB), cryptotanshinone, isocryptotanshinone, miltirone, tanshinol (I and II), and salviol. It also contains a small amount of vitamin E.

The structure of some of these principles is shown as follows:

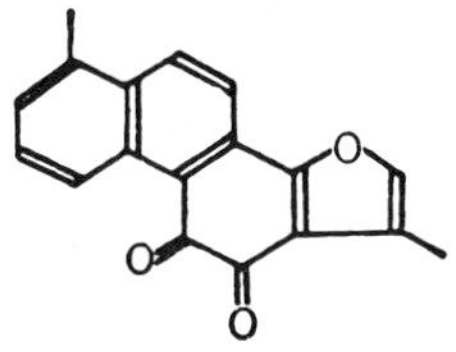

Tanshinone I

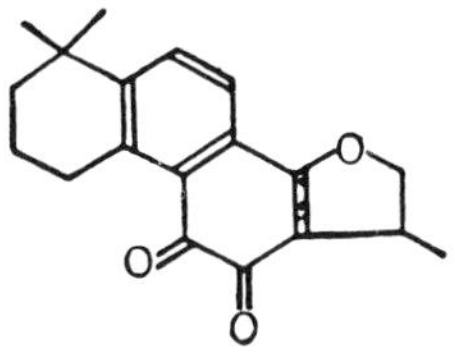

Cryptotanshinone (red crystal)

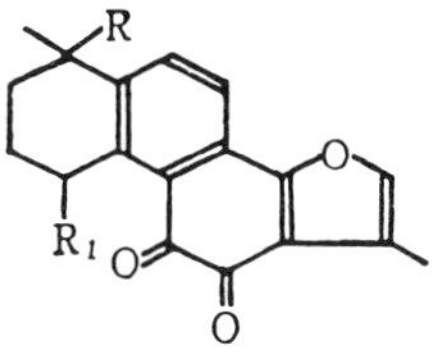

		R =	R_1 =
Tanshinone	II_A	CH_3	H
	II_B	CH_2HO	H
Hydroxytanshinone	II_A	CH_3	OH

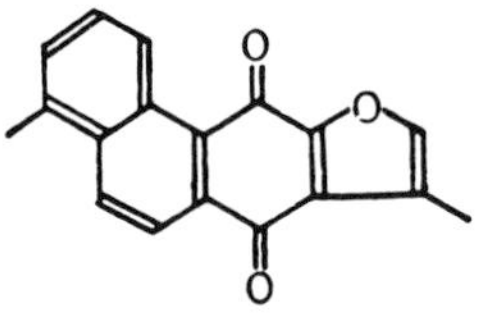

Isotanshinone I

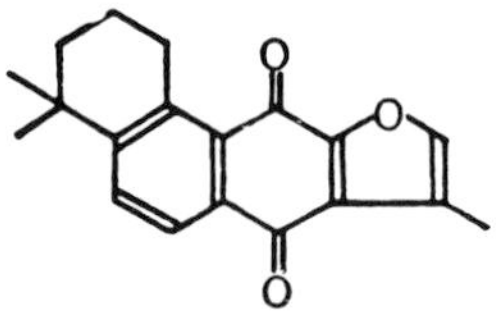

Isotanshinone II

Structure of some other *Tan Seng* principles:

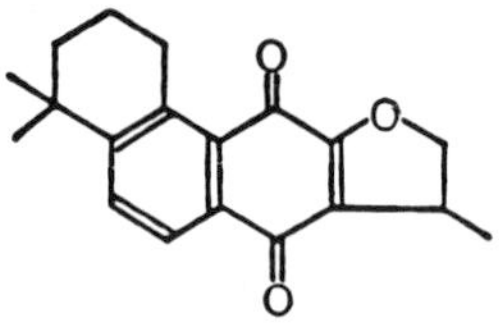

Isocryptotanshinone

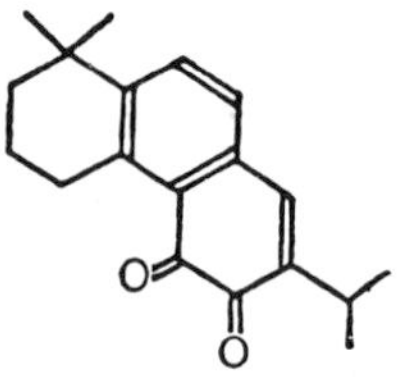

Miltirone

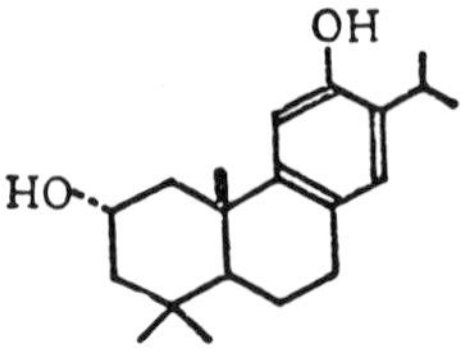

Salviol

Actions

Cardiovascular System

Injection of *Tan Seng* extract into the isolated heart of guinea pigs or rabbits (normal or atherosclerotic) causes relaxation of coronary vascular smooth muscle and an increase in coronary circulation. In anesthetized dogs, the intravenous injection of *Tan Seng* extract in a dose of 4 g/kg increases coronary flow by 70.5% and reduces resistance by 46%. Clinical trials with this extract in patients with coronary artery disease reported a definite increase in coronary circulation index.

The *Tan Seng* extract has an inhibitory effect on the conduction system of the heart, resulting in a slowing of the heart rate. Cardiac contractility is initially inhibited and then increases.

The herb has a protective effect in acute myocardial ischemia. In experiments with animals in which myocardial ischemia was induced either by intracardial injection of norepinephrine (NE) or by partial ligation of a coronary branch, administration of *Tan Seng* extract reduced ischemic symptoms, improved the pathological electrocardiogram (ECG) pattern, and stimulated the regeneration of myocardial tissue. Tanshinone IIA, one of the active ingredients of the herb, and its water-soluble sulfonate salt were used clinically and showed an acceleration of the coronary blood flow and collaterals by blocking Ca^{2+} entry.

In studies of the slow action potential of guinea pig heart muscles, Liu and Tsei[3] reported that tanshinone IIA can inhibit the Ca^{2+} channel and at higher concentrations (50 to 100 μmol/l) it also blocks the Na^{+} channel.

Fan et al.[2] studied the effects of tanshinone IIA on the isolated heart muscles of guinea pigs, rabbits, and pigs. They found that this agent can inhibit cardiac contractility and reduce automaticity by decreasing the Ca^{2+} influx. Chen et al.[1] reported that *Tan Seng* extract can inhibit the ATPase activity of red blood cell membranes.

Tan Seng extract can dilate peripheral blood vessels, resulting in a fall in blood pressure. The effect can be blocked by atropine. However, *Tan Seng* extract cannot reduce hypertension induced by NE.

Tan Seng and tanshinone IIA have an inhibitory effect on several coagulation factors in the plasma. They are antihemolytic and anticoagulant. *Tan Seng* also has a fibrinolytic effect and can inhibit platelet aggregation. *In vitro* experiments showed that the water extract of *Tan Seng* significantly prolonged blood clotting time and fibrin formation time. The mechanisms of this action are not completely understood.

Tan Seng can also reduce cholesterol and triglyceride levels in the plasma, and has been widely used as an anticholesterol agent. *Tan Seng* can increase oxygen-deprivation tolerance of mice, reducing oxygen consumption, especially in the heart muscle. Experiments showed that when mice were kept in a low oxygen chamber, animals treated with *Tan Seng* had higher survival times than control animals.

Central Nervous System

The herb is an analgesic and sedative. Its water extract can prolong the sleeping time induced by chloraldehyde or barbiturates.

Antibacterial Effects

Tan Seng extract or tanshinone I and IIA have an inhibitory effect on many bacteria, including staphylococci, *Escherichia coli*, dysentery bacteria, and typhoid bacteria. The alcoholic extract of the herb or tanshinol has a marked inhibitory effect on the growth of the tubercule bacillus.

Pharmacokinetics

In rats, tanshinone IIA was found to be uniformly distributed into nerve tissue and visceral organs after intravenous injection. The liver and gallbladder had the highest concentrations of this compound. The biological half lives were $t_{1/2\alpha}$ = 26 min and $t_{1/2\beta}$ = 108 min for ^{35}S-labeled tanshinone IIA, and $t_{1/2\alpha}$ = 27 min and $t_{1/2\beta}$ = 199 min for ^{3}H-labeled tanshinone IIA.

Of the administered dose, 70% was excreted in the feces via the bile, while the remainder appeared in the urine.

This herb is relatively nontoxic. When the water extract of *Tan Seng* was administered intraperitoneally to rabbits in a dose of 2 to 3 g/kg/d for 14 d, no toxic signs developed. In mice, the LD_{50} was 80.5 ± 3.1 g/kg (i.p.).

Therapeutic Uses

Traditional Chinese physicians prescribed this herb to stabilize the heart and calm nerves, "lighten" blood, and remove "stagnant" blood. More modern uses include the treatment of angina pectoris, cerebral atherosclerosis, diffusive intravascular clotting, and thrombophlebitis.

Shanghai clinical reports showed that in 323 angina patients who received *Tan Seng* tablets for 1 to 9 months, the effectiveness was 81% and an improvement in the ECG pattern was observed in 57.3% of all cases. Some of the patients had a fall in plasma phospholipids and a significant increase in coronary circulation index. Some patients who also had hypertension complications showed a fall in blood pressure after continued administration of the herb.[4]

In other studies, 65 patients suffering from stroke due to cerebral atherosclerosis received the *Tan Seng* extract by intravenous infusion or intramuscular administration. Of the 65, 9 patients recovered completely from the paralysis and language difficulties caused by the stroke, 48 patients showed significant improvement, and only 8 cases showed no progress.

In cases of diffusive intravascular clotting, the herb can dilate blood vessels, improve circulation, and slow the clotting time of blood; this is especially true in the early stages of administration. In cases of thrombophlebitis, oral administration of *Tan Seng* tincture produced a significant improvement in 113 patients with thrombophlebitis, 28 of whom were completely cured.

Other uses include treatment of hepatitis, acute surgical infectious diseases, mastitis, erysipelas, otitis media, tonsillitis, and bone marrow infection. Reports indicate that all of these diseases respond well to treatment with the herb. It has also been reported that *Tan Seng* is effective in treating many skin diseases, such as shingles, neurological dermatitis, and psoriasis.

References

1. **Chen, C. et al.,** *New Pharmacol. J.*, 4, 45, 1970.
2. **Fan, C. et al.,** *Acta Pharmacol. Sin.*, 7, 527–533, 1986.
3. **Liu, Q. Y. and Tsei, T. D.,** *Acta Physiol. Sin.*, 42, 254–261, 1990.
4. **Shanghai Hospital Team,** *Chin. Med. J.*, 11, 68, 1976.

CHUAN XIANG (川芎) — the dried rhizome of *Ligusticum chuanxiang* Hort.

Chemistry

Three alkaloids have been isolated from this plant: tetramethylpyrazine (TMP), leucylphenylalanine anhydride, and perlolyrine. The herb also contains several essential oils.

An oily alkaloid ($C_{27}H_{37}N_3$) has been isolated from another species of *Ligusticum*, *L. wallichii* Franch. This particular species also contains ferulic acid, cnidilide, neocnidilide, and ligustilide.

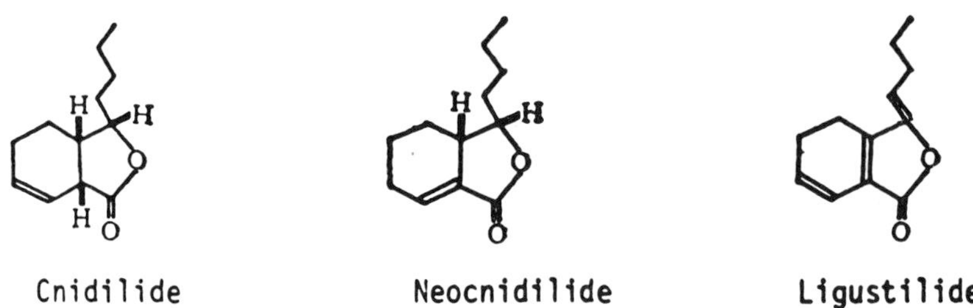

Actions

The herb increases myocardial contractility and slows the heart rate; vagotomy does not affect this action. Additionally, it improves coronary circulation and reduces oxygen consumption in the heart muscle; TMP is the active principle producing this effect.

Other observed effects include vasodilatation of peripheral vessels and lowering of blood pressure. The total alkaloids or TMP alone can lower vascular resistance, resulting in an increase in blood circulation in the aorta and lower extremities. *In vitro* experiments with rabbit aorta strips showed that TMP can inhibit contractions induced by NE or KCl.

Chuan Xiang extract has an inhibitory effect on ADP-induced platelet aggregation and on thromboxane A_2 synthesis. It acts as an antiatherosclerosis and anticholesterolemia agent. It can increase uterine contraction. Additionally, extracts of the herb can prolong the sleeping time induced by barbiturates and reduce the central stimulatory effect of caffeine.

Pharmokinetics

TMP is rapidly absorbed from the gastrointestinal tract and uniformly distributed into the cortex and cerebellum. The peak effect is observed within 1 to 3 h after oral administration. The biological half-life, $t_{1/2}$, is 29 min.

The drug is eliminated primarily by metabolism after absorption. Metabolites are excreted both in the urine and bile.

Toxicity

This herb is relatively nontoxic. The LD_{50} in mice is 65.9 ± 31.3 g/kg for intraperitoneal administration and 66.4 ± 3.2 g/kg for intramuscular administration. TMP has an LD_{50} in mice of 239 mg/kg (intravenous administration).

Female patients taking this herb occasionally show early menstruation, and it is therefore not recommended for use by females suffering from dysmenorrhea or other hemorrhagic diseases.

Therapeutic Uses

Chinese herbal lore claimed that this herb would mobilize and promote the flow of *qi* (bodily gas), promote blood flow, remove blood stasis and relieve pain. It was used in the treatment of stroke, headache, and amenorrhea.

In the treatment of angina pectoris, TMP has shown an efficacy rate of 88%. For this treatment, 100 to 200 mg of the herb are given by intravenous infusion after dilution.

Chuan Xiang is also used to treat cerebral ischemia. In clinical trials with 545 patients, TMP produced 80 to 90% improvement in the acute stage of cerebral ischemia. TMP injections at 40 mg/2 ml are used in this treatment. The daily dose is 40 to 80 mg, diluted with isotonic saline or glucose solution for intravenous infusion. The treatment continues for 10 d.

Other uses of the herb include the relief of postpartum pain and to promote discharge of the placenta or dead tissue from the uterus. *Chuan Xiang* administration can also significantly improve symptoms of acute hepatic icterus due to toxic bacterial infection or microcirculation stagnation; children show particularly positive response to this treatment.

Preparations of the herb include an injection ampule in which each ml is equivalent to 10 g of raw material; this is administered by intramuscular injection, at 2 ml daily. A *Chuan Xiang* alkaloid injection is also available, in which each milliliter is equivalent to 5 g of crude herb containing 4 mg of total alkaloids; the standard dose is 10 ml diluted with glucose solution for intravenous infusion.

The herb is also a primary component of *Kuan Seng II* (冠心二), a combination preparation. This preparation contains the following herbs: *Chuan Xiang, Tan Seng*, "red flower", red peonies, and *Kong Seng* (降香). It is water extracted and filtered. Ampules contain 2 ml equivalent to 2 g of each herb, to be used directly for intramuscular injection or diluted for intravenous infusion. It has also been made into tablets for oral administration.

YEN XING LEAF (銀杏叶) or PEI GO SU LEAF (白果树叶) — the dried leaves of *Ginkgo biloba* L.

Chemistry

The leaf contains kaempterol-3-rhamnoglucoside, ginkgetin ($C_{32}H_{22}O_{10}$), isoginketine ($C_{32}H_{22}O_{11}$), and bilobetin ($C_{31}H_{20}O_{10}$). Several other substances have been isolated, including isorhamnetin, shikimic acid, D-glucaric acid, and anacardic acid.

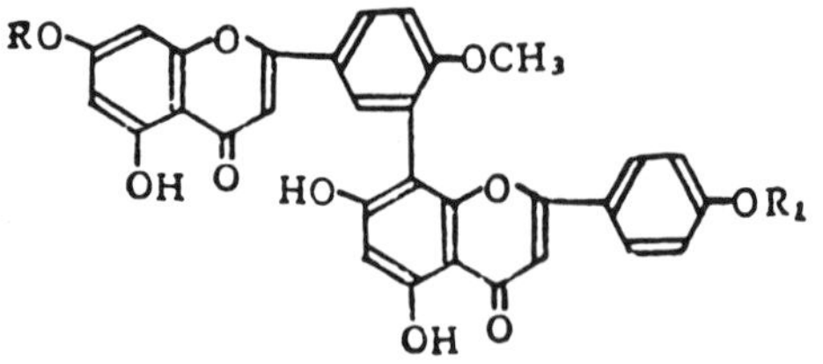

	R=	R_1=
Ginkgetin	CH_3	H
Isoginkgetin	H	CH_3
Bilobetin	H	H

Actions

The herb stimulates vasodilatation, resulting in an increase in blood flow and a lowering of blood pressure. It helps to lower plasma cholesterol concentrations. It relaxes smooth muscles and antagonizes muscular contractions induced by $BaCl_2$. *Yen Xing* leaf extract or ginkgetin can also aid in bronchodilatation, antagonizing histamine-induced bronchial constriction.

Adverse effects include nausea and vomiting, increased salivation, and loss of appetite. Patients may also complain of headache, dizziness and tinnitus, and hypersensitivity reactions, such as skin rash. Orthostatic hypertension may occur if large doses are administered.

Therapeutic Uses

In folk medicine, this herb has been used as an antitussive, antiasthmatic, and anodyne. Its effects on the circulatory system allow its use in the treatment of coronary artery disease, angina pectoris, hypercholesterolemia, and Parkinson's disease.

In the treatment of coronary artery disease and angina pectoris, the onset of action is slow. It usually takes 3 to 10 d of continuous therapy to produce a significant effect. Often, a total of 30 to 40 d of continuous therapy is required. In the treatment of Parkinson's disease, the herb can increase cerebral circulation and improve symptoms significantly.

Common preparations of the herb include four types of tablets and two injection solutions.

1. *Kuan Sen Tone* tablets (冠心酮片) contain 1.14 mg of the combined ginkgetin. The normal oral dose is four tablets t.i.d.
2. *Shu Sen Tone* tablets (舒心酮片) contain 0.5 mg of ginkgetin. The standard dose is two tablets t.i.d. This preparation can be used sublingually.
3. *Shu Sen Tin* tablets (舒心寧片), or 6911, contains 2 mg of the active principle of the herb. The normal dose is one to two tablets t.i.d.
4. *Song Huang Tone* tablets (双黄酮片), at 0.6 mg, are administered sublingually.
5. *Shu Sen Tone* injection contains 0.3 mg active principle/ml and is used intramuscularly. As a precaution, hypersensitivity should be tested before clinical use.
6. *Shu Sen Tin* injection contains 1.3 to 1.4 mg/2ml of the active principle, and is intramuscularly administered.

GE GEN (葛根) — the dried root of *Pueraria lobata* Ohwi or *P. pseudohirsuta* Tang et Wang

Chemistry

Four glucosides have been isolated as the active principles of this herb: daidzin ($C_{21}H_{20}O_9$), diadzin-4,7-diglucoside ($C_{27}H_{30}O_{14}$), puerarin, and xylopurarin. A synthetic daidzein is already on the market.

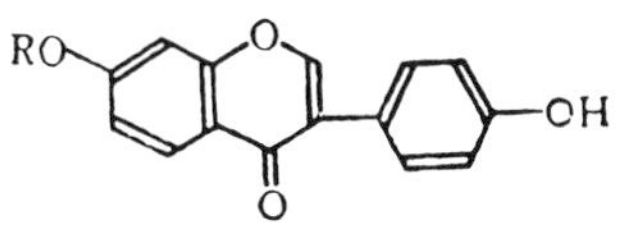

Daidzein R = H

Daidzin R = glucose

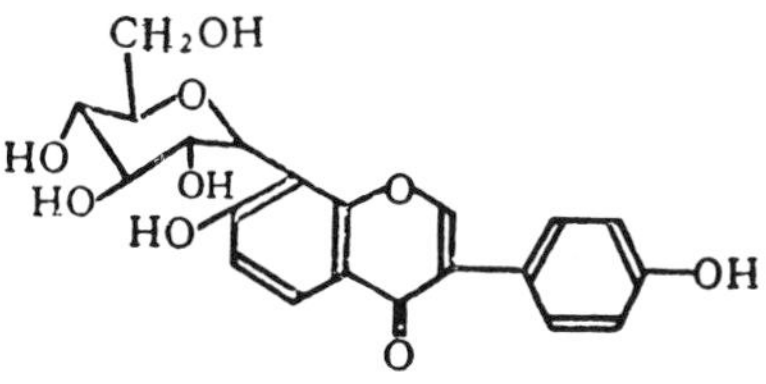

Puerarin

An alkaloid, kassein, has been isolated from the root. It has a muscarine-like action.

Actions

This herb can improve coronary circulation and lower myocardial oxygen consumption. The glycosides can cause muscular relaxation, especially in the left coronary vessel, and a decrease in heart rate. The most potent of the herb's glycosides is puerarin.

Ge Gen glycosides can also protect the myocardium against pitressin-induced ischemia. Studies *in vivo* and *in vitro* conducted by a group of investigators from Shangtung demonstrated that puerarin exhibits a competitive β-adrenergic blocking effect on rabbit atrial muscles and tracheal smooth muscles. This agent can lower blood pressure and heart rate in spontaneously hypertensive rats (SHR), but has little or no effect on normal Wistar rats. It also reduces plasma renin activity in SHR.

The herb has been observed to improve cerebral circulation, mainly by lowering vascular resistance of the cerebral vessels; it should be noted that this effect is dose dependent. Other observed effects include reduction in blood pressure, especially in renal hypertension. Extracts of the herb have shown antipyretic and anticonvulsant effects.

Absorption, Metabolism, and Excretion

Puerarin, given orally to volunteers, was completely absorbed from the intestine. After absorption, it was bound (42%) to plasma albumin. It was distributed primarily in the liver and kidney, and less extensively in the brain. It was eliminated mainly by metabolism in the liver; only 10% of the absorbed dose was excreted in the urine in its unchanged form.

The toxicity of the herb is relatively low. Mice receiving the herb at 2 g/kg/d for 2 months showed no significant pathological abnormalities.

Patients receiving the herb or glycosides may occasionally complain of a slight discomfort in the upper abdomen. No other side effects have been reported.

Therapeutic Uses

This herb is used by traditional Chinese physicians to dispel ''pathogenic factors'' from the superficial muscles, to reduce ''heat'', to reinforce *yang*, to promote ''eruption'', to promote salivation, and to relieve thirst.

It has been used in the treatment of angina pectoris and hypertension. It is also reported to be effective in the treatment of earlier stages of deafness and in optic nerve atrophy or retinitis; no explanation is available on the possible mechanism of effectiveness.

Tablets prepared by alcohol extract of the herb contain 0.2 g, equivalent to 1.5 g of crude herb. The standard dose is one tablet taken three times per day. *Ge Gen* glucosides tablets or injection ampules contain 10 mg of the active principle per tablet, or 200 mg/2 ml in the ampule.

References

1. **Lu, X. R. et al.,** *Acta Pharmacol. Sin.*, 7, 537–538, 1986.
2. **Song, X. P. et al.,** *Acta Pharmacol. Sin.*, 9, 55–58, 1988.

SAN QI (三七) — the dried root of *Panax zingiberensis*

FIGURE 2. The root of *San Qi*.

Chemistry

Of the herb content, 12% is saponins. These have been isolated and identified as arasaponins A, B, C, D, E, and R. After water hydrolysis, the saponins give genin products and sugars. The genins of arasaponin are panaxadiol and panaxatriol, as described in Chapter 1 on ginseng.

Actions

On the cardiovascular system, the herb exerts a major effect in dilating the coronary vessels by reducing vascular resistance. This results in an increase in coronary flow and a lowering of blood pressure. The herb also reduces the oxygen consumption of heart tissue by indirectly reducing myocardial metabolic rate. In addition, it can improve coronary collateral circulation.

The herb has a direct effect on blood, as well. It can accelerate blood clotting time and increase the probability of blood coagulation. It also lowers blood cholesterol levels.

Toxicity

The LD_{50} for *San Qi* extract is 2.5 to 3 g/kg in rabbits (intravenous), 5 to 7.5 g/kg in rats (intraperitoneal), and 7.5 to 10 g/kg in mice (intraperitoneal). Arasaponin A has an MLD of 450 to 460 mg/kg in mice.

Patients taking the herb may sometimes experience dryness of mouth, high skin temperatures, nervousness, and insomnia. Nausea and vomiting are also common side effects.

Therapeutic Uses

Chinese physicians used this herb to arrest bleeding, remove blood stasis, and relieve pain. Recent studies have shown that in treatment of angina pectoris, this herb can produce a 95.5% improvement in symptoms and 83% improvement in the ECG pattern. Treatment of hemorragic diseases has also proven positive; the herb can usually stop bleeding in cases of hemoptysis and hematemesis. The effectiveness of the herb in lowering plasma cholesterol, however, is still under debate, as insufficient data exist to confirm reported results.

The standard dosage for *San Qi* is a total of 1 to 1.5 g, administered orally in three divided doses per day. A 30% injection solution is also available; the dosage for intramuscular injection is 2 ml one to two times daily.

GUA LOU (瓜蒌) — the dried fruit of *Trichosanthes kirilowii* Waxia and *T. uniflora* Hao

Chemistry

The herb contains saponin, organic acids, and resin.

Actions

Gua Lou dilates the coronary vessels and increases myocardial tolerance to oxygen deprivation. It is also an expectorant and a potent laxative.

Clinical trials with this herb in intravenous infusion have shown no significant toxicity. Occasionally, a mild hypotensive effect, chilling, and headache have been reported. These may be due to impurities in the injected extract.

Therapeutic Uses

Traditional Chinese medicine prescribes this herb to remove ''heat'', reduce phlegm, relieve stuffiness of the chest, and as a laxative to relieve constipation.

It is used in angina pectoris, with an effective rate of 78.9%, depending upon the length of drug administration. Water extracts of the herb have been made into a tablet form, which is given in doses of four tablets (equal to 30 g of raw material) t.i.d. for 1 month. An injection solution is also available, in which 2 ml is equivalent to 10 g of raw herb. This is given by intramuscular or intravenous injection.

The herb has been used as an antitussive and as an expectorant. It has also been used in the treatment of acute mastitis.

SONG JI SHANG (桑寄生) — the dried leaf and stem of *Loranthus parasiticus* (L.) Merr.

Chemistry

The active principles are saponins, including avicularin ($C_{20}H_{18}O_{11}$) and quercetin.

Actions

Wu's pharmacology placed *Song Ji Shang* and *Hu Ji Shang* (see below) together in discussing their pharmacological actions and therapeutic uses. This was probably an error: although the two share a common name, *Ji Shang*, they are derived from different plants.

Song Ji Shang dilates the coronary vessels, reduces myocardial ischemic symptoms, and decreases the oxygen demand of cardiac tissue. It can antagonize vasopressin-induced vasoconstriction of the coronary vessel. The herb has also been observed to lower peripheral blood pressure; this effect can be slightly antagonized either by atropine or vagotomy.

The herb's central inhibitory effects include an ability to antagonize caffeine stimulating effects on the central nervous system (CNS). It also has a diuretic effect, but is less potent and more toxic than theophylline. Researchers have also attributed an antibacterial effect to the herb.

In mice, the LD_{50} is 11.24 g/kg of crude herb (i.p.) and 1.17 g/kg of saponin (i.p.).

Patients taking this herb may sometimes complain of headache, dizziness, loss of appetite, abdominal distension, and thirst.

Therapeutic Uses

Chinese folk medicine used this herb to support the liver and kidney, strengthen the muscles and bones, and to prevent miscarriage. Modern medical practitioners use the herb to treat angina pectoris, cardiac arrhythmias, and hypertension. The herb is also used to treat frostbite: the water extract of the plant is made into an ointment or mixed with glycerin, which is then applied to the affected area.

The herb is most widely available via *Song Ji Shang* packages, which contain 39 g of raw herb per package. It has also been water extracted to produce a beverage, which is served as a tea drink twice a day. An injection solution is available, with ampules containing 2 ml (equivalent to 4 g of the raw herb) for intramuscular or intravenous injection.

Reference

Wu, P. C., *Pharmacology of Zhong Cai Yao,* Peoples' Health Publisher, Beijing, 1983, 90–91.

HU JI SHANG (槲寄生) — dried leaf and stem of *Viscum coloratum* Nakai and *V. album* L.

Chemistry

The herb contains oleanolic acid, β-amyrin, and mesoinositol. The leaf of the plant also contains the following substances: flavoyadorinin A (alboside), flavoyadorinin B, homoflavoyadorinin B, lupeol, and myristic acid. The seed of the plant contains agglutinins, alkaloids, quercitol, querbrachitol, and vitamins E and C.

The species of the plant found in the province of Kwangsi, *V. album* L., contains three toxic substances: viscotoxin A2, A3, and B. All three are 45-amino acid peptides with three S-S linkages.

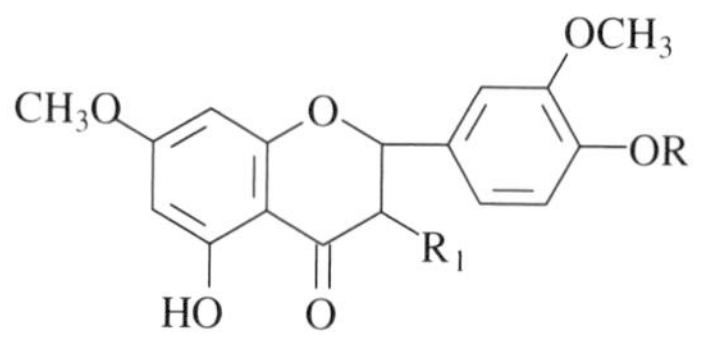

	R =	R_1 =
Flavoyadorinin A	H	O-glucose
Flavoyadorinin B	-glucose	H
Homoflavoyadorinin B	-glu-O-apiose	H

Actions

Chinese medical literature frequently interchanged the herbs *Song Ji Shang* and *Hu Ji Shang*. It should be noted, though, that while these herbs have similar pharmacological actions and therapeutic uses, they come from different plants.

Hu Ji Shang has its chief action on the cardiovascular system, with some anticancer properties exhibited. In recent studies, the water or alcoholic extract of *Hu Ji Shang* was given to rabbits by intravenous injection. The result was a drop in blood pressure to 32% of the original level, for a dose of 1 ml/kg (equivalent to 0.32 g/ml). The mechanism of this action is not yet understood, but possibly involves a central effect. The herb increases coronary flow and produces a decrease in the heart rate, which can be blocked by atropine. Atropine, however, does not alter the antihypertensive effect of this extract.

Therapeutic Uses

Hu Ji Shang is primarily used as an antihypertensive agent. It is usually given as a 20% tincture, 1 to 2 ml t.i.d. Patients taking the tincture report feeling an improvement, with few side effects.

In other studies, a group of patients with later-stage stomach cancer was given the extract intramuscularly for several weeks. This resulted in a prolongation of the patients' survival time. A preparation called Iscador, which is manufactured by a Chinese pharmaceutical company, has been used in these cases. Patients were tracked for 5 years and reported excellent results.

YIN YANG HUO (淫羊藿) — the dried plant of *Epimedium brevicorum* Maxim., *E. horeanum* Nai, or *E. sagitatum* (Sieb. et Zucc.) Maxim.

Chemistry

The active principles are the glycosides icariin and noricariin. The herb also contains ceryl alcohol and some essential oils and fatty acids.

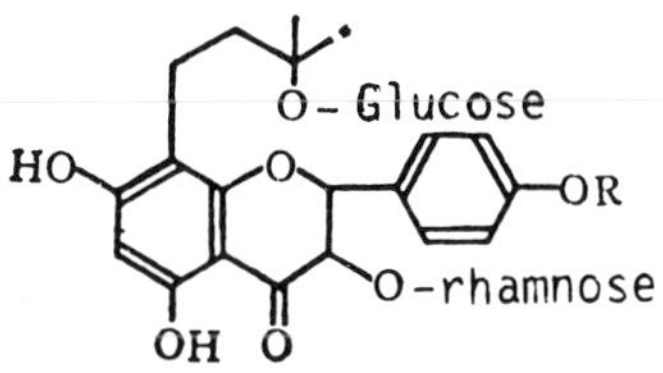

HO, O-Glucose, OR, O-rhamnose, OH, O

	R =
Icariin	CH_3
Des-O-methylicariin	H

Actions

The herb's primary effects are on the cardiovascular system. It dilates the coronary vessels and increases the coronary flow by reducing vascular resistance. It lowers the blood pressure, demonstrating a longlasting action. Tolerance to the herb, however, can develop after continuous use.

Yin Yang Huo extract is also reputed to have a sexual stimulatory effect on males. The herb can stimulate growth of the prostate, testes, and anus rector muscle. Studies have reported that persons taking this herb show an increase in sperm production and in the urinary excretion of 17-ketosteroids.

The herb acts as an antitussive and expectorant; its antitussive effect is due to a central inhibitory action. It also has antibacterial and antiviral properties; extracts can inhibit the growth of staphylococci, streptococci, and pneumococci in cultures. Additionally, the herb displays anti-inflammatory actions.

The herb is relatively nontoxic. Patients taking *Yin Yang Huo* may experience some gastrointestinal disturbances, such as nausea and vomiting, and occasionally dryness of mouth.

Therapeutic Uses

Yin Yang Huo was described in the *Herbal Classic of the Divine Plowman* as a body-building agent, *yang* supporter, an agent to reinforce muscles and bones, and to help the liver and kidney.

It is commonly used in the treatment of angina pectoris, chronic bronchitis, and neurasthenia. Studies of the herb's effectiveness in 1066 cases of chronic bronchitis have shown a 74.6% efficacy rate, documenting the herb's antitussive and expectorant effects. Additionally, intramuscular injection of a 10% *Yin Yang Huo* solution has been reported to significantly improve symptoms in cases of child paralysis.

In 0.3-g tablet form (equivalent to 2.7 g of raw material), the herb is administered in doses of four to six tablets b.i.d. for 1 month; administration is stopped for 7 to 10 d, then resumed in a second series if required.

Yin Yang Huo pills are also available, made from a condensed 80% extract. In a 20% tincture, the herb is administered in doses of 5 ml t.i.d. before meals. Injection solutions are prepared in ampules of 2 ml, equivalent to 1 g of raw material, used intramuscularly.

MAR DONG (麦冬) — the dried root of *Ophiopogon japonicus* (Thunb.) Ker-Gawl

Chemistry

The active principles include: β-sitosterol, stigmasterol, and ophiopogenin B.

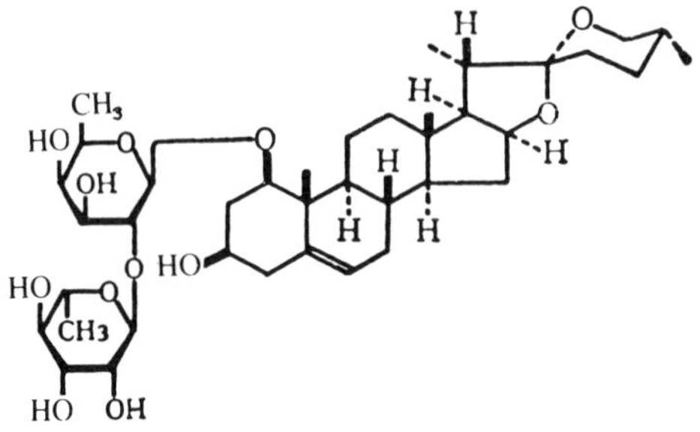

Ophiopogonin B

Actions

Mar Dong increases coronary blood flow, produces a strophanthin-like inhibitory action on Na-K ATPase, and increases myocardial contractility. It has also been observed to increase the tolerance to oxygen deprivation, to slowly elevate blood pressure, and to exhibit antibacterial properties.

Clinical trials on 101 patients with angina showed a 74 to 84% effective response rate to the herb. Symptoms improved and the ECG pattern changed. Common side effects are abdominal distension, gas, and loose bowel movements.

Therapeutic Uses

This herb is used in Chinese medicine as a *yin* supporter, to smooth lung functions, and to stop coughing. It is generally prepared as a decoction of 10 ml, equivalent to 15 g of raw material; this is administered three times a day for 3 to 18 months.

It has also been made into an injection solution, with 2 ml equivalent to 4 g of raw material. This is administered via intramuscular or intravenous injection.

WEI MAO (卫矛) or GUI JIAN YU (鬼箭羽) — the dried young branch, leaf, and fruit of *Euonymus alatus* (Thunb.) Sieb.

Chemistry

The leaf of the plant contains quercetin, dulcite (1.1.%), epifriedelinol (0.3%), friedelin, and some resin. The seeds contain saturated fatty acid and some organic acids.

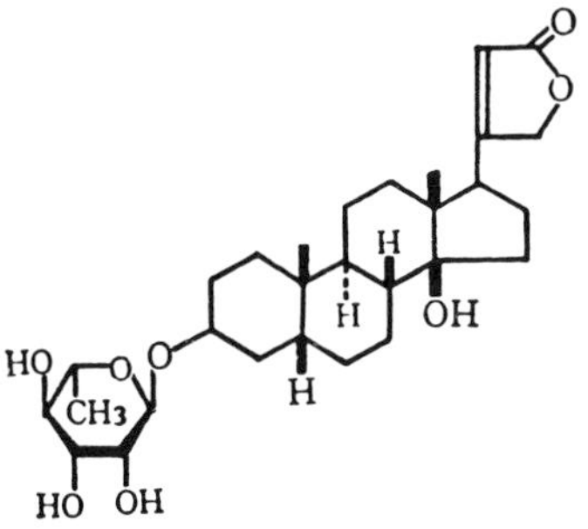

Evomonoside

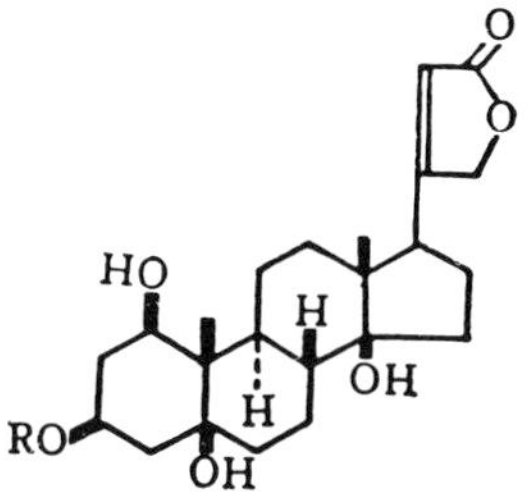

Glucoevonoside R= β-D-glu

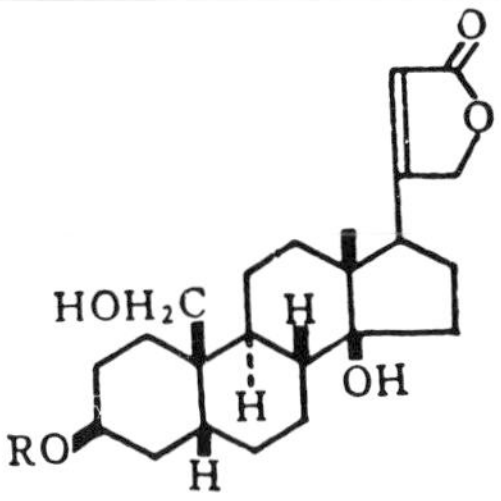

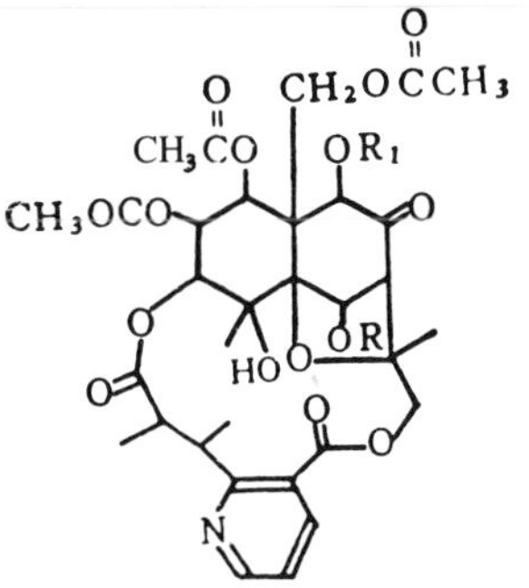

	R =
Evonoloside	α-L-rhamn
Glucoevonoloside	α-L-rhamn-O-β-D-glu

	R=	R_1 =
Evorine	H	$COCH_3$
Evozine	H	H
Evonine	$COCH_3$	$COCH_3$

The seeds from the species *Euonymus europaea* L. contain several cardiac glucosides and alkaloids. The glucosides include: evomonoside (which gives the genin glucoevonogenin and rhamnose after water hydrolysis), glucoevonoloside, and evonoloside. Alkaloids isolated from the seeds include evonine, evorine, and evozine.

Several alkaloids have also been isolated from the *E. alatus formastriates* (Thunb.) Mak. species. These include evonymine, wilfordine, and alatamine.

Actions

This herb slows the heart rate and protects against myocardial ischemia. It can increase tolerance to oxygen deprivation, and has a significant, albeit temporary, hypotensive effect.

It acts as a depressant on the CNS and can lengthen barbiturate-induced sleeping times. Its effects on metabolism include a reduction of blood sugar levels via stimulation of the beta cells of pancreatic islets. Additionally, quercetrin has been found to be a good expectorant.

Therapeutic Uses

This herb was used in folk medicine to regulate *qi* (bodily energy) and blood, relieve pain, eliminate stagnant blood, and treat dysmenorrhea.

It is administered in the form of a syrup, with 1 ml equivalent to 1 g of raw material. In the treatment of angina pectoris, the standard oral dose is 10 to 30 ml t.i.d.

JI MU (檵木) — the dried plant of *Loropetalum chinense* (Rbr.) Oliv.

Chemistry

The plant contains flavone and glycosides, including quercitrin and isoquercitrin.

Actions

Ji Mu increases cornary circulation and reduces oxygen consumption of heart tissue. It slows the heart rate and increases myocardiac contractility. The herb also displays a hemostatic action, shortening blood clotting time.

Other observed actions include the ability to contract uterine muscles in both pregnant and nonpregnant women. It has an antibacterial effect against organisms such as staphylococci and *Escherichia coli*.

Therapeutic Uses

This herb was formerly used as an antipyretic, a detoxicant, and a hemostatic. Today, it is used to treat angina pectoris, bronchitis in elderly persons, alimentary indigestion, and bleeding — including duodenal ulcer bleeding, uterine bleeding, and skin infections.

Preparations of the herb include an injection solution made from the *Ji Mu* flavone, with 2 ml containing 0.25 mg of flavone (equivalent to 3.12 g of leaf) for intramuscular administration, twice a day. An injection solution is also made from the white crystalline substance of the extract, with 2 ml containing 40 mg of crystal substance for intramuscular or intravenous injection.

For cases of chronic bronchitis, food poisoning, and indigestion, the herb is administered in tablet form. For the former, a 0.3-g tablet is used. For the latter two, a 200-mg tablet is available, with a dose of one tablet t.i.d.

CHANG BAI RUI XIANG (長白瑞香) — the dried root and stem of *Daphne koreane* Nakei

Chemistry

A substance called daphnetin has been isolated from this plant.

Actions

It can dilate the coronary vessels and reduce the oxygen consumption of heart muscle and improve its function.

Therapeutic Uses

The ancient Chinese pharmacopoeia recommended the herb to ''unplug'' the stagnation of blood in the circulatory system. Recent studies indicate a 92% effective rate in the treatment of angina pectoris and simple arthritis. The herb is dispensed in the form of an injection solution, with 2 ml equivalent to 3 g of raw material, for intramuscular administration.

MAO DONG QING (毛冬青) — the dried root or leaf of *Ilex pubescens* Hook. et Arn.

Chemistry

The plant contains flavone and ursolic acid. The root of a related species, *Ilex pubescens* var. *glabra* Chang, has been shown to contain seven crystalline substances. These include compounds such as 3,4-dihydroxyacetophenone, hydroquinone, scopoletin, and vomifliol. Their pharmacological actions have not been investigated.

Actions

This herb's primary actions affect coronary vasodilatation. It reduces the heart rate and the oxygen consumption of the heart muscle. The onset of action is slow and the duration of action is long. Experimental data show that the compound 3,4-dihydroxyacetophenone can effectively block slow inward current and delay the inactivation of fast coronary potassium ion current of the heart muscle.

Other actions include the lowering of blood pressure — this can be lessened by atropine. The herb has antitussive and expectorant effects, and displays some antibacterial properties.

The LD_{50} of this herb was 920 mg/kg in mice (intravenous).

Therapeutic Uses

This herb is used to treat angina pectoris, acute myocardiac infarction, cerebral thrombosis, and thrombophlebitis. It has also been reported to be effective in the treatment of central angiospastic retinitis: a *Mao Dong Qing* injection was administered intramuscularly 12 to 52 d, resulting in a 90% efficacy rate.

Standard preparations of the herb include a tablet equivalent to 6 g of raw material; the normal dose in this form is 5 to 10 tablets t.i.d. A decoction package is also available, containing 120 g raw material per package. In injection form, 2 ml of solution containing 20 mg of the active principle (equivalent to 8 g raw material) is administered intramuscularly. Additionally, a syrup can be prepared, with each ml equal to 3 g of raw material, and a normal dose of 20 to 30 ml t.i.d.

SI JI QING (四季青) — the dried leaf of *Ilex chinensis* Sims

Chemistry

The herb contains protocatechuic acid, protocatechuic aldehyde, ursolic acid, and a large amount of tannic acid.

Actions

Si Ji Qing reduces vascular resistance in coronary vessels, resulting in an increase in blood flow. It reduces the oxygen consumption of heart tissue and has a protective effect against myocardial ischemia. Additionally, it exhibits both antibacterial and anti-inflammatory activity.

The herb is almost completely absorbed from the intestine. It is eliminated mainly by the kidney.

Its toxicity is very low. The LD_{50} for the raw plant was 133 $\pm$ 11.6 g/kg in mice (oral) and 104 $\pm$ 6.5 g/kg (intravenous). The LD_{50} for protocatechuic aldehyde was 0.51 $\pm$ 0.014 g/kg (intramuscular) in mice.

Therapeutic Uses

The primary use of this herb is in the treatment of angina pectoris and thrombophlebitis. It is also given internally to treat burns, or may be externally applied on burn wounds; treatment in cases of extremity ulceration is also reported. It has been known to accelerate wound healing and to exert an antibacterial effect. This antibacterial effect allows it to be effective in the treatment of infectious diseases such as acute and chronic bronchitis, pneumonia, dysentery, acute pancreatitis, cholecystitis, nephritis, pelvic infection, urethritis, and cervicitis.

For the treatment of burns and wounds, the herb is prepared as a water extract, with 1 ml equivalent to 1 g of raw material. This solution can be sprayed on the wound surface, applied as a soaked gauze, or rubbed directly on the skin.

From the water extract, two concentrations of emulsion are derived. In the first, oil is added to the water extract to form an emulsion in which 1 ml is equivalent to 1.8 g of raw material. In the second, 1 ml is equivalent to just 0.6 g of raw material.

The herb is also available as a syrup (1 ml equivalent to 1 g of raw material, with a dose of 20 ml t.i.d.), tablets (each equivalent to 4 g of raw material), or injection solution (1 ml equivalent to 2 g of raw material, intramuscular dose of 2 to 4 ml).

SHE XIANG (麝香) — the dried secreta from the male pocket of *Moschus sifanicus* Prezewalki or *M. moschiferus* L.

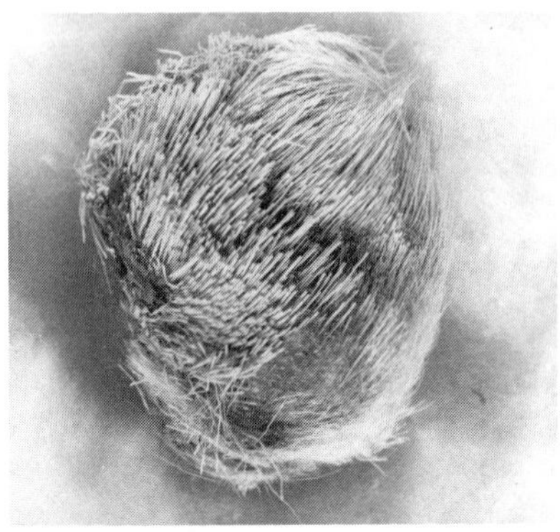

FIGURE 3. The male pocket of *Moschus sifanicus*.

Chemistry

Between 0.5 to 2% of the herb is muscone, which is the active principle of the secreta and produces a very distinctive aroma. The natural herb has now been largely replaced by synthetic products. Muscone is a colorless liquid with a melting point of 142 to 145°C. In addition to muscone, *She Xiang* contains several kinds of androstanes, which have male-stimulating effects, and muscopyridine.

The secreta from *Viverricula indica* Desmarest contains a substance very similar to muscone — zibetone. Zibetone has an odor similar to muscone, and has the following structure:

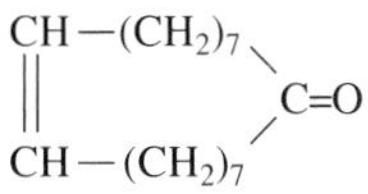

CH—$(CH_2)_7$
C=O
CH—$(CH_2)_7$

$(CH_2)_{12}$—CH—CH_3
CO——CH_2

N
—CH_3

Zibetone 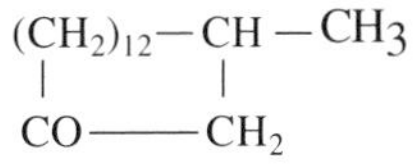 Muscone 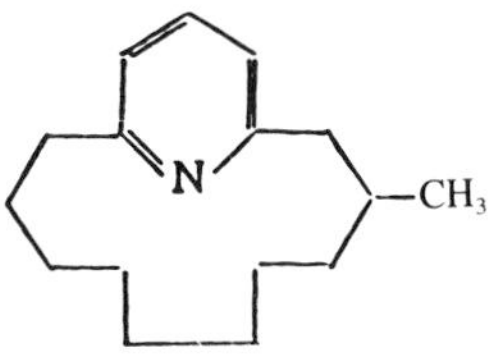 Muscopyridine

Actions

The herb's effects on the heart include an increase in both myocardiac contractility and tolerance to oxygen deprivation. It also improves blood flow.

On the CNS, small doses of either natural and synthetic muscone has a stimulatory effect, and can reduce barbiturate-induced sleep. In large doses, however, they can depress the CNS and prolong sleeping time. Chinese herbal lore also claimed that the herb would stimulate the CNS and increase the intelligence quotient of children.

She Xiang has anti-inflammatory properties. Muscone can decrease capillary permeability, with a greater potency than that seen in salicylic acid. It decreases prostaglandin E and F tissue concentrations.

The herb is also reputed to stimulate male sex hormones, and to increase uterine contractions.

This agent is rapidly metabolized in the body. The half-life is 1.4 min. The LD_{50} of synthetic muscone is 290.7 mg/kg in mice (intraperitoneal).

Patients taking this substance generally show few side effects. Occasionally, they may experience dizziness, nausea and vomiting, and loss of appetite.

Therapeutic Uses

This herb was used in folk medicine as a remedy for stroke, convulsions, shock, and sudden circulatory collapse. Many ancient pharmaceutical recipes or formulas contain *She Xiang* — for example, the *Lu Shien* pill (六神丸) and the *An Gon Niu Huang* pill (安宫牛黄丸) have *moschus secreta* as a major ingredient. These pills are used to stabilize the mind and nerves, to relieve swelling, and to relieve pain. These products are sold on the street or in the market without prescription, for the treatment of poison-induced swelling, trauma, and bone fractures.

In the treatment of angina pectoris, the herb generally requires 2 to 5 min to show effect. It is contraindicated to pregnant women. Generally, synthetic muscone is more potent than the natural product.

The herb is also reported effective as an anticancer agent. Injection of muscone in patients suffering from esophagal or gastrointestinal cancer produced significantly symptomatic improvement, especially in the early stages of the disease.

The herb has been used in the treatment of vascular migraines. In these cases, prophylactic use gives best results.

General preparations of the herb include sublingual tablets, which contain 30 mg of synthetic muscone. These are used as a substitute for nitroglycerin in the treatment of angina. Injection solutions are prepared at 1.5 mg/ampule for intramuscular or intravenous use. An aerosol preparation is also available, containing 180 mg muscone/vial.

She Xiang cardiotonic pills, a mixture of *She Xiang* and ginseng, are widely used in China. It is claimed that this preparation has an effective rate of about 32% in the treatment of angina.

JU HUA (菊花) — the dried flower of *Chrysanthemum morifolium* Ramat.

Chemistry

Several types of substances have been isolated from this flower. These include bornol (an essential oil); chrysanthenone and camphor; the alkaloid stachydrine ($C_7H_{13}O_2N$); and several glucosides: acacetin-7-rhamnoglucoside, cosmosiin and acacetin-7-glucoside, diosmetin-7-glucoside. Also found are adenine, choline, vitamin B and A-like substances.

The *Ju Hua* root contains the following substances: chrysartemin A and B; chrysandiol and chlorochrymorin

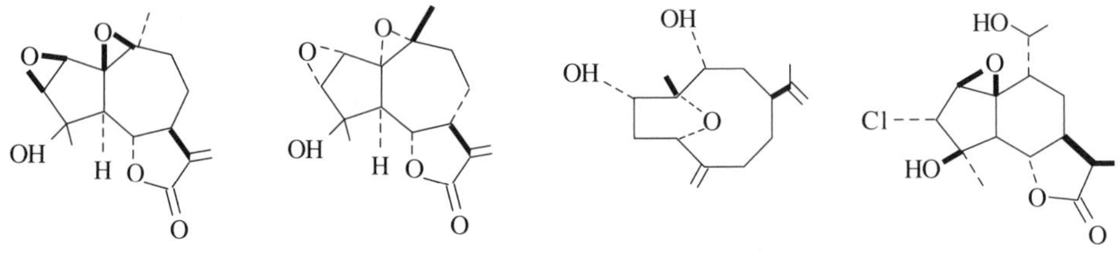

Chrysartemin A Chrysartemin B Chrysandiol Chlorochrymorin

Actions

This herb increases coronary vasodilatation and coronary blood flow, but has little effect on cardiac contractility or oxygen consumption. It reduces the capillary permeability induced by histamine, and operates as an antibacterial and antipyretic agent.

Therapeutic Uses

In southern China, this herb is very popular during the summer time, when it is consumed as a tea. In folk medicine it was used as an antipyretic, to clear the eye and the mind, and as an antitoxin. It is widely used as a remedy for the common cold, headache, dizziness, red eye, swelling, and hypertension.

It is commonly used in cases of angina pectoris and hypertension. To make a water extract, 300 g of the herb is condensed into 500 ml; the standard dose is 25 ml twice a day, or taken as a tea.

Chapter 6

ANTIHYPERCHOLESTEROLEMIC HERBS

In Chinese markets and pharmacies there are many foods and beverages (mainly teas) which are touted as substances that lower blood cholesterol levels. The effectiveness and mechanisms of action of these products are not yet fully understood.

For example, *Tuo Cha* (陀茶), a well-known tea grown in the southwestern provinces of China, Yunnan and Szechuen, is reputed to lower plasma cholesterol and is widely used by Chinese both locally and abroad. The white fungus *Pei Mu Er* (白木耳), a dried sporophore of *Tremella fuciformis* (Tremellaceae), is widely used as a delicate dessert and is said to lower cholesterol. These substances most likely contain some type of resin that can bind cholesterol or bile acids in the intestine and promote their excretion in the feces.

Ginseng, as described in Chapter 1, is well known and widely used in Chinese and Japanese folk medicine as a treatment for arteriosclerosis. But ginseng is extremely expensive and, in ancient times, was considered to be the Emperor's private property. Thus, its use as an anticholesterol agent is not economically practical, despite its medicinal value. The following less costly herbs have been described in the Chinese pharmacopoeia as useful in the treatment of hypercholesterolemia.

LING ZHI (灵芝) — the dried fructification of the fungus *Ganoderma lucidum* (Polyporaceae)

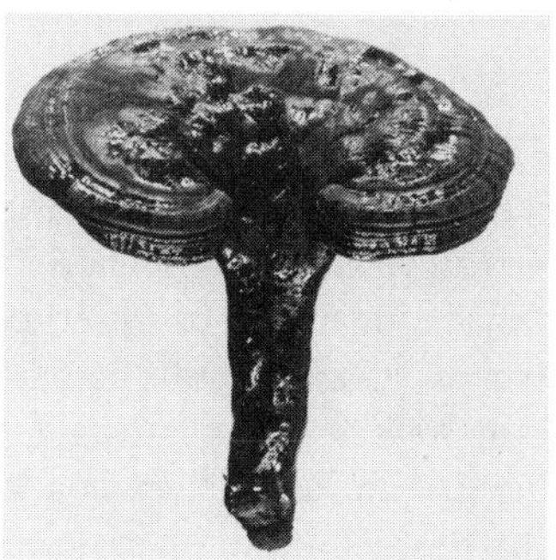

FIGURE 1. *Ling Zhi.*

Chemistry

The herb contains ergosterol, fungal lysozyme, proteinase, several amino acids, and organic acids. Recently several polysaccharides have been isolated from *Ganoderma lucidum* Leyss. ex Fr.

HO

Ergosterol

Actions

Ling Zhi has been shown to lower plasma cholesterol and phospholipid levels, helping to prevent atherosclerotic changes in the arterial wall. It increases myocardiac contractility and systolic volume, and can improve cardiac function and reduce O_2 consumption.

The herb affects the central nervous system (CNS) acting as an anticonvulsant, sedative, analgesic, and antitussive agent. It can also relax smooth muscles of the intestine and uterus, antagonizing spasmodic contractions induced by histamine or pitressin. Additionally, it lowers blood sugar levels and protects hepatic functions.

Zhan et al.[4] gave this herb to mice and found it can prolong the sleeping time induced by barbiturate and reduces body temperature.

Peng and Wang[3] studied the pharmacological actions of this plant and found that in rats it can increase the capillary circulation and the number of capillaries under microscopic observation. The herb also increases the cAMP concentration of the heart muscles and platelets, but not cGMP. In an experiment with mice they have found that this herb can reduce O_2 consumption.

Lin and colleagues[1,2] suggested that the active principles of the herb are their polysaccharide components. They reported that these polysaccharides antagonize significantly the suppressive effect of hydrocortisone on lymphocyte proliferation of mouse spleen induced by injection of convanavalin A. The polysaccharides also enhance the reactive ability of aged mice and potentiate the activity of DNA polymerase α in intact mouse spleen cells. Based on their data, Lin and colleagues then concluded that *Ling Zhi* has the property to restore the immunodeficiency and an antiaging effect.

The toxicity of this herb is low. The LD_{50} for *Ling Zhi* syrup is 69.6 ml/kg in mice, and 4 ml/kg in rabbits (by stomach).

Therapeutic Uses

In Chinese, the name *Ling Zhi* means "spirit plant". In traditional Chinese medicine, it is used as a tonic, roborant, sedative, and stomachic. The *Ben Cao Kong Mu* explained that "continued use of *Ling Zhi* will lighten weight and increase longevity."

The herb is currently used in the treatment of hyperlipemia, angina pectoris, chronic bronchtis, hepatitis, and leucopenia. Clinical trials in 2000 cases of chronic bronchitis reported an effective rate of 60 to 90%.

Several preparations of *Ling Zhi* are used. Common general preparations are syrups and tinctures. As a syrup, the herb is administered in doses of 4 to 6 ml daily. As a 20% tincture, a dose of 10 ml t.i.d. is given.

In the treatment of insomnia and neurasthenia, the herb is prescribed in 1-g tablets, with a dose of three tablets t.i.d. Cases of chronic bronchitis are treated with a fermented extract, at a dose of 25 to 60 ml t.i.d. In the treatment of leukopenia, 0.5-g capsules are prepared, with an equivalent of 4.16 g of cultured extract; the dose is four capsules t.i.d.

References

1. **Lei, S. W. and Lin, Z. B.,** *J. Beijing Med. Univ.*, 23, 329–333, 1991.
2. **Xia, D., Lin, Z. B. et al.,** *J. Beijing Med. Univ.*, 21, 533–537, 1989.
3. **Peng, H. M. and Wang, L. J.,** *Chin. Traditional Herbal Drug*, 17, 21–25, 1986.
4. **Zhan, S. W. et al.,** *Chin. J. Hosp. Pharm.*, 11, 55–58, 1991.

SHAN ZHA (山楂) — the dried fruit of *Crataegus pinnatifida* Bge. var. *major*, *C. pinnatifida*, or *C. cuneata* Sieb. et Zucc.

Chemistry

The active principles are chlorogenic acid, caffeic acid, citric acid, crataegolic acid, maslinic acid, ursolic acid, and some saponins.

Actions

The herb lowers blood cholesterol by increasing cholesterol's catabolism. This helps the surface of the atherosclerotic area in the arterial wall to shrink and become smoother. It improves coronary circulation, and increases the ^{86}Rb uptake of the heart muscle via an increase of blood flow. This reduces O_2 consumption and protects against myocardial ischemia. Additional effects are an increase in myocardial contractility and a lowering of blood pressure. It also acts as an antibacterial agent.

Shan Zha is virtually nontoxic. Nausea and vomiting may be observed if large quantities are consumed. The LD_{50} for the 10% *Shan Zha* extract is 33.8 ml/kg for rat and 18.5 ml/kg for mice.

Therapeutic Uses

Shan Zha is used not only for medicinal purposes, but is also sold on street corners as a popular snack food. Chinese consider it useful in reducing food stagnancy and blood stasis. As a medicine, it is used to treat hypercholesterolemia, angina pectoris, and hypertension.

The herb is prepared in three forms. The 0.3-g sugar-coated *Kuo-san-nin* tablet is equivalent to 3 g of raw material, and is prescribed in doses of five tablets t.i.d. *Kuo-san-nin* extract is administered in doses of 1 tablespoon t.i.d., equivalent to 15 g of the raw herb. In syrup form, 1 ml is equivalent to 0.65 g of raw material, with normal dosage of 20 ml t.i.d.

SHOU WU (首乌), also called HE SHOU WU (何首乌) — the dried tuberous root of *Polygonum multiflorum* (Polygonaceae)

Chemistry

Approximately 1.1% of the plant comprises active principles. These include chrysophenol ($C_{15}H_{10}O_4$), emodin ($C_{15}H_{10}O_5$), emodin methyl ester, rhein ($C_{15}H_8O_6$), and the glycoside rhaphantin ($C_{21}H_{24}O_9$). In addition, a large quantity of lecithin and other glycosides is found.

Actions

The herb lowers plasma cholesterol levels by reducing intestinal absorption of dietary cholesterol. It has also been suggested that the lecithin contained in the herb can block cholesterol uptake from the plasma into the liver, preventing the deposition of cholesterol into plaques in the arterial wall.

The herb has also been observed to reduce heart rate and to slightly increase coronary circulation. Emodin derivatives can increase peristalsis in the large intestine and produce a laxative effect. The herb also exhibits antibacterial properties.

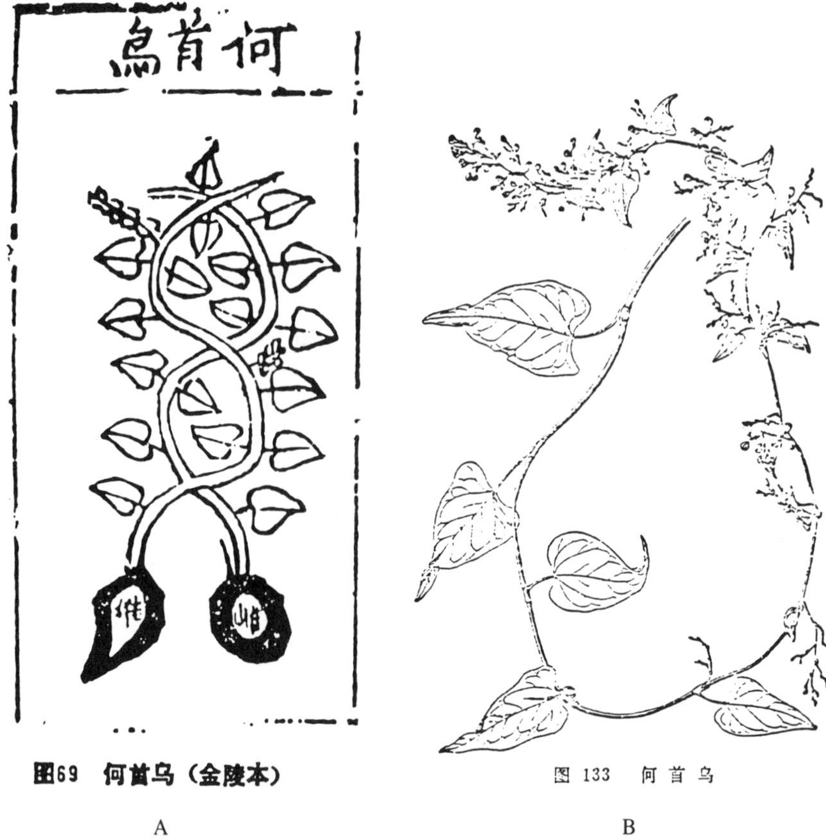

何首烏
图69 何首乌（金陵本）
图 133 何首乌
A
B

FIGURE 2. Drawings of *He Shou Wu*. [(A) From *Ben Cao Kong Mu* (Jian-nin print); (B) from *Zhong Cao Yao Yue*, Vol. 2, p. 161.]

This herb is relatively nontoxic. The common adverse effects include nausea, vomiting, abdominal distention, and loose bowel movements.

Therapeutic Uses

Traditionally, *Shou Wu* is used as a laxative, detoxicant for boils, to replenish the "vital essence" of the liver and kidneys, to "resupply" blood, to treat early graying of the hair, to treat backache and neurasthenia, and to lower plasma cholesterol. In 813 A.D., Li Ao described the action of *Shou Wu*. Li's work noted (as reported by Unschuld) that the herb made an impotent man sexually active, helping him to regain his reproductive powers (in other words, to father a son), rejuvenizing him to a lifespan of 160 years (Unschuld, *Medicine in China*, pp. 230–234). Not surprisingly, this story has never been confirmed.

This herb received no attention in Chinese medical-pharmaceutical circles until the 4th century A.D. Now, the herb is commonly used in the treatment of hypercholesterolemia, with a 62 to 82% effective rate reported. Cholesterol levels rise again, however, after administration of the herb is stopped. Other uses include treatment of neurosis, neurasthenia, and insomnia. For the latter two, the onset of action is slow, with at least 2 weeks of administration required before positive effects are seen. Some cases have been reported in which the herb was used to treat malaria and the whooping cough.

This herb is generally prescribed in the form of 0.8 g tablets, at a dose of five to six tablets three times a day. It is also prepared as a 20% injection solution, each ampule containing 2 to 4 ml for intramuscular injection. Tablets are also prepared in combination with other herbs, including *Shan Wu, Wu Mei Zhu, Tan Sen,* and *Huang Ling*.

JUE MING ZI (决明子) — the dried seed of *Cassia obtusifolia* or *C. tora* (Leguminosae)

Chemistry

The herb contains many active substances, including chrysophenol, emodin, aloe-emodin, rhein, physcion, obtusin, aurantio-obtusin, chrysobtusin, rubrofusarin, norrubrofusarin, and toralactone. Emodin is the ingredient responsible for the herb's laxative effects.

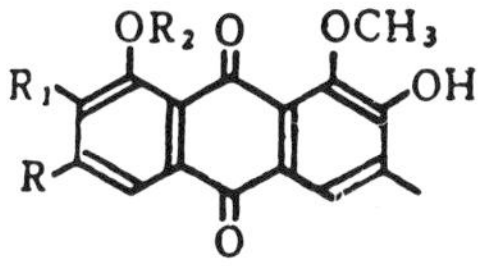

	R=	R_1=	R_2=
Obtusin	OCH_3	OCH_3	H
Aurantio-obtusin	OH	OCH_3	H
Chrysobtusin	OCH_3	OCH_3	CH_3

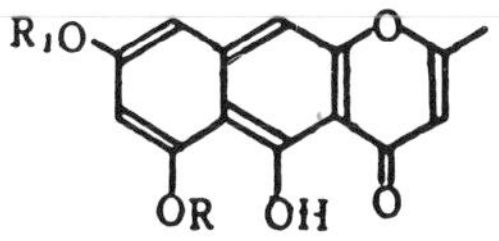

	R=	R_1=
Rubrofusarin	H,	CH_3
Nor-rubrofusarin	H	H
Rubrofusarin-6-β-gentiobioside	β-gentiobiose	CH_3

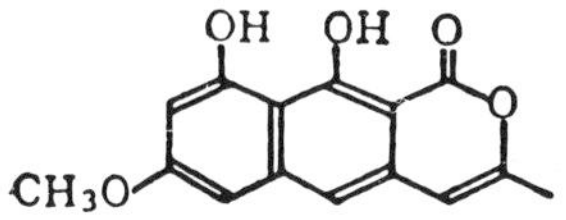

Toralactone

Actions

Jue Ming Zi lowers plasma cholesterol and prevents the formation of atherosclerotic plaque in the arterial wall. It also acts as an antihypertensive, antibacterial, and laxative.

Gastrointestinal disturbances such as nausea, abdominal distension, and loose bowel movements are the most commonly observed adverse effects.

Therapeutic Uses

Traditionally, the herb was used to remove "heat" from the liver, improve visual acuity, and as a laxative. Modern physicians use the herb to treat hypercholesterolemia and hypertension.

In the treatment of hypercholesterolemia, it has been reported that normal plasma cholesterol levels can be achieved in 80% of cases if the herb is taken continuously for 2 weeks; if treatment is continued for a longer period, a 96% success rate has been observed.

The herb can be prescribed as a syrup of 0.75 g raw material per ml at an oral dose of 20 ml three times a day. It is prepared in tablet form (3 g of herb per tablet), with a dose of five tablets t.i.d. A water decoction of 50 g is also used, administered in divided doses twice a day.

The herb is also used to treat vaginitis. A water extract of the herb is used to wash the vulva and vagina for 15 to 20 min daily for 10 d.

WU TONG (梧桐) — the dried leaves, seeds, flowers, barks, and roots of *Firmiana simplex* (L.) W. F. Wright

Chemistry

The herb contains betaine, choline, β-amyrin, β-amyrin acetate, lupenone, heutriacontane, rutin, octacosanol, and β-sitosterol.

The seeds also contain small quantities of caffeine, fatty acids, sterculic acid, and lupenone. Their structure is as follows:

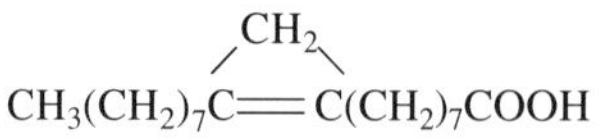

$$CH_3(CH_2)_7C=C(CH_2)_7COOH \text{ (with } CH_2 \text{ bridging the two C atoms)}$$

Sterculic acid

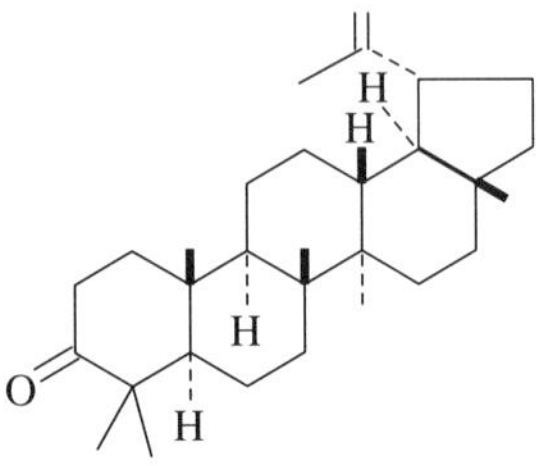

Lupenone

Actions

The alcoholic extract of the herb can reduce plasma cholesterol levels but does not affect triglyceride levels. It has a vasodilatating effect, increases coronary flow, and lowers peripheral blood pressure.

The toxicity is very low. The water extract of this herb had an LD_{50} in mice of 8.3 g/kg (intravenous). Clinically, a slightly discomfort in the gastrointestinal tract was reported. Occasionally, patients also complained of dryness of mouth and fatigue.

Therapeutic Uses

Chinese physicians believed that *Wu Tong* could remove ‘‘heat’’, act as a detoxicant, smooth lung functions, and increase the appetite. It was also reputed to stimulate hair pigmentation.

The herb is now used to treat hypercholesterolemia and hypertension. It is also applied externally, to treat burns.

ZE XIE (泽泻) — the dried root and stem of *Alisma orientalis* (Sam.) Juzep.

Chemistry

The major active principles are alisol A and B, alisol monoacetate, and the essential oil epialisol A.

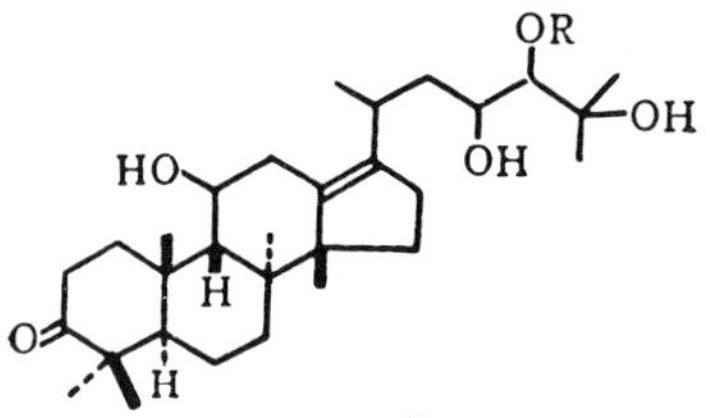

	R =
Alisol A	H
Alisol A mono-acetate	OC-CH_3

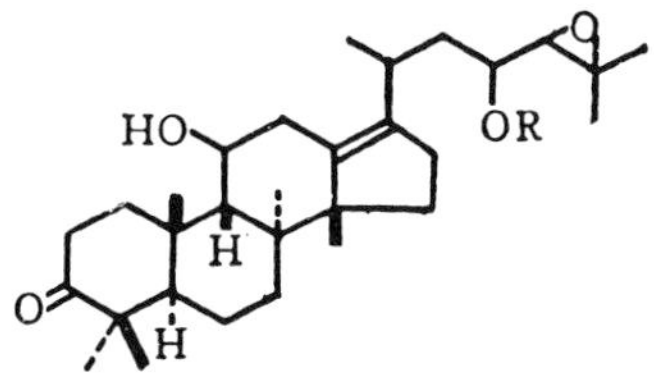

	R =
Alisol B	H
Alisol B monoacetate	-OC-CH_3

Actions

Ze Xie was claimed to be effective as a diuretic for the removal of edemic fluid. Clinical trials showed that the herb can lower plasma cholesterol levels, protect hepatic function, and increase the urinary excretion of Na, Cl, and urea.

Therapeutic Uses

Clinical trials to lower hypercholesteremia showed a better result than in cases of hypertriglyceride. Subjective symptoms were improved.

It has also been used in edema due to kidney malfunction. Herb, in doses of 3 to 12 g, was administered in decoction form.

PU HUANG (蒲黄) — the dried pollen of *Typha augustifolia* L. or *T. orientalis* Presl.

Chemistry

The active principles are isothamnetin, α-typhasterol, and oligosaccharide.

Actions

The herb lowers plasma cholesterol by inhibiting the intestinal absorption of dietary cholesterol or the reabsorption of biliary cholesterol. This helps prevent the development of atherosclerosis. Its direct effects on the cardiovascular system include a slowing of the heart rate, an increase in coronary circulation, and a lowering of peripheral blood pressure.

The herb has also been shown to increase intestinal and uterine smooth muscle contraction. It acts as a hemostatic, reducing the clotting time of blood. Additionally, it acts as an anti-inflammatory agent.

The herb has a very low toxicity. Patients may occasionally complain of nausea and vomiting, loss of appetite, abdominal distension, constipation, dryness of mouth, and fatigue.

Therapeutic Uses

The *Ben Cao Kong Mu* described this herb as a drug to mobilize stagnant blood and circulation. It also noted that the herb displayed hemostatic and analgesic actions.

It is used today in the treatment of hypercholesteremia, angina pectoris, exudative eczema, postdelivery bleeding, and to stop bleeding in hematemesis and hematuria.

HU CHANG (虎杖) — the dried root and stem of *Polygonum cuspidatum* Sieb. et Zucc.

Chemistry

The active principles include several glucosides: polygonin, glucofragulin, emodin, polydatin, and flavonoids.

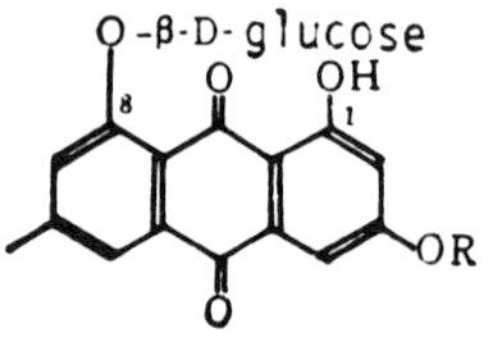

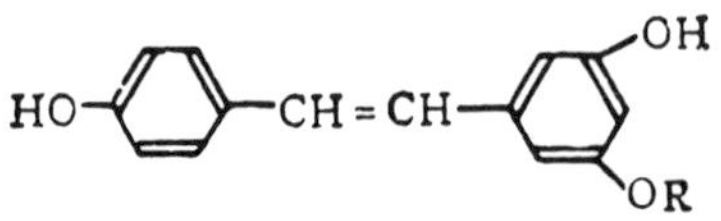

	R=		R=
Emodin-8-β-glucoside	H	Resveratrol	H
Physcion-8-β-D-glucoside	CH_3	Polydatin	β-D-glucose

Actions

Polydatin can lower blood cholesterol levels, and increase myocardial contractility and coronary circulation.

Therapeutic Uses

Clinical trials to use this herb in 136 patients showed 83 of 94 cases of hypercholesterol lowering their cholesterol level.

Hu Chang tablet, prepared from extracts equivalent to 2 g raw material per tablet, was given as three tablets orally t.i.d. Polydatin tablet (20 or 40 mg/tablet) was given as one tablet t.i.d.

Chapter 7

ANTISHOCK HERBS

Shock is a clinical syndrome caused by inadequate organ perfusion. The causes of shock are many. Commonly observed causes include cardiovascular collapse, toxic shock, bacterial septicemic shock, and dehydration shock. Because the pathophysiologic abnormalities of these processes are quite different, the results of treatment are also varied.

This fact was not recognized by the traditional Chinese Tai-fu, who generally treated shock symptomatically. It is thus difficult to evaluate whether Chinese antishock herbs are effective or beneficial. Some of the herbs mentioned in this chapter have recently been categorized by Chinese pharmacologists and clinicians on the basis of experimental and clinical studies. Not all of these, however, were mentioned in classical Chinese medical lore or the *Ben Cao Kong Mu* as antishock agents.

ZHI SHI (枳实) — the dried fruit of *Citrus aurantium*

Chemistry

Several sympathomimetic amines have been isolated from this fruit, including synephrine and *N*-methyltyramine.

Two flavones have been fractionated. They are tangeratin and nobiletin, with structures as follows:

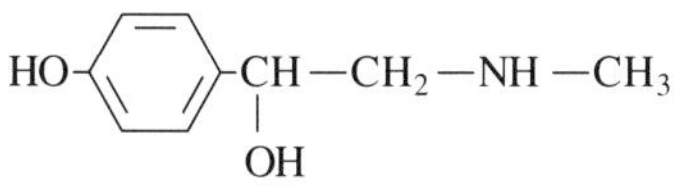

$HO-C_6H_4-CH(OH)-CH_2-NH-CH_3$

Synephrine

$HO-C_6H_4-CH_2-CH_2-NH-CH_3$

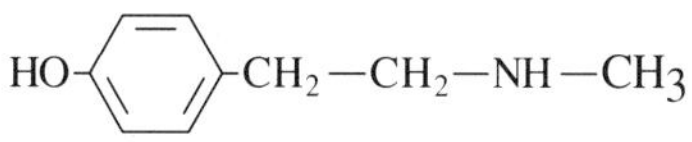

N-Methyltyramine

	R_1 =	R_2 =
Tangeretin	H	H
Nobiletin	H	OCH_3

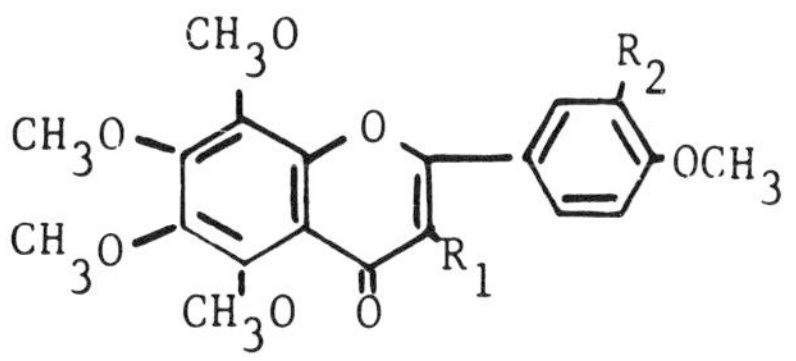

Actions

Synephrine is an α_1-adrenergic agonist, and can elevate blood pressure by constriction of arterioles. It is less potent than epinephrine, but has a longer duration of action. *N*-methyltyramine produces a similar vasoconstricting effect by depleting norepinephrine from the terminal storage granules.

The herb also increases cardiac contractility, improves coronary circulation and cerebral blood flow, and inhibits intestinal smooth muscle contraction.

Both tangeratin and nobiletin were found to exhibit an inhibitory effect on histamine release from mast cells.

Therapeutic Uses

In Chinese folk medicine, this herb was commonly used to treat indigestion. It was also prescribed to relieve abdominal distension and ptosis of the anus or uterus.

Today, the herb is used in the treatment of shock, particularly toxic shock and anaphylactic shock. It is also used in the treatment of heart conditions and cardiac exhaustion. Generally, the herb is prepared as an injection solution, with each ml equivalent to 4 g of raw material; this is diluted with glucose solution for intravenous infusion.

SAN LONG ZHI (山莨菪) — the root of *Scopolia tangutica* Max.

Chemistry

The active principles include hyoscyamine, scopolamine, anisodamine, and anisodine which are cholinergic blocking alkaloids.

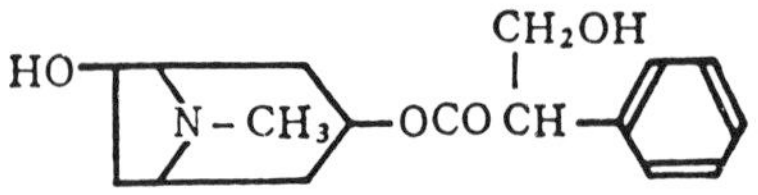

Anisodamine 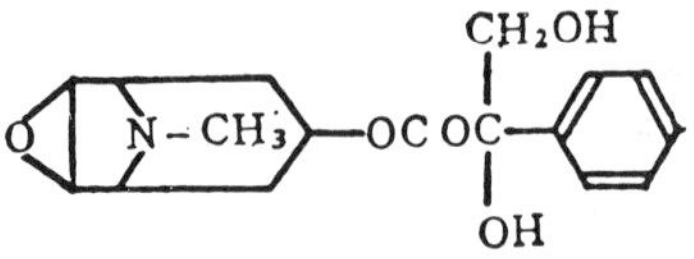　　　　Anisodine

Actions

San Long Zhi has a similar effect on the cardiovascular system to atropine. It can accelerate the heart rate and increase cerebral blood flow. It also has sedative and anticonvulsant effects on the central nervous system (CNS) and was the major ingredient in the ancient Chinese "anesthetic recipe". It inhibits the cortex and stimulates the respiratory center. In animals, it can reduce observed tremors.

The herb also has some peripheral effects. Scopolamine, which is more potent than either anisodamine or anisodine, can relieve intestinal spasms, dilate the pupils, inhibit salivation, and antagonize organophosphorous intoxication.

This herb is quite toxic. It has the same adverse effects as other cholinergic blocking agents, including dryness of the mouth, facial flushing, pupillary dilatation, blurred vision, and difficulties in urination and defecation. The LD_{50} of anisodine and anisodamine to mice is 482 to 595 mg/kg and 123 to 140 mg/kg, respectively.

The herb is contraindicated in persons suffering from glaucoma.

Therapeutic Uses

In folk medicine, the herb was used as an anticonvulsant, an analgesic, and in the relief of blood stasis. More recent uses include:

1. Treatment of shock caused by acute infectious diseases or intoxication — patients usually respond satisfactorily to *San Long Zhi* alkaloids, which can reduce mortality and improve overall physical condition. It is, however, important to continuously monitor patients' circulation (particularly retinal microcirculation), blood pressure, and respiration.
2. Cerebral thrombosis and acute spinal cord inflammation.
3. As an adjuvant in anesthesia — in combination with general anesthetic agents, anisodamine and anisodine can exert an analgesic and cerebral inhibitory effect while stimulating respiration.
4. Ocular distress — anisodamine or anisodine, in combination with 1% procaine, has been used as a local anesthetic in the eye in the treatment of retinal spasms and central retinitis.
5. Biliary duct or duodenal spasms.

Preparations of the herb are in either tablet or injection form. Anisodamine is administered in tablets containing 5 or 10 mg of the alkaloid at doses of one tablet t.i.d. It is also administered intravenously (5-, 10-, or 20-mg/ampule) at doses of 10 to 40 mg for acute toxic shock, or 0.3 to 2 mg/kg for children; in the treatment of cerebral thrombosis, 30 to 40 mg is diluted with glucose solution for intravenous infusion.

Anisodine is prepared in tablets of 1- or 3-mg strength. Intramuscular injection is prepared in ampules of 2 or 5 mg/ml.

Section IV
Nervous System

Chapter 8

ANESTHETIC AND MUSCLE-RELAXING HERBS

ANESTHETIC HERBS

Surgery, which has historically been dominated in China by Western-trained surgeons, was not categorized as a special field in Chinese traditional medicine. There was, however, one surgeon whose name is highly praised in Chinese medical literature. Hua Tau (華陀), who practiced surgery around 141 to 203 A.D., is credited with introducing the use of *Ma Fu* decoction (麻沸湯) for general anesthesia. Thanks to this prescription, he was able to perform many major operations, including abdominal and gastrointestinal surgery. (Historical records do not, however, indicate what percentage of those operations were successful.)

During the Ming Dynasty, another Chinese physician named Zhang Jin York (張景岳, 1563 to 1640 A.D.) introduced the use of *Mon Han Yao* (蒙汗药) as an anesthetic agent. This compound combined aconite root, rhododendron, olibanum, and other herbal substances.

Since then, many anesthetic formulas have been listed in Chinese medicine. Most are taken with wine to produce general anesthesia. The majority of these formulas are mixtures of herbs, with aconite and datura as the major ingredients. From a modern scientific viewpoint, these preparations are obsolete. Modern research has shown that datura is an adjuvant in anesthesia, but by itself is not an anesthetic agent.

It should be noted here that the herb *San Long Zhi* (山莨菪) described in the last chapter as an antishock agent, contains similar active principles to *Man Tao Luo*, and was also used as an anesthetic adjuvant.

MAN TAO LUO (曼陀罗) or YANG JIN HUA (洋金花) — the dried leaf or seed of *Datura stramonium* (Solanaceae) and the flower of *D. metal* L.

Ben Cao Kong Mu described this herb's shape and growth in detail.

Chemistry

The active principles are scopolamine and hyoscyamine. The seed also contains other substances, such as daturodiol and daturolone.

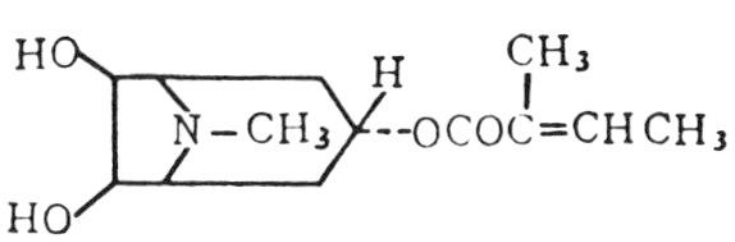

HO, HO, N–CH_3, H, CH_3, --OCOC=CHCH$_3$

Meteloidine

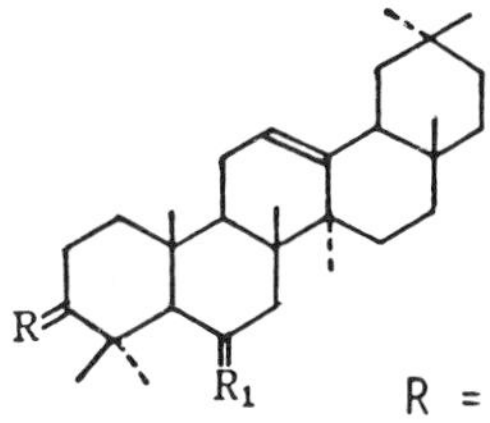

R, R_1

	R =	R_1 =
Daturodiol	β-OH,α-H	β-OH,α-H
Daturolone	O	OH, H

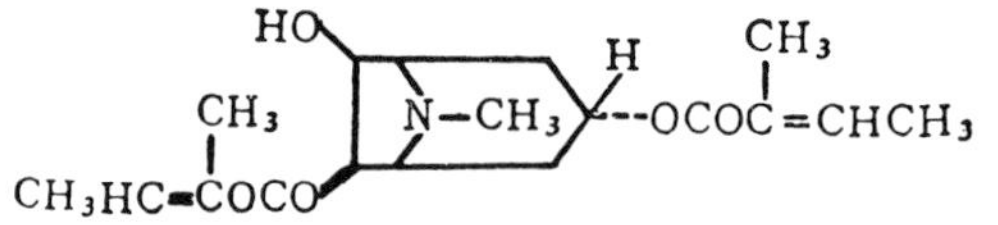

6,7-Di-OH-(1)-tropane-3α,6β-diol ditigloyl ester

Actions

Datura alkaloids are cholinergic blocking agents, specifically blocking muscarinic receptors of the cholinergic nerve ending. In the CNS, scopolamine has an inhibitory effect, reaching only an analgesic stage without producing complete muscular relaxation.

The *Datura* alkaloids can antagonize arteriolar and venular spasms induced by acetylcholine or 5 hydroxytryptamine (5HT), resulting in an increase in blood flow.

This herb has an antiasthmatic effect and can increase plasma immunoglobulin A (IgA) level and phagocytic activity. It dilates the bronchioles and reduces pulmonary secretions.

Toxicity

Administration of *Datura* alkaloids to dogs in a dose of 75 mg/kg — 37 to 40 times larger than the minimum effective dose — produces general anesthesia and convulsions of the extremities.

The commonly observed side effects of this herb in humans are dryness of mouth, facial flushing, dry skin, increase in heart rate, elevation of body temperature, and difficulties in urination.

Therapeutic Uses

Man Tao Luo has been used as an spasmolytic, analgesic, antiasthmatic, and antirheumatic agent. It is used as a general anesthetic for major operations on the head, extremities, and vertebrae.

In injection solutions, it is prepared at 0.5 or 5 mg/ml; 0.08 mg/kg is given subcutaneously. The herb is also administered orally in a dose of 0.3 to 0.45 g, mixed with wine.

CHAN SU (蟾酥) — dried toad venom, or the dried secretion of the skin glands of *Bufo bufo gargarizans* Cantor or *B. melanotictus*

Chemistry

The *bufo secreta* contains several potent substances which are genin derivatives and alkaloids. The active ones include bufadienolide, bufotalin, cinobufagin, cinobufotalin, cinobufaginol, desacetylcinobufagin, resibufogenin, and cinobufotoxin.

The cardiac genin group:

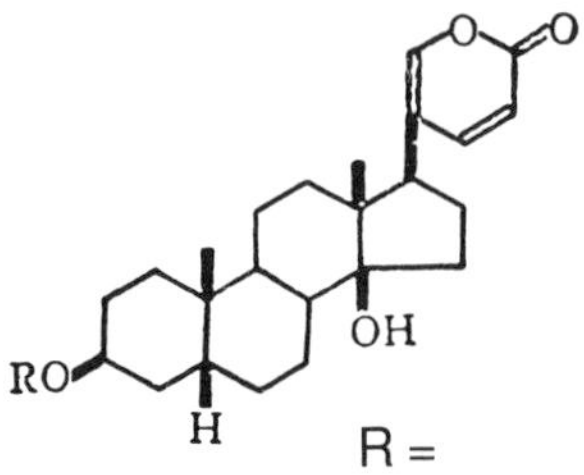

	R =
Bufalin	H
3-Bufolyl-suberic acid	CO $(CH_2)_6COOH$

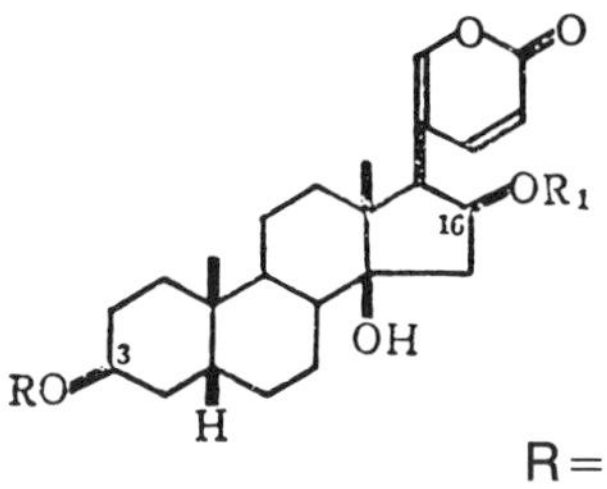

	R =	R_1 =
Bufotalin	H	$COCH_3$
Desacetylbufotalin	H	H

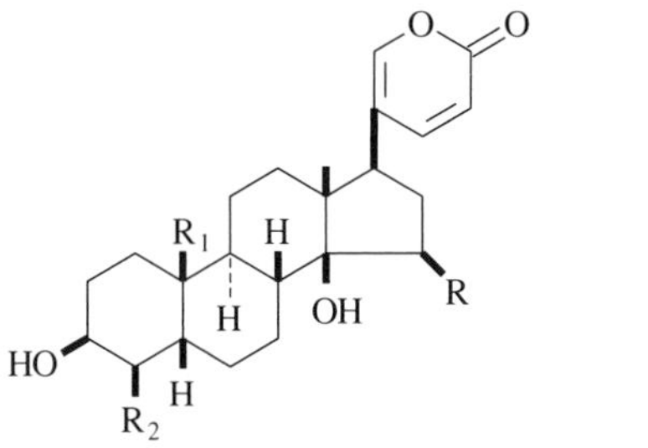

	R=	R_1=	R_2=
4β-OH-bufalin	H	CH_3	OH
15β-OH-bufalin	OH	CH_3	H
19β-OH-bufalin	H	CH_2OH	H

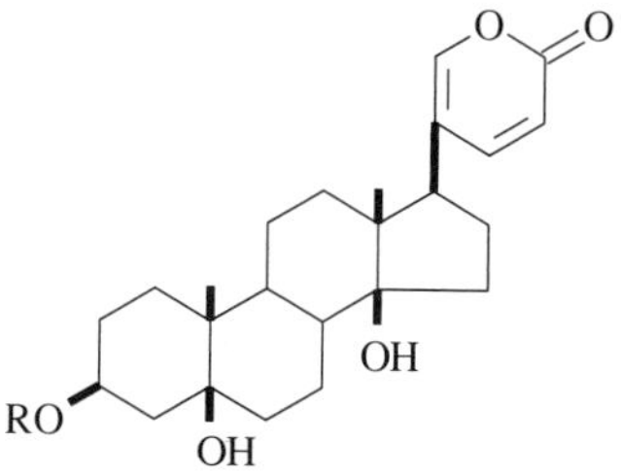

	R=
Telocinobufagin	H
3-Telocinobufagyl-suberic acid	$CO(CH_2)_6COOH$

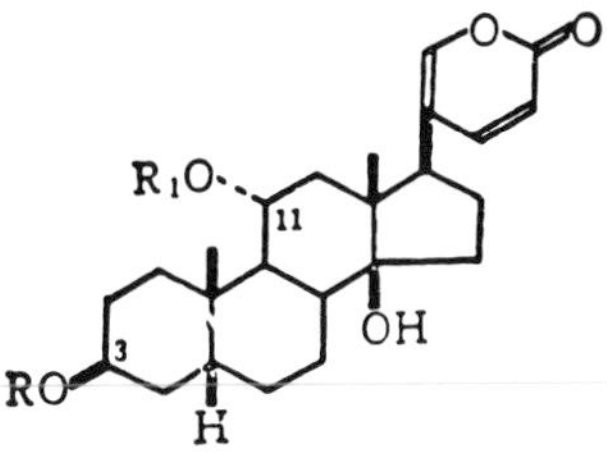

	R=	R_1=
Gamabufotalin	H	H
3-Gamabufotalyl-suberic acid	$CO(CH_2)_6COOH$	H
3-(11-*O*-Acetyl-gamabufotalyl) suberic acid	$CO(CH_2)_6COOH$	$COCH_3$

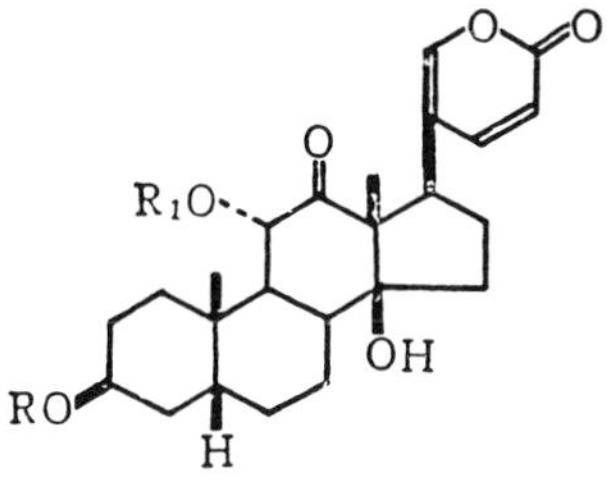

	R=	R_1=
Arenobufagin	H	H
3-Arenobufagyl-suberic acid	$CO(CH_2)_6COOH$	H

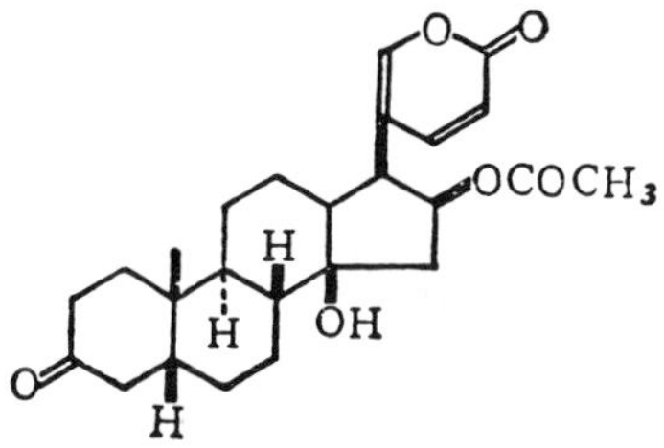

Bufotalon

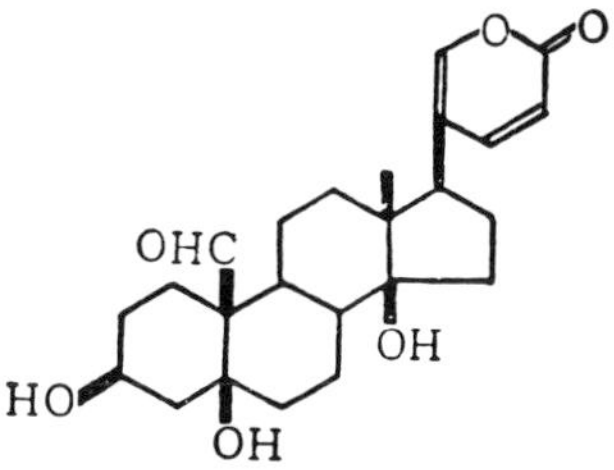

Bufotalidin

The genin group which does not have the C-14 OH radical:

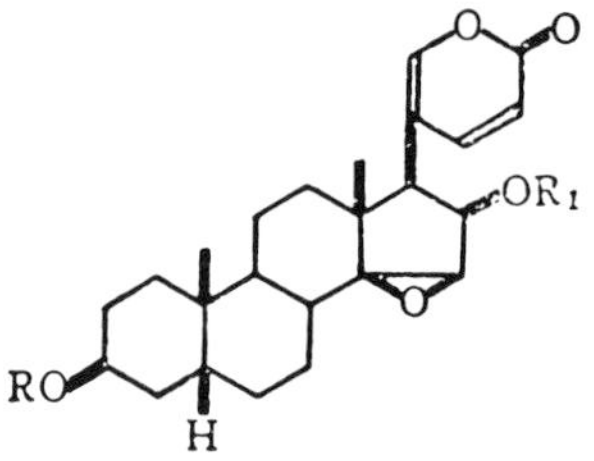

	R=	R_1=
Cinobufagin	H	$COCH_3$
Desacetylcinobufagin	H	H

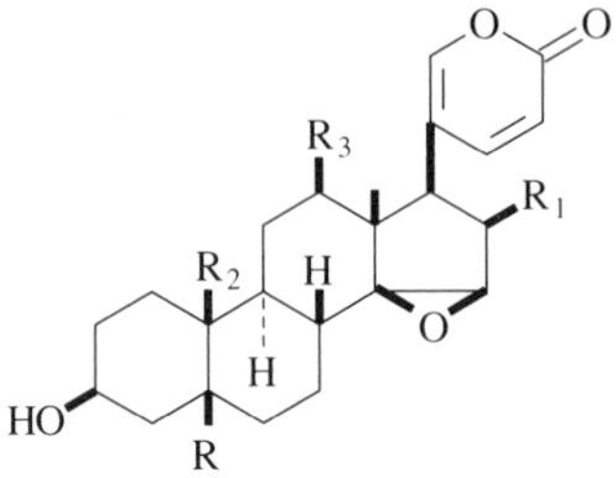

	R=	R_1=	R_2=	R_3=
12β-OH-cinobufagin	H	$OCOCH_3$	CH_3	β-OH
Desacetylcinobufaginol	H	OH	CH_2OH	H
10β-OH-19-CH_3-cinobufagin	H	$OCOCH_3$	OH	H

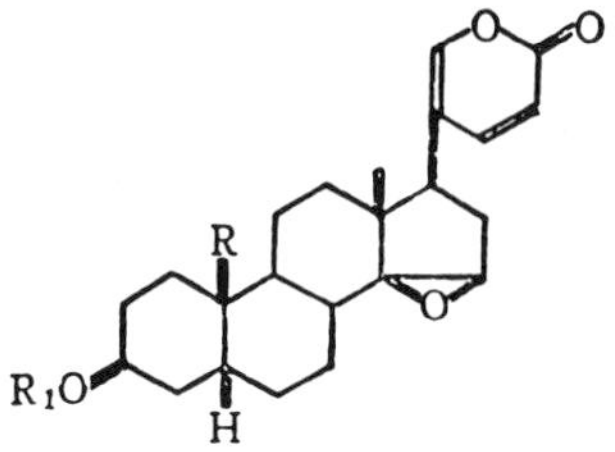

	R=	R_1=
Resibufogenin	CH_3	H
3-Resibufogenyl-suberic acid	CH_3	$CO(CH_2)_6COO$
Resibufogin	CHO	H

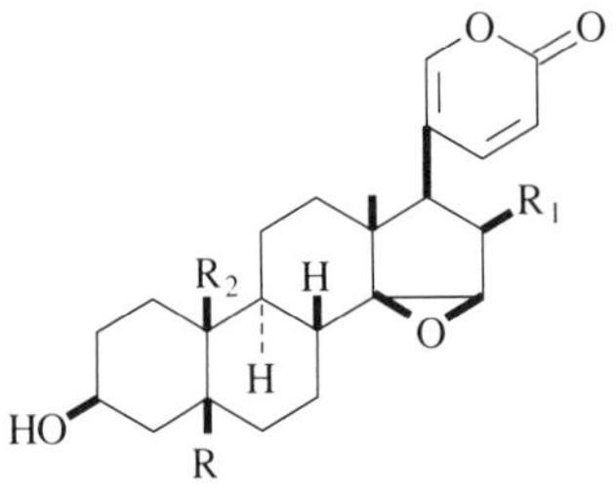

	R=	R_1=	R_2=
Cinobufotalin	OH	$OCOCH_3$	CH_3
19-Oxdesacetyl-cinobufotalin	OH	OH	CHO
19-OH-cinobufotalin	OH	$OCOCH_3$	CH_2OH
Bufotalinin	OH	H	CHO
19-OH-marinobufagin	OH	H	CH_2OH

The alkaloids: bufotenine and bufotenidine

Bufotenine

Bufotenidine

Others: adrenaline and cholesterol

Actions

Chan Su is the major ingredient of a Chinese formula called the *Lau Shan pill* (六神丸), which has been used as a revitalizing drug, to stimulate the feeble heart and stop serious pain. In the early 1930s, Chen and his co-workers found that *Bufo secreta* was a potent cardiotonic agent.[1] The structure of bufalin is similar to that of other digitalis glycosides. Therefore, it is not surprising that this herb was listed in the cardiotonic chapter.[2] But on the other hand, bufotenine has a structure which is also similar to that of 5HT and has effects on the CNS.

1. Local anesthetic — both cinobufagin and bufotalin have a potent surface anesthetic effect. In terms of its anesthetic effect on the cornea it is 10 times more potent than cocaine and 100 times more potent than procaine. Its duration of action is relatively long.
2. Cardiotonic — bufalin can increase myocardial contractility and improve the circulation.
3. Diuretic — reduces the tubular reabsorption of Na and Cl ions.
4. Stimulates the respiratory center.

Toxicity

Chan Su is quite toxic. The major toxic effects include gastrointestinal symptoms such as nausea and vomiting, diarrhea, and abdominal pain. Cardiovascular toxicity includes heart block, irregular heart beat, which can be blocked by atropine, vasodilatation, and a tendency to bleed. CNS toxicity includes dizziness, numbness of the extremities, and, in large doses, coma.

Treatment of *Chan Su* intoxication includes gastric lavage with 0.2 to 0.5% potassium permanganate solution and a subcutaneous injection of atropine in a dose of 0.5 to 1 mg.

The LD_{50} of *Chan Su* in mice is 41 mg/kg (intravenous), 96.6 mg/kg (subcutaneous), and 36.2 mg/kg (intraperitoneal).

Therapeutic Uses

This herb is described in Chinese pharmacopoeia as a detoxicant and anodyne. As a surface anesthetic, it is applied to the throat, nose, or mouth. A combination of *Chan Su* (9 g), aconite root (9 g), datura (16 g), and peppermint (6 g) is made into an alcoholic extract at 30% concentration; this is applied locally to mucosal surfaces.

Chan Su, in combination with raw aconite root, raw *Ban Sha* (半夏) and *Shi Shin* (细辛), is ground into a powder for surface anesthetic use.

The herb has also been used in the treatment of esophageal cancer.

References

1. **Chen, K. K. et al.,** *J. Biol. Chem.*, 87, 741, 1930; *J. Pharmacol. Sci.*, 56, 1535, 1967.
2. **Jang, C. S.,** [*Recent Research on Chinese Herbs*], China Science Library and Equipment, Shanhai, 1954, 55.
3. **Liu, C. Y.,** *Jiangsi Med. Drug*, 5, 665, 1965.

JIU LI XIANG (九里香) — the dried leaf and young foliferous branches of *Murraya paniculata* (Rutaceae)

Chemistry

The leaf contains several essential oils, including L-cadinene, methylanthranilate, bisabolene, β-caryophyllene, gerariol, carene, 5-guaizulene, osthol, paniculatin, and coumurrayin. The bark of the stem contains mexoticin ($C_{16}H_{20}O_6$) and scopoletin ($C_{18}H_8O_4$) and some exoticin analogs.

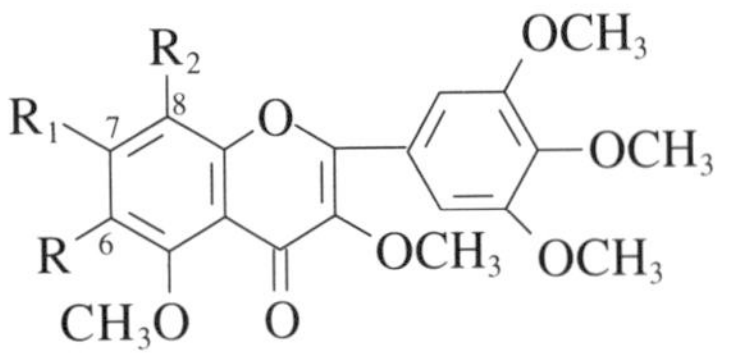

	R=	R_1=	R_2=
Exoticin	OCH_3	OCH_3	OCH_3
6-Demethoxyl-exoticin	H	OCH_3	OCH_3
7-Demethoxyl-exoticin	OCH_3	H	OCH_3
8-Demethoxyl-exoticin	OCH_3	OCH_3	H

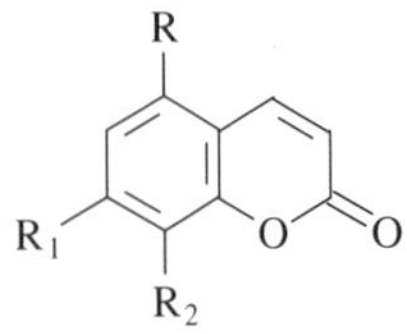

	R =	R_1 =	R_2 =
Osthol	H	OCH_3	$CH_2—CH = C(CH_3)_2$
Coumurrayin	OCH_3	OCH_3	$CH_2—CH = C(CH_3)_2$
Paniculatin	H	OCH_3	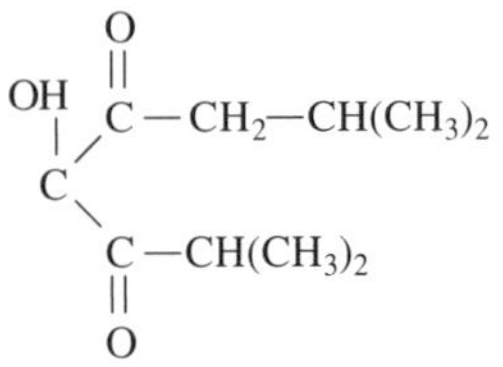 $C(OH)(C(=O)—CH_2—CH(CH_3)_2)(C(=O)—CH(CH_3)_2)$

Actions

In folk medicine, the herb was used to promote the flow of *qi*, to relieve pain, and to remove toxic substances. It is an antispasmodic and can antagonize muscular spasms induced by histamine or $BaCl_2$.

The water extract of this herb (12.5%) produces a local anesthetic effect, but can be irritating.

MUSCLE-RELAXING HERBS

Relaxation of skeletal muscles is a necessity during surgical operations. This effect can be achieved with peripherally acting drugs such as curare-like agents, which block nicotinic receptors at the neuromuscular junction, or with succinylcholine-like agents, which produce a depolarizing action at the neuromuscular junction (NMJ). Most Chinese herbs which produce a muscle-relaxing effect are curare-like substances.

FANG JI (防己) or HAN FANG JI (汉防己) — the dried tuberous root of *Stephania tetrandra* S. Moore (Menispermaceae)

Chemistry

The total alkaloid content of this herb is between 1.5 to 2.3%. The primary alkaloids found are *d*-tetrandrine (approximately 1%), fangchinotine (0.5%), and cyclanoline (0.1%).

It should be pointed out once again that Chinese herbal terminology is very confusing. Chinese *Fang Ji*, as described here, is generally called *Han Fang Ji* (汉防己) and differs considerably from Japanese *Fang Ji* (日本防己), which is derived from the root of *Sinomenium acutum* Rehder et Wilson. The latter contains the alkaloids sinomenine and disinomenine, which are phenanthrene derivatives similar to morphine.

Japanese *Mu Fang Ji* (木防己) is the root of *Cocculus trilobus*, which contains the alkaloids triboline and isotrilobine. Both have structures very similar to tetrandrine, a *d*-tubocurarine-like substance (see Table 1).

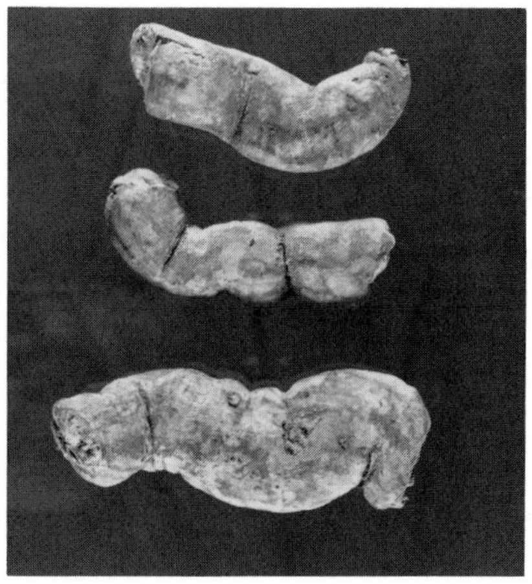

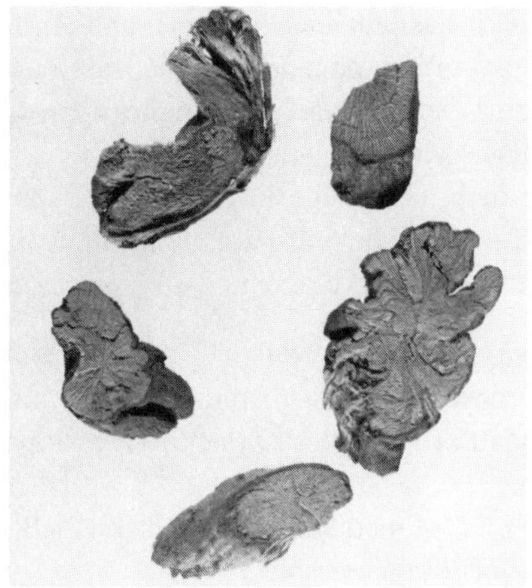

A B

FIGURE 1. (A) *Fan Ji*; (B) *Mu Fang Ji*.

TABLE 1
Known Components Isolated from Different *Fang Ji*

Name of *Fang Ji*	Major components	Chemical formula	Melting point (°C)
Han Fang Ji	Tetrandrine	$C_{38}H_{42}O_6N_2$	217
Stephania tetrandra	Demethyltetrandrine	$C_{37}H_{40}O_6N_2$	241–242
Mu Fang Ji (Kong Fang	MufangchineA	$C_{32}H_{24}O_{13}N_2$	278–280
Ji, Ting Fang Ji)	Mufangchine B	$C_{14}H_{33}O_{11}N$	232–233
Cocculus thunbergii	(Mufangchine)		
Japanese *Han Fang Ji*	Sinomenine	$C_{19}H_{23}O_4N$	162, 182
Sinomenium acutum	Disinomenine	$(C_{19}H_{22}O_4N)_2 \cdot 2CH_3OH$	222
	Sinactine	$C_{20}H_{21}O_4N$	174
	Acutumine	$C_{20(21)}H_{27}O_8N$	240
	Diversine	$C_{20}H_{27}O_5N$	80–93
	Tuduranine	$C_{18}H_{29}O_3N$	125
Japanese *Mu Fang Ji*	Trilobine	$C_{36}H_{36}O_3N_2$	235
Cocculus trilobus	Isotrilobine	$C_{36}H_{36}O_3N_2$	215

R =
d-Tetrandrine CH_3
Fangchinoline H

Cyclanoline

Actions

The herb has a curare-like action. Methylated tetrandrine and metetrandrine iodine, were found to be four times more potent than *d*-tubocurarine in their ability to block the depolarizing action of acetylcholine on the NMJ.

Fang Ji has anti-inflammatory and antihypersensitivity actions. The herb has a direct stimulatory effect on adrenocorticosterone secretion. Peng et al. recently reported that sinomenine exerts a marked immunosuppressive effect and significantly decreases the ratio of cGMP/cAMP of plasma in mice.

The herb has analgesic properties. Japanese *Fang Ji* and sinomenine are as effective as morphine in relieving pain. It displays antiarrhythmic effects (see Chapter 3) and acts as an antihypertensive agent.

Toxicity

Overdose may cause respiratory paralysis. Therapeutic doses, however, have little effect on the heart or respiratory system. Occasionally, patients may develop cyanosis and excess sweating.

The LD_{50} of metetrandrine in mice is 1.3 mg/kg (intravenous), which is ten times greater than that of *d*-tubocararine.

Therapeutic Uses

Popular prescriptions of this herb include use as a diuretic, antiphlogistic, and anti-rheumatic.

As an adjuvant in anesthesia for abdominal operations, metetrandrine has been used in combination with acupuncture to obtain good anesthetic effects. The dose used is 0.8 mg/kg administered intramuscularly, or diluted to 5 to 10 ml with isotonic glucose solution for intravenous injection. Usually it takes 2 to 5 min to produce complete muscle relaxation, which will last for 40 min. In some cases, a fall in blood pressure has been observed.

In the treatment of arthritis and neuralgia, the standard dose is 6 to 12 g daily; if *d*-tetrandrine tablets (0.02 g) are used, the dose is 1 to 2 tablets t.i.d. Tetrandrine may also be used as an antiarrhythmic agent, displaying a quinidine-like action on the heart (see Chapter 3).

Reference

Peng, H. M. et al., *Acta Pharmacol. Sin.*, 9(4), 377, 1988.

MU FANG JI (木防己) — the dried root of *Cocculus thunbergii*; Japanese MU FANG JI (日本木防己) — the dried root of *Cocculus trilobus* (Thunb).

Chemistry

The root or rhizome contains several alkaloids. The major ones are magnoflorine ($C_{20}H_{24}O_4N+$, approximately 0.41%), trilobine, homotrilobine, etc. Their structures are shown as follows:

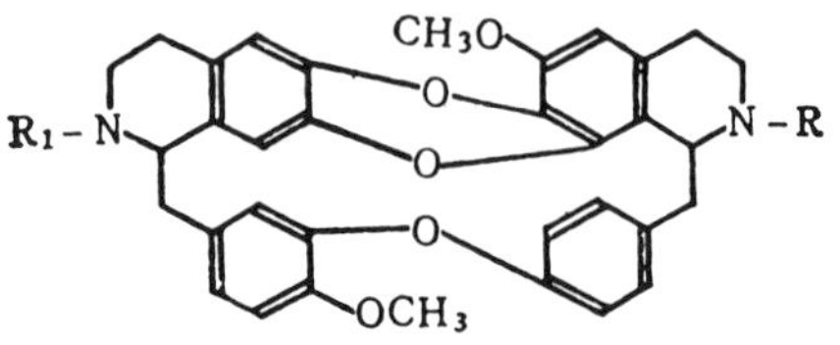

	R =	R_1=
Trilobine	H	CH_3
Homotrilobine	CH_3	CH_3

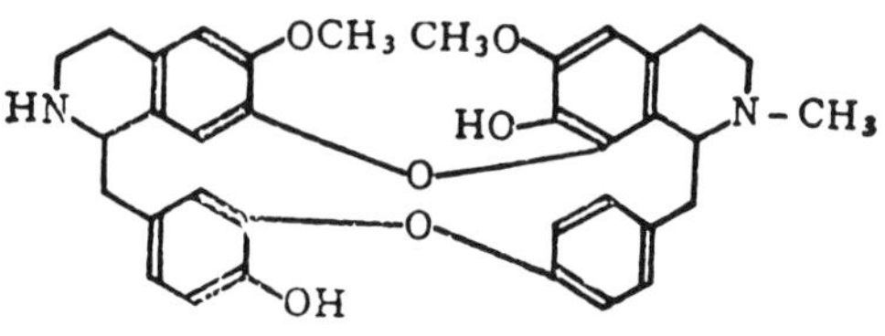

Trilobamine

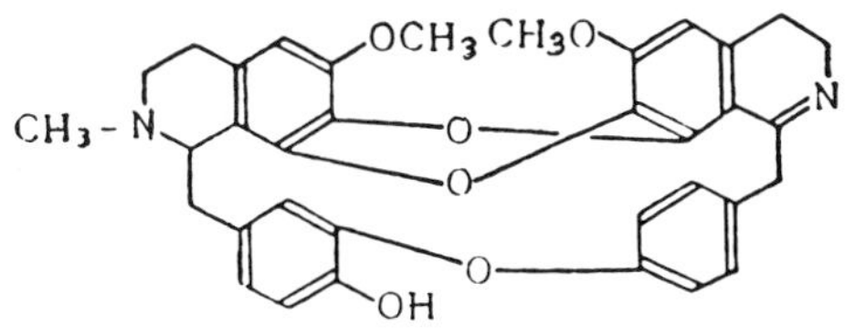

Normenisarine

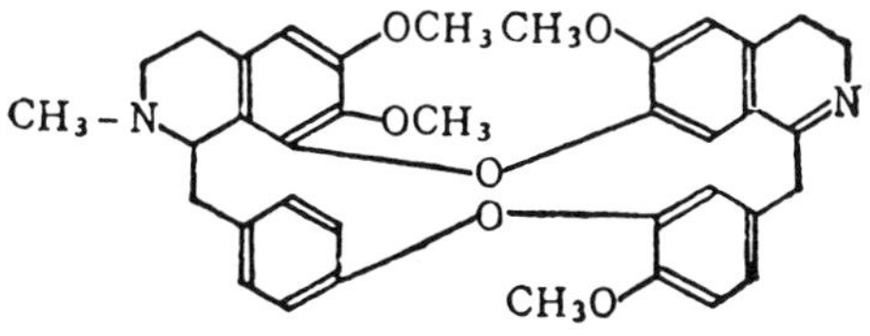

Coclobine

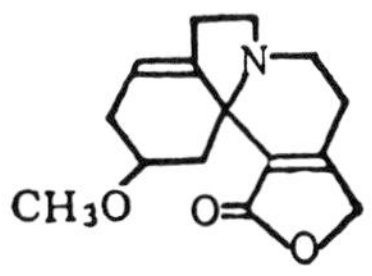

Cocculolidine

Actions

This herb possesses an analgesic effect and can reduce swelling. It was used to relieve arthritis pain and neuralgia. It is also effective in the treatment of pulmonary and cardiac edema.

Cocculolidine has an insecticidal effect.

JAPAN HAN FANG JI (日本汉防己) or QING TENG (青藤) dried root and rhizome of *Sinomenium acutum* (Thunb.) Rehd. et Wils.

This herb can be found in Yunnan, Geizhow, and Szechuan. Different names were given according to its habitat.

Chemistry

The root and rhizome contain several alkaloids. They constitute the active principles of the herb and have the phenanthrene structure similar to that of morphine derivatives.

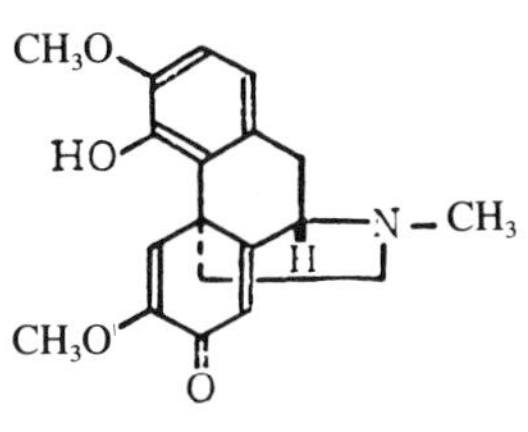

Sinomenine

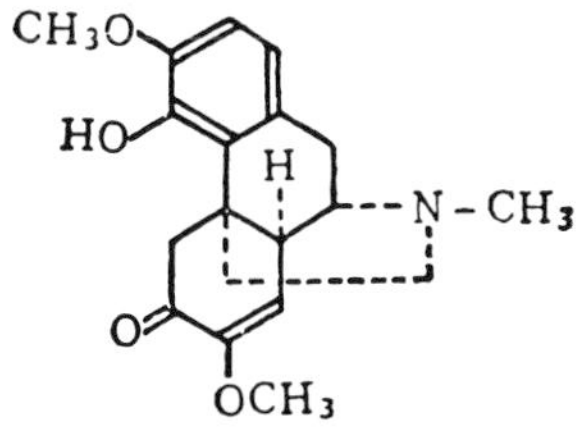

Sinoacutine

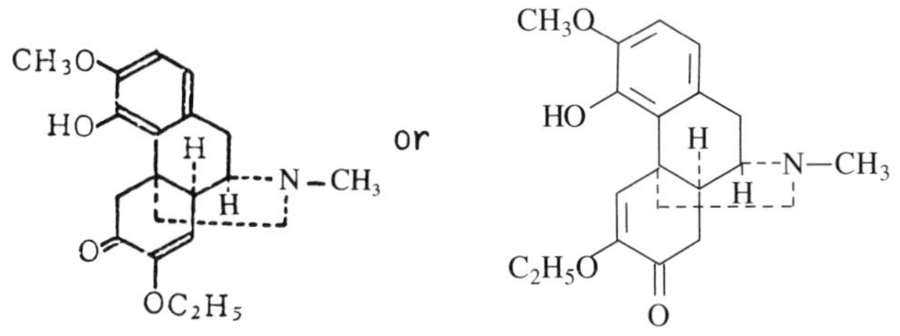

or

Isosinomenine

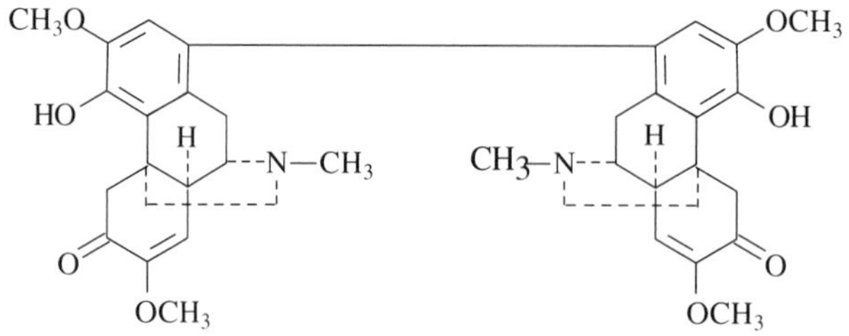

Disinomenine

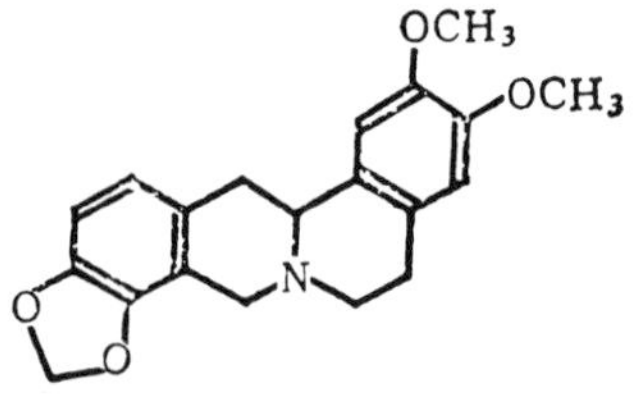

Sinactine

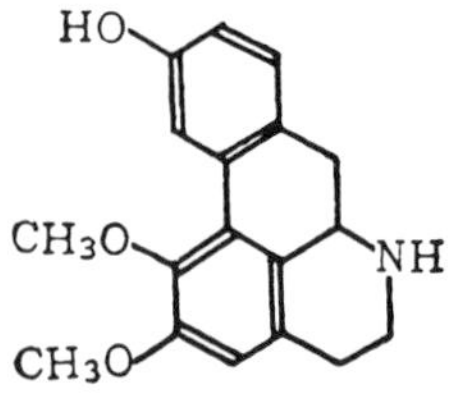

Tuduranine

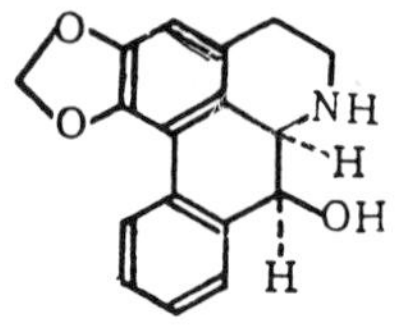

Michelalbine

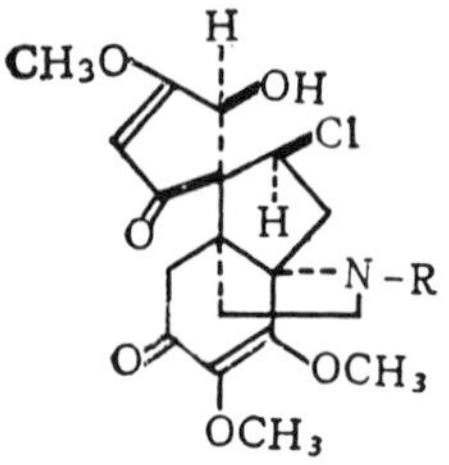

	R=
Acutumine	CH_3
Acutumidine	H

Actions

The pharmacological action of sinomenine is similar to that of morphine but much less potent. Its analgesic effect is about 1/25 of morphine. It is sedative and antitussive. It can reduce the swelling of joints and shows an anti-inflammatory effect. The LD_{50} to mice is 580 mg/kg (by stomach).

Folk medicine used it to treat arthritis and traumatic swelling of extremities. It was given in a dose of 6 to 8 g of root taken with wine or in decoction.

The following Chinese herbs belong to the genus *Stephania*, although their names are not similar to *Fang Ji*. They contain several alkaloids, some of which are similar in structure to tetrandrine. Their therapeutic uses are not as well described as those of *Fang Ji*.

JIN XIAN DIAO WU GUI (金钱吊乌龟) — the root of *Stephania cepharantha* Hayata

Chemistry

Alkaloids which have been identified in this herb include cepharanthine ($C_{37}H_{35}O_8N_2$), isotetrandrine ($C_{38}H_{42}O_6N_2$), cycleanine ($C_{38}H_{42}O_6N_2$), cepharanoline ($C_{36}H_{36}O_6N_2$), berbamine ($C_{37}H_{40}O_6N_2$), cepharamine ($C_{19}H_{23}O_4N$), and homoaromoline ($C_{37}H_{40}O_6N_2$).

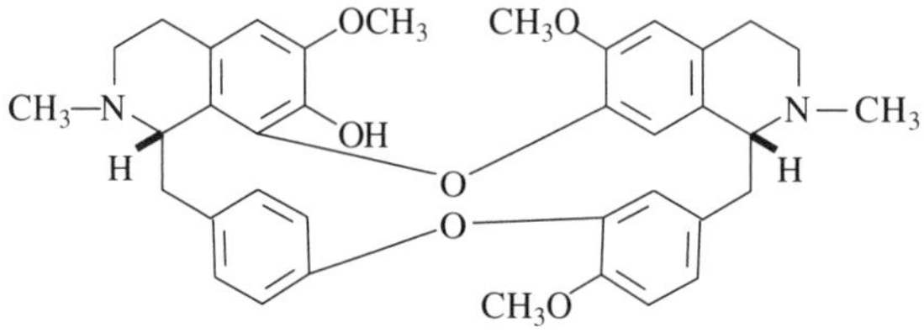

Homoaromoline

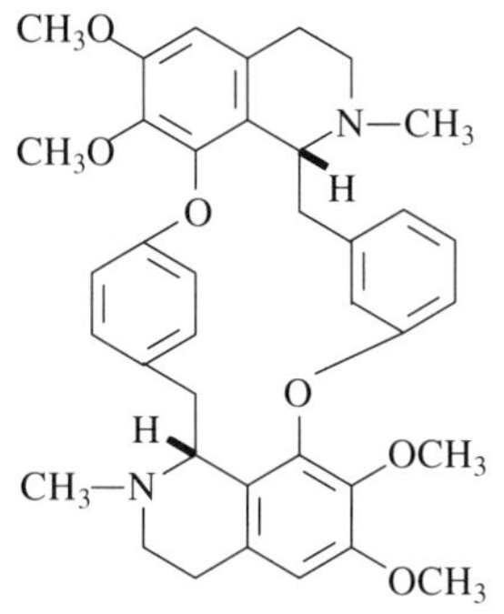

Cycleanine

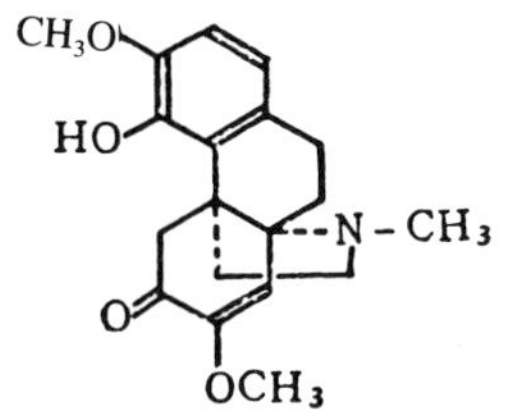

Cepharamine

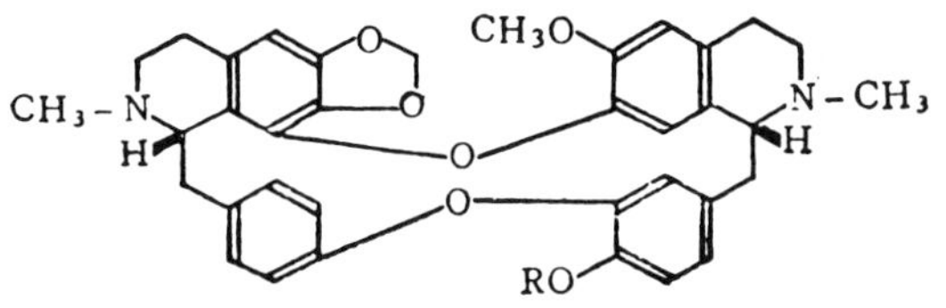

	R =
Cepharanthine	CH_3
Cepharanoline	H

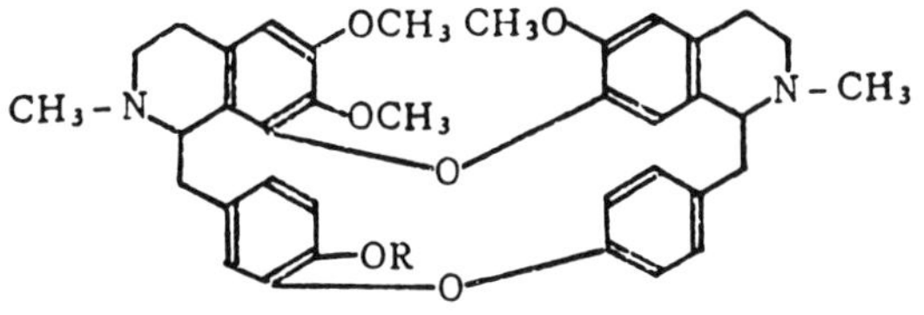

	R =
Isotetrandrine	CH_3
Berbamine	H

The herb and its alkaloids have anti-inflammatory effect to reduce joint swelling. They also have antipyretic and analgesic action. Ju et al. reported that berbamine exhibits a protective effect on ischemic heart.

Reference

Ju, H. S. et al., *Chin. Circ. J.*, 6, 227–229, 1991.

QIAN JIN TENG (千金藤) — from *Stephania japonica* (Thunb.)

Chemistry

Alkaloids comprise approximately 2.2% of the herb. These include stephanine ($C_{19}H_{19}O_3N$), protostephanine ($C_{21}H_{27}O_4N$), epistephanine ($C_{31}H_{38}O_6N_2$), hypoepistephanine ($C_{36}H_{36}O_6N_2$), homostephanoline ($C_{30}H_{35}O_5N$), metaphanine ($C_{19}H_{23}O_5N$), prometaphanine

($C_{20}H_{25}O_5N$), hasubanonine ($C_{21}H_{27}O_5N$), insularine ($C_{38}H_{40}O_8N_2$), cyclanoline ($C_{20}H_{24}O_4N$), stephanoline ($C_{31}H_{42}O_7N_2$), stepinonine ($C_{36}H_{34}O_7N_2$), and steponine ($C_{20}H_{24}O_4N$).

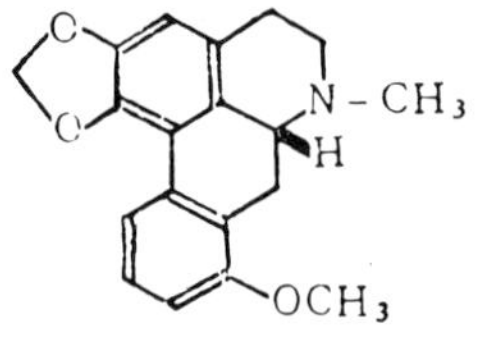

Stephanine

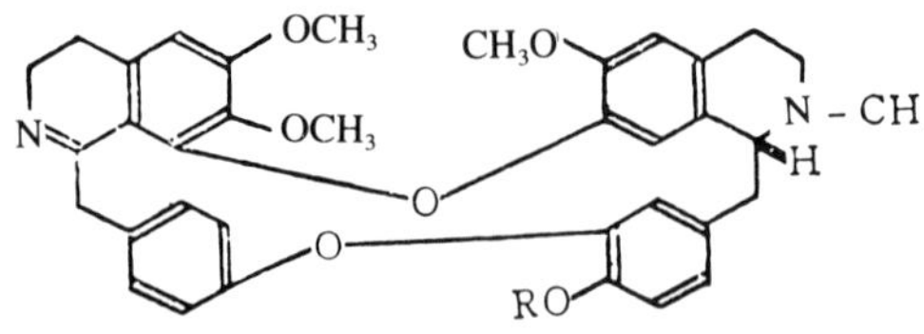

	R=
Epistephanine	CH_3
Hypoepistephanine	H

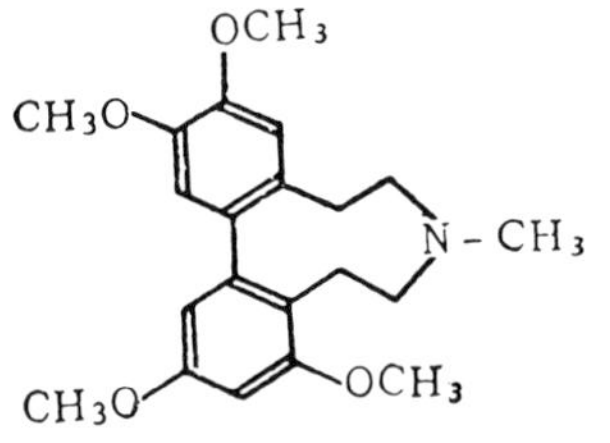

Protestephanine

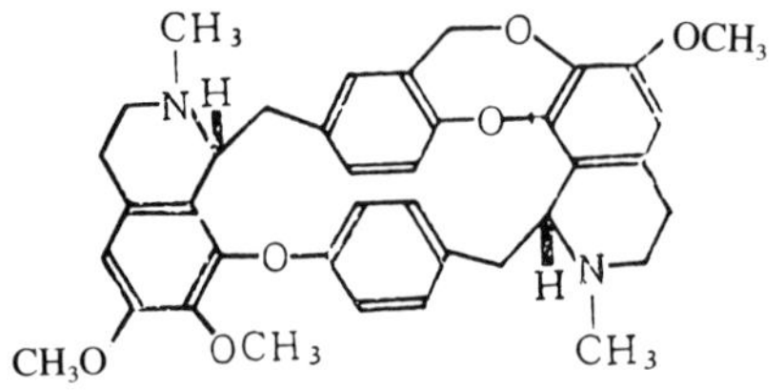

Insularine

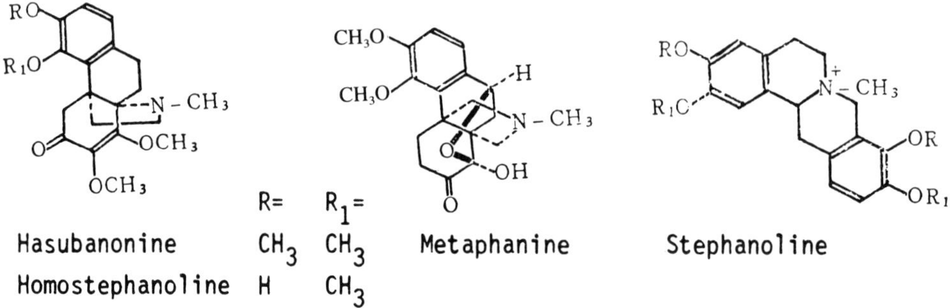

	R=	R_1=
Hasubanonine	CH_3	CH_3
Homostephanoline	H	CH_3

Metaphanine

Stephanoline

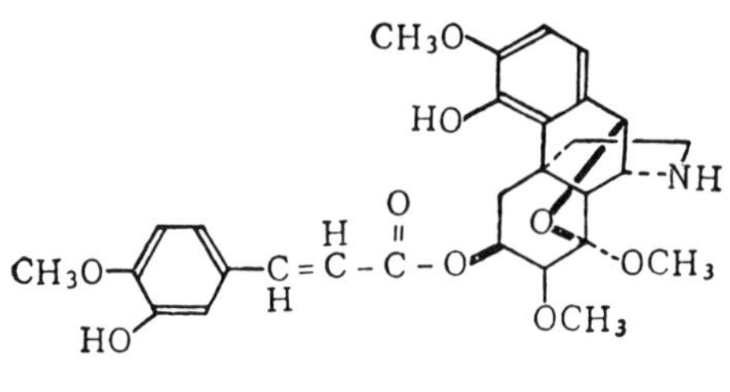

Stephisoferuline

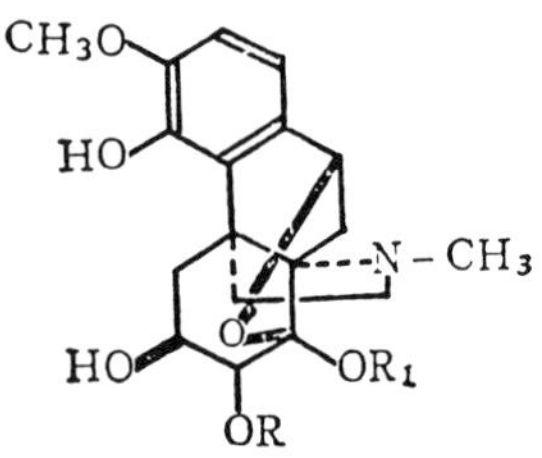

	R=	R_1=
Hernandine	H	CH_3

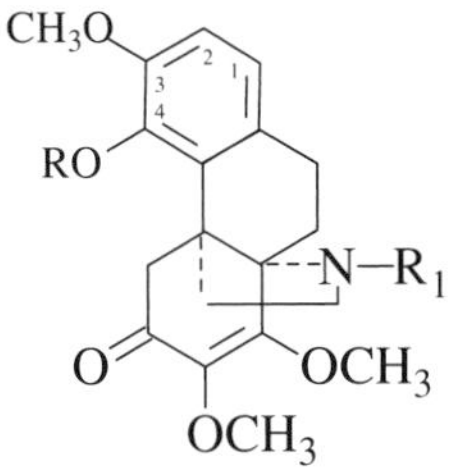

	R=	R_1=
4-Demethylhasubanonine	H	CH_3
4-Demethyl-norhasubanonine	H	H

Qian Jin Teng can also be derived from *Stephania hernendifolia* (Willd.) Walp., which contains the following alkaloids: *dl*-tetrandrine, fangchinoline ($C_{37}H_{40}O_6N_2$), 4-demethylhasubanonine ($C_{20}H_{25}O_5N$), isochondrodendrine ($C_{36}H_{35}O_6N_2$), hernandine ($C_{19}H_{25}O_6N$), stephisoferuline ($C_{29}H_{33}O_9N$), hernandoline ($C_{20}H_{25}O_5N$), hernandolinol ($C_{20}H_{27}O_5N$), and 3-*O*-demethylhernandifoline.

Therapeutically this herb is primarily used to treat nephritis edema, urinary tract infection, rheumatic arthritis and sciatic neuralgia. It is given in dose of 12 to 25 g in decoction.

HUA QIAN JIN TENG (华千金藤) — from *Stephania sinica* Diels.

Chemistry

The herb contains the following alkaloids: *l*-tetrahydropalmatine, stepharotine, stepharine, and tuduranine.

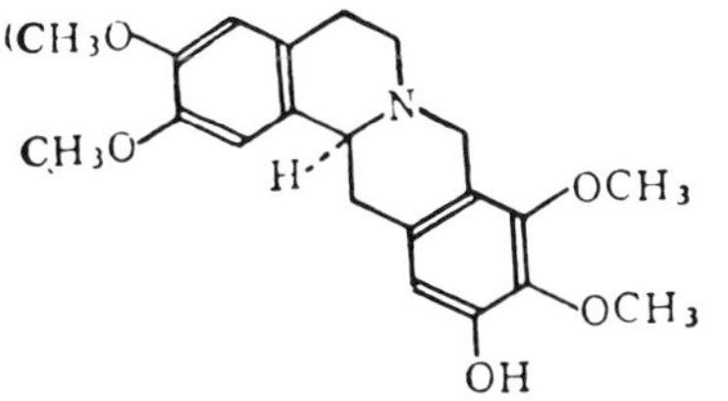

Stepharotine

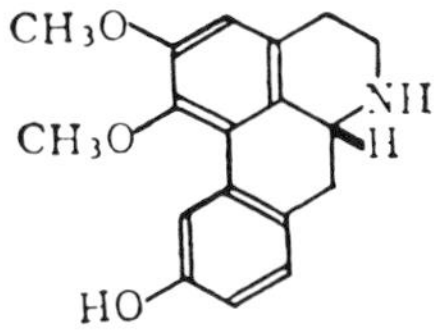

Tuduranine

Actions

It has an analgesic effect and is used in the treatment of stomach ache, neuralgia, and toothache.

YE DOU GEN (野豆根) or SHAN DOU GEN (山豆根) — the dried root and rhizome of *Menispermum dauricum* D.C., a species belonging to the same family as *Fang Ji, Stephania*

Chemistry

This is another herb whose name is commonly confused. Chinese pharmacopoeia interchange this herb with *Shan Dou Gen* or *Kong Dou Gen*, an antipyretic which is derived from the root or rhizome of *Sophora subprostrata. Ye Dou Gen* or *Shan Dou Gen*, as mentioned in the text, *Zhon Cao Yao Xue* (中草药学), is the rhizome of *Menispermum* and is used as a muscle relaxant.

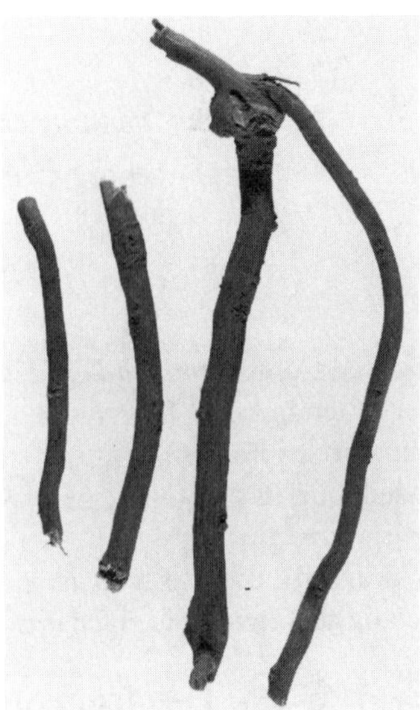

FIGURE 2. *Ye Dou Gen* or *Shan Dou Gen* (*Menispermum dauricum*).

The herb contains several alkaloids, the quantity of which varies depending upon the region in which the herb is grown. These alkaloids include dauricine, daurinoline, dauricoline, dauricinoline, stepharine, stepholidine, magnoflorine, menisperine, sinomenine, acutumine, and a small quantity of tetrandrine. The last three are the chief alkaloids also obtained from *Stephania*. The leaf of the plant contains acutuminine.

	R =	R_1 =	R_2 =
Dauricine	CH_3	H	CH_3
Daurinoline	CH_3	H	H
Dauricoline	H	H	H
Dauricinoline	H	H	CH_3

Cheilanthifoline

	R =
Menisperine	CH_3
Magnoflorine	H

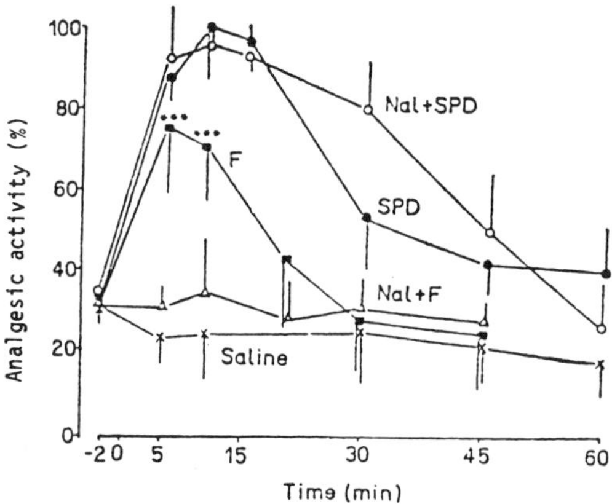

FIGURE 3. The analgesic effect of stepholidine in mice (SPD) *l*-stepholidine, 60 mg/kg; (F) fantanyl, 150 μg/kg; (Nal) naloxone, 2 mg/kg. (From Zhang, Z. D., et al., *Acta Pharmacol. Sin.*, 7, 522, 1986. With permission.)

CH_3O, HO, N, H, OCH_3, OH

Stepholidine

CH_3O, CH_3O, NH, H, O

Stepharine

Actions

Studies of the pharmacological action of this herb are based on animal experimentation with *l*-stepholidine, the main active principle of the herb. Zhang et al.[6] reported that stepholidine produces an analgesic effect in mice which is not antagonized by naloxone. Their results are illustrated in Figure 3. They also claimed that stepholidine is a dopamine antagonist which can block apomorphine-induced vomiting in dogs and reduce blood pressure.

Xia et al.[4] reported that *l*-stepholidine can lower the intraocular pressure in the rabbit's eyes. A 0.5% solution produced an effect slightly less potent than 0.5% of Timerol solution.

Xiong et al.[5] administered *l*-stepholidine to dogs and rats at a dose of 2.5 and 0.5 mg/kg, respectively. This resulted in a hypotensive effect which can be antagonized by phentolamine, an alpha blocker, or yohimbine, an α_2-blocker. They claimed that this agent is an α_2-adrenergic agonist.

Gu et al.[3] found that the administration of *l*-stepholidine to spontaneously hypertensive or Goldblatt's kidney rats can lower blood pressure and decrease plasma prolactin levels. They proposed that the agent acts on central dopamine receptors.

Clinical trials on 20 patients at a dose of 50 mg *l*-stepholidine t.i.d. for 3 months showed a significant relief of the symptoms of headache and insomnia. Patients also indicated that the symptoms of nightmares and loss of concentration were substantially reduced.[1]

References

1. **Dong, L. L. and Cai, N.,** *New Drugs Clin. Remedies* (in Chinese), 9, 83–85, 1990.
2. **Feng, H. M. et al.,** *Acta Pharmacol. Sin.*, 9, 377–380, 1988.
3. **Gu, T. H. et al.,** *Acta Pharmacol. Sin.*, 11, 445–449, 1990.
4. **Xia, X. P. et al.,** *Acta Pharmacol. Sin.*, 11, 137–140, 1990.

5. **Xiong, Z. L. et al.,** *Acta Pharmacol. Sin.*, 8, 497–502, 1987.
6. **Zhang, Z. D. et al.,** *Acta Pharmacol. Sin.*, 7, 522–526, 1986.
7. *Zhong Cao Yao Yue* (中草药学) Vol. 2, p. 289 on *Shan Dou Gen, Menispermum dauricum* and p. 499 on *Kong Dou Gen* or *Shan Dou Gen, Sophora subprostrata,* 1976.

BA JIAO FENG (八角枫) — dried root of *Alangium chinense* (Lour.) Harns

Chemistry

The active ingredient of this root is the alkaloid *dl*-anabasine or neonicotine ($C_{10}H_{14}N_2$).

From another species of *Alangium, A. lamarckii*, several additional alkaloids have been isolated. They are cephaeline, emetine, psychotrine, tubulosine, isotubulosine, demethyltubulosine, demethylpsychotrine, alangicine ($C_{28}H_{36}O_5N_2$), deoxytubulosine ($C_{29}H_{37}O_2N_3$), demethylcephaeline ($C_{27}H_{36}O_4N_2$), alamarckine ($C_{25}H_{33}O_4N$), alangimarckine ($C_{20}H_{37}O_3N_3$), and ankorine ($C_{19}H_{29}O_4N$). The glucoside alangiside has also been isolated.

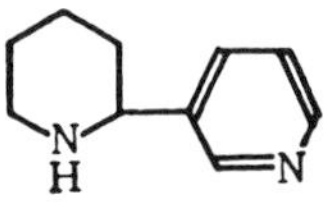

dl-Anabasine or Neonicotine

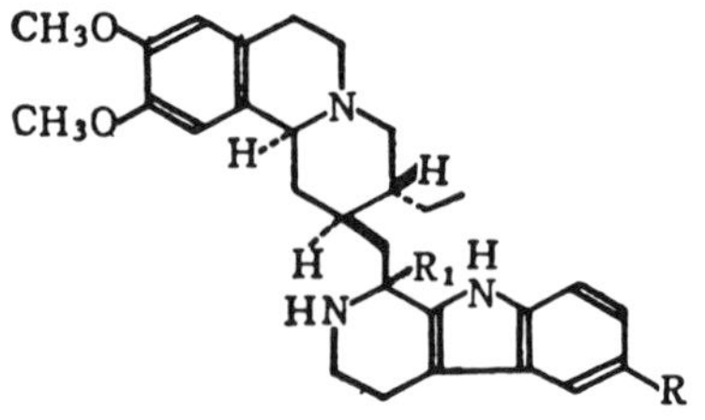

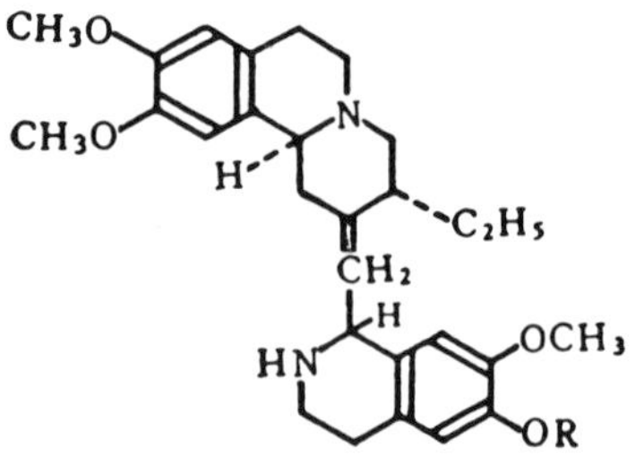

	R=	R_1=
Tubulosine	OH	β-H
Isotubulosine	OH	α-H
Deoxytubulosine	H	β-H

	R=
Emetine	CH_3
Cephaeline	H

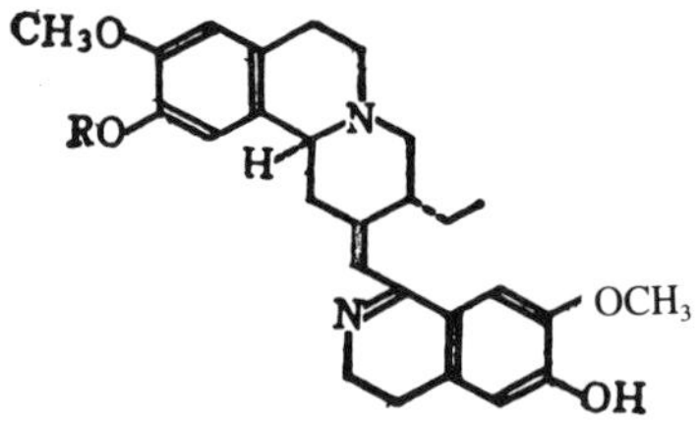

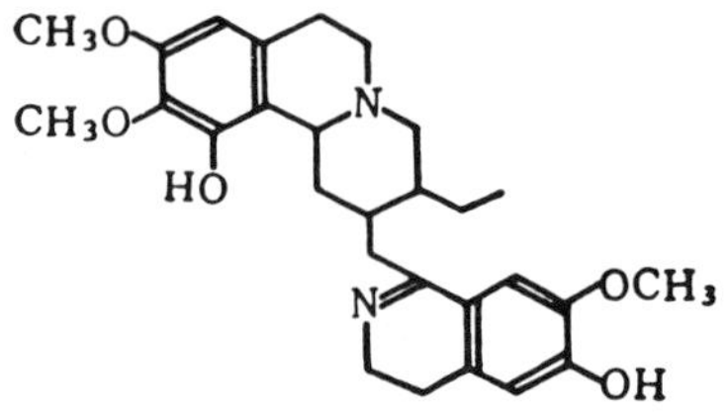

	R=
Psychotrine	CH_3
Demethylpsychotrine	H

Alangicine

Alangimarckine

	R =	R_1 =
Alangiside	H	CH_3
or	CH_3	H

Actions

The alkaloids are the active principles of the herb and have a curare-like blocking effect on the NMJ. This action can be antagonized by neostigmine. The alkaloids cause myocardial stimulation, increase contractility, and may cause fibrillation. They can also increase blood pressure.

Intravenous administration of the alkaloids causes an initial stimulation of the respiratory system, followed by depression; the latter is due mainly to depression of respiratory muscles. A similar effect is seen on the CNS. The alkaloids rapidly penetrate the blood brain barrier, causing CNS stimulation followed by depression.

Ba Jiao Feng is used as an adjuvant in anesthetic. It is also combined with local anesthetics, such as procaine, to produce greater muscle relaxation. The dose of alkaloid used is 0.3 to 0.6 mg/kg.

The herb is also used in the treatment of rheumatic arthritis. For this purpose, 3 to 6 g of root is prepared as a decoction. It is contraindicated in pregnant women.

XI SHENG TENG (錫生藤) — the plant of *Cissampelos pareira* Linn.

Chemistry

This plant contains the following alkaloids: cissampareine, hayatine, hayatinine, *dl*-beheerine, *dl*-curine, D-guereitol, *d*-isochondrodendrine, hayatidine, cissamine, and menisnine.

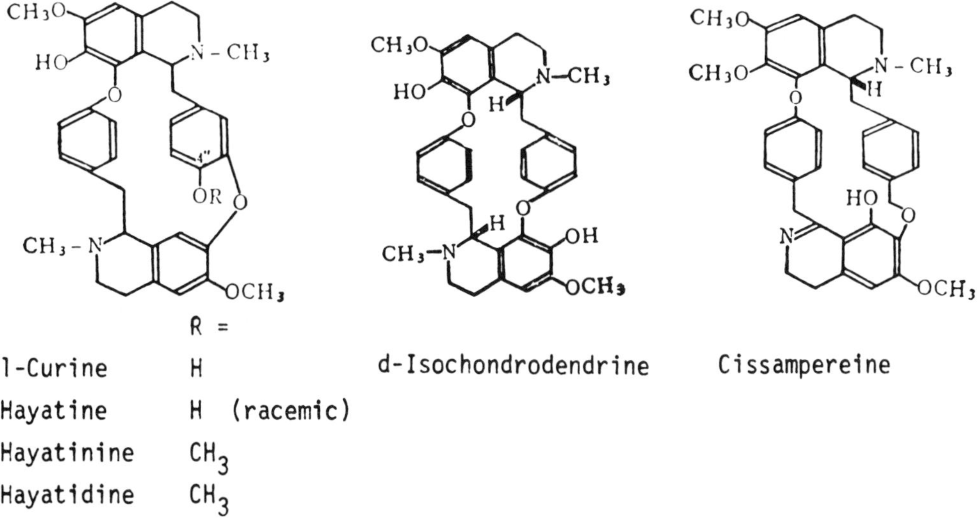

Actions

The herb's pharmadynamic effect is similar to that of *d*-tubocurarine and involves a blockade of NMJ depolarization. The onset of action is fast and the duration short. It can be used externally on wound surfaces to relieve pain, or can be prepared as an ointment for the treatment of traumatic injuries and rheumatitis. Folk medicine also used it to treat asthma and cardiac diseases.

Like *d*-tubocurarine, hayatine can liberate histamine after parenteral administration, but the amount of histamine being released is approximately half of that from *d*-tubocurarine.

The LD_{50} of hayatine to mice is 0.446 mg/kg (i.p.).

LUO SHI TENG (络石藤) — dried leafy stem of *Trachelospermum jasminoides*

Chemistry

This herb contains arctiin and several glucosides, including tracheloside, nortrachelo-side, matairesinoside.

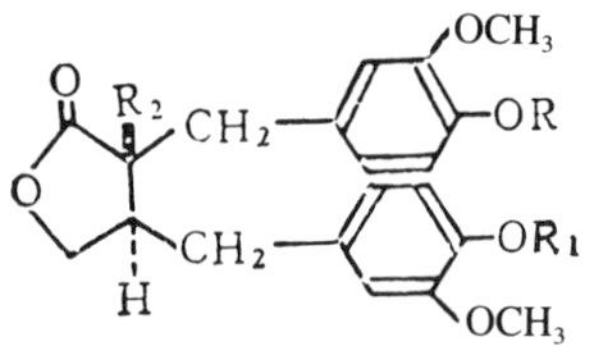

	R =	R_1 =	R_2 =
Arctiin	–glucose	CH_3	H
Matairesinoside	–glucose	H	H
Tracheloside	–glucose	CH_3	OH
Nortracheloside	–glucose	H	OH
Matairesinol-4,4′-di-*O*-β-D-glucopyranoside	–glucose	–glucose	OH
Nortrachelogenin-4,4′-di-*O*-β-D-glucopyranoside	–glucose	–glucose	OH
Arctigenin-4′-β-gentiobioside	–gentiobiose	CH_3	H

Therapeutic Uses

It is used to relieve muscle rigidity, to ensure the normal flow of *qi* and blood in the collaterals, to remove blood stasis, and to stop bleeding.

Chapter 9

SEDATIVE AND HYPNOTIC HERBS

Traditional Chinese medicine believes that nervousness and irritation are caused either by "emptiness" of the blood and *qi* or an excess of "fire" in the heart, liver and spleen. These result in symptoms of insomnia, heavy dreaming, convulsions, or delirium. Ancient recipes for *An Shen Yao* (安神药), meaning herbal mixtures to calm or stabilize the mind, was popularly prescribed to reduce anxiety and irritation, and to "balance" mental activity. The mechanism of these brews were quite simple: they cooled the "fire" in the heart, liver, spleen, and kidney. In traditional medicine, this is called "smoothing of the middle *jiao* (中焦) (or the visceral organs) to suppress evil energy."

SUAN ZAO REN (酸枣仁) — the dried seed of *Ziziphus spinosa* (Rhamnaceae) or *Z. jujuba* Mill.

Chemistry

This herb contains betulin, betulic acid, and the glycosides jujuboside A and B, which on hydrolysis produce jujubogenin ($C_{30}H_{48}O_4$). The seed also contains some vitamins and organic acids.

Recently, Han and Park reported that several other alkaloids had been isolated from the seeds and fruit of this herb. The alkaloids isolated from the seeds are peptide alkaloids which have been named sanjoinines — 14 in all. The fruit contains 12 different alkaloids with a phenanthrene radical; these have been named daechu alkaloids.

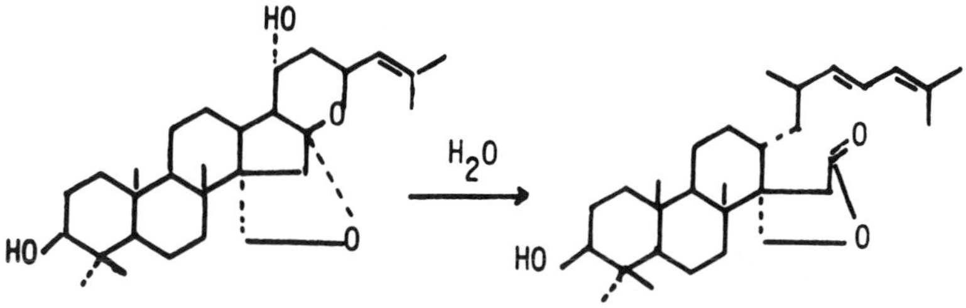

Jujubogenin $\xrightarrow{H_2O}$ Edelin lactone

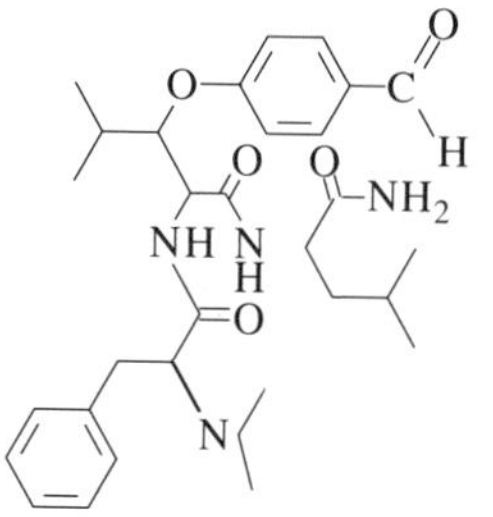

Sanjoinine G2

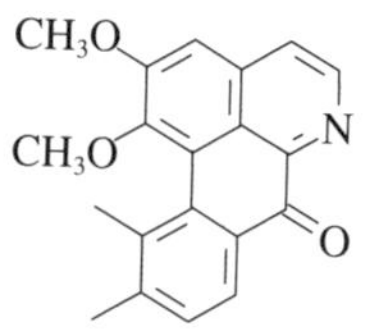

Daechu alkaloid C (oxonuciferine)

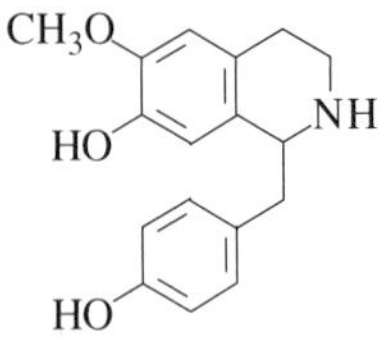

(+)-Coclauline 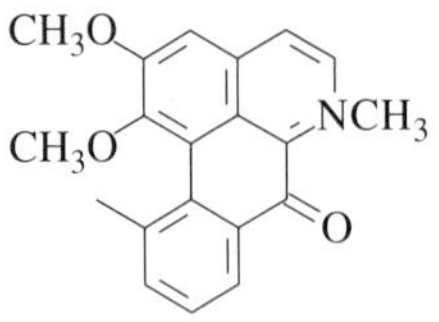 Nuciferine 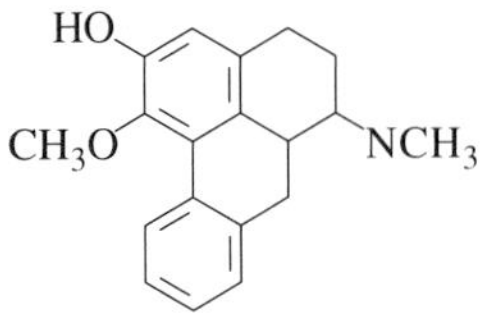 N-Methylasimilobine 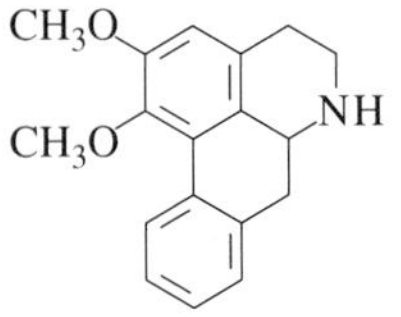 Nornuciferine

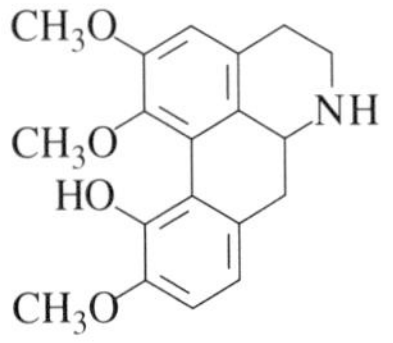

Norisocorydine 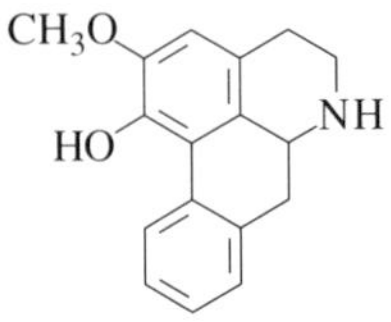 Caaverine 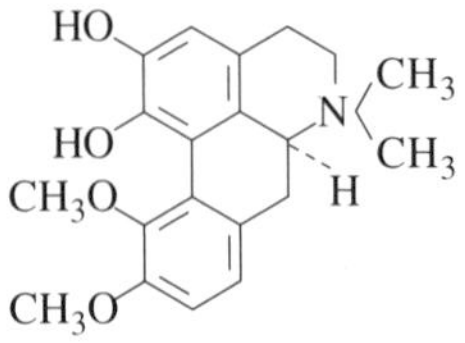 Zizyphusine

The alkaloid fractions from the seeds or fruit are the active principles in the herb's sedative action; the peptide alkaloids isolated from the stem bark are not effective.

Actions

The Chinese medical classics list this herb as one that can calm the mind, preserve *qi*, nourish muscles, and enrich bone marrow. In recent laboratory investigations, the herb showed a tranquilizing and hypnotic effect. It is also an analgesic and anticonvulsive.

The toxicity of the herb is very low. The LD_{50} in mice is 13.3 g/kg (i.p.).

Therapeutic Uses

The herb is usually used to treat neurasthenia, irritation, and insomnia. The normal dose is 9 to 18 g in a decoction.

Reference

Hans, B. H. and Park, M. H., in *Folk Medicine,* Steiner, Ed., American Chemical Society, Washington, D.C., 1986, 205–215.

YE HUANG HUA (野黄花) — the dried stem or whole plant of *Patrinia scabiosaefolia* Fisch.

Chemistry

This herb contains essential oils and the glycoside patrinoside.

Actions

It has sedative and antibacterial actions, and can protect hepatic function.

Adverse effects include a slight respiratory depression and loosening of bowel movements.

Therapeutic Uses

Clinical trials have shown that this herb is effective in the treatment of insomnia caused by neurasthenia or acute infections.

FU LING (茯苓) — the dried sclerotium of the fungus *Poria cocos* (Polyporaceae)

Chemistry

It contains several organic acids, including pachymic acid ($C_{33}H_{52}O_5$), tumulosic acid ($C_{31}H_{50}O_4$), eburicoic acid ($C_{31}H_{50}O_3$), and pinicolic acid ($C_{31}H_{45}O_3$). It also contains the polysaccharide pachymarose.

Actions

Fu Ling has been shown to be effective in its tranquilizing effects, as a diuretic, and as a cardiotonic. It also lowers blood sugar levels. As an antibacterial and anticancer agent, it contributes to an increase in the immune response of the body to cancer cells.

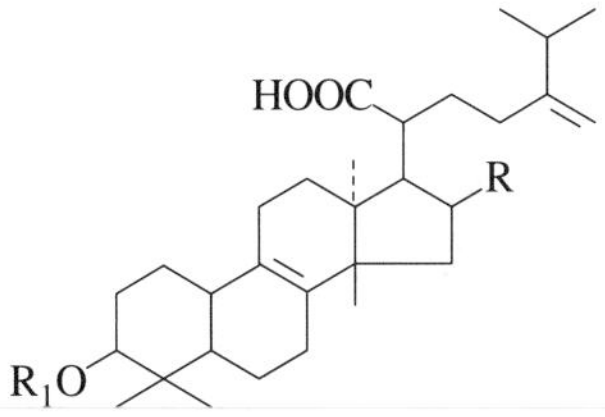

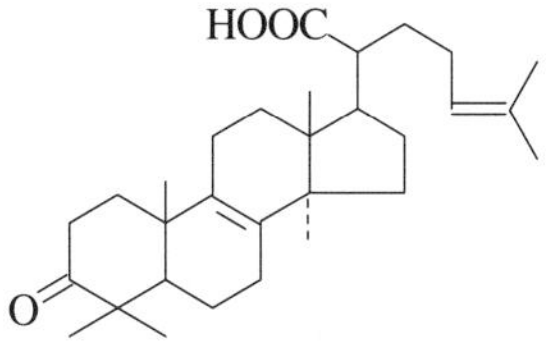

	R=	R_1=
Pachymic acid	OH	$COCH_3$
Tumulosic acid	OH	H
Eburicoic acid	H	H

Pinicolic acid

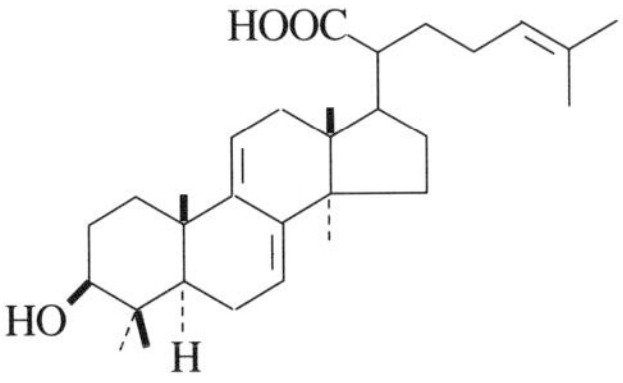

24-Trien-21-oic acid

HAN XIOU CAO (含羞草) — dried stem of *Mimosa pudica* L.

Chemistry

The herb contains the glycoside mimoside ($C_{14}H_{20}O_9N_2$).

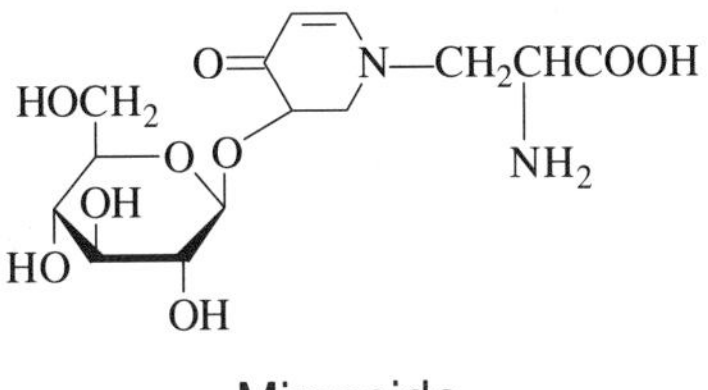

Mimoside

Therapeutic Uses

It was used to treat neurosis and has a tranquilizing effect. It was also used to treat trauma wound and hemoptysis. The dose is 5 to 7 g in decoction. It is contraindicated to pregnant women. The decoction solution also can be used externally — applied to a wound.

ZHU ZI TOU (豬屎豆) — the plant of *Crotalaria mucronata*

Chemistry

It contains the following alkaloids:

- Mucronatine, $C_{18}H_{25}O_6N$
- Mucronatinine
- Retroresine
- Usaramine
- Nilgirine, $C_{17}H_{23}O_5N$
- Vitexin, $C_{21}H_{20}O_{10}$

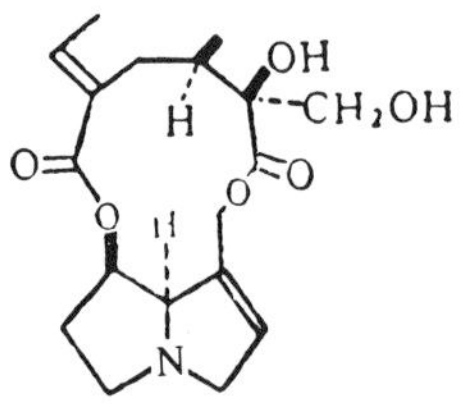

Mucronatine

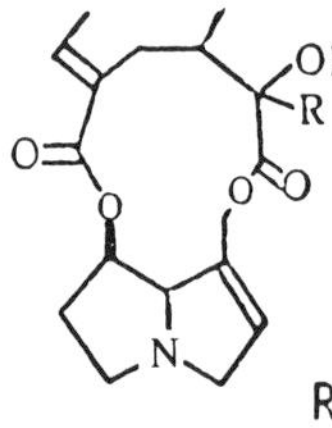

	R =
Mucronatinine	CH_2OH
Nilgirine	H

Two glucosides are also found: vitrexin-4-*O*-xyloside ($C_{26}H_{28}O_{14}$) and apigenin ($C_{15}H_{19}O_5$).

Therapeutic Uses

This herb was used in folk medicine to treat frequent urination in children and to produce a calming effect in cases of neurosis.

Chinese medicine recommends this herb to promote water metabolism and remove "dampness". It is also used to reinforce the function of the spleen, and to relieve mental stress.

This herb is also used to treat edema, chronic diarrhea, and pelvic infections. In the treatment of neurasthenia and insomnia, it is administered as a decoction, at a dose of 9 to 18 g in combination with *Suan Zao Ren.*

LING YANG JIAO (羚羊角) — antelope's horn (*Saiga tatarica*)

Chemistry

The primary contents of the herb are calcium phosphate, proteins, and insoluble inorganic salts.

Actions

The water extract of the powdered horn exerts a sedative effect in animals and can prolong the sleeping time induced by barbiturates. Intravenous administration of the water extract also produces an antipyretic effect in rabbits, at a dose of 40 mg/kg.

Therapeutic Uses

The old Chinese classics stated this herb could check hyperactivity of the liver and relieve convulsions. It was also reputed to remove "heat" from the liver, improve eyesight, and produce antipyretic and detoxicant effects.

A decoction from 50 g of *Ling Yang Jiao* was taken orally to treat dysentery or neurosis. Externally the agent was mixed with wine and applied on the inflamed mastitis.

DAO DOU (刀豆) — dried ripe seed of *Canavalia gladiata* (Jacq.)

Chemistry

Approximately 20% of the herb is canavaline, canavanine, and urease. Other substances that have been isolated from the seed include gibberelin A_{21} and A_{22}, canavalia gibberellin I ($C_{19}H_{22}O_7$), and canavalia gibberellin II ($C_{19}H_{22}O_6$).

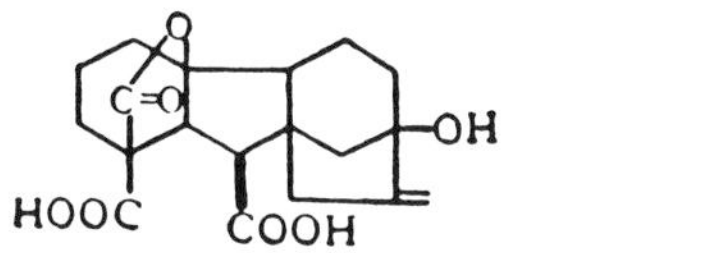

Canavalia gibberellin-I

C=O
OH
HOH₂C COOH

Canavalia gibberellin-II

Therapeutic Uses

Traditional medical lore claimed that this herb "warmed" the viscera and suppressed evil *qi*.

It was given in dose of 7 to 12 g in decoction, or ground into powder form and taken orally.

SHA YUAN ZI (沙苑子) — dried seed of *Astragalus complanatus* R. Br.

Chemistry

This herb contains the glucoside astragalin ($C_{20}H_{20}O_{11}$) and two additional substances: canavanine and homoserine.

HO O OH
O – glucose
OH O

Astragalin

Therapeutic Uses

Traditional medicine used this herb to reinforce the liver and kidney, and to improve visual acuity. It was given in a dose of 8 to 14 g in decoction.

TABLE 1
Other Central Depressant Herbs

Name	Source	Actions			
		Hypotensive	Sedative	Anticonvulsive	Others
Di Long (地龙)	*Pheretima aspergillam* (earthworm)	+	+	+	Diuretic, antipyretic, and uterus contraction
Wu Gong (蜈蚣)	Scolopendridae	+	+		
Quan Xie (全蝎)	*Buthus martasi* (scorpion) which contains katsutoxin	+	+	+	Antispasmodic

Chapter 10

ANTICONVULSIVE HERBS

Convulsions are general muscle spasms. The causes are many, most originating from stimulation of motor centers in the CNS. Convulsions may be idiopathic (as in many epileptic convulsions), may result from a toxic effect (as in tetanus toxin exposure), may be due to fever or drugs, or may be simply due to water or electrolyte imbalance.

In traditional Chinese medicine, herbs were used mostly in the treatment of convulsions caused by fever, which often occurred in children during acute infections. Thus, the herbs described here are also effective in reducing body temperature.

GOU TENG (鈎藤) — the dried hooked stem of *Uncaria rhynchophylla, U. macrophylla, U. hissata, U. sessilifructus,* or *U. sinensis* (Rubiaceae)

Chemistry

The active principles of this herb are alkaloids. These include rhynchophylline, isorhynchophylline, corynoxeine, isocorynoxeine, corynantheine, hirsutine, and hirsuteine.

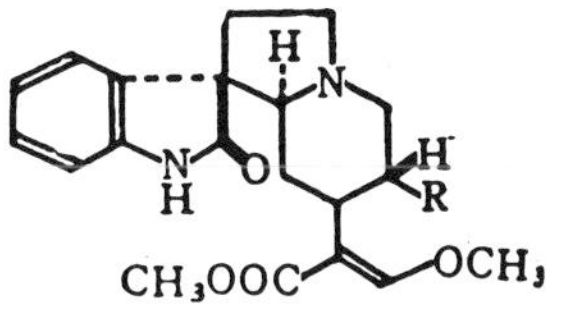

	R =
Rhynchophylline	CH_2CH_3
Corynoxeine	$CH{=}CH_2$

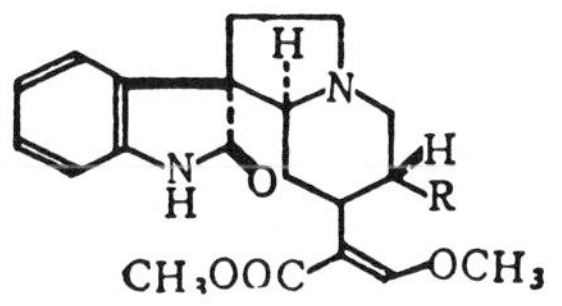

	R =
Isorhynchophylline	CH_2CH_3
Isocorynoxeine	$CH{=}CH_2$

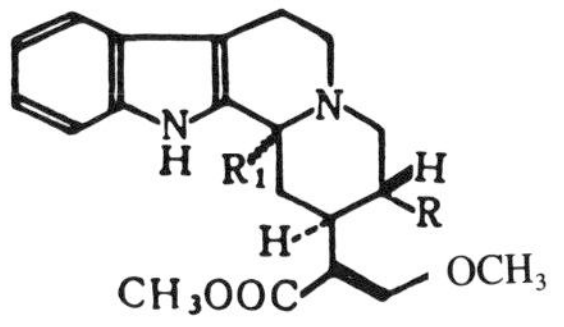

	R =	R_1 =
Hirsutine	CH_2CH_3	β-H
Hirsuteine	$CH{=}CH_2$	β-H
Corynantheine	$CH{=}CH_2$	α-H
Di-hydrocorynantheine	CH_2CH_3	α-H

Actions

As a sedative and anticonvulsive, the herb is used to treat childhood epilepsy (febrile convulsions). In adults, it is administered as an antispasmodic and sedative. It was also used in traditional medicine during the 8th month of pregnancy to reduce fetal movement and postpartum spasms. It has an antispasmodic effect on smooth muscles.

In experiments with animals, the herb was observed to antagonize caffeine-induced central stimulation and to reduce cortical excitation.

The herb also lowers blood pressure. It has a triphasic effect: there is an early fall in blood pressure, followed by a rapid return to the original level, and then a gradual second decrease which lasts for a longer period.

The LD_{50} of the total alkaloids in mice is 514.6 ± 29 mg/kg (oral) and 144.2 ± 3.1 mg/kg (i.p.).

Therapeutic Uses

This herb was described in a medical classic as having the ability to remove ''heat'', check hyperfunction of the liver and subdue ''endogenous wind'', and to relieve dizziness, tremors, and convulsions.

Gou Teng is given in doses of 6 to 15 g, prepared as a decoction, in cases of pediatric febrile convulsions. For hypertension, it is administered in doses of 6 to 9 g as a decoction. Frequently, it is combined in equal amounts with other herbs such as chrysanthemum (菊花), *Xia Ku Cao* (夏枯草), or mulberry leaf (桑叶).

In tablet form, the herb is prepared with 5 or 10 mg of total alkaloids; the standard dose is 10 to 20 mg t.i.d.

TIAN MA (天麻) — the dried tubes of *Gastrodia elata* Blume; during development of the plant, *Gastrodia* is symbiotic with a fungus

Chemistry

The active principles of the herb are vanillyl alcohol, vanilin, vitamin A, and small quantities of the glycoside, gastrodin.

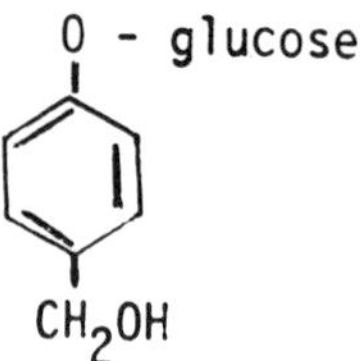

Gastrodin

Actions

This herb has anticonvulsive, sedative, and analgesic effects. In addition, it can increase coronary and cerebral blood flow, and lower peripheral blood pressure.

Therapeutic Uses

The *Ben Cao Kong Mu* stated that this herb was useful in subduing ''exuberant *yang*'' of the liver, calming internal wind, and relieving convulsions and fainting.

It is used in the treatment of numbness or spasms of the extremities. It is particularly effective in relieving facial muscle spasms and trigeminal neuralgia. Some reports claim that it is effective in relieving anginal pain.

An injection solution is prepared from the herb. Each ampule contains 2 ml equivalent to 5 g of raw material; the standard dose is 2 to 4 ml, administered by intramuscular injection.

NAN XING (南星) — the dried stem of *Arisaema consanguineum* Shott., *A. heterophyllum* Blume, and *A. murense* Maxim.

Chemistry

Because of its high toxicity, only ''prepared'' or roasted *Nan Xing* is prescribed in Chinese apothecaries. It contains alkaloids and saponin.

Actions

Ben Cao Kong Mu described this herb as an effective agent in the treatment of tetanus, spasms, epilepsy, and neuralgia. It is frequently used as a sedative and expectorant.

As an anticonvulsive, 3 to 9 g of the herb is decocted as a standard dose. Experiments with animals have shown its effectiveness against strychnine-induced convulsions.

Nan Xing is used as an anticancer agent, with 15 to 45 g prepared as a decoction. This brew is taken daily as a tea. It is also prescribed as a wine extract, and has been applied externally, especially on the uterine cervix.

Toxic effects include loss of feeling and taste sensations, salivation, ulceration of the mucosa, pharyngeal edema, and, in severe cases, loss of voice. The LD_{50} in mice is 13.5 g/kg (i.p.).

NIU HUANG (牛黄) — the gallstone of *Bos taurus domesticus* (Bovidae)

Chemistry

The main ingredients of the stone are cholic acid, bilirubin, cholesterol, ergosterol, vitamin D, Ca^{2+} salts, Cu, and Fe. Recent investigations showed that the stone contains substances which encourage smooth muscle contractions.

Actions

As an anticonvulsive and sedative agent, this herb can antagonize the central stimulation produced by cocaine or caffeine; but it does not affect central stimulation produced by strychnine.

The herb displays antipyretic, erythropoietic, and choleretic effects. In addition, it lowers blood pressure and stimulates myocardial contraction.

Therapeutic Uses

The *Ben Cao Kong Mu* described *Niu Huang* as a drug which could remove ''heat'' from the heart and liver, induce resuscitation, eliminate phlegm, relieve convulsions, and remove toxic substances.

The herb is frequently decocted with other herbs for treatment of high fevers, coma, or convulsions. The standard recipe is 0.15 to 0.3 g of *Niu Huang*, 1.5 g of *Huang Lin* (黄連), 4.5 g of *Huang Qin* (黄芩), and 9 g of *Zhi Zi* (梔子).

It is also prepared as an injection solution for intravenous administration, for the treatment of convulsions.

HU JIAO (胡椒) — the dried fruit of *Piper nigrum* L.

Chemistry

Hu Jiao contains several alkaloids, including piperine,chavicine, and piperamine. By weight, the fruit comprises 0.8% essential oils, which include piperonal, dihydrocarveol ($C_{10}H_{18}O$), caryophyllene, and cryptone.

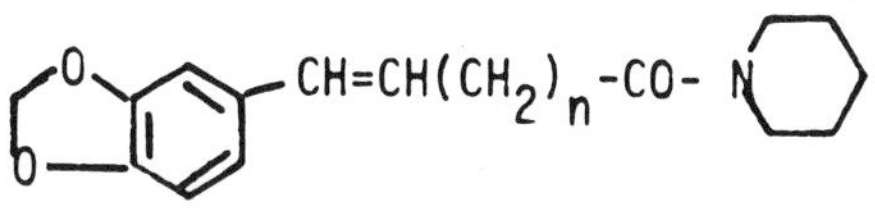

CH=CH(CH$_2$)$_n$-CO-N

Piperoleine A n= 4

Piperoleine B n= 6

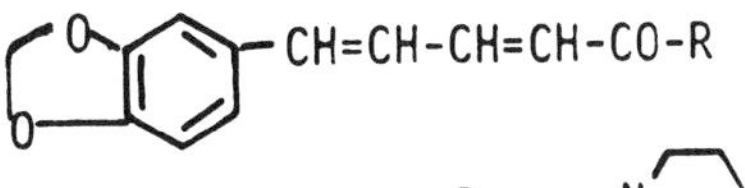

CH=CH-CH=CH-CO-R

Piperine R =

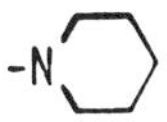

Piperyline R =

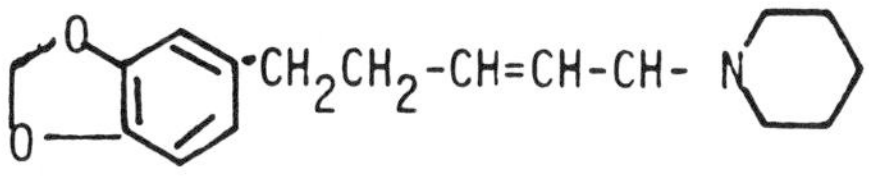

CH$_2$CH$_2$-CH=CH-CH-N

Piperamine

Actions

Traditional Chinese medicine prescribed this herb to warm the stomach and remove "coldness" from the body. Its known actions are as an anticonvulsive and sedative. *Hu Jiao* can stop convulsions induced by electrical shock. The onset of action is, however, relatively slow. Generally, it takes 2 to 3 d of continuous administration to show an anticonvulsant effect on patients. Woo stated that the *Piper* extract can prolong hexobarbital-induced sleeping time and markedly reduce the strychnine mortality.

Hu Jiao can be irritating to the system. Thus, it is recommended to be administered after meals. Prolonged administration can result in withdrawal syndrome. The herb should not be given together with alcohol.

The LD_{50} of piperine in rats was 348.6 mg/kg (i.p.).

Reference

Woo, W. S., in *Advanced Chinese Medicinal Material Research,* Chang, H. M. et al., Ed., World Scientific Publishing Co., Singapore, 1985, 129–146.

JIAN CAN (僵蚕) — dried larvae of the silkworm, *Bombyx mori* L., which have died due to infection by the fungus *Beauveria bassiana* (Bals.) Vuill.

Chemistry

The herb contains proteinases and other enzymes, including chitinase. The *Beauveria* fungus itself contains an active substance called bassianins.

Actions

In folk medicine, this herb was used as an anticonvulsant and spasmolytic agent, to reduce phlegm, and to resolve masses.

It has anticonvulsive, sedative, and hypnotic properties. It is an antibacterial agent, and has been used in the treatment of epilepsy, acute upper respiratory infections, epidemic parotiditis, and diabetes mellitus. Some reports indicate that the herb may have some anticancer effects.

CI JI LI (刺蒺藜) — dried fruit of *Tribulus terrestris*

Chemistry

This herb contains the glycosides tribuloside ($C_{30}H_{26}O_{13}$) and astragalin. Two alkaloids are also found — harmane ($C_{12}H_{10}ON_2$) and harmine ($C_{13}H_{12}ON_2$).

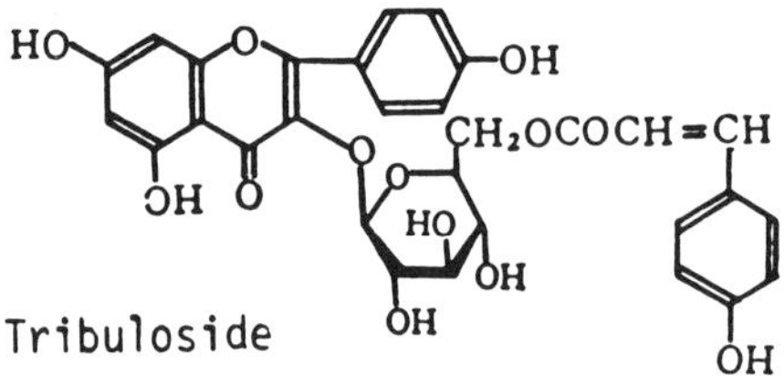

Tribuloside

Actions

The water extract has a slight hypotensive and diuretic effect. It is used as an anticonvulsant and to improve visual acuity.

It is given in dose of 5 to 7 g as decoction and should be given cautiously in pregnant women.

Chapter 11

ANALGESIC HERBS

Traditional Chinese medicine believes that the root of pain is an incompatibility between blood and *qi*. Thus, in Chinese medical classics it was stated that ''stagnation causes pain'' and that ''opening the stagnancy will remove that pain''.

When discussing analgesic agents, physicians and pharmacologists usually consider opium, morphine, and their analogs. Contrary to popular opinion, opium did not originate in China. Instead, it was introduced to China more than ten centuries ago, and the opium poppy *Papaver* was planted in the southwestern part of the country. Some other species of *Papaver* which do grow naturally in China, however, are used in analgesic prescriptions listed in the ancient Chinese pharmacopoeia.

YAN HU SUO (延胡索) — the dried tuber of *Corydalis turtschaninovii* Bess f. *yanhusu* (Papaveraceae)

FIGURE 1. The tuber of *Corydalis turtschaninovii.*

Chemistry

Approximately 0.65% of the herb consists of alkaloids, more than ten of which have been identified and found to be phenanthrene derivatives. The major ones are *d*-corydaline, *dl*-tetrahydropalmatine, and corydalis H, I, J, K, and L. All of these have analgesic effects; *dl*-tetrahydropalmatine is the most potent.

	R_1=	R_2=	R_3=	R_4=	R_5=
d-Corydaline	CH_3	CH_3	CH_3	CH_3	CH_3
dl-Tetrahydropalmatine	CH_3	CH_3	CH_3	CH_3	H
l-Tetrahydrocoptisine	$-CH_2-$		$-CH_2-$		H
dl-Tetrahydrocoptisine	$-CH_2-$		$-CH_2-$		H
l-Tetrahydrocolumbamine	CH_3	H	CH_3	CH_3	H
d-Crybulbine	H	CH_3	CH_3	CH_3	CH_3
Corydalamine	CH_3	CH_3	CH_3	H	H

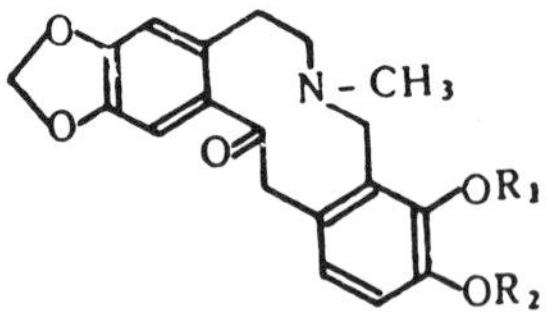

	$R_1=$	$R_2=$
Protopine	$-CH_2-$	
α-Allocryptopine	CH_3	CH_3

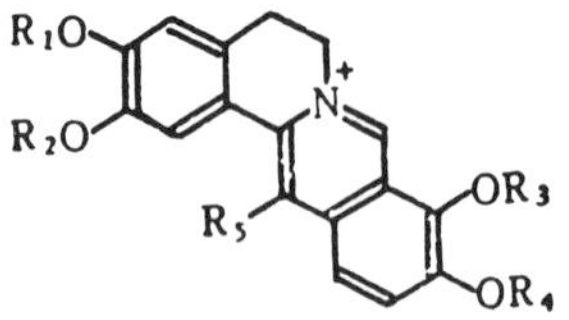

	$R_1=$	$R_2=$	$R_3=$	$R_4=$	$R_5=$
Coptisine	$-CH_2-$		$-CH_2-$		
Dehydrocorydaline	CH_3	CH_3	CH_3	CH_3	CH_3
Columbamine	CH_3	H	CH_3	CH_3	H
Dehydrocorydalmine	CH_3	CH_3	CH_3	H	H

An *l*-form derivative named rotundine has been synthetized from *dl*-tetrahydropalmatine. It was reported that this synthetic rotundine is more potent and less toxic than the parent substance.

Actions

Like morphine, the *Yan Hu Suo* alkaloids exert an analgesic action by inhibiting the reticular activating system in the brain stem. They are, however, less potent than morphine. Continuous use will result in tolerance, and is likely to lead to a cross tolerance to morphine.

The alkaloids have sedative and hypnotic effects, and are synergistic with barbiturates. They also inhibit myocardial contraction, slow the heart rate, and increase coronary flow.

Overdosage intoxication leads to central nervous system (CNS) depression and muscle relaxation. The LD_{50} of *dl*-tetrahydropalmatine in mice is 151 mg/kg (i.v.).

Therapeutic Uses

The *Ben Cao Kong Mu* described this herb as an agent to stimulate the circulation of blood, promote *qi*, and relieve pain. It is used in the treatment of neuralgia, dysmenorrhea, and gastrointestinal spasms.

Chinese pharmaceutical companies have produced several preparations from the *Yan Hu Suo* alkaloids for use as analgesics. Available preparations include a 30-mg tablet containing all of these alkaloids, a 10% tincture prescribed in doses of 5 ml t.i.d., and an ampule for subcutaneous injection (60 mg/2 ml). Additionally, the drug may be prescribed in powder form, with 3 to 9 g taken as a decoction.

Several species of *Corydalis* are listed in the Chinese pharmacopoeia; they contain phenanthrene derivatives, but differ from those of *Yan Hu Suo*. The pharmacological actions of these alkaloids have not been completely investigated. Two similar herbs are described as follows:

XIA TIAN WU (夏天无) — from *Corydalis decumbens* (Thunb.) Pers.

The active principles of this plant are protopine, bulbocapnine ($C_{19}H_{19}O_4N$), and *d*-tetrahydropalmatine.

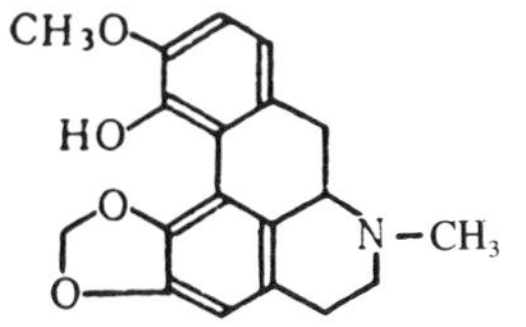

Bulbocapnine

This herb is effective in relieving pain in deep tissues, such as neuralgia or bone fractures. It also exhibits antihypertensive and antirheumatic properties. Overdosage can result in spinal cord stimulation and convulsions. It is generally administered as a decoction of 5 to 7.5 g.

CHUAN DUAN CHANG CAO (川断腸草) — from *Corydalis incisa* (Thunb.) Pars.

Chemistry

More than ten alkaloids have been isolated from this plant. They are protopine ($C_{20}H_{19}O_5N$), pallidine ($C_{19}H_{21}O_4N$), sinocecatine ($C_{19}H_{21}O_4N$), corynoline ($C_{21}H_{21}O_5N$), isocorynoline, acetylcorynoline ($C_{23}H_{23}O_6N$), corynoloxine ($C_{21}H_{19}O_5N$), corycavine ($C_{21}H_{21}O_5N$), *l*-corypalmine ($C_{20}H_{23}O_4N$), corydalic acid methyl ester ($C_{22}H_{23}O_6N$), corydamine ($C_{20}H_{18}O_4N_2$), coptisine ($C_{19}H_{14}O_4N$), *l*-tetrahydrocorysanine ($C_{20}H_{19}O_4N$), corysanine ($C_{20}H_{16}O_4N$), *l*-cheilanthifoline ($C_{19}H_{19}O_4N$), *l*-scoulerine, coreximine ($C_{19}H_{21}O_4N$), and reliculine ($C_{19}H_{23}O_4N$).

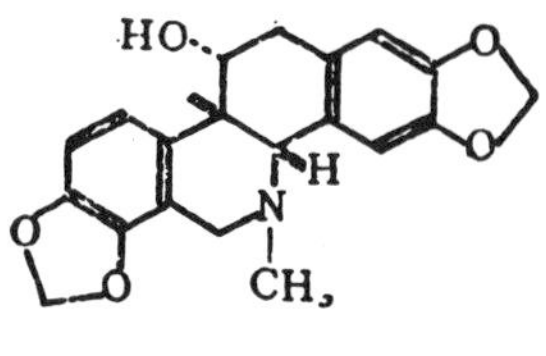

Corynoline

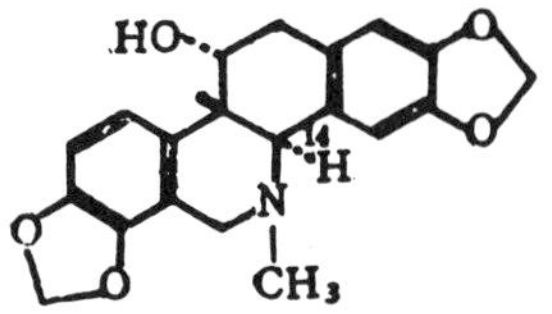

d-14-Epicorynoline

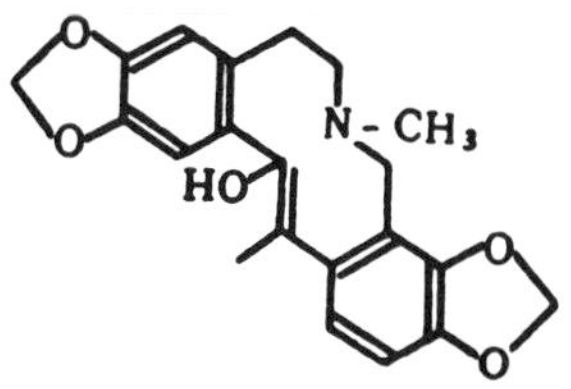

Corycavine

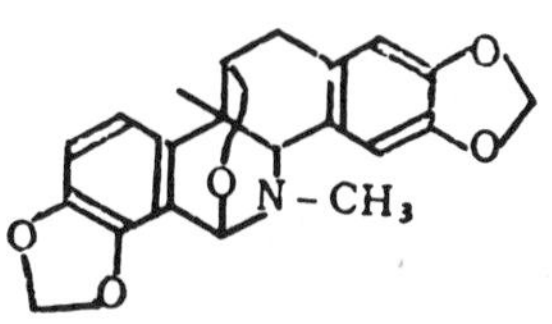

Corynoloxine

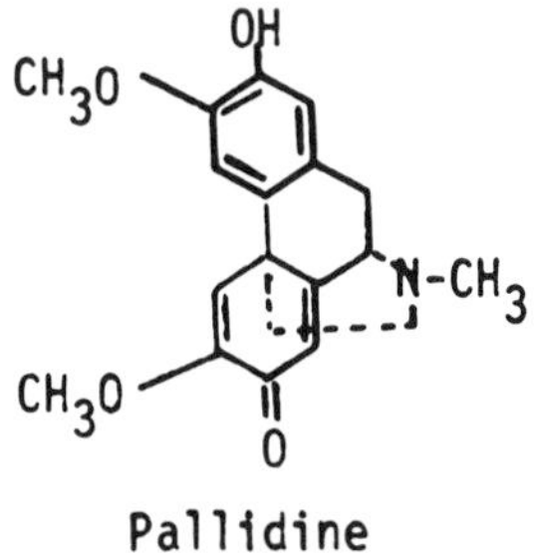

Pallidine

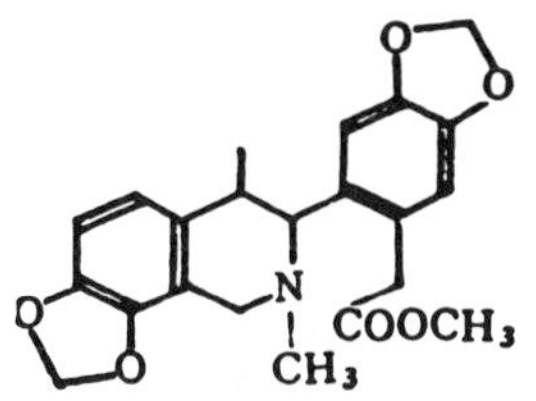

Corydalic acid methyl ester

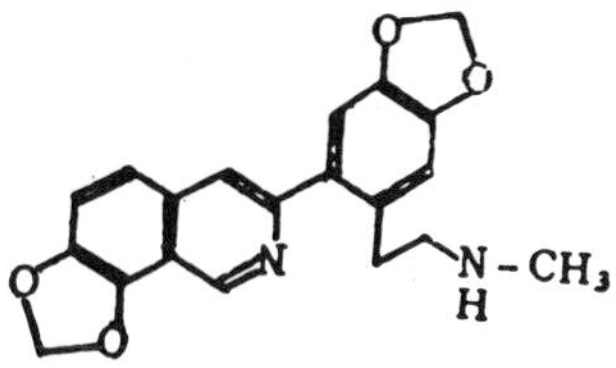

Corydamine

BAI QU CAI (白屈菜) — the dried whole plant of *Chelidonium majus* (Papaveraceae)

Chemistry

The plant contains several alkaloids, including chelidonine, protopine, stylopine, allocryptopine, chelerythrine, sparteine, and coptisine.

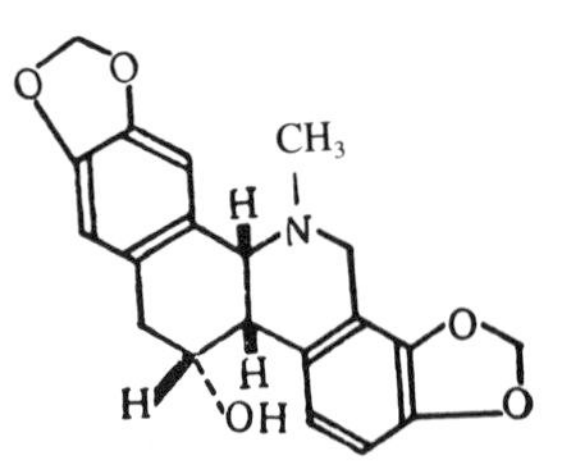

Chelidonine

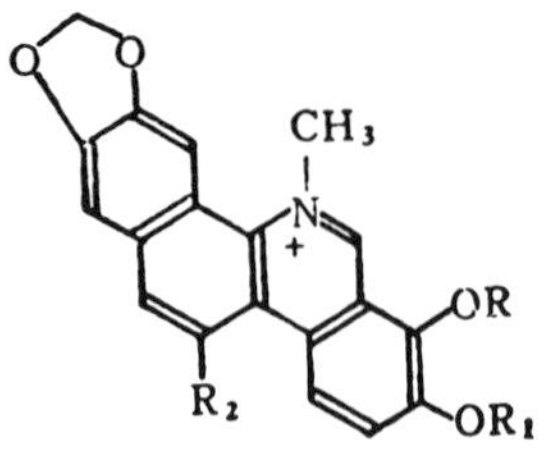

	R=	R_1=	R_2=
Chelerythrine	CH_3	CH_3	H
Chelilutine	CH_3	CH_3	OCH_3
Sanguinarine	$-CH_2-$		H
Chelirubine	$-CH_2-$		OCH_3

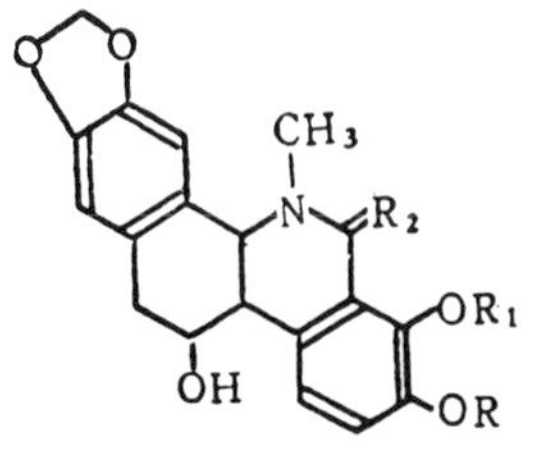

	R=	R_1=	R_2=
α-Homochelidonine	CH_3	CH_3	H_2
Oxychelidonine	$-CH_2-$		O

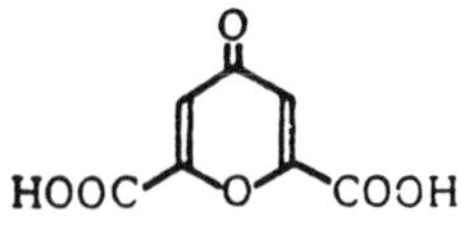

Chelidonic acid

Actions

The alkaloids of this herb produce an analgesic effect similar to that of morphine, with an action lasting from 4 to 48 h. The alkaloids also have an anesthetic effect on sensory nerve endings. Additionally, they are antitussive and cause relaxation of gastrointestinal muscles, and have antibacterial activity.

In clinical trials, doses above 1 g produced dizziness, headache, sweating, nausea and vomiting, and, occasionally, hypotension. The LD_{50} of the alkaloids in mice is 300 mg/kg (subcutaneous).

Therapeutic Use

This herb has been used as an anodyne, analgesic, antitussive, diuretic, and detoxicant. It is used in the treatment of abdominal pain, peptic ulcers, chronic bronchitis, and whooping cough.

It is prescribed in combination with other herbs in the form of a tincture or water extract to relieve gastrointestinal spasms. It may also be applied externally to the skin in the form of an ointment or paste, to treat insect bites and infections.

XI XIN (细辛) — the dried whole plant of *Asarum heterotropoides mandshuricum* or *A. sieboldii* (Aristolochiaceae)

Chemistry

The major components of this plant are essential oils, including eucarvone, safrole, β-pinene, asoryl-ketone, and asarilin.

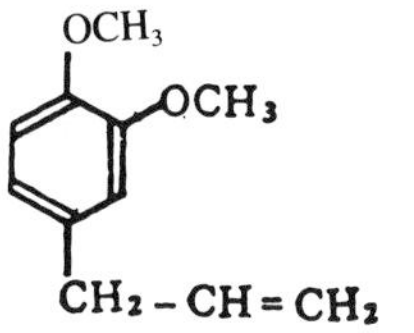

OCH₃ / OCH₃ / $CH_2-CH=CH_2$

Methyleugenol

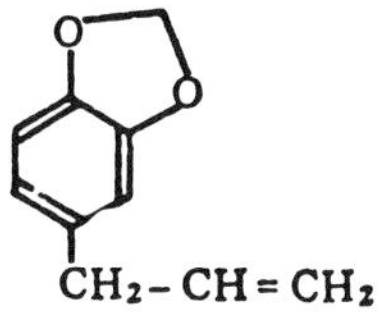

$CH_2-CH=CH_2$

Safrole

Actions

Xi Xin is an analgesic and sedative, and displays both antipyretic and anti-inflammatory properties. It is effective as a local anesthetic. Additionally, it has been used to stimulate respiration.

Adverse effects include headache, sweating, irritation, dyspnea, and coma. Large doses may result in death, due to respiratory depression. The LD_{50} in mice is 123.7 mg/kg (by stomach) and 7.78 mg/kg (intravenous).

Therapeutic Uses

The Herbal Classic of the Divine Plowman describes this as an herb which can induce diaphoresis, dispel cold and "wind", relieve pain, warm the lungs, and reduce phlegm.

As a treatment for the common cold and influenza, the herb is administered in doses of 0.9 to 3 g. It can be used to relieve a toothache, by direct application of the powder on the gingival surface. A 3% solution can be injected subcutaneously as a local anesthetic; the anesthetic effect lasts for $1^1/_2$ h.

CAO WU (草乌) — the dried tuberous root of *Aconitum kusnezoffii*, *A. chinense* Paxt., *A. vilmorinianum* Kom. and *A. pariculigerum* Nakei

Chemistry

The major components are the alkaloids hypaconitine, aconitine, aconine, mesaconitine, and talatisamine. Other substances whose structures have not yet been elucidated have also been found.

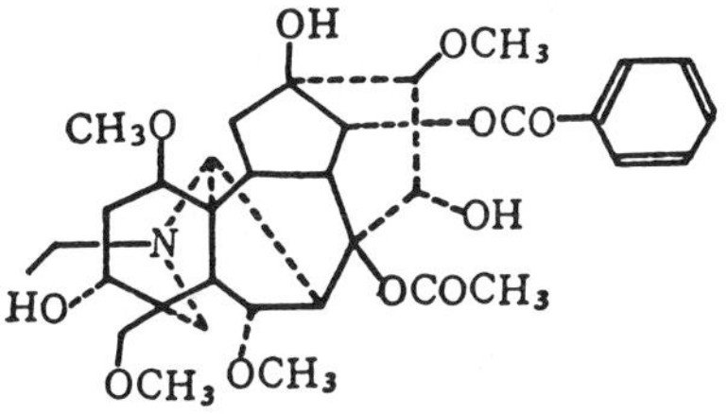

OH, OCH_3, CH_3O, OCO, OH, N, HO, $OCOCH_3$, OCH_3, OCH_3

Aconitine

Actions

The alkaloids are analgesic and sedative. They have a vagal-stimulating effect which reduces the heart rate and causes vasodilatation of coronary vessels. They also have a local anesthetic effect, especially on muscosal membranes, but only limited penetration to deep layers.

They are quite toxic. Overdosage results in paresthesias, dryness of mouth, speech and visual difficulties, bradycardia, a fall in blood pressure, and, in extreme cases, coma.

Therapeutic Uses

The great Chinese surgeon, Hua Tau (華佗, 145 to 203 AD) introduced a famous anesthetic recipe combining *Cao Wu* and *Datura stramonium*. This was mixed with wine and given to patients to relieve the pain associated with surgical operations.

QING FENG TENG (青风藤) — the dried stem of *Sinomenium acutum* (Thunb.) or *S. acutum cinereum* (Menispermaceae) or *Sabia japonica* Maxim.

This herb is also called Japan *Han Fan Ji* (see Chapter 8).

Chemistry

The stem contains several alkaloids, including sinomenine, disinomenine, magnoflorine, acutumine, sinactine, isosinomenine, tuduranine, and sinoacutine. The active principle responsible for analgesia is sinomenine.

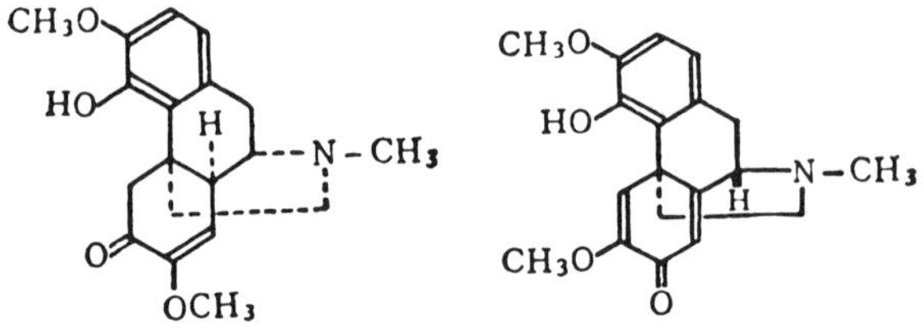

Sinomenine

Sinoacutine

Actions

Structurally, sinomenine can be considered a morphine analog. Its analgesic effect is weaker than that of morphine and its duration of action is shorter. Tolerance to this alkaloid develops only after continuous use for a long period; no addiction syndrome has been reported.

The herb is more potent than aspirin as an anti-inflammatory agent. It has been suggested that this action is due to stimulation of the hypothalamus-pituitary axis on adrenocorticotropic-hormone secretion.

Other observed actions include a lowering of the blood pressure with little effect on the myocardium. The herb acts on sympathetic ganglia and the central vasomotor center. Additionally, it relaxes smooth muscle and displays an antihistaminic effect.

Therapeutic Uses

The *Ben Cao Kong Mu* described this herb as an antirheumatic and analgesic agent. It is used mainly to relieve rheumatic arthritis pain. For treatment of this condition, 6 to 9 g of the herb is prepared in a decoction for drinking.

ZU SI MA (祖司麻) — the dried root bark of *Daphne giraldii* Nitsche or *D. retusa* Hemsl., *D. tangutica* Maxim.

Chemistry

The root bark contains daphnetin.

Therapeutic Uses

This herb is used as an analgesic, anti-inflammatory and antibacterial agent. It is prescribed in the form of an injection ampule (250 mg/amp), which is diluted with glucose solution for intravenous infusion.

WU YAO (乌药) — dried tuberous root of *Lindera strychnifolia* (Sieb. et Zucc.)

Chemistry

The active principles are essential oils, including lindestrene ($C_{15}H_{18}O$), liderane ($C_{15}H_{16}O_4$), linderene ($C_{15}H_{18}O_2$), linderalactone ($C_{15}H_{16}O_3$), isolinderalactone, isofuranogermacrene ($C_{15}H_{20}O$), linderoxide ($C_{16}H_{20}O_2$), isolinderoxide ($C_{16}H_{20}O_2$), lindestrenolide ($C_{16}H_{18}O_2$), and neolinderalactone ($C_{15}H_{16}O_3$).

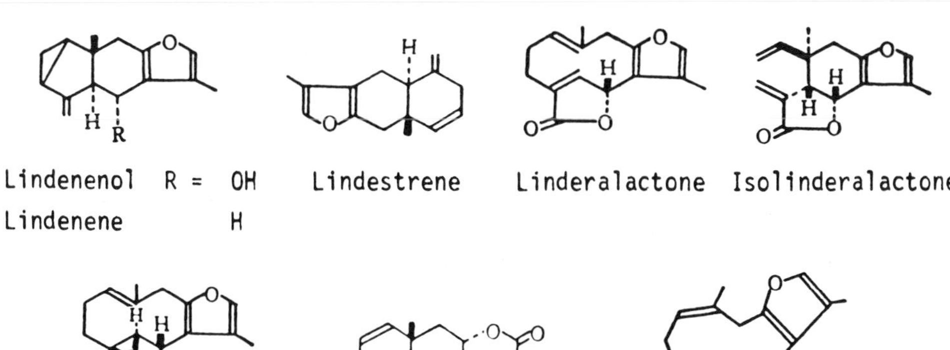

Lindenenol R = OH
Lindenene H

Lindestrene

Linderalactone

Isolinderalactone

Linderane

Lindestrenolide

Neolinderalactone

Therapeutic Uses

This herb is used mainly to promote the circulation of *qi* and to ease pain. The standard dose for the herb is 6 to 9 g prepared as a decoction.

MO YAO (没药) — a gum resin obtained from the stem of *Commiphora myrrha* Engler and other related species

Chemistry

The resin contains several essential oils, including myrcene, α-camphorene, Z-guggulsterol, guggulsterol I ($C_{27}H_{44}O_4$), makulol ($C_{20}H_{34}O$), cembrene A, and *d*-sesanin.

Therapeutic Uses

It was claimed that this resin can activate blood flow, relieve pain, and promote tissue regeneration.

LIAN MIAN ZHEN (兩面針) — dried root of *Zanthoxylum nitidum*

Chemistry

This root contains nitidine ($C_{21}H_{18}O_4N$), oxynitidine ($C_{21}H_{17}O_5N$), and vitexin ($C_{21}H_{20}O_{10}$).

Nitidine　　Oxynitidine

Therapeutic Uses

The herb is used as an analgesic. Traditional Chinese physicians also prescribed it to increase blood flow and promote the circulation of *qi*.

GOU MIN (鉤吻) — dried root and stem of *Gelsemium sempervirens* Ait. or *G. nitidum* Mich.

Chemistry

Several alkaloids have been isolated from this herb. They are gelsemine ($C_{20}H_{22}O_2N_2$), gelsemicine ($C_{20}H_{25}O_4N_2$), and sempervirine. Gelsemicine is the active principle and is very toxic.

The Chinese *Gou Min*, as listed in *Ben Cao Kong Mu*, actually belongs to the species *Gelsemium elegans* Benth. Zhou et al. reported that the alkaloids isolated from this Chinese species are different from those isolated from *C. sempervirens*. These include koumine ($C_{20}H_{22}ON_2$), kouminine, kouminicine, and kouminidine.

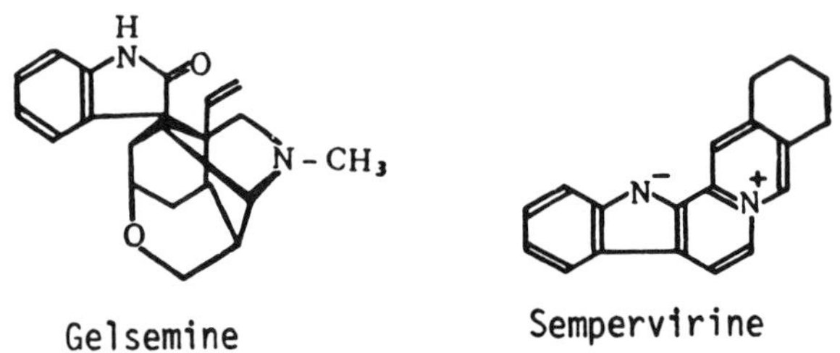

Gelsemine　　Sempervirine

Actions

Gelsemicine can cause salivation, tremors, and skeletal muscle paralysis. The latter, as shown in animal studies, can result in death from respiratory arrest.

Both gelsemine and gelsemicine can cause pupillary dilatation when applied locally on the eye. Koumine and kouminine, however, do not produce such effects.

Gelsemicine is very toxic. The MLD in mice is 0.1 to 0.12 mg/kg (subcutaneous or interperitoneal); in rabbits, it is 0.05 to 0.06 mg/kg (intravenous); in dogs, 0.5 to 1.0 mg/kg (intravenous). Koumine and kouminine are much less toxic.

Reference

Zhou, S. G. et al., *Chin. J. Physiol.*, 5, 131–140, 1931; *Chin. J. Physiol.*, 5, 345–352, 1931.

TABLE 1
Other Analgesic and Anti-Inflammatory Herbs

Name	Source	Actions: Analgesic	Actions: Anti-inflammation	Actions: Others
Du Huo (独活)	*Angelica pubescens* root	+	+	Sedative
Wu Jia Pi (五加皮)	*Acanthopanax gracilistylus* W. W. Smith	+	+	Ginseng-like effect, used as tonic and aphrodisiac
Hai Tong Pi (海桐皮)	*Erythrina variegata, E. aborescens*	+	+	Antirheumatic, diuretic, relieves edema
Xi Xian Cao (豨莶草)	*Siegesbeckia orientalis, S. pubesceae*	+	+	Antirheumatic, sedative
Hu Gu (虎骨)	Tiger bone	+	+	Sedative, promotes bone healing
Bai Hua She (白花蛇)	*Agkistrodon ocutus*	+		Sedative, vasodilatation

Chapter 12

ANTIPYRETIC HERBS

An antipyretic effect can be achieved by vasodilatation and sweating. This may be useful in treating febrile conditions, helping to lower body temperature towards normal levels. Two terms often confused, *Jie Re Yao* (解热药) and *Chin Re Yao* (清热药), are used in Chinese herbal medicine to describe antipyretic herbs. *Jie Re Yao* is a herb that actually lowers body temperature, while *Chin Re Yao* is one that allows the patient to experience a subjective feeling of coolness. The latter is used when the patient's body temperature is normal, but he feels hot; the patient may not necessarily suffer any infectious disease. These types of herb are popular items sold in Chinese drug stores or street corners, especially during the summer months, when stalls selling these herbs proliferate like lemonade stands in the U.S.

When the Divine Plowman tasted hundreds of plants in search of treatments for fever, he did not know that the primary cause of fever is infection. Therefore, neither *The Herbal Classic of the Divine Plowman* nor the *Ben Cao Kong Mu* distinguishes between herbs which are really antipyretic or those which lower temperature secondarily through an antibacterial effect.

In the past century, Chinese investigators have been mostly interested in examining plants that appear to contain active alkaloids or glycosides. Therefore, little is known about plants which contain simple salicylates or acetaminophen-like derivatives.

CHAI HU (柴胡) — the dried root of *Bupleureum chinense* D.C. or *B. scorzonerifolium*

Chemistry

This root contains the saponins saikoside Ia, Ib and II. In addition, it contains bupleurumol, essential oils, and fat. The genins of these saikosides are products with a sterol structure.

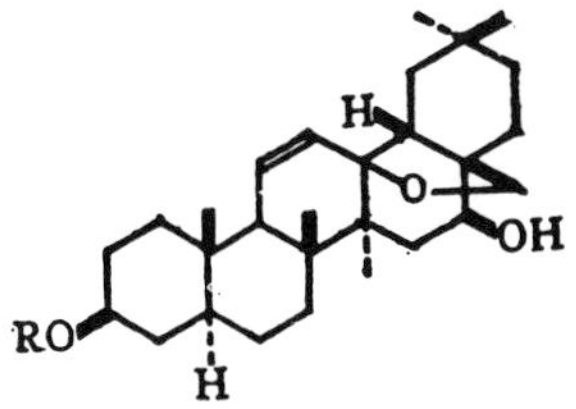

H O OH RO H

Saikosaponin C R = -glu-O-glucose (O) rhamnose

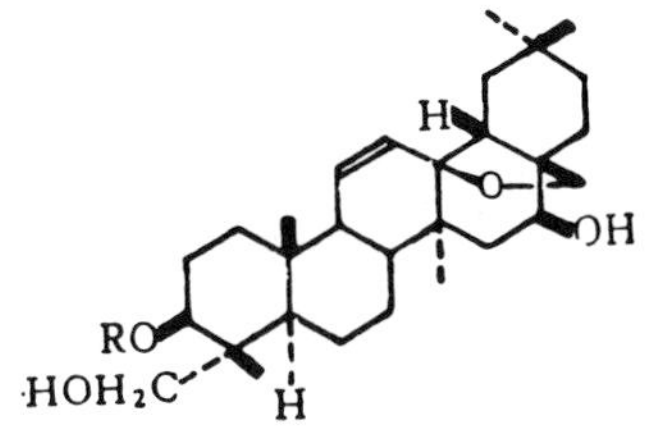

H O OH RO HOH_2C H

Saikosaponin A R = -fucose-O-glucose

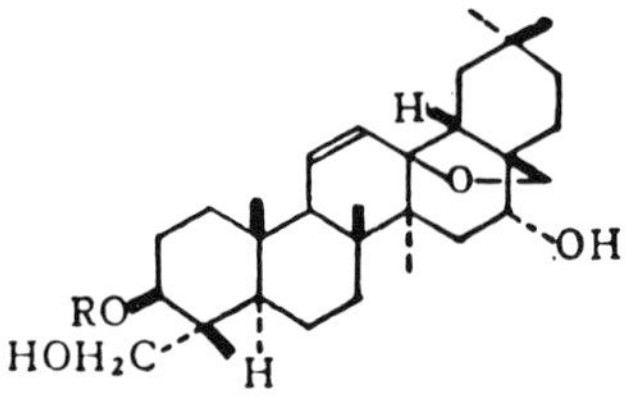

	R =
Saikosaponin D	-fucose-O-glucose

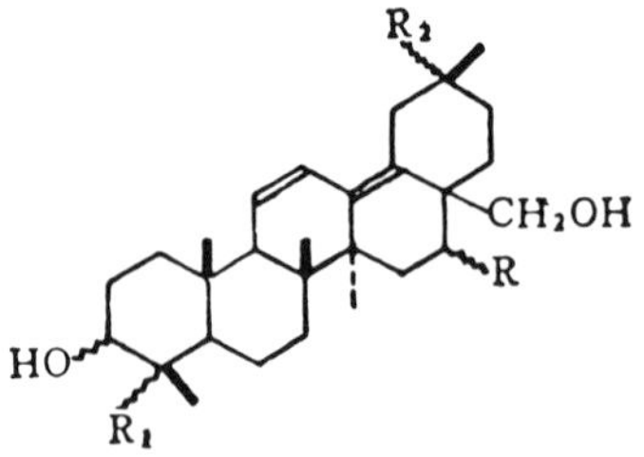

	R=	R_1=	R_2=
Saikosapogenin A	β-OH	α-CH_2OH	α-CH_3
Saikosapogenin C	β-OH	α-CH_3	α-CH_3
Saikosapogenin D	α-OH	α-CH_2OH	α-CH_3

Actions

This herb is primarily an antipyretic agent, exhibiting an inhibitory effect on the body's temperature center. It has sedative, analgesic, and antitussive actions. It has been used as both an antibacterial and as an antimalaria agent.

As an anti-inflammatory agent, *Chai Hu* can block the increase in capillary permeability induced by histamine or 5 hydroxytryptamine (5HT). As an antilipemia agent, it can lower plasma cholesterol and triglycerol levels. It has also been reported to protect liver functions and act as a choleretic.

Toxicity

Saikosides are potent hemolytic agents. They are very irritating if injected into subcutaneous tissue. Clinical trials with large doses of *Chai Hu* (2 g/d) showed symptoms of gastrointestinal irritation, abdominal distension, loss of appetite, constipation, and, occasionally, facial and extremity edema.

The LD_{50} in mice is 47 g/kg (oral), 1.9 g/kg (subcutaneous), 112 mg/kg (interperitoneal), and 70 mg/kg (intravenous).

Therapeutic Uses

Chai Hu has been used as a medicating agent and febrifuge to relieve depression of *qi* of the liver, and to boost *yang qi*.

The herb is an effective antipyretic. For almost any febrile condition (e.g., bacterial infection), injection of *Chai Hu* solution can bring body temperature down to normal levels within 24 h. No recurrences have been observed. It is commonly prescribed for children and pregnant women, especially in cases of common cold, influenza, and malaria. The herb is also used as an adjuvant with other herbs in the treatment of several infectious diseases.

For the treatment of dysmenorrhea, the herb is used in combination with the herbs *Dang Gui* (当归) and *Bai Shao* (白芍), with 12 g of each prepared in a decoction. Other uses of the herb include treatment of acute pancreatitis, pleuritis and neuralitis, gastritis, acute cholecystitis, and hepatitis.

FANG FENG (防风) — the dried root of *Ledebouriella divaricata* Hiroe (Umbelliferae)

Chemistry

Approximately 0.3 to 0.6% of the herb's content consists of essential oils, alcohol derivatives, and organic acids.

Actions

The herb has antipyretic, analgesic, and antibacterial effects. It has been used in the treatment of migraine headache, the common cold, and rheumatoid arthritis.

It is usually prescribed in a decoction together with the herb *Bai Zhi* (白芷), in proportions of 9 g and 6 g, respectively. The decoction is given orally.

JING JIE (荆芥) — the dried aerial parts of *Schizonepeta tenuifolia* (Labiatae)

Chemistry

Approximately 1 to 2% of the herb comprises essential oils. The main active principles are *d*-menthone and *d*-limonene.

Actions

This herb causes diaphoresis, lowers body temperature, and has an anticonvulsive effect. It has an antibacterial action and can increase blood coagulation.

Classic Chinese medical literature claimed that this herb was useful to expel ''wind'' and promote ''eruption''. It is used to lower body temperature in the treatment of the common cold. In addition, it is used to treat sore throats, conjunctivitis, and measles, and to stop bleeding.

SHENG MA (升麻) — dried rhizome of *Cimicifuga heracleifolia, C. dehurica,* or *C. foetida*

Chemistry

The rhizome contains acids and sterol derivatives, including ferulic acid ($C_{10}H_{10}O_4$), isoferulic acid, cimigenol, khellol ($C_{13}H_{10}O_5$), aminol ($C_{14}H_{12}O_6$), cimifugenol, and cimitin.

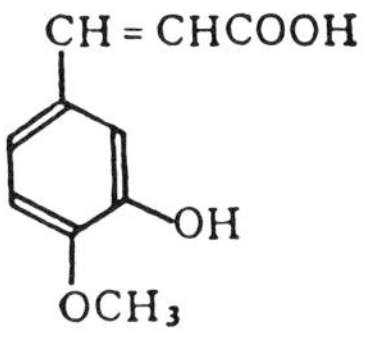

Isoferulic acid 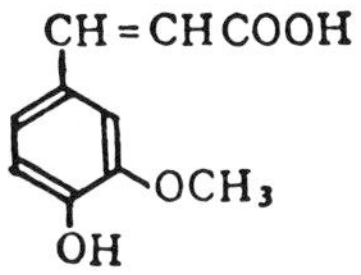　　Ferulic acid 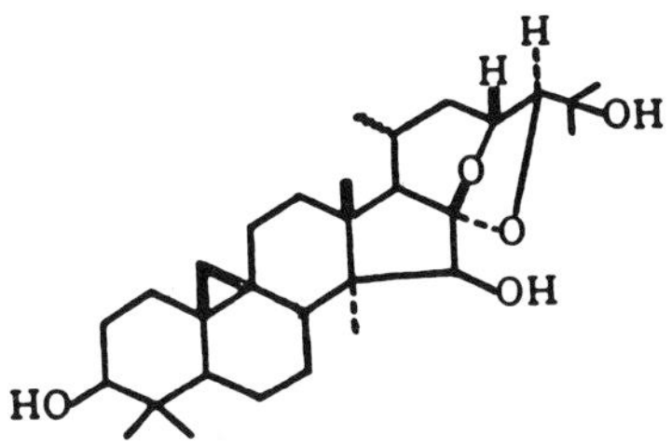　　Cimigenol

Actions

This herb was used in folk medicine to induce diaphoresis and promote skin eruption, remove ''heat'', counteract toxins, invigorate vital functions, and prevent ptosis.

A classic prescription formula for the herb is *Sheng Ma Ge Gen Tang* (升麻葛根湯), which is a soup made from *Sheng Ma, Ge Gen*, and licorice root. It is given to children at the onset of measles, particularly in cases with severe skin eruption.

HE YE (荷叶) — dried leaf of *Nelumbo nucifera* Gaertn.

Chemistry

The leaf contains several alkaloids, including nuciferine ($C_{19}H_{21}O_2N$), roemerine ($C_{18}H_{17}O_2N$), *O*-nornuciferine ($C_{18}H_{19}O_2N$), anonaine ($C_{17}H_{15}O_2N$), lirodenine ($C_{16}H_{19}O_2N$), dihydronuciferine ($C_{19}H_{19}O_2N$), pronuciferine ($C_{19}H_{21}O_3N$), anneparine ($C_{19}H_{23}O_3N$), *N*-methylcoclaurine ($C_{18}H_{21}O_3N$), and *N*-methylisococlaurine ($C_{17}H_{21}O_3N$).

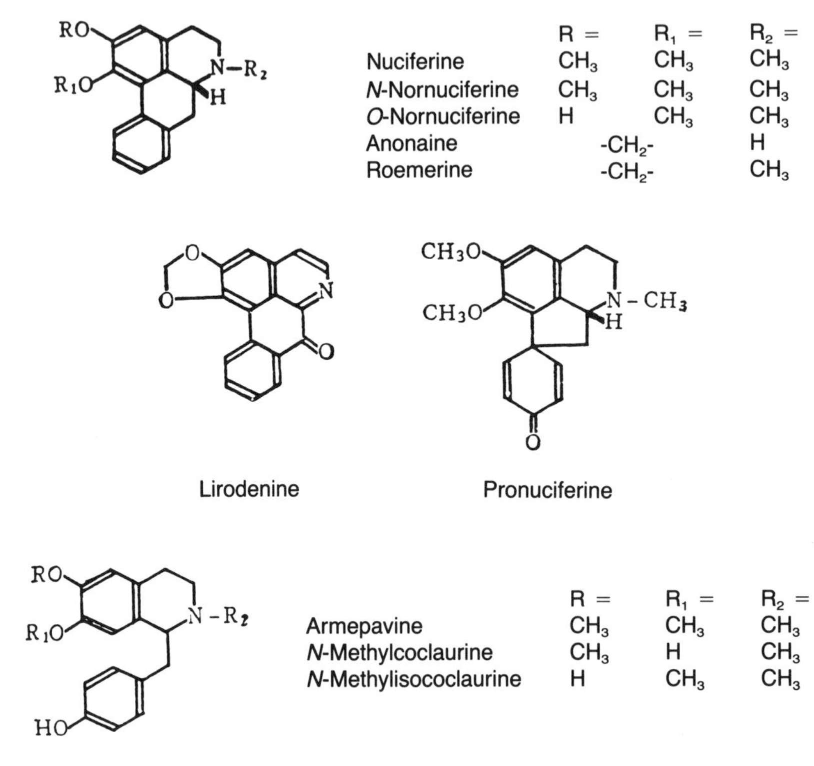

	R =	R_1 =	R_2 =
Nuciferine	CH_3	CH_3	CH_3
N-Nornuciferine	CH_3	CH_3	CH_3
O-Nornuciferine	H	CH_3	CH_3
Anonaine	-CH_2-		H
Roemerine	-CH_2-		CH_3

Lirodenine Pronuciferine

	R =	R_1 =	R_2 =
Armepavine	CH_3	CH_3	CH_3
N-Methylcoclaurine	CH_3	H	CH_3
N-Methylisococlaurine	H	CH_3	CH_3

Actions

The herb is used to disperse body heat during summers and is said to increase essential body energies, in particular those of the defensive systems. Rice wrapped in *He Ye* is a southern Chinese dish commonly served in the summertime.

The alkaloids of this leaf have a relaxing effect on smooth muscles.

LIAN ZI XIN (蓮子心) — the dried plumule and radicle in the seed of *Nelumbo nucifera* Gaertn.

Chemistry

The herb contains the following alkaloids: liensinine ($C_{37}H_{42}O_6N_2$), isoliensinine, neferine ($C_{28}H_{44}O_6N_2$), lotusine ($C_{19}H_{24}O_3N$), methylcorypalline ($C_{12}H_{17}O_7N$), and demethylcoclaurine ($C_{16}H_{17}O_3N$).

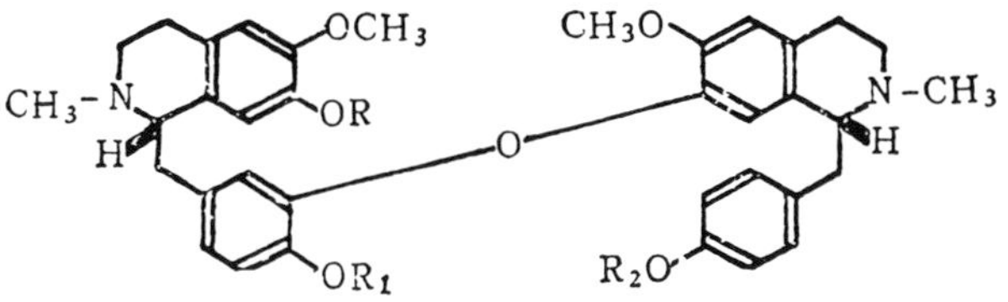

	R=	R_1=	R_2=
Liensinine	CH_3	H	H
Isoliensinine	H	H	CH_3
Neferine	CH_3	H	CH_3

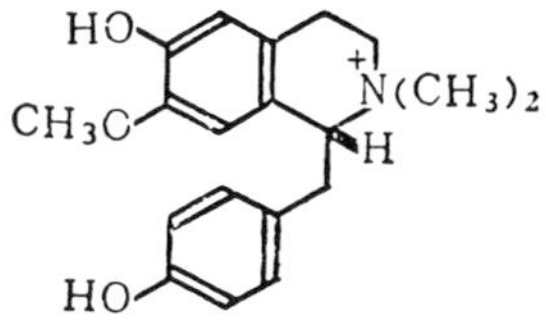

Lotusine

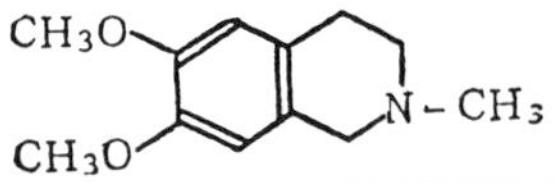

Methyl-corypalline

Actions

This herb was used to remove ‘‘heat’’. It has tranquilizing and antihypertensive effects.

HUANG TENG (黄藤) — dried stem of *Fibraurea recisa*

Chemistry

The stem contains the following alkaloids: palmatine ($C_{21}H_{22}O_4N$), jatrorrhizine ($C_{20}H_{20}O_4N$), fibramine ($C_{25}H_{28}O_7N$), fibraminine ($C_{18}H_{19}O_7N$), fibralactone ($C_{27}H_{28}O_9$), and sterol ($C_{27}H_{45}OH$).

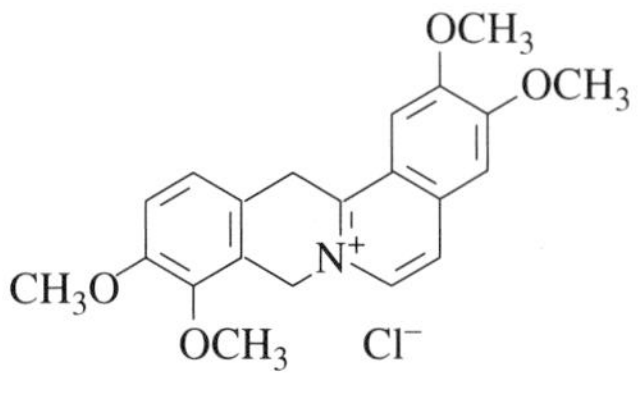

Palmatine Cl

Actions

The herb is used as an antipyretic and detoxicant. It is used therapeutically in the treatment of tonsillitis and pharyngitis.

HUANG LU (黄栌) — dried leaf and twig of *Cotinus coggygria cinerea*

Chemistry

The herb contains myricetin ($C_{15}H_{10}O_8$), myricitrin ($C_{21}H_{20}O_{12}$), fisetin ($C_{15}H_{10}O_6$), and fustin ($C_{15}H_{12}O_6$).

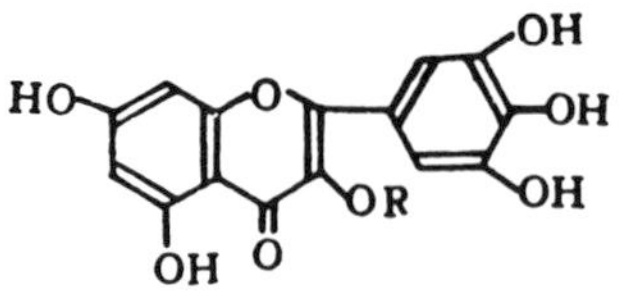

	R =
Myricetin	H
Myricitrin	-rhamnose

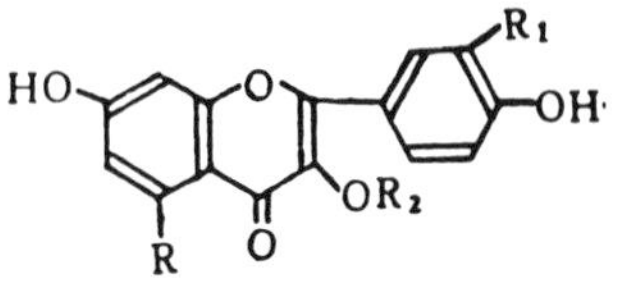

	R=	R_1=	R_2=
Isoquercetin-glucoside	OH	OH	-glucose
Fisetin	H	H	OH

Actions

Chinese traditional medicine prescribed the herb to eliminate "dampness" and "heat", and as an antipyretic.

LUO DE DA (落得打) or JI XUE CAO (积雪草) — dried whole plant of *Centella asiatica* (L.) Urbar.

Chemistry

The plant contains several saponin compounds, including asiaticoside ($C_{48}H_{78}O_{19}$), madecassoside ($C_{48}H_{98}O_{20}$), brahmoside ($C_{47}H_{78}O_{19}$), and brahmissoside ($C_{53}H_{88}O_{24}$). On hydrolysis, the following substances are produced: thankunic acid ($C_{30}H_{48}O_8$), isothankunic acid, and asiatic acid ($C_{30}H_{48}O_6$). The plant also contains the alkaloid hydrocotyline ($C_{23}H_{33}O_8N$).

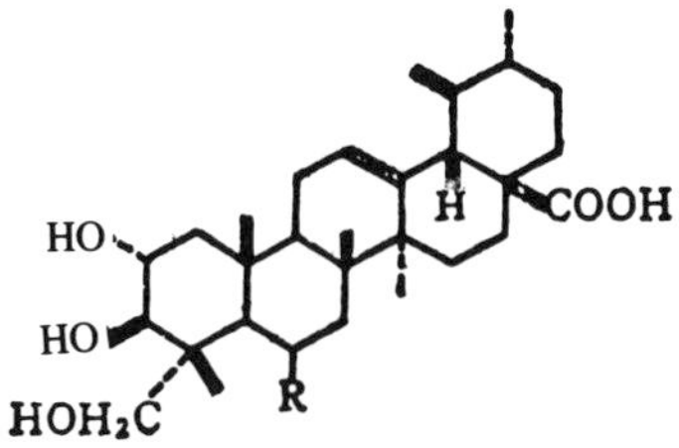

	R =
Asiatic acid	H
Brahmic acid	OH

Actions

This herb displays antibacterial properties. It is also able to lower blood pressure and slow the heart rate. It is used as an antipyretic, detoxicant, and diuretic.

BAI ZHI (白芷) — dried root of *Angelica dahurica* (Fisch. ex Hoffm.) or *A. taiwaniana* Boiss.

It was listed as one of the upper-class herbs in the *Herbal Classic*. In his article, Hu gave a detailed analysis of how to determine the botanical name of this herb as *Angelica*, which is closely related to *Dang Gui* (当归), *Angelica sinensis* (Chapter 27).

Chemistry

The whole plant contains essentials oils and the root contains various furocoumarin derivatives. Among them there are 0.2% Byak-angelicin, 0.2% Byak-angelicol and oxypeucedanine, imperatorin, isoimperatorin, phellopterin, xanthotoxine, marmesin, scopoletin, neobyakangelicol, etc. The structures of these compounds are illustrated.

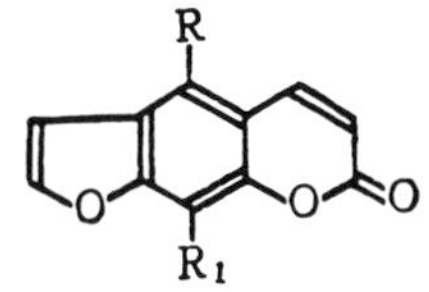

	R =	R_1 =
Byak-angelicine	OCH_3	$O-CH(OH)-CH(OH)-CH(CH_3)_2$
Byak-angelicol	OCH_3	$O-CH_2-CH(-O-)C(CH_3)_2$
Oxypeucedanine	$O-CH_2-CH(-O-)C(CH_3)_2$	H
Imperatorin	H	$O-CH_2-CH=C(CH_3)_2$
Isoimperatorin	$O-CH_2-CH=C(CH_3)_2$	H
Phellopterin	OCH_3	$O-CH_2-CH=C(CH_3)_2$
Xanthotoxine	H	OCH_3
Anhydrobyakangelicin	OCH_3	$O-CH_2-C(=O)-CH(CH_3)_2$
Neobyakangelicol	OCH_3	$O-CH_2-CH(OH)-C(CH_3)=CH_2$

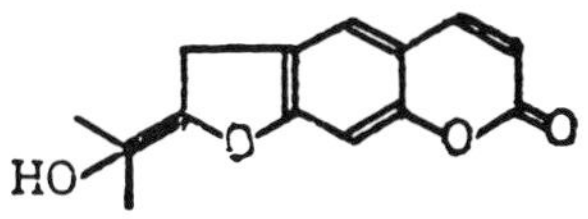

Marmesin

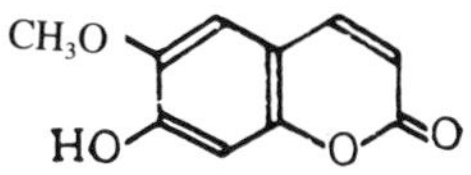

Scopoletin

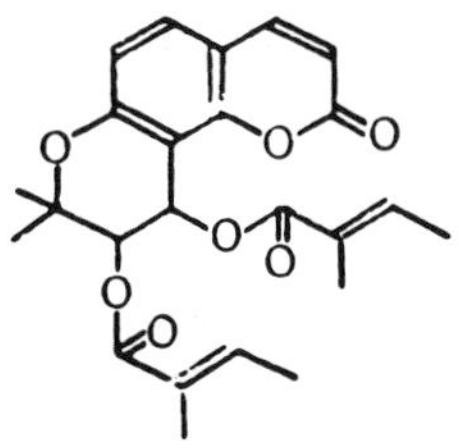

Anomalin

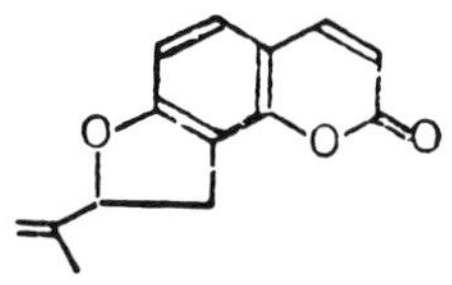

Angenomalin

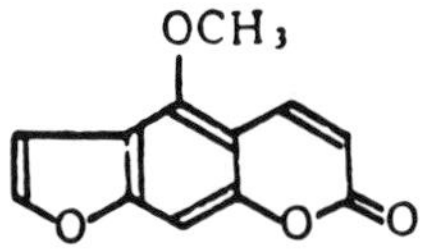

Bergapten

Actions

This herb possesses antipyretic and analgesic action. It is also antibacterial. The decoction of the herb can inhibit the growth of *Escherichia coli*, *B. dysenteriae*, and typhoid bacilli. It was reported that Byak-angelicin can dilate the coronary vessels and increase the coronary circulation.

Therapeutic Uses

Bai Zhi was used in folk medicine to treat toothache and headache. It was used externally in cases of mastitis and wound infection.

Reference

Hu, S. Y., in *Advances in Chinese Medicinal Materials Research,* Chang, H. M. et al., Eds., World Scientific Publishing Co., Singapore, 1985, 17–33.

GAO BEN (藁本) — dried rhizome and root of *Ligusticum sinense* or *L. jeholense*

Chemistry

This herb contains nothosmyrnol ($C_{11}H_{14}O_2$).

Actions

Gao Ben was used to induce diaphoresis, dispel cold and "wind", to relieve headaches, and to treat dermatitis.

Table 1 shows other herbs that were described in the Chinese pharmacopoeia as able to remove intensive heat and dyphoria and to quench thirst.

TABLE 1
Other Antipyretic Herbs

Agent	Source	Part	Ingredient	Effect and uses
Gui Zhi (桂枝)	*Cinnamomum cassia*	Twig	Cinnamic aldehyde, cinnamyl acetate	Antibacterial, vasodilatation
Shi Gao (石膏)	*Gypsum fibrosum*		$CaSO_4 \cdot 2H_2O$	
Dan Zhu Je (淡竹叶)	*Lophatherum gracile*	Aerial part	Arundoin, cylindrin, friedelin	Antipyretic, diuretic, antibacterial
Zhi Zi (栀子)	*Gardenia jasminoides*			Antipyretic
Zhi Mu (知母)	*Anemarrhena asphodeloides* Bge.	Root	Sarsasapogenin, markogenin, neogitogenin	Antipyretic, antibacterial
Zi Su (紫苏)	*Perilla frutescens* L.	Leaf	*l*-Perilla aldehyde, alcohol	Antibacterial, antitussive
Giang Huo (羌活)	*Notopterygium incisum* or *N. forbesii*	Root, rhizome		Antirheumatic
Sheng Jiang (生姜)	*Zingiber*		Essential oil	Anti-inflammatory, stimulates gastric secretion
Niu Bang (牛蒡)	*Arctium lappa*	Fruit		Antibacterial, relieve sore throat
Ju Hua (菊花)	*Chrysanthemum morifolium* Ramat. or *C. boreale*	Flower	Chrysanthemin, stachydrine, acactin	Antibacterial, coronary dilatation (see Chapter 5)

Chapter 13

ANTIRHEUMATIC HERBS

Rheumatism, in ancient Chinese terminology, was referred to as ''wind'' and ''dampness'', and was treated symptomatically. The medical classics described the group of herbs discussed in this chapter as drugs that can dispel ''wind'', repel ''dampness'', dispose of coldness, help reverse paralysis, and relieve pain. We classify these antirheumatic herbs together with agents which act on the nervous system because their primary actions are to relieve pain; they can be considered analgesic agents.

XU CHANG QING (徐長卿) — the dried root of rhizome of *Cynanchum paniculatum*

Chemistry

This herb contains paeonol, glycosides, and organic acids.

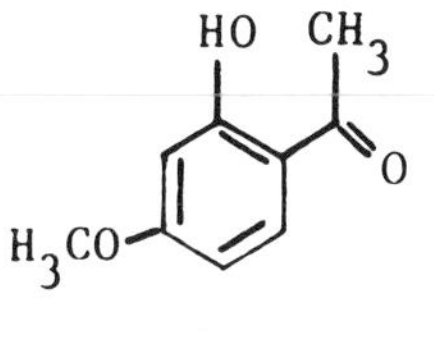

Paeonol

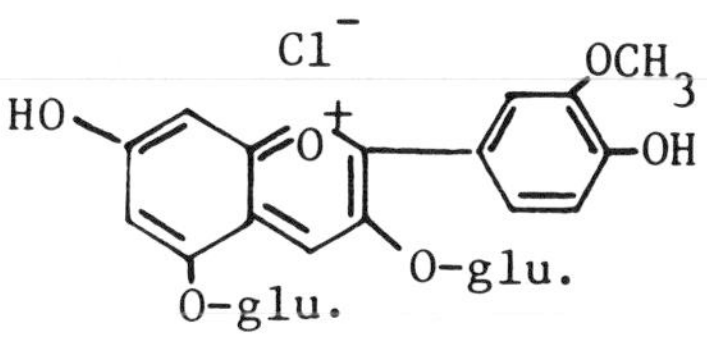

Paeonin

Actions

The herb is a sedative and analgesic, with a very long duration of action. Its effects on the cardiovascular system include a slowing of the heart rate, an increase of the PQ interval, improvement of myocardial circulation, and a lowering of blood pressure. Additionally, it lowers plasma cholesterol levels, and exhibits some antibacterial properties.

Therapeutic Uses

According to *The Herbal Classic of the Divine Plowman*, this herb ''antagonizes toxic devils and repels ugly *qi*.'' It was used in folk medicine to dispel ''wind'', relieve pain, and induce diuresis. It was mainly used to treat rheumatic arthritis and to relieve pain in cases of surgical wounds, trauma, toothache, and dysmenorrhea. It has also been used in cases of mastitis, skin infection, eczema, neurodermatitis, contagious dermatitis, and shingles.

The standard preparation of the herb is 30 to 60 g made into a decoction or extracted with wine, for oral administration; the extract can also be applied externally to the skin to relieve pain. An injection solution of the herb is also used, containing 50 mg paeonal per ml; this is administered in doses of 1 to 2 ml by intramuscular injection in the treatment of arthritis.

CANG ER ZI (苍耳子) — dried fruit of *Xanthium sibiricum* (Compositae)

Chemistry

The fruit contains the glycoside xanthostrumarin, the alcohol derivatives xanthanol and isoxanthanol, the fat xanthumin, and some proteins, alkaloids, and vitamins.

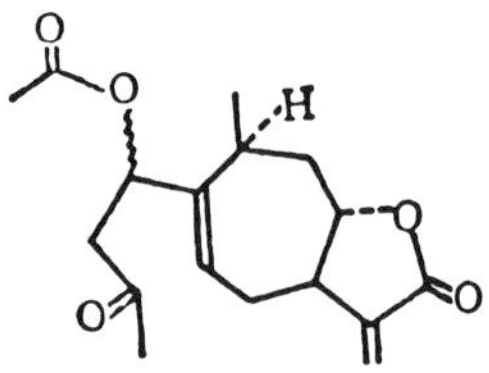

Xanthinin

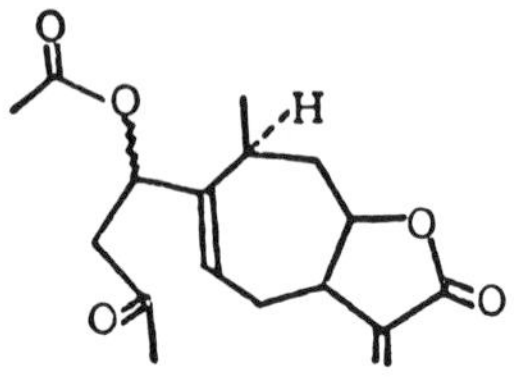

Xanthumin

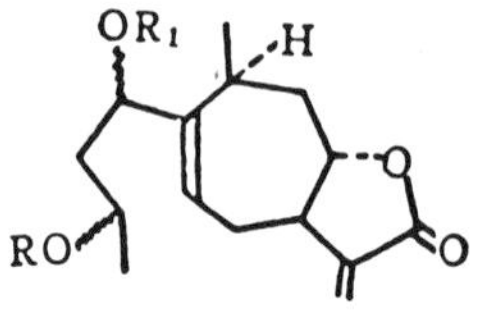

	R=	R_1=
Xanthanol	H	$COCH_3$
Isoxanthanol	$COCH_3$	H

Actions

This herb has antibacterial and antitussive properties. It has a respiratory stimulating effect and can lower blood pressure and blood sugar levels.

Cang Er Zi is quite toxic. The LD_{50} in mice is 0.93 g/kg (i.p.). The toxic symptoms include irregular respiration, dyspnea, and coma. This herb can cause abnormalities in hepatic and renal function. Patients taking the herb may complain of malaise, headache, and gastrointestinal disturbances. In severe overdosage, it causes cardiac arrhythmias, swelling of the liver, hematuria and oliguria, and coma.

Therapeutic Uses

The Herbal Classic of the Divine Plowman stated that this herb had "major effects in the relief of headaches, dispelling "wind" and "dampness", treatment of paresthesias and necrotic muscle . . . " It is used to treat hypersensitivity rhinitis, neurotic headache, and rheumatic arthritis. It is effective in treating pain in the extremities, sciatic neuralgia, eczema, pruritis, chronic sinus infections, otitis media, and paroditis.

QIN JIU (秦艽) — the dried root of *Gentiana macrophylla, G. straminea, G. crassiaulis*, or *G. dahurica*

Chemistry

The root contains essential oils and several alkaloids, including gentianine, gentianidine, and gentianol.

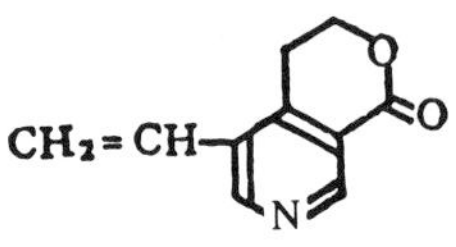

CH_2=CH–

Gentianine

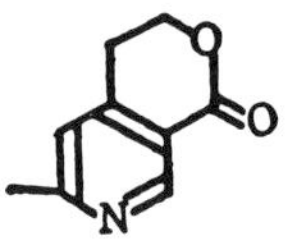

Gentianidine

CHO

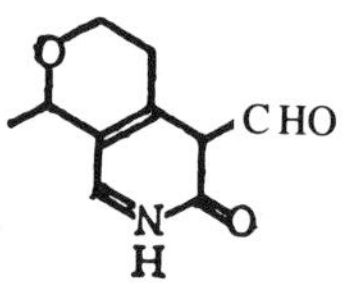

Gentianol

Actions

According to *The Herbal Classic of the Divine Plowman*, this herb was used in the treatment of rheumatism and fever due to deficiency of *qi* and blood and to cause diuresis and remove water.

The herb has been shown to be an antipyretic analgesic, and to produce an anti-inflammatory effect. Gentianine is about twice as potent as aspirin. Its anti-inflammatory effect is due to stimulation of adrenocorticotropic hormone (ACTH) secretion, resulting in an increase of corticosterone.

In addition, the herb has antihypersensitivity and antihistaminic effects. It can produce sedation and hypnosis, and lowers both the heart rate and blood pressure. It also has an antibacterial effect.

At a regular oral dose of 100 mg, gentianine often causes nausea and vomiting, and other symptoms of gastrointestinal disturbance. The LD_{50} of this alkaloid in mice is 480 mg/kg (oral), 250 mg/kg (intravenous), and 350 mg/kg (interperitoneal).

WEI LING XIAN (威灵仙) — the dried root and rhizoma of *Clematis chinensis, C. hexapetala*, or *C. manshurica* (Ranunculaceae)

Chemistry

This herb contains anemonin, anemonol, and other saponins.

Actions and Therapeutic Uses

The actions of the herb include analgesia, diuresis, and antibacterial effects. It was prescribed in traditional medicine to dispel "wind" and remove "dampness", and to relieve pain. It is used in the treatment of the common cold, tonsillitis, acute icteric hepatitis, rheumatic arthritis, and laryngitis.

The herb is decocted in doses of 15 to 60 g, taken as a drink or applied externally to swollen tissues.

XUE SHANG YI ZHI GAO (雪上一枝蒿) — the dried root of *Aconitum brachypodum* (Ranunculaceae)

Chemistry

This herb contains bullatine A, B, C, and D, and aconitine.

Actions

The root and aconitine have an analgesic effect. The onset of action is slow, but the duration of action is long. This herb was used to relieve pain, activate blood circulation, and reduce swelling. It is commonly used to relieve traumatic and arthritic pain.

The herb is toxic. The LD_{50} in mice is 1.02 ± 0.18 g/kg. It is contraindicated in pregnancy and in children.

SAN HAI TON (山海棠) or ZI JIN PI (紫金皮) — the whole plant of *Tripterygium hypoglaucum* (Levl.) Hutch.

Chemistry

The plant contains several alkaloids as well as triptolide.

Actions

It produces anti-inflammatory and antiswelling effects through stimulation of the ACTH-corticosterone axis. It is used in the treatment of rheumatic arthritis. Standard preparations include a 0.2-g tablet, which is equivalent to 13 g of raw material; a 20% tincture; and a 10% ointment, for external application.

CHUAN SHAN LONG (穿山龙) — the dried rhizome of *Dioscorea niponica* Makino

Chemistry

The rhizome contains dioscin, which on hydrolysis gives diosgenin and trillin.

Actions

The saponin components of this herb have an anti-inflammatory action similar to cortisol. It also has antitussive, expectorant, and antiasthmatic effects.

Therapeutic Uses

In traditional medicine, this herb was used to dispel "wind" and remove "dampness", to promote blood circulation, relieve pain, expel phlegm, and arrest coughing. It is claimed to be effective in the treatment of arthritic pain, rheumatic endocarditis, sciatic neuralgia, and chronic bronchitis.

The herb is available in tablet, tincture, and injection form. For the treatment of chronic bronchitis, 0.5-g tablets are prescribed in a dose of 4 tablets t.i.d. For other ailments, a tincture of 60 g in 500 ml of wine is prepared, taken in doses of 10 to 15 ml t.i.d. Standard injection solution of the herb is 2 ml per ampule, equivalent to 1 g of raw material; the normal dose is 2 to 4 ml by intramuscular injection once a day.

GUO GANG LONG (过岗龙) — dried stem of *Entada phaseoloides* (L.) Merr.

Chemistry

The stem contains entageric acid ($C_{30}H_{48}O_6$), which has a sterol structure.

Actions

It is used as an antirheumatic, to promote collateral flow, and to relieve blood stasis.

TOU GU CAO (透骨草) — the whole plant of *Impatiens balsamina* L.

Chemistry

This herb contains gentisic acid ($C_7H_6O_4$), ferulic acid ($C_{10}H_{10}O_4$), *p*-coumaric acid ($C_9H_8O_3$), sinapic acid ($C_{11}H_{12}O_5$), caffeic acid ($C_9H_8O_4$), scopoletin ($C_{10}H_8O_4$), and lawsone ($C_{10}H_6O_3$).

Actions

Tou Gu Cao is used to treat arthritis and relieve pain.

Chapter 14

CENTRAL STIMULATING HERBS

MA QIAN ZI (马錢子) — the dried seed of *Strychnos pierriana* or *S. nux-vomica* (Loganiaceae)

Chemistry

By weight, the seed contains approximately 2 to 5% alkaloids. The major ones are strychnine and brucine (approximately 1 to 1.4% each). It also contains small amounts of vomicine, pseudostrychnine, pseudobrucine, *N*-methylsecpseudobrucine, and novacine.

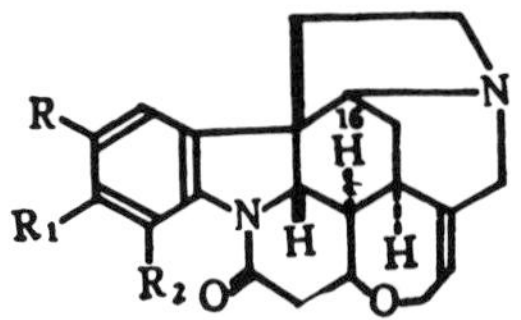

	R=	R_1=	R_2=
Strychnine	H	H	H
Brucine	OCH_3	OCH_3	H
α-Colubrine	H	OCH_3	H
β-Colubrine	OCH_3	H	H

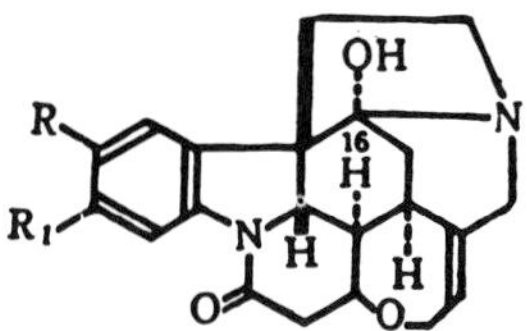

	R=	R_1=
Pseudostrychnine	H	H
Pseudobrucine	OCH_3	OCH_3
16-Hydroxy-α-colubrine	H	OCH_3
16-Hydroxy-β-colubrine	OCH_3	H

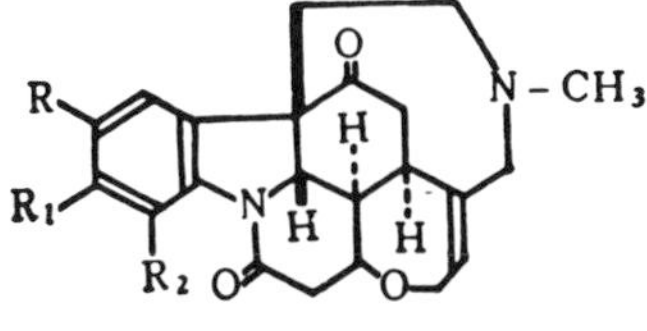

	R=	R_1=	R_2=
Vomicine	H	H	OH
Novacine	OCH_3	OCH_3	H
Icajine	H	H	H

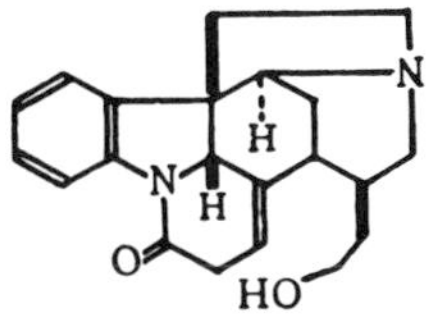

Isostrychnine

Actions

This herb inhibits Renshaw cells, the interneuronal cells, of the spinal cord. This results in an increase of central nervous system (CNS) reflex stimulation. The herb also stimulates the cortex, increasing visual, auditory, and smelling sensations.

It has been observed to stimulate gastric secretion, acting as a stomachic. It is also effective as an expectorant, and as an antibacterial agent.

This herb is toxic. In children, 50 mg of the seed, or 1 g in adults, can cause intoxication. This is manifested by orthotonic contractions of the extremities and shallow respiration. In extreme cases, paralysis, convulsions, and coma can occur.

The LD_{50} of the seed in mice is 155 to 300 mg/kg (oral). The LD_{50} of brucine in mice is 3.03 mg/kg (oral).

Therapeutic Uses

Strychnine nitrate has been shown to improve patient conditions in cases of childhood paralysis, sciatic neuralgia, and neurasthenia; the standard dose is 1 to 3 mg administered subcutaneously.

"Resurrection pills" (九转回生丹), which contain deep-fried *Strychnos* seeds plus *Di Long* (地龙), are used to treat hemiplegia, muscular weakness, and impotence. Additionally, the herb has been used in cases of hypertrophic vertebral inflammation and chronic anemia.

YI YE CHAU (一叶萩) — the leaf, flower, and young branch of *Securinega suffruticosa* (Pall.) Rehd.

Chemistry

The herb contains securinine (0.12 to 1%) and dihydrosecurinine and others.

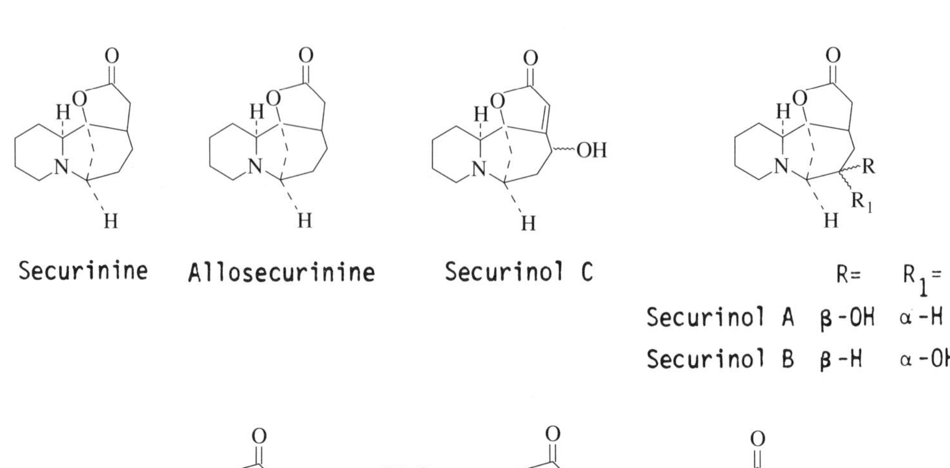

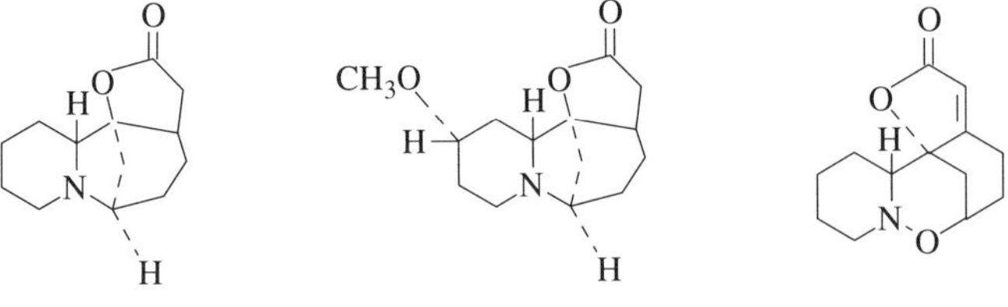

Actions

Securinine is similar to strychnine in that it can stimulate the spinal cord, but is less potent and has a shorter duration. It can also stimulate the midbrain and cerebral cortex. Small doses can increase myocardial contractility and increase the heart rate. Inhibition of cholinesterase results in an increase in salivary secretion; in this action, it is less potent than physostigmine.

Securinine is rapidly deposited in the intestine and only about 20% of the administered dose is absorbed into the body fluids. After absorption, the compound is swiftly destroyed and plasma concentration falls rapidly.

Toxic symptoms include tremors of the extremities and rigid contractions. Death can occur from orthotonic convulsions. The LD_{50} of securinine in mice is 19.3 mg/kg (intramuscular), 20.4 mg/kg (subcutaneous), 31.8 mg/kg (interperitoneal), and 270 mg/kg (oral).

Therapeutic Uses

The herb is used to treat paralysis, neuralgia, recurrent neuritis, chronic sciatic nerve inflammation, neurasthenia, depression, and schizophrenia. It is occasionally used in the treatment of menopausal syndrome.

Preparations of the herb include 4 mg securinine tablets and injection ampules containing either 4 mg/ml or 16 mg/ml.

SHI SUAN (石蒜) — the dried rhizome of *Lycoris radiata* (L'Her.) Herb. or *L. longituba* Y. Han et Fan and *L. aura* (L'Her.) Herb.

Chemistry

This herb contains more than 10 different alkaloids, including galanthamine, lycoremine, lycorine, lycoramine, lycorenine, tazettine, pseudolycorine, dihydrolycorine, homolycorine, lycoricidine, and lycoricidinol.

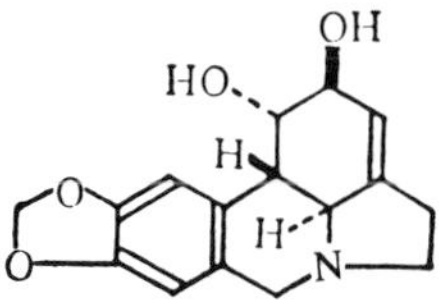

Lycorine 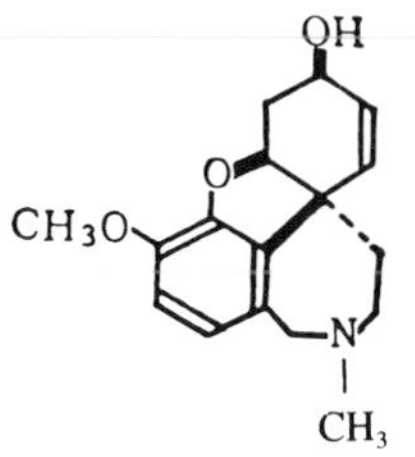 Galanthamine 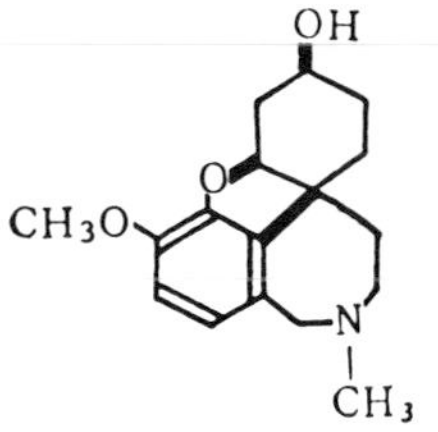 Tazettine

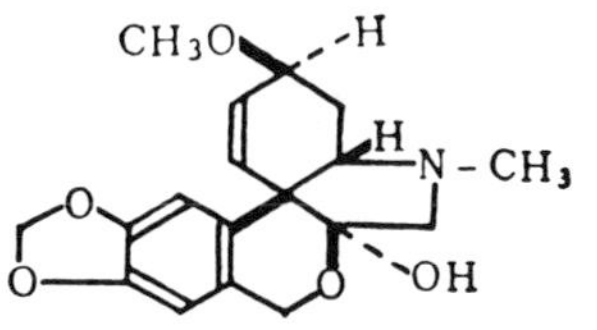

Lycoramine 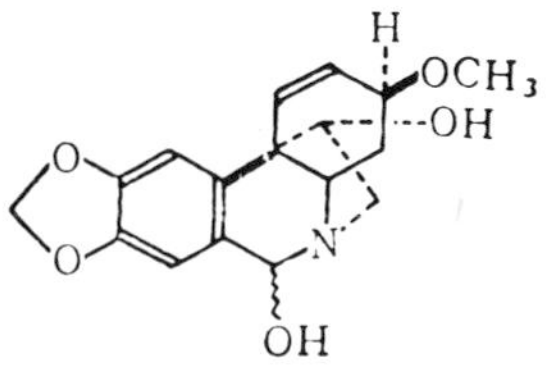 Haementhidine

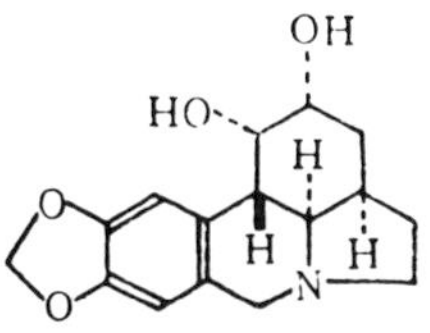

Zephyranthine 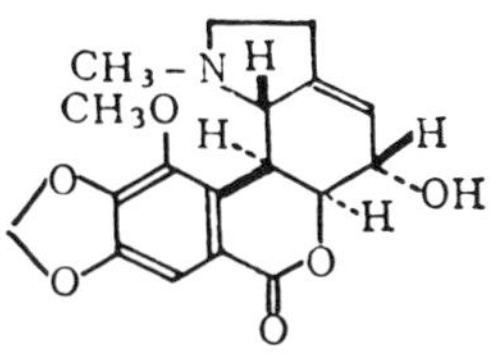 Nerinine

Actions

Galanthamine and lycorine are cholinesterase inhibitors, but are less potent than physostigmine. They can increase skeletal muscle contraction. They inhibit the CNS and produce sedative and hypnotic effects. These alkaloids also have an analgesic effect and potentiate the action of morphine.

Wang et al showed that lycorenine can lower blood pressure through a central effect. The oxidation product of lycorine and oxylycorine (AT-1840), shows anticancer properties. It can inhibit the growth of sarcoma S-180 and ascites cells, and has increased the survival rate of mice inoculated with these cancer cells.

The herb can stimulate secretion from the pituitary gland and the adrenal cortex, resulting in an increase of antidiuretic hormone and corticosterone secretion. It has anti-inflammatory properties, and has been observed to stimulate uterine contraction. It has an emetic effect slightly weaker than that of morphine.

Toxicity

This herb is irritating. When in contact with the skin, it causes itching and swelling. If it gets into the respiratory tract, it can result in nose bleeds.

Symptoms of overdosage include salivation, nausea and vomiting, slowing of the heart, tremors, and convulsions. Death can occur due to depression of the respiratory center. Immediate gastric lavage and injection of an antidote, such as atropine, are necessary to counteract intoxication.

Therapeutic Uses

This herb was used to treat childhood paralysis, muscle weakness, and rheumatic arthritis. It is also used as a rat killer and larva killer.

Standard preparations include 5-mg galanthamine tablets, administered in doses of two tablets t.i.d. Galanthamine can be injected intramuscularly in solutions of 1 or 2.5 mg/ml, with a total dose ranging from 2.5 to 10 mg. A *Shi Suan* powder is also available for use as an emetic; the normal dose for this purpose is 30 to 50 g, taken orally.

References

Wang, H., Gu, T. H., et al., *Acta Pharmacol. Sin.,* 1, 30–34, 1980.

CHANG NAO (樟脑) — camphor, the root, branch, and leaf of *Cinnamomum camphora* L. Presl.

Chemistry

The purified active substance isolated from this plant is *d*-camphor.

Actions

The primary action of camphor is the stimulation of higher centers of the CNS. Large doses might provoke convulsions, though the herb has little effect on the respiratory center. Its metabolic product, oxacamphora, has a remarkable cardiac stimulating effect and can raise blood pressure.

The herb also relaxes gastrointestinal muscle contractions and cools the skin, relieving itching and pain. It has some antibacterial properties.

Therapeutic Uses

Camphor is generally used to stop vomiting. The standard dose is 0.1 to 0.2 g administered orally, in capsule form. A 20% camphor injection solution is also used, administered subcutaneously or intramuscularly to stimulate respiration. Camphor ointment or lotion is commonly used to rub swollen areas, producing a cooling effect.

CHA (茶) or TEA — the dried leaf of *Camellia sinensis* O. Kze.

The natural habitat of this plant is south of the Yangi River, east from the Zhejiang Province of China, to Assam-Burma in the west, including the hills of Thailand and Vietnam.

History shows records of the use of *Cha* in China as far back as 2700 B.C. It is speculated, though not confirmed, that *Sheng Nung*, the Divine Plowman, may have been the first to taste *Cha* and introduce the practice of boiling it in water for health reasons. In any case, *The Herbal Classic of the Divine Plowman*, written in 100 B.C., certainly listed the plant among its many herbs.

China's tea-drinking culture grew up gradually from the 3rd century A.D. Both Buddhists and Taoists promoted the use of the drink, the former to keep awake during long meditations and the latter because of a belief that *Cha* could improve health and longevity.

Japan began to grow tea in its remote islands around 800 A.D., and around 1840, the Duchess of Bedford introduced the afternoon tea to British society. After that, the British brought the seeds of the tea plant to India and Ceylon (Sri Lanka). Under the supervision of the British, using modern scientific cultivation techniques, production was enlarged substantially.

At present, *Cha* production in China is the second largest in the world, at about 600 million lb/year, only slightly below India's about 672 million lb. Only 70 million lb of *Cha* are exported from China, however. The rest (88%) is consumed by the Chinese, not as a medicinal herb, but as a daily beverage.[3]

Commercially, there are three kinds of *Cha*: black, green, and oolong. The Chinese, however, use much more sophisticated classification systems to identify from 165 kinds of *Cha* (1985 data) up to 330 kinds (1988 data). A tea store in China generally carries over 100 different kinds of *Cha*.

Chemistry

Tea leaf comprises 1 to 5% alkaloids, including caffeine, theophylline, theobromine, xanthine. Tannic acids constitute another 9.5 to 21% of the leaf. Tea leaf also contains other elements, including metals of nutritional import. Figure 1 summarizes data on different elements which have been analyzed from tea leaf.

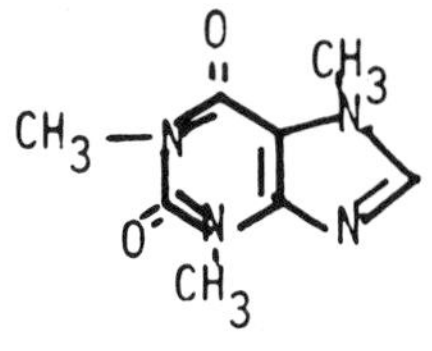

Caffeine 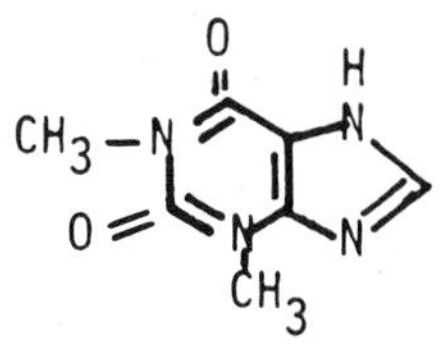 Theophylline 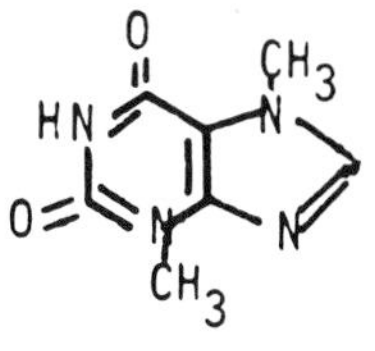 Theobromine

Actions

Cha stimulates the CNS, first at the cortical level, and then continuing to the medulla as the dose is increased. Theophylline and caffeine, which is less potent, have an inhibitory effect on phosphodiesterase, indirectly increasing cAMP level in the cytoplasm. This increase in intracellular cAMP results in a stimulation of myocardial contraction and an increase in heart rate.

The herb has a diuretic effect. Theophylline is the most potent of the three xanthine derivatives in the inhibition of tubular Na- and Cl-ion reabsorption. In addition, it increases

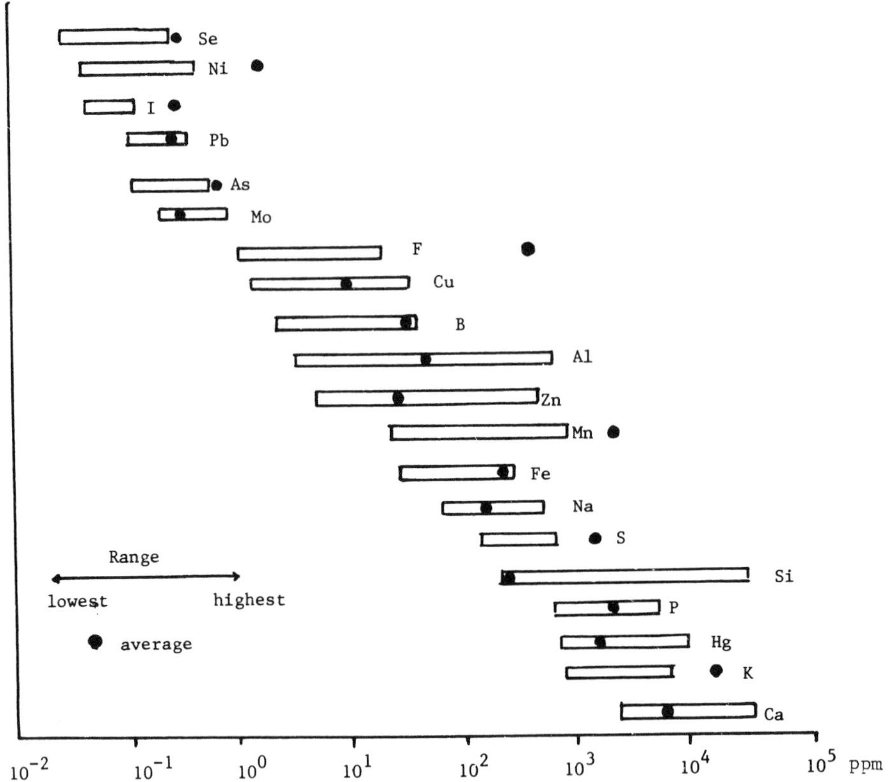

FIGURE 1. Elements from the tea leaf. (From Chen, C. S., *Tea Bull.*, 4, 1–10, 1990. With permission.)

renal blood flow. Theophylline is also the most potent of the xanthines in its ability to relax smooth muscles, causing bronchial dilatation and increasing coronary blood flow.

Table 1 compares the effects of these three xanthine derivatives on the CNS, heart, and kidney.

Several Japanese investigators have studied *Cha's* anticancer actions. They researched the effect of tea extract or tea catechin on ascites tumor or solid tumor in rats and mice. Their work indicated that there is a significant inhibitory effect and prolongation of survival time.

Wang et al.[5] reported that the green tea polyphenol can protect the polycyclic aromatic hydrocarbon-induced skin tumor initiation in mice. Yen et al.[6] showed that the green tea extract does exert a definite inhibitory effect on the stomach carcinoma cells, by inhibiting their growth from the G-1 stage to the S stage. Lou et al.,[4] from the same laboratory, extracted a substance from green tea, T-8750, and showed that the substance exerts a dose-dependent inhibitory or necrotic effect on human stomach carcinoma cells.

The herb has also been found to contain an antioxidant which exerts anti-X-irradiation effects. One species of *Cha, Tuo Cha*, is reputed to lower hypercholesteremia.

The tannic acid of *Cha* is a strong astringent agent. When a strong tea is consumed, the tannic acid can precipitate proteins in the alimentary mucosa, forming a proteinacious membrane which protects the mucosa membrane from ulceration. This also leads to an interference with the intestinal absorption of nutrients and vitamins.

Lastly, *Cha* has even been shown to help in the prevention of tooth decay. The fluoride content of tea leaf varies from 160 to 660 μg/kg dry weight. Studies of the water extract

TABLE 1
Effect of the Three Xanthine Derivatives

	Major source	Stimulation of		Diuretic effect
		CNS	Myocardium	
Theophylline	Tea	Moderate	Potent	Potent
Caffeine	Coffee	Potent	Moderate	Moderate
Theobromine	Coconut	Slight	Minimum	Slight

from ten different commercial teas showed that there is a definite antibacterial and anticaries activity.

Therapeutic Uses

Purified caffeine is given to stimulate the CNS, especially in cases of overdose of hypnotic or anesthetic agents. Theophylline is effective in the treatment of coronary disease, asthma, and biliary colic.

Cha has been used in combination with vitamin C to protect against X-irradiation. Clinical trials showed that it has an effective rate of 94% in the treatment of bacterial dysentery. This is probably due to its high content of tannic acid.

Soaked tea leaf or strong tea extracts are used widely in Chinese households to treat burn wounds, because of its astringent effects.

References

1. **Chen, C. S.,** *Tea Bull.* (in Chinese), 4(1), 1–10, 1990.
2. **Chow, K. and Kramer, I.,** *All the TEA in China,* China Books and Periodicals, San Francisco, 1990.
3. **Kramer, I.,** *China Today,* 40(3), 44, 1991.
4. **Lou, M. C. et al.,** *Tea Sci.* (in Chinese), 9(2), 172–174, 1989.
5. **Wang, Z. Y. and Klan, W. A. et al.,** *Chem. Abstr.,* 110, 18232b, 1989.
6. **Yen, Y. S. et al.,** *Tea Sci.,* 9(2), 161–166, 1989.

AN XI XIANG (安息香) — balsam obtained from *Styrax tonkinensis* or *S. hypoglaucus*

Chemistry

This herb contains sumaresinolic acid ($C_{30}H_{48}O_4$), coniferyl alcohol ($C_{18}H_{12}O_3$), styracin ($C_{18}H_{16}O_2$), vanillin ($C_8H_8O_3$), α-phenylpropyl cinnamate ($C_{18}H_{18}O_2$), and balsamic acid.

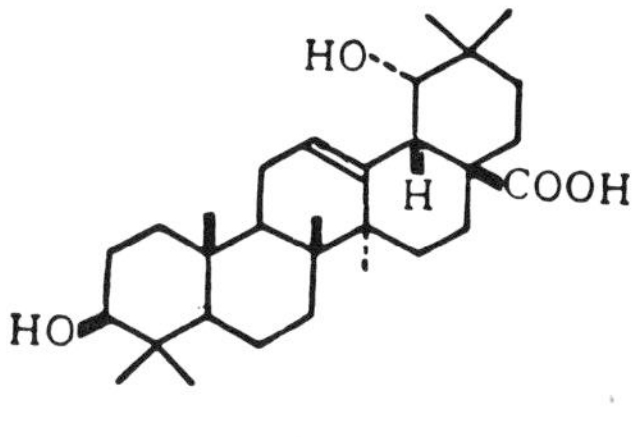

Siaresinolic acid

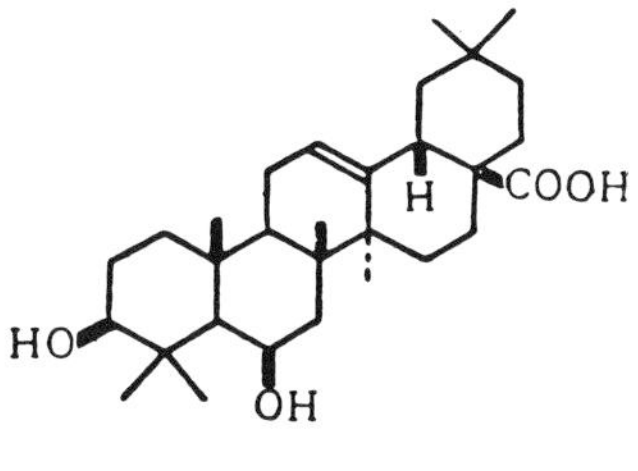

Sumaresinolic acid

Therapeutic Uses

In folk medicine, this herb was used as an aromatic stimulant for the resurrection of dying persons. It was also recommended to promote circulation and to relieve pain.

APPENDIX: PSYCHOTOMIMETIC HERBS

GU KO YI (古柯叶) — dried leaf of *Erythroxylum coca* Laur.

This plant is grown mainly in southern China, in the provinces of Kwangung and Kwangsi. It is not listed in the old pharmacopoeia, and the Chinese have not learned how to use it as either an addictive drug or an antihunger agent.

Chemistry

The plant contains several alkaloids, including *l*-cocaine ($C_{17}H_{21}O_4N$), cinnamylococaine ($Cl_9H_{23}O_4N$) α,β-trevilline ($C_{28}H_{46}O_8N_2$), ecgonine ($C_9H_{15}O_3N$), benzoylecgonine ($C_{16}H_{19}O_4N$), ecgonine-methylester ($C_{17}H_{17}O_3N$), ecgonidine-methylester ($C_{10}H_{15}O_2N$), nor-egonine ($C_8H_{13}O_3N$), tropacocaine ($C_{16}H_{19}O_2N$), hygrine ($C_8H_{15}ON$), hygroline ($C_8H_{17}ON$), and cuskhugrine ($C_{18}H_{24}ON_2$).

Actions

l-Cocaine is the main active principle of this plant. It has a potent local anesthetic effect when applied to mucosal membranes. It is rapidly absorbed from the mucosal surface. Additionally, it has a vasoconstriction effect.

This substance stops hunger sensations. It has a psychotomimetic effect, causing visual hallucinations and psychic delusions.

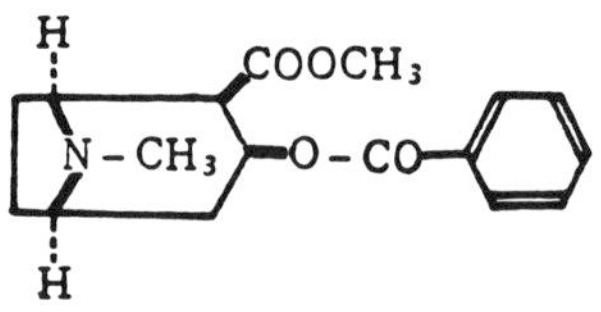

Cocaine

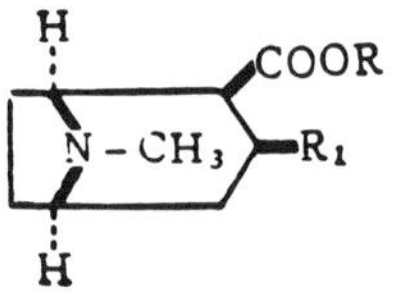

	R=	R_1=
Benzoyl ecgonine	H	$CO\text{-}C_6H_5$
Ecgonine-methylester	CH_3	H

Section V
Alimentary System

Chapter 15

STOMACHIC AND "WIND"-DISPELLING HERBS

In earlier times, the term "nutrition" had a somewhat different meaning than it does today. People believed that food was not the only source of nutrition. They thought that both healthy and ill persons benefit from herbs which could augment nutrition. These herbs were thought to increase the absorption of nutrients and strengthen body functions. Thus, stomachics and tonics were immensely popular and were more commonly taken than other herbs. They were taken before meals to stimulate the appetite and taste buds. They were also given to patients recuperating from illnesses.

"Wind"-dispelling (驱风) herbs are spices and peppery agents. When taken orally, they produce an increase in salivary and gastric secretions, and stimulate gastrointestinal motility, resulting in the removal of gas from the lumen. These types of herbs were very popular for cooking and used in medicine to treat abdominal distention due primarily to indigestion.

LONG DAN (龙胆) — the dried root and rhizome of *Gentiana manschurica, G. scatura, G. triflora,* or *G. regescens*

Chemistry

The herb contains saponins, gentiopicrin, the sugar gentianose, and several alkaloids, including gentianine.

Actions

This herb tastes bitter, and can increase gastric secretions and stimulate the appetite. It has anti-inflammatory and choleretic effects. It is a sedative and can slow the heart rate and lower blood pressure. It also has antibacterial activity. It was used popularly as a stomachic and in the treatment of acute icteric hepatitis. It was also found effective in treating urinary infections.

Overdosage results in nausea and vomiting.

ZHI QIAO (枳壳) — the dried unripe fruit of *Citrus aurantium* L. or *C. wilsonii* Tanaka

Chemistry

The fruit contains several essential oils and flavone.

Actions

It tastes bitter and sour, and can enhance gastrointestinal peristalsis. Additionally, it stimulates uterine contraction. It is used in the treatment of indigestion and to correct mild ptosis of the uterus.

HOU PO (厚朴) — the dried bark of *Magnolia officinalis* or *M. officinalis biloba* (Magnoliaceae)

Chemistry

The main active components found in the bark are essential oils and some alkaloids. The essential oils include machiolol, magnolol, tetrahydromagnolol, isomagnolol, and konokiol. The alkaloids include magnocurarine and tubocurarine.

Magnolol Liriodenine Konokiol Crytomeridiol

Actions

The bitter taste of the bark can stimulate salivation, gastric secretions, and reflexive intestinal peristalsis. Decoctions of the herb can increase uterine contraction.

The alkaloids of *Hou Po* have a neuromuscular blocking effect and cause relaxation of skeletal muscles. They can also lower blood pressure; the effect is not due to the histamine release which is usually caused by tubocurarine. Additionally, the herb exhibits some antibacterial properties.

Overdosage can cause respiratory paralysis. The LD_{50} in mice is 6.12 ± 0.04 g/kg (interperitoneal) and 4.25 ± 1.5 g/kg (intravenous) in cats.

BO HE (薄河) — peppermint, dried aerial parts of *Mentha haplocalyx* (Labiatae)

Chemistry

This plant contains the essential oils menthol (70 to 90%), menthone (10 to 20%), and menthyl acetate.

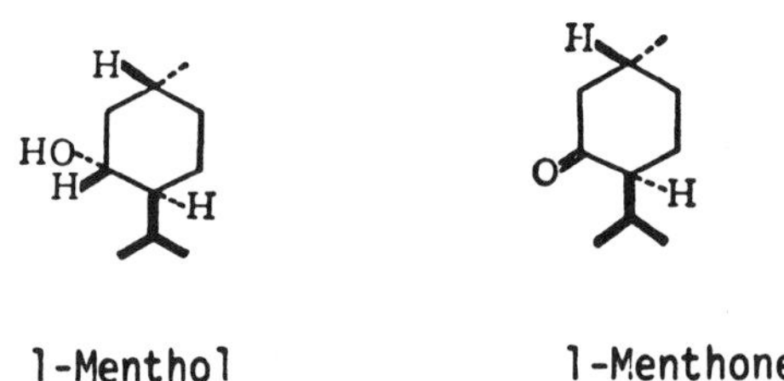

1-Menthol 1-Menthone

Actions

Menthol can stimulate gastrointestinal tract motility, promote the forward movement of luminal contents, and dispel gas. It also stimulates the central nervous system, dilates peripheral blood vessels, and increases sweat gland secretion. When applied locally to mucosa or skin, it produces a sensation of coolness. It can slowly penetrate the dermal layer of the skin and improve circulation in the local region where it is applied.

Therapeutic Uses

Chinese physicians use this herb to dispel ''wind'' and ''heat'', and to promote eruption. It is given orally in the form of a water extract, with a total dose of 3 to 5 ml taken two or three times a day, to dispel abdominal distention. A menthol ointment is rubbed into the skin in the treatment of the common cold or headache. *Bo He* is also given in combination with chrysanthemum, in doses of 3 to 9 g made in a decoction and served as beverage.

BA JIAO HUI XIANG (八角茴香) — dried ripe fruit of *Illicium verum* Hook.

Chemistry

The fruit contains several essential oils, including anethol ($C_{10}H_{12}O$), anisaldehyde ($C_8H_8O_7$), safrole ($C_{10}H_{10}O_2$), and anisic ketone ($C_{10}H_{12}O_2$).

The plant *Illicium lanacedatum* A. S. Smith is quite different from *I. verum*. The former is toxic and contains some different ingredients: anisatin ($C_{15}H_{28}O_8$), neoanisatin ($C_{15}H_{20}O_7$), pseudoanisatin ($C_{15}H_{22}O_6$), and shikimic acid ($C_7H_{10}O_5$).

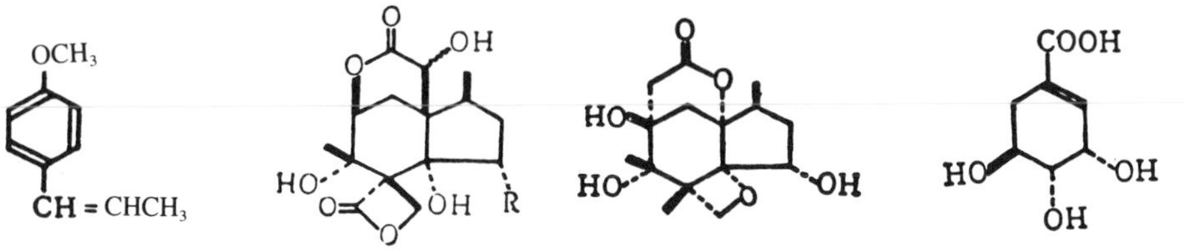

Anisic ketone	Anisatin	R = OH	Pseudoanisatin	Shikimic acid
	Neoanisatin	H		

Therapeutic Uses

Ba Jiao Hui Xiang is popularly used as a cooking spice. It was reputed to be capable of warming the viscera and expelling cold, ensuring normal flow of *qi*, and relieving pain.

XIN YI HUA (辛夷花) — dried flower buds of *Magnolia liliflora*, *M. biondii*, or *M. denudata*

Chemistry

The flower contains eugenol ($C_{10}H_{11}O_2$), safrole ($C_{10}H_{10}O_2$), citrol ($C_{10}H_{16}O$), and anethol ($C_{10}H_{12}O$). The leaf contains two alkaloids: salicifoline ($C_{12}H_{20}O_2N$) and magnocurarine ($C_{19}H_{24}O_3$).

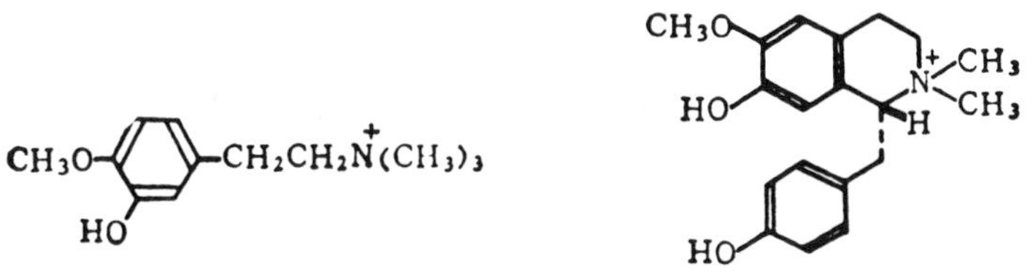

Salicifoline

Magnocurarine

Therapeutic Uses

It was used to dispel wind and cold, and to relieve nasal congestion.

TAN XIANG (檀香) — the dried heartwood of *Santalum album* L. or *S. spicatum*

Chemistry

This wood contains several essential oils, including α,β-santalol ($C_{15}H_{24}O$), α,β-santalene ($C_{15}H_{24}$), santene (C_9H_{14}), α-santenone ($C_9H_{14}O$), α-santenol ($C_9H_{16}O$), santalone ($C_{11}H_{16}O$), santalic acid ($C_{15}H_{22}O_2$), teresantalic acid ($C_{10}H_{14}O_2$), isovaleraldehyde ($C_5H_{10}O$), teresantalol ($C_{10}H_{16}O$), and tricycloekasantal ($C_{12}H_{10}O$). It also contains santalin, deoxysantalin, sinapyl aldehyde, caniferyl aldehyde, and syringic aldehyde.

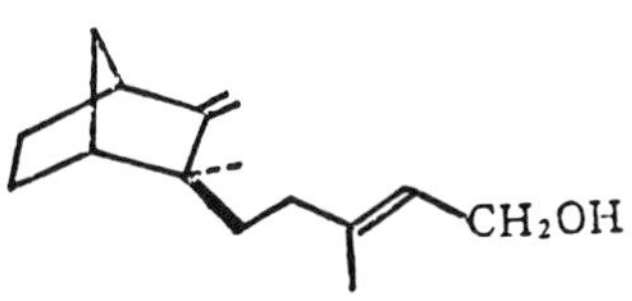

β-Santalol 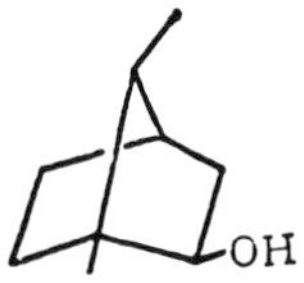 α-Santenone

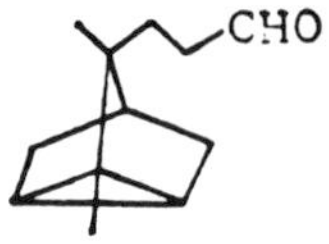

Tricycloekasantal

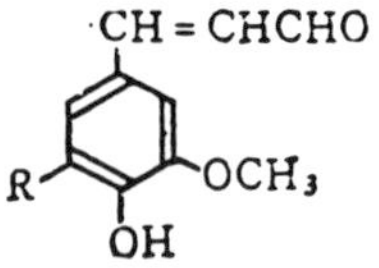

	R=
Sinapyl aldehyde	OCH_3
Ferulaldehyde	H

Therapeutic Uses

The herb was used to regulate the flow of *qi* and stomach functions. The essential oils served as stomachics.

XIAO HUI XIANG (小茴香) — the dried fruit of *Foeniculum vulgare* Mill.

Chemistry

The herb contains several essential oils, including anethol ($C_{10}H_{12}O$), *d*-fenchone ($C_{10}H_{16}O$), methylchavicol ($C_{10}H_{12}O$), and anisaldehyde ($C_8H_8O_2$).

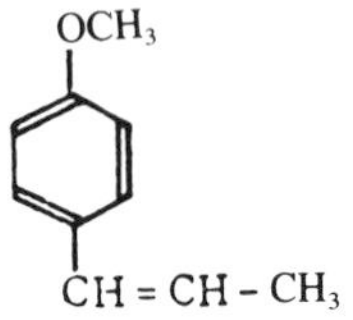

Anethol 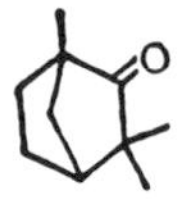 d-Fenchone 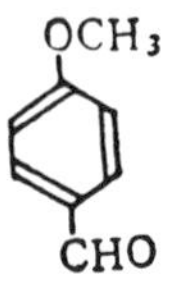 Anisaldehyde 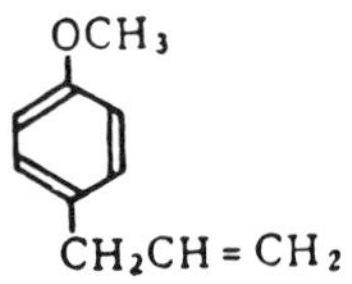 Methylchavicol

Therapeutic Uses

The herb was used to restore normal functioning of the stomach, warm taste sensations, dispel cold, normalize the flow of *qi*, and relieve pain.

TABLE 1
Other Stomachics and "Wind"- Dispelling Agents

Name	Source	Part of plant	Components	Effects and uses
Chen Pi (陈皮)	*Citrus reticulata* Blanco	Skin of fruit	Essential oil, vitamin B, glucosides	Increases gastric secretion and peristalsis, used as antiemetic, antihiccough, to dispel abdominal distension; dose: 3–9 g
Sheng Jiang (生姜)	*Zingiber officinale* Rose	Rhizome	Essential oil, amino acids	Increases gastric secretion and peristalsis; dose: 3–9 g
Dou Kou (豆蔻)	*Alpinia katsumadii* Hayata	Seed	Alpinetin	Warms stomach, dispels abdominal distension, increases appetite; dose: 3–6 g
Mu Xiang (木香)	*Aucklandia lappa* Decne.	Root	Essential oil	Dispels wind, stomachic as antispasmodic, treats abdominal distention; dose: 3–9 g
Cang Zhu (苍术)	*Atractylodes lancea* or *A. chinensis*	Rhizome		Lowers blood pressure; antibacterial, antiviral
Huo Xiang (藿香)	*Pogostemon cablin* Benth.	Leaf	Patchouli alcohol, essential oil, estragole, and isoanethele, limonene	Increases gastric secretion, antifungus, antispirochaeta
Pei Lan (佩兰)	*Eupatorium odoratum*			Antitoxin
Cao Guo (草果)	*Amonum tsaoko*	Fruit		Expectorant, antimalarial
Sha Ren (砂仁)	*Amomum villosum* or *A. longiligulare*	Fruit		Arrests diarrhea, prevents miscarriage
Bi Ba (荜茇)	*Piper longum*	Fruit spike		Dispels cold, relieves pain, stomachic, dispels wind and *qi*
Gao Liang Jiang (高良姜)	*Alpinia officinarum*	Rhizome		

Chapter 16

HERBS PROMOTING DIGESTION

This group of herbs is mainly used to promote the digestion of food in the gastrointestinal tract. They work by resupplying the digestive enzymes or by stimulating their activity. In Chinese medicine, it was believed that stagnation of food in the gastrointestinal tract is due to weakness of the stomach and spleen. Therefore, herbs were suggested as a remedy to strengthen the spleen, dispel cold, and aleviate weaknesses of the stomach.

MAI YA (麦芽) — germinated barley, the dried germinated fruit of *Hordeum vulgare*

Chemistry

The fruit contains several enzymes, including invertase, amylase, and proteinase. Additionally, it contains fat, vitamins B and C, maltose, and dextrose.

Therapeutic Uses

Because of its content of digestive enzymes, this herb can help to digest carbohydrates and protein. It is used in cases of indigestion and loss of appetite. The standard dose is 15 to 30 g made into a decoction, taken for 3 consecutive days.

The herb is prescribed to stop milk spitting by babies during feeding. For this purpose, 6 to 12 g of *Mai Ya* is extracted with warm water, and fed to the baby. The herb is also given to stop lactation in nursing mothers; the mechanism of this effect is not clear.

DING XIANG (丁香) — the dried flower bud of *Eugenia caryophyllata* Thunb.

Chemistry

The herb is grown mainly in southern Asia, Malaysia, Indonesia, and China's Kwangtung Province. About 15 to 20% of the herb consists of essential oils. The main principles of these oils are eugenol ($C_{10}H_{12}O_2$), acetyleugenol, α- and β-caryophylline, ylangene ($C_{15}H_{24}$), chavicol ($C_9H_{10}O_6$), and humulene.

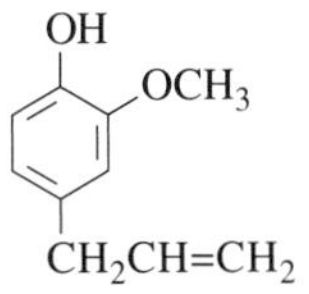

OH, OCH_3, $CH_2CH{=}CH_2$

Eugenol

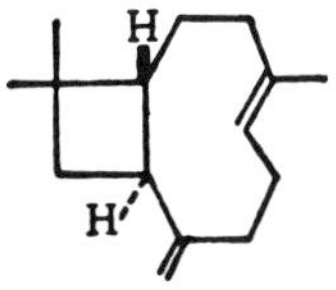

β-Caryophyllene

The flower bud also contains some flavonoids, including rhamnetin, kaempferol, oleanolic acid, eugenitin, and isoeugenitin. The bark contains the trimethylether of ellagic acid, β-sitosterol, and mairin.

Actions

The extract of the herb can stimulate gastric secretions, with a greater effect on pepsin than acid, resulting in an increase in digestion and a dispelling of gases. The herb is also a potent ascaricide. At a dose of 0.1 to 1.0 g/kg, the alcoholic extract or the essential oil can expel ascaria from the host with few adverse effects.

Ding Xiang has both antibacterial and antifungal effects. It is active against staphylococci, diphtheria, and typhoid bacilli. At a concentration of 1:8,000 to 1:16,000, extracts of the herb can inhibit the growth of many fungi, including *Candida albicans*.

The LD_{50} of *Ding Xiang* oil in mice is 1.6 g/kg (oral); 5 g/kg in dogs, and 19.3 g/kg in rats. Animals given this herb can develop the symptoms of hematuria and mucosal necrosis and can die from paralysis.

Therapeutic Uses

Ding Xiang is prescribed as a remedy for hiccoughs and as an antiemetic agent. An alcoholic extract of the herb is used externally to treat fungal infections and various skin infections.

LA LIAO (辣蓼) — dried whole plant of *Polygonum hydropiper* or *P. flaccidum*

Chemistry

The plant contains persicarin ($C_{16}H_{11}O_7 \cdot SO_3K$), rhamnazin ($C_{17}H_{14}O_7$), quercimeritrin ($C_{21}H_{20}O_{12}$), tadeonal ($C_{16}H_{22}O_2$), and isotadeonal ($C_{16}H_{22}O_2$).

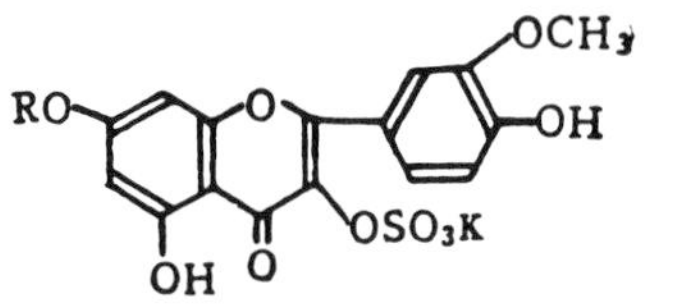

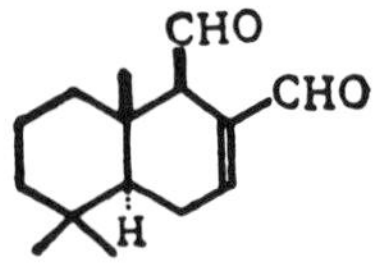

	R =
Persicarin	H
Persicarin-7-methyl ester	CH_3

Tadeonal

Therapeutic Uses

This herb is used to remove dampness and food stagnancy, and is commonly prescribed to combat indigestion. It is also effective in the treatment of dysentery and enteritis.

GOU GI (枸桔) — the ripe fruit of *Poncirus trifoliata* (L.) Raf.

Chemistry

Several substances have been isolated from this fruit, including poncirin ($C_{28}H_{34}O_{14}$), limonin ($C_{26}H_{30}O_8$), imperatorin ($C_{16}H_{14}O_4$), bergapten ($C_{12}H_8O_4$), neohesperidin ($C_{28}H_{34}O_{15}$), citrifoliol ($C_{16}H_{14}O_5$), myrcene, camphene, and γ-terpinene.

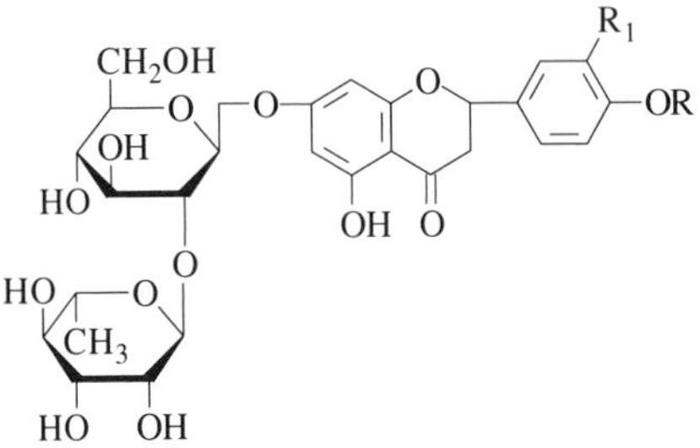

	R =	R_1 =
Citrofolioside	CH_3	H
Neohesperidin	CH_3	OH

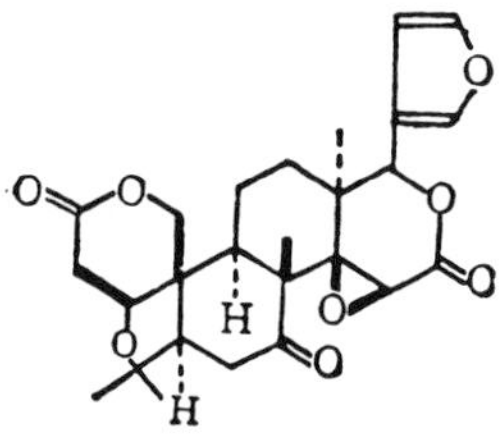

Limonin 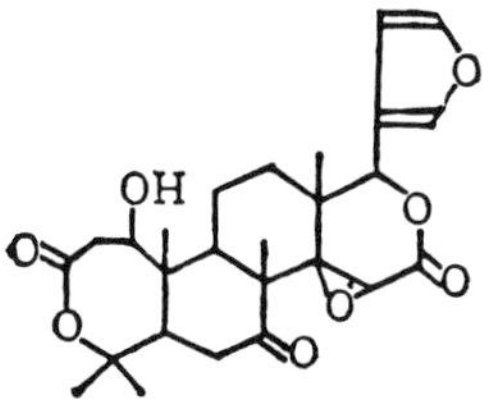 Deacetylnomilin 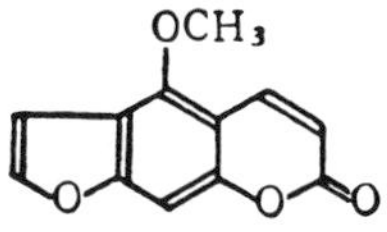 Bergapten

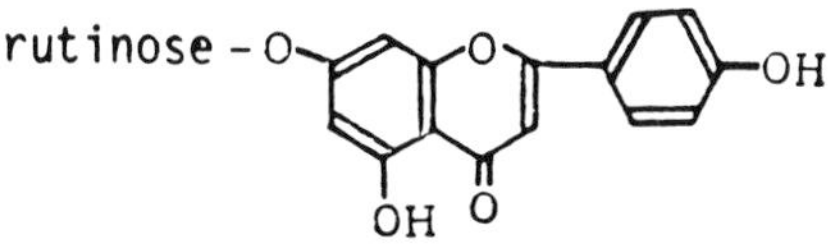

Rhoifolin 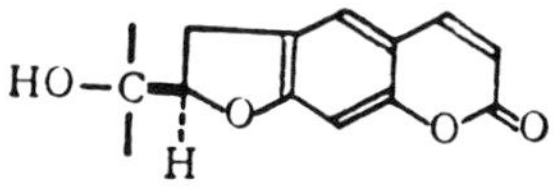 Marmesin

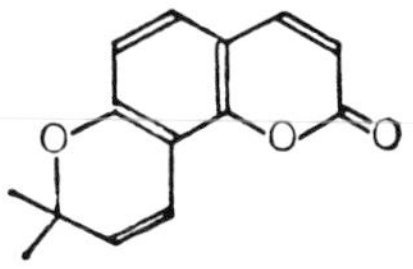

Seselin 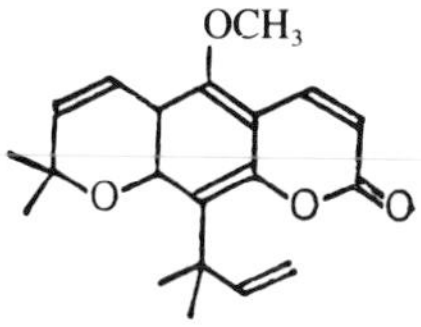 Poncitrin

Therapeutic Uses

Gou Gi is used to treat gastric pain and constipation. It has been used with success to treat prolapse of the uterus or rectum.

SUO LUO ZI (娑罗子) — dried ripe fruit of *Aesculus chinesis* Bunge or *A. arlsonii*

Chemistry

The fruit contains several saponin products, hydrolyzed to give genin and sugars. The genins include protoescigenin and escigenin ($C_{30}H_{48}O_5$); the sugars include oligosaccharides and amylose.

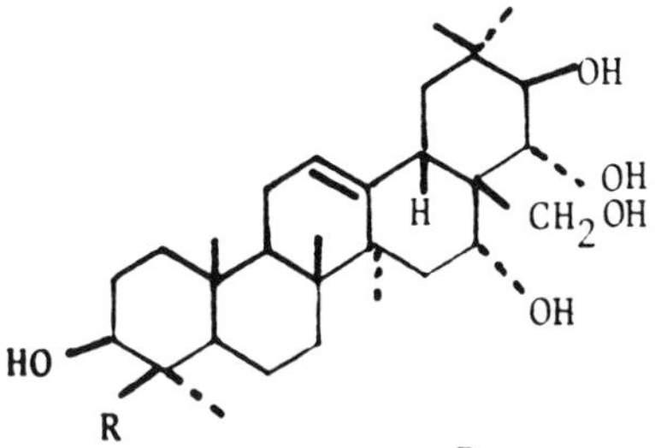

	R =
Protoescigenin	CH_2OH
Barringtogenol C	CH_3

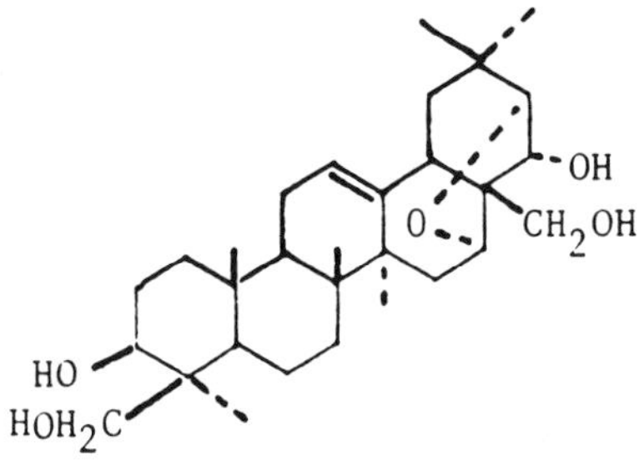

Escigenin

Therapeutic Uses

It is used to promote the circulation of *qi*, to relieve fullness and pain in the epigastrium, and to promote digestion.

TABLE 1
Other Herbs Used to Help Digestion

Name	Source	Part of the organism	Component	Effects and uses
Shan Zha (山楂)	*Crataegus pinnatifida* Major	Fruit	Chlorogenic, crataegolic acid	Increases gastric and pancreatic secretion, treats children's indigestion; dose: 9–30 g in decoction or in form of snack-candy (see Chapter 6)
Ji Nei Jin (鸡内金)	*Gallus gallus domesticus* Brisson	Mucous membrane	Ventriculin, vitamins	Increases gastric secretion, used in childhood indigestion; antiemetic; antidiarrhea agent; dose 3–9 g

Chapter 17

ANTIACID AND ANTIULCER HERBS

HAI PIAO XIAO (海螵蛸) — dried internal shell of *Sepia esculenta* Hoyle or *Sepiella maindroni* de Rochebrune

Chemistry

The major components of the shell are $CaCO_3$ (76.7%), gelatin (11.8%), some organic substances, NaCl, and $Ca_3(PO_4)_2$.

Actions

The "herb" is an effective antacid. Its $CaCO_3$ content can neutralize the HCl in gastric juices, and the gelatin reacts with the juices to form a membrane, preventing further bleeding and relieving pain. Thus, it may promote early healing of ulcerated mucosa.

The herb also has a hemostatic effect. It is used to treat spontaneous emission and leucorrhagia, and to relieve gastric hyperacidity and pain.

It is generally prepared as a powder, and administered orally in a dose of 3 g t.i.d.

WA LENG ZI (瓦楞子) — shell of *Arca subcrenata*, *A. granosa*, or *A. inflata*

Chemistry

The components of this "herb" are $CaCO_3$, $Ca_3(PO_4)_2$, and CaO.

Therapeutic Uses

It is used to arrest acid regurgitation, reduce phlegm, and soften and resolve masses. The herb is prescribed in doses of 3 g t.i.d., administered orally.

MU LI (牡蛎) — shell of *Ostrea gigas*, *O. talcenuhanensis*, or *O. rivularis* Gould.

Chemistry

Approximately 80 to 95% of the "herb" consists of $CaCO_3$. Other components include $CaSO_4$, aluminum, and Fe_2O_3.

Therapeutic Uses

The herb has antacid and astringent effects. *The Herbal Classic of the Divine Plowman* described it as useful to check hyperfunction of the *yang* of the liver, to soften and resolve hard masses, and to induce astringency.

Chapter 18

LAXATIVE HERBS

A Chinese proverb states that "sickness comes from the top, while rescue comes from the bottom." Thus, to treat any illness, removal of the bowel contents was advised. Because of this, most Chinese prescription formulas include a laxative as an adjuvant such as *Tai Huang* which appears in two out of ten prescriptions. The mechanism of most laxatives involves stimulation of the intestinal mucosa, resulting in an increase in peristalsis. Their major adverse effect is to increase blood congestion in the abdomen.

TAI HUANG (大黄) — the dried root or rhizome of *Rheum palmatum* L., *R. officinale* Baill., or *R. tanguticum* Maxim. ex Balf.

Chemistry

The major components isolated from the herb are the glycosides rhein-8-monoglucoside, physcion monoglucoside, aloeemodin monoglucoside, emodin monoglucoside, chrysophenol monoglucoside, and sennoside A, B, and C. Rheum tannic acids, gallic acid, and calechin are also present.

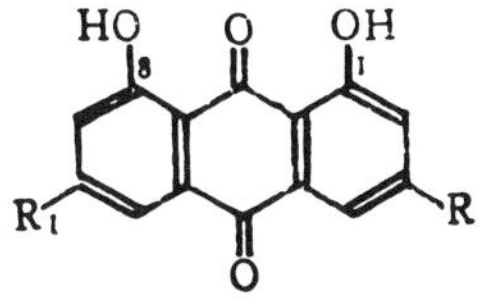

HO O OH, 8, 1, R_1, R, O

	R =	R_1 =
Rhein	COOH	H
Emodin	OH	CH_3
Chrysophenol	CH_3	H
Aloeemodin	CH_2OH	H
Physcion	OCH_3	CH_3

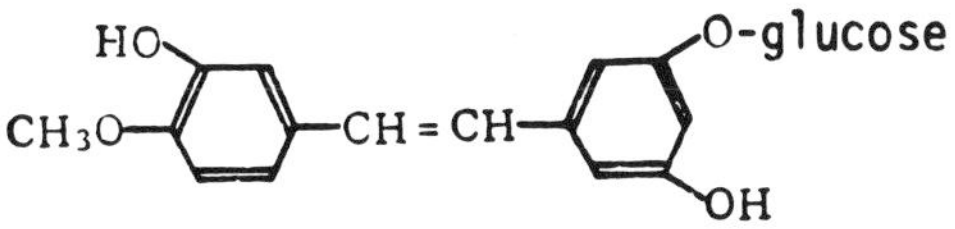

HO, CH_3O, CH = CH, O-glucose, OH

Rhapontin

Actions

This herb is a potent laxative. The hydrolysis products of the glucosides emodin and sennidin are the active principles. They stimulate the large intestine and increase the movement of luminal contents toward the anus, resulting in defecation. Small doses of *Tai Huang* (0.03 to 0.3 g), however, can cause constipation, because its tannic acid content exerts an astringent effect on the mucosa.

The herb has antispasmodic effects, and is about four times more potent than papaverine, an alkaloid isolated from opium. It exhibits choleretic and hemostatic actions. As a diuretic, it increases urinary excretion of Na^+ and K^+, and alkalinizes urine to a pH as high as 8.4.

It has been observed to lower blood pressure and blood cholesterol levels. It is a stomachic, stimulating the appetite. In addition, it displays antibacterial, anthelmintic, and anticancer properties.

Chronic administration of this herb causes pathological changes in the liver, thyroid, and stomach, with hypertropy of the tissue cells. The common adverse effects include nausea, vomiting, diarrhea, and abdominal pain.

Therapeutic Uses

The therapeutic uses of this herb are very broad, and apply to every field of medicine, either as a major remedy or adjuvant therapy. For example, the old Chinese medical formula *San Chin Qi* decoction (三承气湯), which uses *Tai Huang* as the major principle, is used to treat constipation, especially when due to high body temperature. It is also used in small doses, 0.05 to 0.2 g, to treat chronic diarrhea, indigestion, and acute intestinal infections, such as appendicitis and peritonitis, ileitus, and acute hepatic jaundice. It is also effective in gallstone removal and in cholecystitis.

As an adjuvant agent, *Tai Huang* is used as a hemostatic in hempotysis, ulcer hematemesis, ulceration of the oral mucosa, ulcerative wounds, etc.

MANG XIAO (芒硝) — crystalline sodium sulfate ($Na_2SO_4 \cdot 10H_2O$)

Actions

The sulfate ion is not absorbed from the intestinal mucosa. Therefore, the ingested salt will remain in the lumen and produce an osmotic effect to prevent the absorption of fluid. The luminal contents will increase and stimulate the peristaltic movement of the intestine, resulting in a purgative effect.

The salt acts reflexively to induce a choleretic effect, by contracting the gall bladder and relaxing the Vateri orifice. Because of this, it is used in the treatment of cholecystitis and jaundice.

It is generally prescribed in doses of 6 to 15 g, and is taken with warm water.

QIAN NIU ZI (牵牛子) — dried seed of *Pharbitis nil* (L.) or *P. purpurea* (Convolvulaceae)

Chemistry

The herb contains the glycoside pharbitin (approximately 2%). The unripe seed contains several substances called gibberellin I, II, III, IV, V, VI, and VII. The hydrolysis products of pharbitin are pharbilic acid and five sugars.

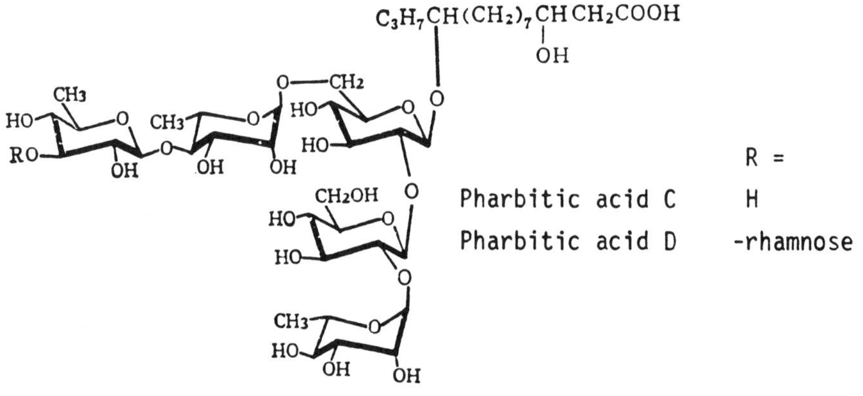

$C_3H_7CH(CH_2)_7CH(OH)CH_2COOH$

	R =
Pharbitic acid C	H
Pharbitic acid D	-rhamnose

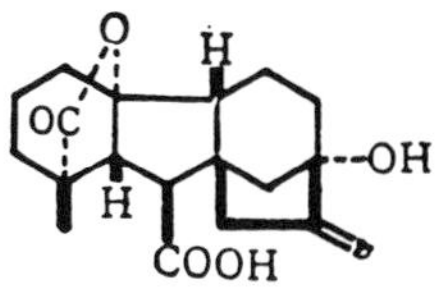

Gibberellia A_{20} 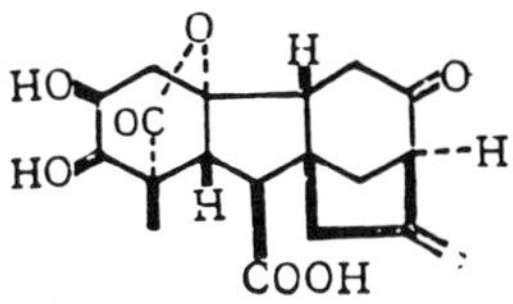 Gibberellia A_{26} 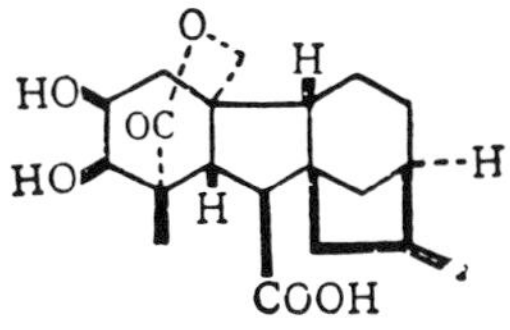 Gibberellia A_{27}

Actions

Pharbitin is a potent purgative. In Chinese medicine, it was stated that small doses of the herb could cause defecation, while large doses would produce water-like diarrhea. It acts by irritating the intestinal mucosea, increasing secretions and peristalsis. The onset of the laxative effect occurs approximately 3 h after oral administration. After absorption into the blood stream, the herb principle reaches the kidney and acts on the renal tubules to produce a diuretic effect by reducing the tubular reabsorption.

This herb can also purge parasites, ascaris and taenia from the intestine. It is commonly used to treat constipation and edema. The normal dose is 3 to 6 g.

It should not be used frequently, and is contraindicated in pregnancy. Overdosage results in hematuria, abdominal pain, nausea and vomiting, and blood in the feces. This herb should not be given together with croton oil.

HUO MA REN (火麻仁) — (Hemp or Marihuana) — the dried fruit and seed of *Cannabis sativa* L.

Chemistry

The fruit contains fat, vitamins B_1 and B_2, muscarine, and choline. It also contains several active substances, such as trigonelline, *l*(*d*)-isoleucine betaine, cannabinol, tetrahydrocannobinol, and cannabidiol.

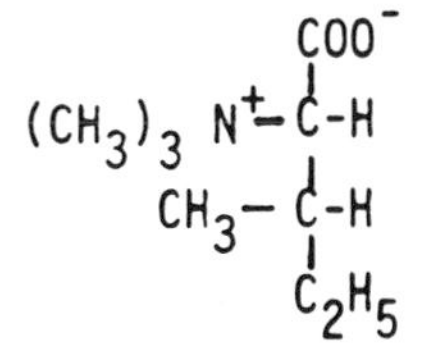

L(d)-Isoleucine betaine 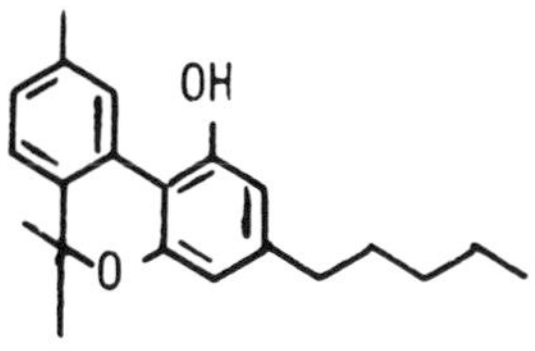 Cannabinol 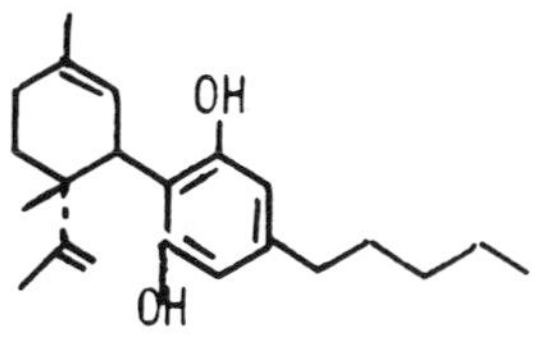 Cannabidiol

Therapeutic Uses

The herb was listed in the *Herbal Classic* as a mild purgative which acts by stimulation of the intestinal mucosa, causing an increase in secretions and peristalsis. It was used to treat constipation of debilitated or elderly persons. The normal dose in these treatments is 9 to 15 g of seed.

Neither the *Herbal Classic* nor other Chinese pharmacopoeias had ever mentioned that this herb has the hallucinogenic effect on the CNS.

Because of its content of muscarine, administration of large doses of *Huo Ma Ren* would cause cholinergic intoxication, manifested as nausea, vomiting, diarrhea, convulsions and coma.

BA DOU (巴豆) — the seed of *Croton tiglium* L.

Chemistry

The seed contains 53 to 57% oil (croton oil) and 18% protein. The oil contains the croton resin, phorbol, and crotonic acid A and D. From the oil, 11 kinds of carcinogenic substances have been isolated. It also contains crotin, a glycoside called crotonoside, and some alkaloids.

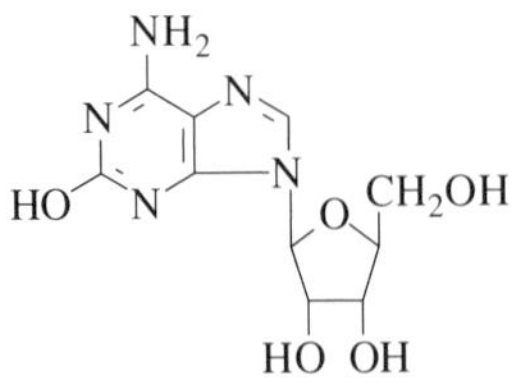
NH2
N
N
N
HO
N
O
CH2OH
HO OH

Crotonside

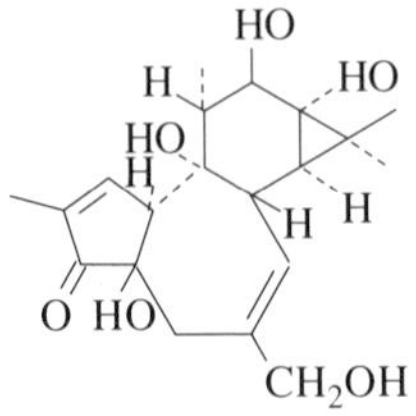
HO
H
HO
HO
H
H
H
O HO
CH2OH

Phorbol

Therapeutic Uses

Ba Dou was listed in the *Herbal Classic* among the ''lower classes of herbs'' because it has a drastic action and is very toxic; it was therefore not recommended for use by upper-class citizens. The herb can produce a drastic purgative effect by irritation of the gastrointestinal mucosa. It is very toxic and carcinogenic. It is generally used only in veterinary medicine.

FAN XIE YE (番泻叶) — the dried leaflets of *Cassia augustifolia* Vahl

Chemistry

Several glycosides have been isolated from this herb, including sennoside A and B ($C_{42}H_{38}O_{20}$), C and D ($C_{42}H_{40}O_{19}$), aloeemodin, dianthrone glucoside, rhein monoglucoside, rhein ($C_{15}H_{18}O_6$), kaempferin, and myricyl alcohol. The herb also contains 0.85 to 2.86% anthraquinone derivatives.

Aloeemodin and rhein are potent purgative agents.

BI MA ZI (蓖麻子) — dried ripe seed of *Ricinus communis*

Chemistry

The major substances isolated from the seed are ricinine ($C_8H_8O_2N_2$), and the fatty acids, ricinolein ($C_{57}H_{104}O_9$), around 80 to 85%, olein, and stearin. A small amount of isoricinoleic acid is present, as is a very small quantity of cytochrome C.

The leaf of the plant contains corilagin ($C_{27}H_{22}O_{18}$).

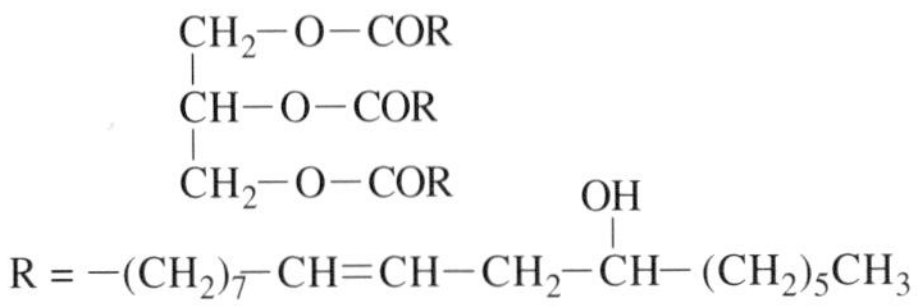
CH2—O—COR
CH—O—COR
CH2—O—COR
OH
R = —(CH2)7—CH=CH—CH2—CH—(CH2)5CH3

Ricinolein

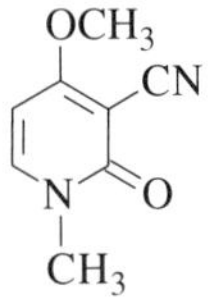
OCH3
CN
N
O
CH3

Ricinine

Actions

Castor oil or ricinus oil, a very commonly used cathartic agent, is extracted from the *Ricinus* seed. The oil is a triglyceride of ricinoleic acid, which, after administration, is hydrolyzed in the lumen of the gastrointestinal tract to form ricinoleic acid. The latter irritates the intestinal wall, producing the desired cathartic action.

TABLE 1
Other Laxative Herbs

Name	Source	Parts	Component	Effect and use
Da Ji (大戟)	*Euphorbia pekinensis* Rupr. or *Knoxia valerianoides*	Root	Euphorbon and euphorbia A, B, C; alkaloids	Purgative, diuretic, incomparable to *Glycyrrhiza* root
Shang Lu (商陸)	*Phytolacca acinosa* Roxb. *P. americana*	Root	Phytolaccine, jaligonic acid, KNO_3, acinosolic acid, spergulagenic acid	Laxative, diuretic, and expectorant; antiasthmatic; antibacterial
Gan Sui (甘遂)	*Euphorbia kansui* Liou	Root	Euphorbone, kanzuiol, α- and β-euphorbol, euphol, euphadienol, tircicallol, 20-deoxyingenol, ingenol, 13-oxyingenol	Laxative, diuretic, incomparable to *Glycyrrhiza* root
Yue Li Ren (榆李仁)	Bush-cherry	Seed		Laxative

Chapter 19

ANTIDIARRHEAL HERBS

Diarrhea can be caused by bacterial infections or by irritation from undigested food, decomposed products, or toxic substances. Actually, diarrhea may be a kind of body defense mechanism, whereby the intestine serves in the removal of unwelcome digested substances from the body before they are absorbed into the blood stream. However, diarrhea is usually accompanied by an excessive loss of body fluids and electrolytes, and the patient feels weak and dehydrated. Most classical Chinese medical books described the treatment of diarrhea very extensively. The following herbs are used mainly as astringents, to protect the intestine by reducing the irritation.

WU BEI ZI (五倍子) — the excrescence produced by an insect on the leaf of *Rhus chinensis* or *R. punjabensis*

Chemistry

Between 50 to 70% of the "herb's" content is gallotannin or gallotannic acid. In addition, it contains some gallic acid, resin, wax, and polysaccharides.

Actions

Due to its large content of tannic acid, this agent exerts its principle action through an astringent effect on protein, coagulating it to form an insoluble protective membrane on the intestinal mucosa. Additionally, the tannin proteinate plugs damaged blood vessels and stops bleeding, and inhibits secretory processes in the mucosa membrane.

The tannic acid can also precipitate toxic substances in the intestinal lumen, converting them into insoluble substances, making it useful as a detoxicant. The herb also has antibacterial activity.

When the tannic acid is absorbed in excessive amounts, it may cause liver damage.

Therapeutic Uses

It is used in the treatment of chronic intestinal infections, hematochezia, proctoptosis, skin infections, and bleeding wounds. It is also used externally to stop bleeding, counteract poisons, and remove "dampness".

The standard dose for the herb is 3 to 6 g, made into a decoction and taken orally.

ZI YU (地榆) — dried root of *Sanguisorba officinalis* L.

Chemistry

The root contains two kinds of glucosides and a large amount of tannic acid. *Zi Yu* glucosides I and II on hydrolysis yield the genin 19-α-hydroxyursolic acid, or promolic acid. The root also contains sanguisorbin A, B, and C; their genin is ursolic acid.

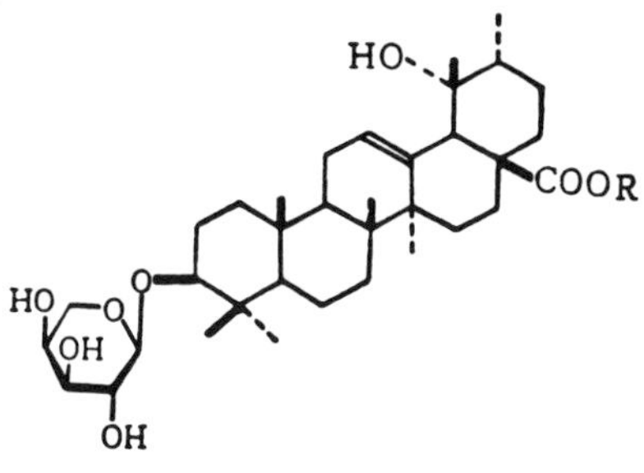

	R =
Zi yu-glycoside I	β-D-glucose
Zi yu-glycoside II	H

Therapeutic Uses

This herb has a strong astringent effect. It is applied externally on burn wounds to reduce the exudation of fluid and prevent subsequent bacterial infection. External use is also prescribed in cases of eczema.

Because of its astringent effect, oral administration of this herb is effective in stopping diarrhea and in treating chronic intestinal infections, duodenal ulcers, and bleeding. It was used to stop uterine bleeding, in menorrhagia, and in the treatment of hematochezia and hemoptysis. The standard dose for these treatments is 4.5 to 15 g, prepared in a decoction.

GOU MEI (构梅) — dried fruit of *Myrica rubra* (Lour.) Sieb. et Zucc.

Chemistry

The herb contains myricetin, a genin hydrolyzed from the glucoside myricitrin.

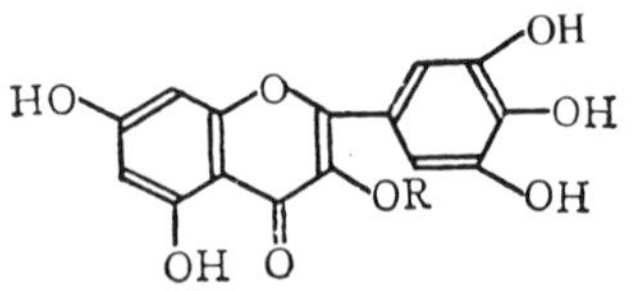

	R=
Myricetin	H
Myricitrin	rhamnose

Therapeutic Uses

It was used to treat gastric pain, diarrhea, and dysentery.

CHUN PI (椿皮) — dried root bark or stem bark of *Ailanthus altissima* (Mill.) Swingle

Chemistry

The herb contains amarolide ($C_{20}H_{25}O_6$), ailanthone ($C_{20}H_{24}O_7$), and afzelin ($C_{12}H_{20}O_{10}$).

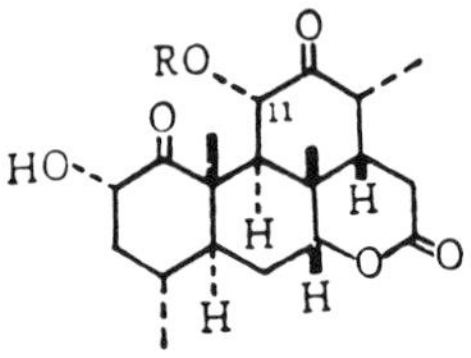

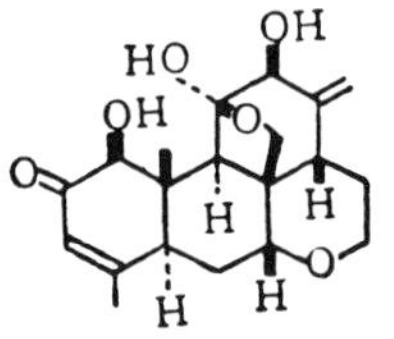

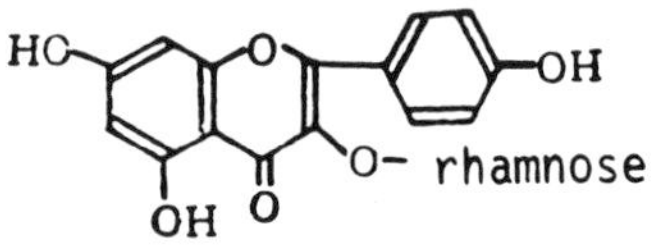

	R =
Amarolide	H
Amarolide-11-acetate	$COCH_3$

Ailanthone

Afzelin

The tree from which this herb is taken originated in China and is also called "the tree of heaven", or Chinese sumac. It was introduced to England in 1751 and brought to the U.S. in 1800 A.D.

Actions

The herb has an antidiarrheal effect, probably due to its tannin content. It also can exhibit a cathartic effect due to its content of resin and volatile oils. It is a hemostatic, emetic, and an antibacterial agent. *Chun Pi* is also a muscle relaxant. Its effect is very similar to that which occurs in beginning smokers while smoking tobacco.

Adverse effects include nausea and vomiting, vertigo, and dizziness.

Therapeutic Uses

It is used to treat diarrhea, dysentery, duodenol ulcers, and as a hemostatic agent. It has also been used to treat gonorrhea and leucorrhea.

Reference

Millspaugh, C. F., *American Medicinal Plants,* Dover, New York, 1974, 132–135.

FAN SHI LIU (番石榴) — dried fruit of *Psidium guajava* L.

Chemistry

The fruit contains several glucosides and other substances, including avicularin ($C_{20}H_{16}O_{11} \cdot H_2O$) and guaijaverin ($C_{20}H_{16}O_{11} \cdot H_2O$); amritoside ($C_{26}H_{26}O_{18}$); crataegolic acid ($C_{30}H_{48}O_4$), luteioic acid, argamolic acid ($C_{30}H_{48}O_5$), and arabinose ester.

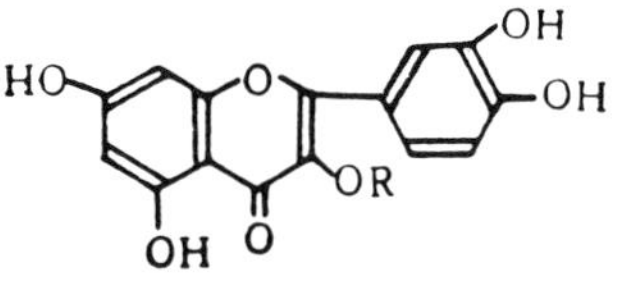

	R =
Avicularin	L-Furanarabinose
Guaijaverin	L-Pyranarabinose

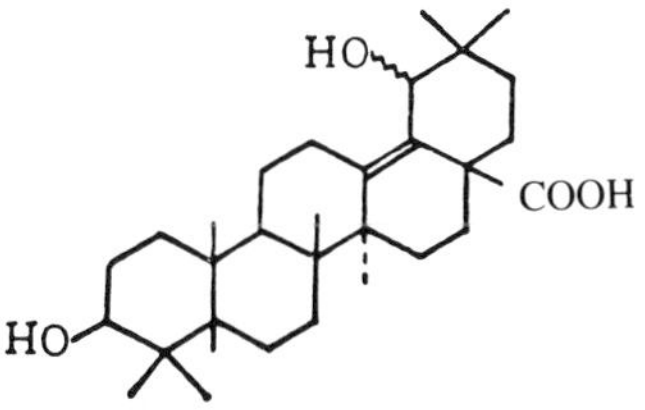

Crataegolic acid

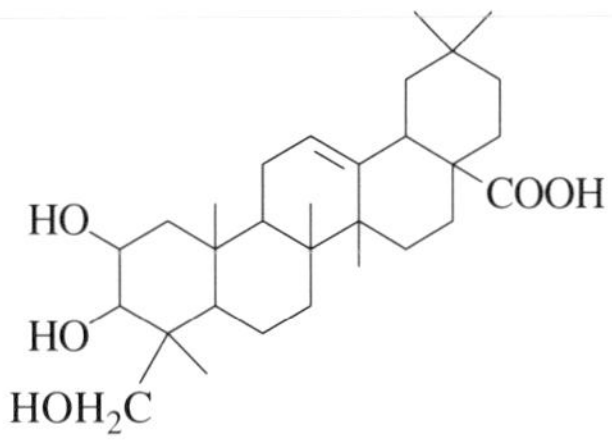

Arjunolic acid

Therapeutic Uses

It was used to treat dysentery and acute gastrointestinal inflammation.

CHUN JIN PI (川槿皮) or MU JIN PI (木槿皮) — dried bark of *Hibiscus syriacus* L.

Chemistry

The bark contains saponarin ($C_{27}H_{32}O_{16}$).

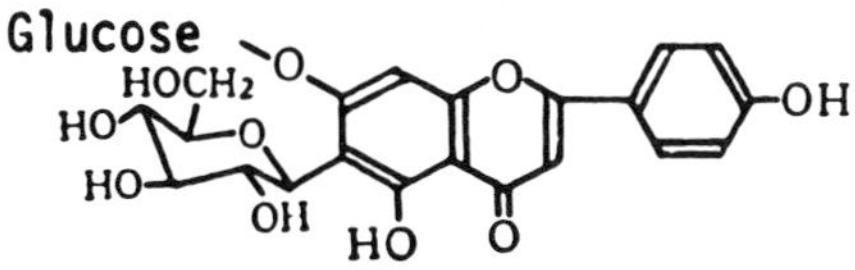

Saponarin

Therapeutic Uses

It was used to clear ''dampness'' and ''heat'', and to treat dysentery, diarrhea, jaundice, eczema, tinea, and scabies. It was also prescribed as a pesticide, to relieve itching.

TABLE 1
Other Astringent Herbs Used as Antidiarrheal Agents

Name	Source	Ingredient	Effect and uses
Chi Shi Zhi (赤石脂)	*Halloysitum rubrum*	Iron oxide, Al^{3+}, and Mg^{2+}, silicic acid	Astringent, styptic, arrests diarrhea and bleeding
Bei Fan (白矾)	Alum	$Al_2(SO_4)_3$	Antiemetic, antidiarrhea, astringent, antibacterial
Rou Dau Kou (肉豆蔻)	*Myristica fragrans* Houtt.	α-Pinene, *d*-Camphene, myristicin	Expels gas, central depressant
He Zi (诃子)	Fruit of *Terminalia chebula* Retz.	Chebulinic acid, corilagin, terchebin, chebulin, quinic acid	Antidiarrhea, treats collapsed anus, antispasmodic
Jin Ying Zi (金樱子)	Dried hip of *Rosa laevigata*		Astringent, stops spontaneous emission
Qian Shi (芡实)	*Euryala ferox* seed		Arrests diarrhea, stops spontaneous emission, and leucorrhagia
Wu Mei (乌梅)	*Prunus mume* Sieb. et Zucc.	Prudomenus tannic acid	Anthelmintic (see Chapter 36)
Bai Guo (白果)	*Ginkgo biloba* L.	Ginkgolic acid, bilobol	Antitussive, antituberculous, antibacterial (see Chapter 5)

Chapter 20

EMETIC AND ANTIEMETIC HERBS

Vomiting is a reflex movement of the stomach to eject its contents backwards into the esophagus and the mouth. It can be caused by direct irritation of the gastric mucosa, or through stimulation of the chemoreceptor trigger zone (CTZ). In cases of drug or food intoxication, it is commonly recommended that gastric lavage be performed and, if necessary, an emetic agent administered to empty the stomach contents before the absorption from the gastrointestinal tract occurs. However, in comatose patients, induction of vomiting may be risky because the vomitus may be aspirated into the lungs.

The antiemetic herbs listed in this chapter act mainly on the central nervous system to prevent continuous vomiting, which can result in a loss of body fluids and electrolytes.

GUA DI (瓜蒂) — the dried melon pedicel of *Cucumis melo* L.

Chemistry

It contains elaterin or melotoxin and cucurbitacin B and E.

Actions

The herb produces emesis after oral administration, due to its irritant effect on the gastric mucosa. However, it has no emetic effect if given parenterally.

The herb has also been observed to improve hepatic function and to increase glucogenesis. It has a protective effect against CCl_4 intoxication. It reduces jaundice and is effective in treating toxic hepatitis.

The LD_{50} of cucurbitacin B in mice is 14 ± 3.0 mg/kg (oral) and 1.0 ± 0.07 mg/kg (subcutaneous injection).

Therapeutic Uses

The *Ben Cao Kong Mu* stated that the herb ''can cause vomiting and salivation''. It was used in Chinese medicine to produce vomiting in the treatment of drug intoxication. It is also used to treat toxic and chronic hepatitis and cirrhosis of the liver.

Preparations of *Gua Di* include a 5% extract, which is administered orally to adults in 3 to 5 ml doses for the treatment of infectious hepatitis. A powder is available for nasal administration, in doses of 0.1 g. Alcoholic extracts of cucurbitacin B and E are also used, with a standard dose of 0.2 to 0.3 mg t.i.d.

LI LU (藜芦) — the dried rhizome of *Veratrum nigrum* L.

Chemistry

The rhizome contains several alkaloids, including jervine, pseudojervine, rubijervine, tienmulilmine, and tienmulilminine.

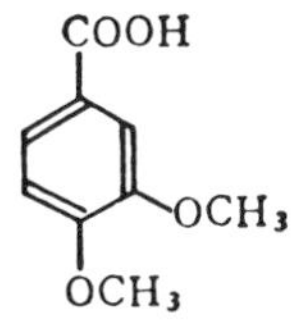

COOH, OCH_3, OCH_3

Veratric acid

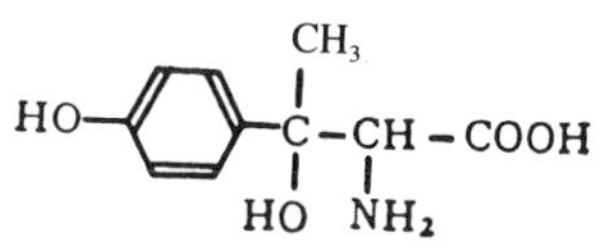

HO–C$_6$H$_4$–C(CH$_3$)(OH)–CH(NH$_2$)–COOH

1-2-Methyltyrosine

HO–C$_6$H$_4$–CH(CH$_3$)–CH(NH$_2$)–COOH

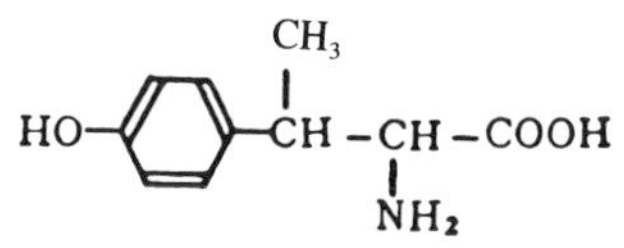

d-2-Oxyl-2-Methyltyrosine

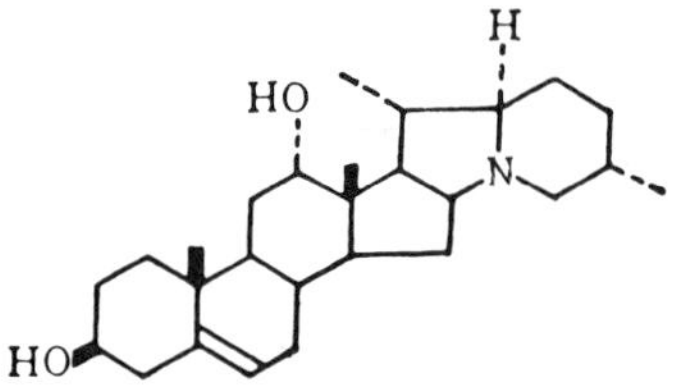

Rubijervine

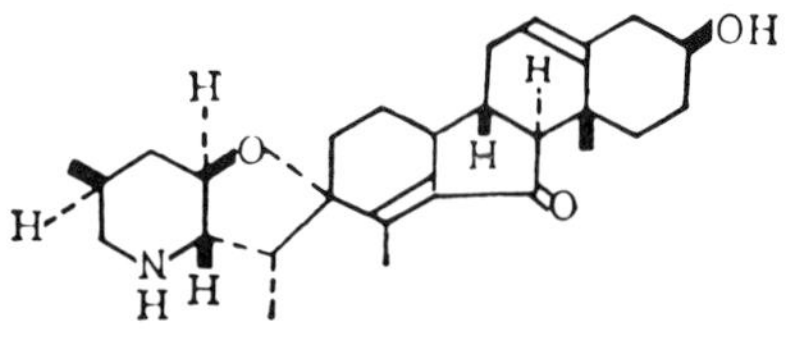

Jervine

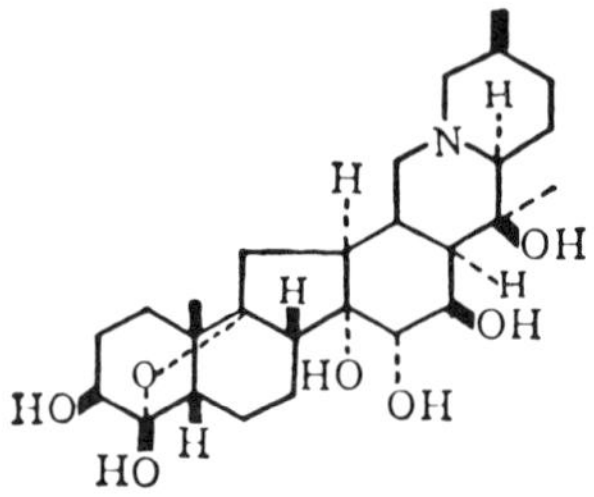

Zygadenine

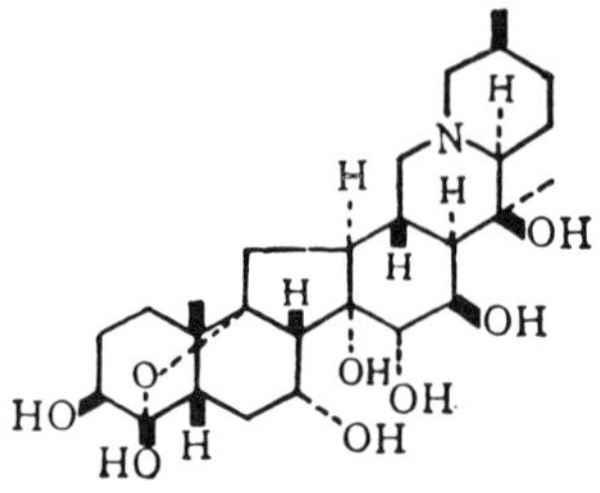

Germine

Actions

The major effect of these alkaloids is mucosal irritation. Oral administration of this herb can cause nausea and vomiting. If it is inhaled into the nasal cavity, it can cause sneezing and coughing.

The herb can lower blood pressure and slow the heart rate, probably through a vagal-stimulating effect. It also has antibacterial and insecticidal effects.

This herb is toxic. Oral administration of 1.8 g/kg to rats can cause death due to cardiac arrhythmias.

Therapeutic Uses

The herb is administered in powder form to produce emesis, with a standard oral dose of 1.5 to 3 g. A 1 to 5% water extract of the powder is commonly used in rural areas to kill flies and larvae found in toilets.

BAN XIA (半夏) — dried tuber of *Pinellia ternata* (Thunb.) Breit.

Chemistry

The tuber contains essential oils, *l*-ephedrine, choline, and several amino acids.

Actions

The herb is an antiemetic, probably due to its inhibitory effect on the CTZ reflex. The principle antiemetic substance is temperature resistant and was isolated from organic solvents. Its complete structure has not yet been identified. This herb also has antitussive and expectorant effects, a cardiac inhibitory effect, and is an antidote for strychnine intoxication.

Therapeutic Uses

Chinese traditional medicine used this herb to remove dampness and resolve phlegm, to smooth the upward adverse flow of *qi*, and arrest vomiting. It is applied externally to wounds to obtain a hemostatic and analgesic effect.

WU ZHU YU (吴茱萸) — the dried ripe fruit of *Evodia rutaecarpa* (Juss.) Benth., *E. rutaecarpa* Benth. var. *officinalis* (Dode) Huang, or *E. rutaecarpa* Benth. var. *bodinieri* (Dode) Huang

Chemistry

The fruit contains several alkaloids, including evodiamine, rutaecarpine, wuchuyine, and rutavine. It also has the essential oils evodene, evodol, and evodin.

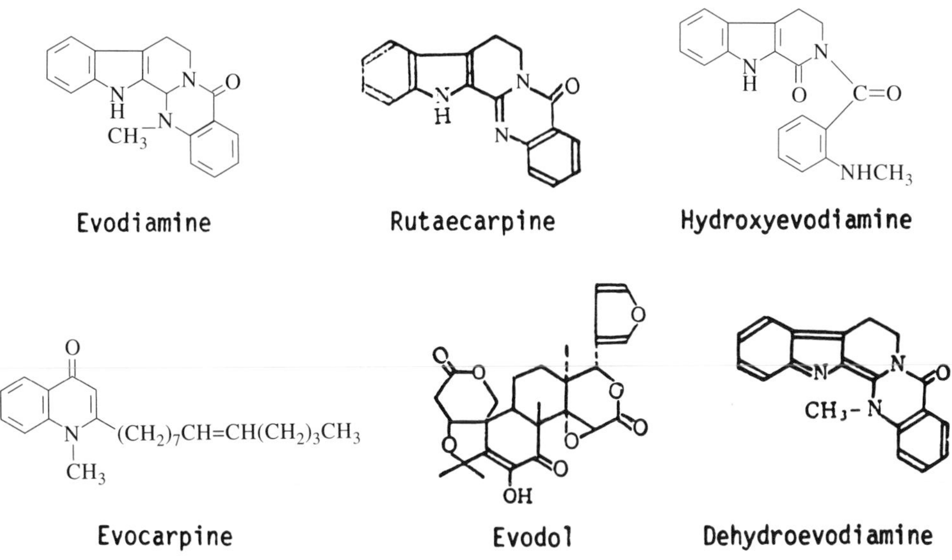

Actions

It has been shown experimentally that this herb and its extract have an antiemetic effect when given orally. It is analgesic and can contract the uterus, lower blood pressure, and exert antibacterial effects.

For thousands of years, this herb was used in the treatment of dysentery. In folk medicine, it was believed that one could repel infection by devils by hanging the herb near the body.

XIANG SI ZI (相思子) — dried seed of *Abrus precatorius* L.

Chemistry

Several alkaloids have been isolated from the seed, including L-abrine ($C_{12}H_{14}O_2N_2$), precatorine ($C_{14}H_{11}O_6N$), hypaphorine ($C_{14}H_{17}O_2N_2$), trigonelline ($C_7H_7O_2N$), cycloarterenol ($C_{30}H_{50}O$), squalene, and 5-β-cholanic acid.

L-Abrine Hypaphorine Precatorine

Therapeutic Uses

This herb was listed in the *Ben Cao Kong Mu* as an antiemetic agent. It is also used as an expectorant and parasiticide.

CHEN XIANG (沉香) — resinous wood of *Aquilaria sinensis* (Lour.) Gilg.

Chemistry

The herb contains several essential oils and resins, including benzylacetone ($C_6H_5CH_2COCH_3$), hydrocinnamic acid ($C_6H_5CH_2CH_2COOH$), agarospirol ($C_{15}H_{28}O$), agarol, α,β-agarofuran ($C_{15}H_{24}O$), dihydroagarofuran ($C_{15}H_{26}O$), norketoagarofuran ($C_{14}H_{22}O_2$), 4-hydroxyagarofuran ($C_{15}H_{26}O_2$), and 3,4-di-OH-dihydroagarofuran.

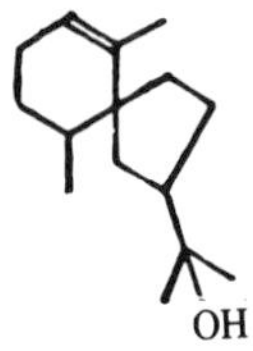

OH

Agarospirol

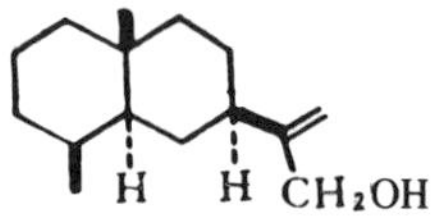

H H CH₂OH

α-Agarofuran

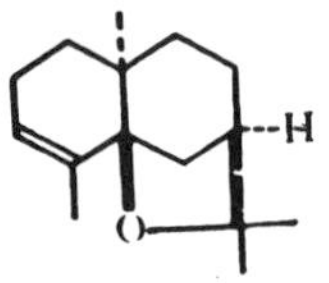

H O

Agarol

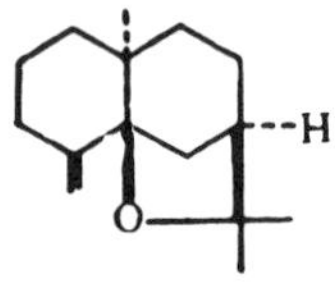

H O

β-Agarofuran

Therapeutic Uses

This herb was used as an antiemetic and to rest the stomach, promote circulation of *qi*, and relieve pain.

Chapter 21

CHOLERETIC AND ANTIHEPATITIS HERBS

The ancient Chinese were familiar with the diagnosis of liver dysfunction, as manifested by symptoms such as jaundice and ascitis. Treatment of this condition was based on the philosophy of balanced *yin* and *yang*: administer herbs clean the stagnation of the organ, circulate *qi*, restore *yang*, and displace ''heat''. While the mechanisms of this *yin-yang* treatment do not fit into western scientific understanding, the herbs used in these treatments have had undeniable effect in relieving these symptoms.

YIN CHEN (茵陈) — dried young shoot of *Artemisia capillaris* Thunb. or *A. scoparia* Waldst. et Kit.

Chemistry

The shoot contains scoparon, chlorogenic acid, caffeic acid, β-pinene, capillin, capillon, capillene, capillarin, and 4-*OH*-acetophenone. It also contains some essential oils.

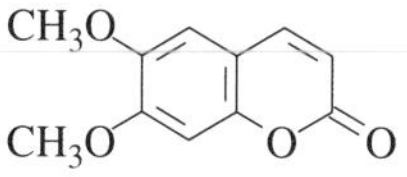

Scoparon 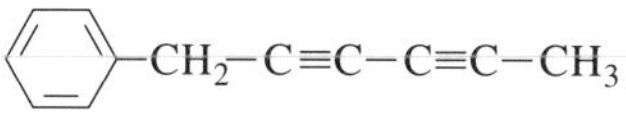   Capillene 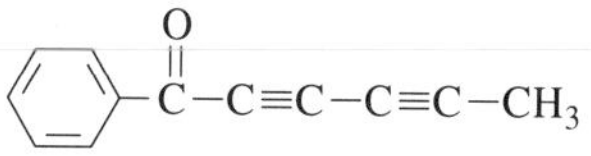   Capillin

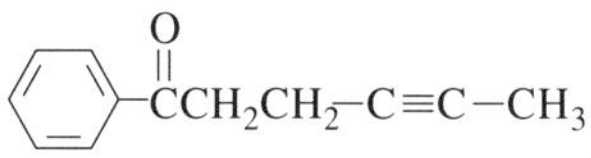

Capillon 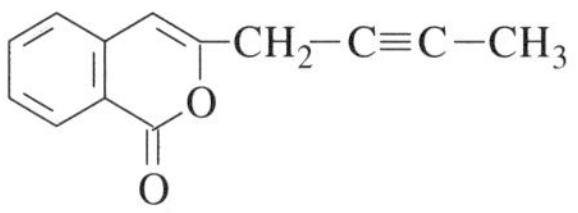   Capillarin 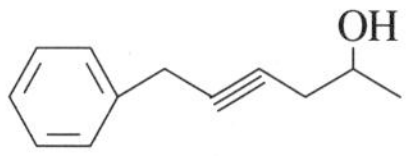   Capillanol

Actions

This herb is primarily used as a choleretic. It is the major ingredient in an old Chinese formula used to treat jaundice. Scoparon, chlorogenic acid, and caffeic acid have been experimentally shown to possess a remarkable ability to stimulate bile secretion and protect the liver against CCl_4 injury. Clinical trials showed that *Yin Chen* is effective in reducing jaundice, curing acute infectious hepatitis, and in treating gallstone-related illnesses. This herb is also effective in treating newborn kernicterus.

The herb is an effective antipyretic, due to its content of essential oils. It has antibacterial and antiviral properties. It has been observed to lower blood pressure and blood cholesterol levels, and to act as an antiasthmatic agent.

Adverse effects include nausea, abdominal distension, and dizziness.

Therapeutic Uses

Yin Chen may be used in hepatitis, infected cholecystitis, and hyperlipemia.

DONG YAO (当药) — the dried whole plant of *Swertia pseudochinensis* Hara

Chemistry

This plant contains the glucosides swertiamarin and swertisin ($C_{22}H_{22}O_{10}$), homoorentin ($C_{21}H_{20}O_{11}$), isovitexin, bellidifolin ($C_{14}H_{10}O_6$), methylbellidifolin, methylswertianin, and decussatin.

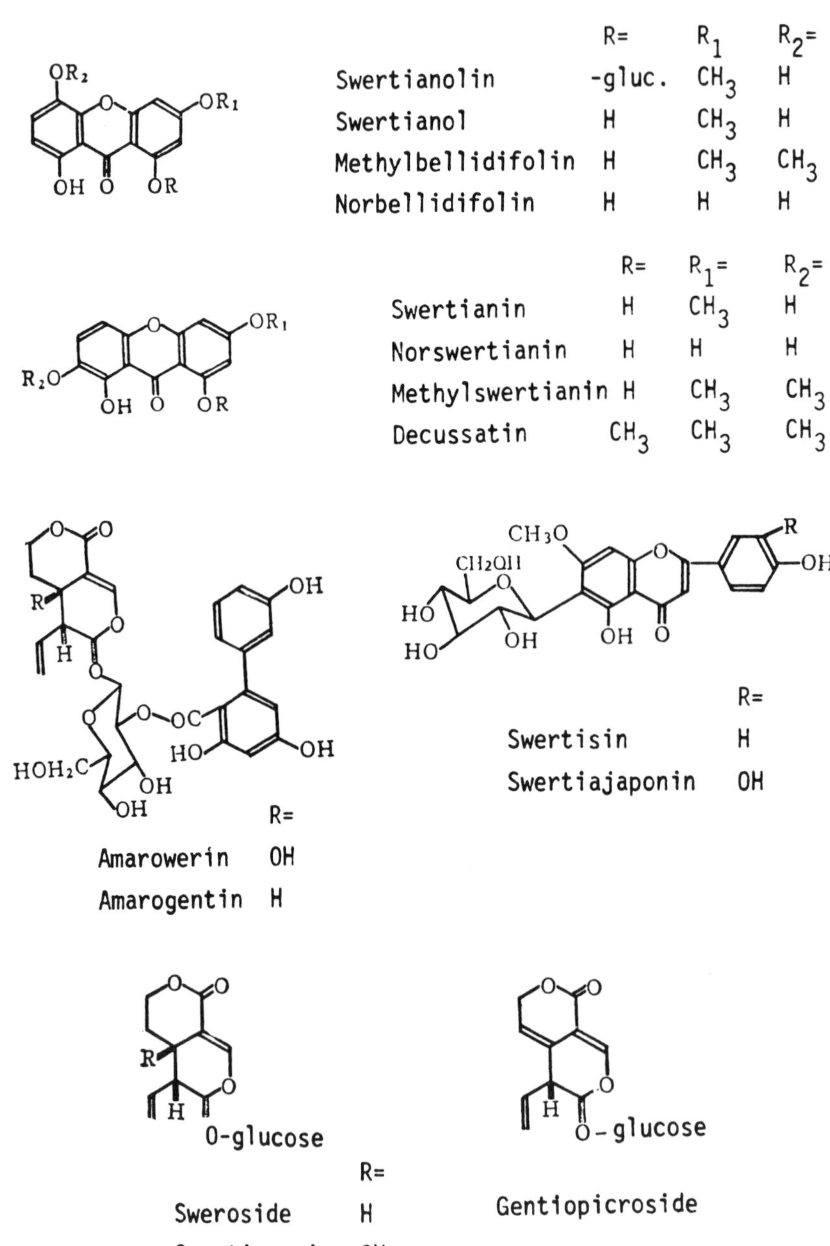

Actions

This herb has a protective effect on the liver and is effective against CCl_4-induced injuries. It is choleretic and can improve hepatic function. Women usually respond to this herb better than men.

Therapeutic Uses

Clinically, the herb is used in the treatment of acute icteric hepatitis and chronic liver diseases. It is generally administered in tablet form (equivalent to 1.2 g), with five tablets taken twice a day. The herb may occasionally cause diarrhea or dizziness, which may be resolved by reducing the dose.

QING YE DAN (青叶胆) — dried whole plant of *Swertia mileensis* (Amaranthaceae)

Chemistry

The herb contains oleanolic acid and some bitter substances.

Therapeutic Uses

It is mainly used in the treatment of acute icteric hepatitis. It is available in tablet form, with each tablet containing 1.57 g of raw material; the standard dose is four to five tablets, taken four times each day.

WU WEI ZI (五味子) — the dried fruit of *Schisandra chinensis* or *S. sphenanthera*

FIGURE 1. *Wu Wei Zi* (or North *Wu Wei Zi*).

Chemistry

This herb contains several hydrocarbon derivatives, including sesquicarene ($C_{15}H_{21}$), β-2-bisabolene, β-chamigrene, α-ylangene, and the glycosides schizandrin, γ-schizandrin, schizandrol, and deoxyschizandrin.

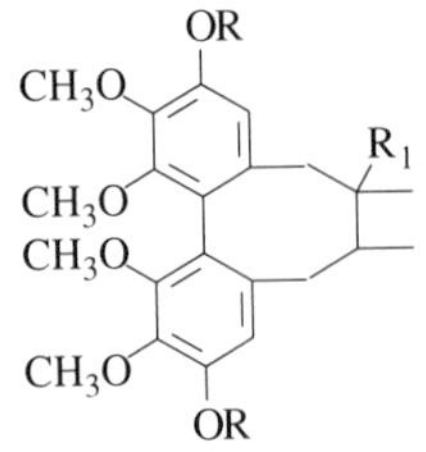

	R=	R_1=
Schizandrin	CH_3	OH
Deoxyschizandrin	CH_3	H
Schizandrol	H	OH

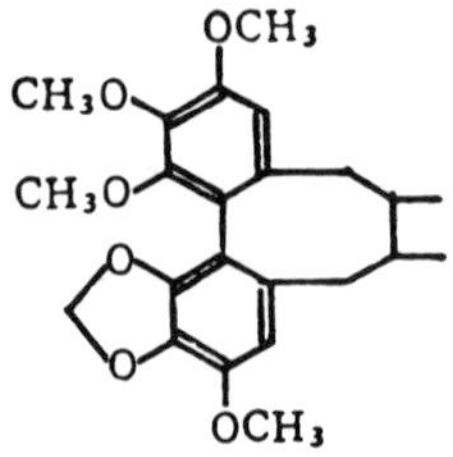

γ-Schizandrin

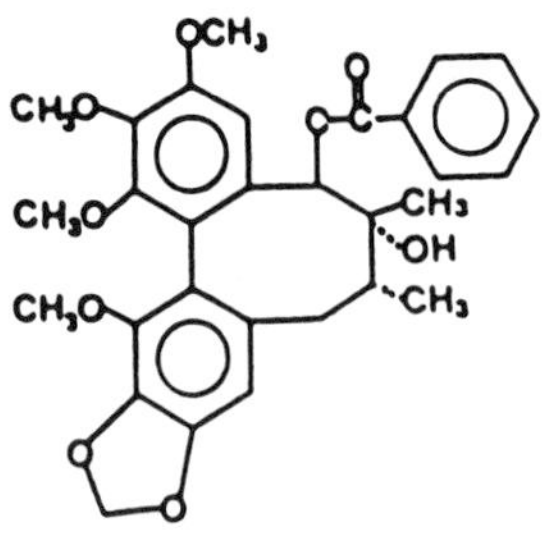

Schizandrer A

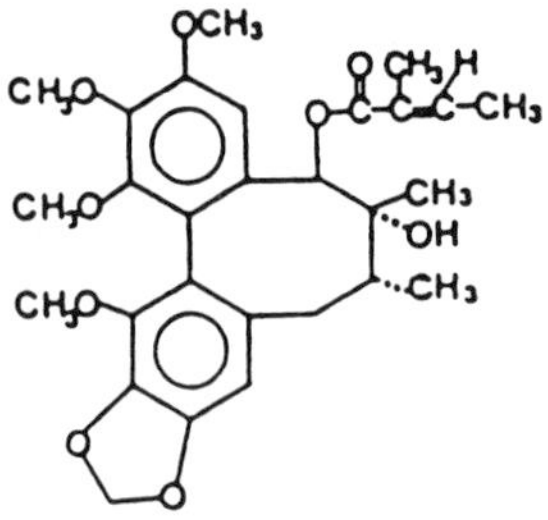

Schizandrer B

Actions

The *Herbal Classic* placed this herb in the upper category.

This herb and its active principle, γ-schizandrin, have a protective effect on the liver and can lower γ-glutamyl transferase activity. In animals exposed to CCl_4, treatment with *Wu Wei Zi* or γ-schizandrin produced a remarkable improvement in liver function and recovery of the degenerated tissue. Protein synthesis and cytochrome P-450 content were greatly increased.

It has also been claimed that *Wu Wei Zi* can improve mental functions and antagonize the central convulsive effective of caffeine. It can, however, potentiate the stimulating effect of strychnine.

The alcohol extract of the herb can prolong barbiturate sleeping time in mice. This effect is believed to be due to inhibition of barbiturate metabolism in the liver.

Wu Wei Zi has a direct stimulating effect on the respiratory center. It increases both the frequency and amplitude of respirations and can effectively antagonize the respiratory depressant effect of morphine.

The water or alcohol extract of the herb can increase myocardial contractility but has little effect on blood pressure. In addition, the herb has antitussive and expectorant properties and can stimulate uterine contraction. It was also claimed to have a potent sperm-stimulating effect.

The herb is relatively nontoxic. Doses up to 5 g/kg do not produce mortality in mice.

Therapeutic Uses

This herb is used in the treatment of nonicteric hepatitis. Clinical trials showed that the herb's efficiency rate in effecting recovery, especially in chronic cases, reached 80% after 9 weeks of administration. A dosage of 3 g t.i.d. for 1 month is recommended. The serum transferase level usually returns to normal after this period.

The herb is also traditionally used to treat coughs, neurasthenia, dysentery, and indigestion. For these conditions, it is taken in the form of a tincture, 2 to 5 ml t.i.d.

Reference

Liu, G. T., in *Advances in Chinese Medical Materials Research,* Chang, H. M. et al., Eds., 1985, World Scientific Publishing Co., Singapore, 257–268.

ZHI ZI (栀子) — the dried ripe fruit of *Gardenia jasminoides* Ellis var. *radicans* Mak.

Chemistry

The fruit contains the following major principles: gardenin, gardenoside, shanzhiside, usolic acid, crocin, and crocetin.

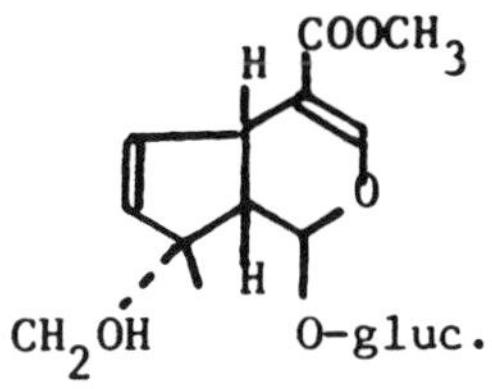

Gardenoside 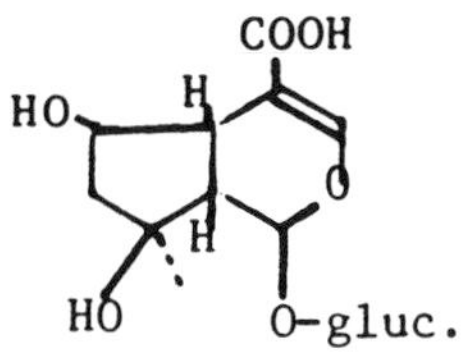 Shanzhiside 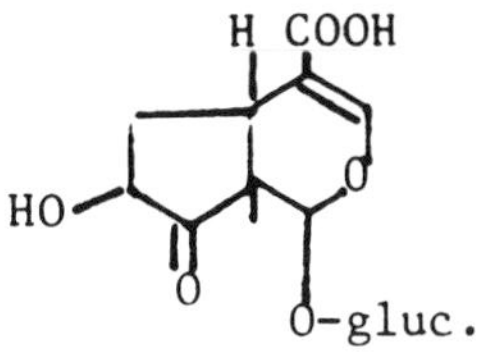 Cardoside

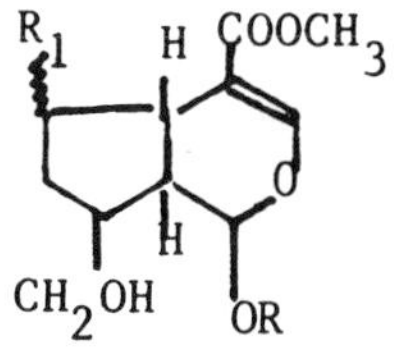

	R =	R_1 =
Geniposide	-glucose	H
Genipingentiobioside	-gentiobiose	H
Scandoside methylester	-glucose	β-OH

Actions

The herb is an effective choleretic. The water extract of the fruit, gardenoside, and gardenin can each stimulate biliary secretion and reduce plasma bilirubin levels.

It displays sedative, hypnotic, anticonvulsant, antibacterial, and anthelmintic properties. It has been observed to lower body temperature; the duration of this effect is prolonged. It can lower blood pressure, probably through an action on the medullary sympathetic center. It is a hemostatic; external application can produce an anti-inflammatory effect. In addition, it can inhibit gastric secretion and mobility.

Therapeutic Uses

This herb is primarily used in the treatment of acute icteric hepatitis. It can also be applied externally to traumatic wounds to produce analgesic, antibacterial, and hemostatic effects.

DI ER CAO (地耳草) — dried whole plant of *Hypericum japonicum* Thunb.

Chemistry

The plant contains quercetin, quercitrin, and some glucosides.

Actions

The herb has antipyretic and detoxificant effects. It is effective in the treatment of acute icteric hepatitis. It also has antibacterial activity, slows the heart rate, and lowers blood pressure.

Therapeutic Uses

Traditional Chinese medicine prescribed this herb as an antipyretic, diuretic, antiphlegm agent, and detoxicant.

YU JIN (郁金) — dried tuberous root of *Curcuma aromatica*, *C. kwangsiensis*, *C. longa*, or *C. zedoaria* Rosc.

Chemistry

Approximately 6% of the herb consists of an essential oil. The major principles of this oil are *l*-curcamene (65.5%), sesquiterpene (22%), camphor (2.5%), and camphene (0.8%).

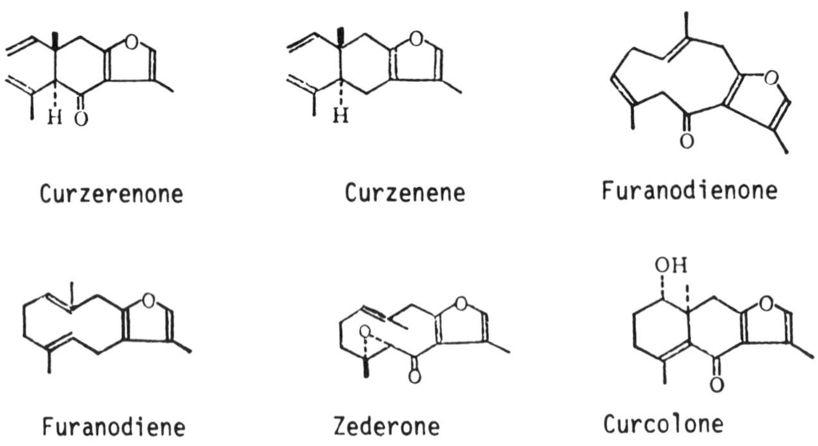

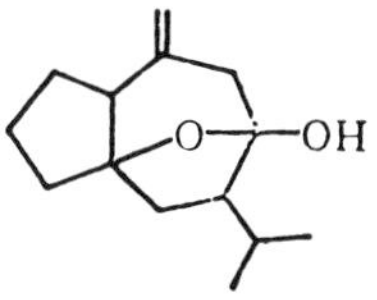

Curcumol 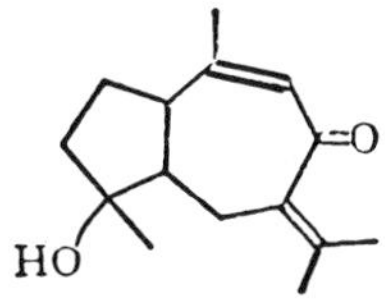   Procurcumenol 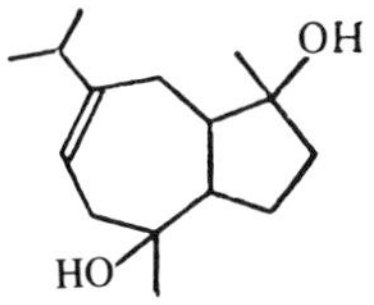   Curcumadiol 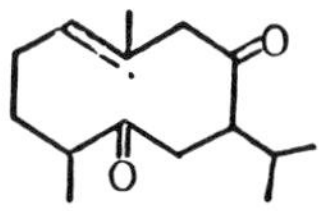   Curdione

Actions

The essential oil of *Yu Jin* has been found to stimulate contraction of the gallbladder and increase the secretion of bile.

Therapeutic Uses

In folk medicine, this herb was used to activate blood flow, remove blood stasis, promote the circulation of *qi*, relieve depression, remove ''heat'' from the heart, resuscitate and normalize gallbladder functions, and cure jaundice. It was claimed that the herb can solubilize gallstones.

CHU PEN CAO (垂盆草) — dried whole plant of *Sedum sarmentosum* Bge.

Chemistry

This plant contains a glucoside which is effective in the treatment of acute icteric hepatitis, but less effective in treating chronic nonicteric hepatitis. It also contains two alkaloids: *N*-methylisopelletierine and dihydro-*N*-methylisopelletierine.

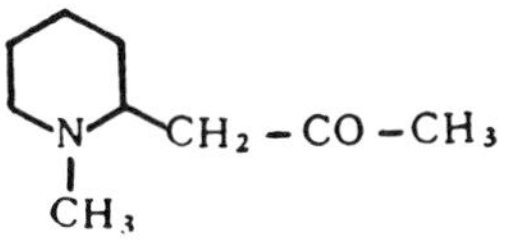

N-Methylisopelletierine 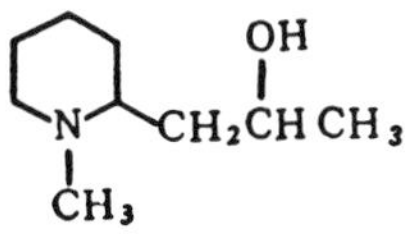   Dihydro-N-methylisopelletierine

Therapeutic Uses

The herb is prescribed as an antipyretic, detoxicant, and diuretic. It is also applied externally in the treatment of mastitis and furuncle, to promote the drainage of pus.

SHUI FEI JI (水飞蓟) — the dried fruit of *Silybum marianum* (L.) Gaertn.

Chemistry

The fruit contains silybin, silymarin, silydiamin, silyckristin, dehydrosilybin, and silybinomer.

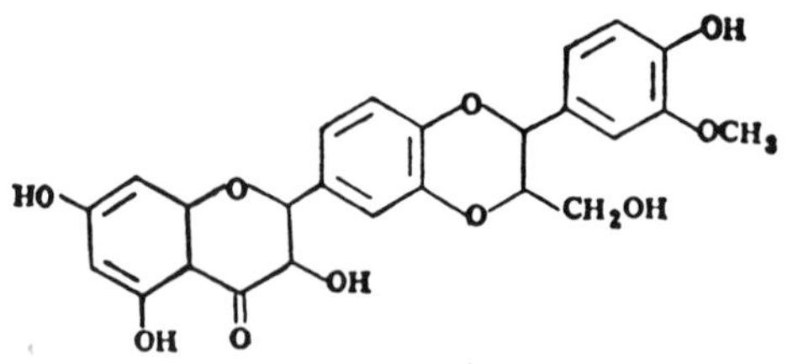

Silymarin

Actions

The herb can help maintain normal functioning of the liver. Its protective action lasts for over 30 h. It promotes the regeneration of injured hepatic cells and increases glycogenesis and nucleic acid metabolism.

It can increase the secretion of bile into the duodenum and has also been observed to have protective effects on brain tissue and against X-irradiation.

Chapter 22

TONICS AND SUPPORTING HERBS

This chapter discusses herbs that were described in Chinese pharmacopoeias as providing nourishment to strengthen patients' *yang* and replenish *yin*. These were classified by *Sheng Nung* in the upper class of herbs, and only the rich and the aristocracy could afford to use them in support of weakening health.

They are subjectively considered effective as sedatives, pain relievers, and agents to treat impotency. They are multifunctional, but are treated in Chinese medical texts primarily as tonics or body-supporting agents. We have opted to place them with herbs that affect the alimentary system.

DONG CHONG XIA CAO (冬虫夏草) — the dried fungus *Cordyceps sinensis*, grown on the larvae of the caterpillar *Hepialus armoricanus* Oberthuer

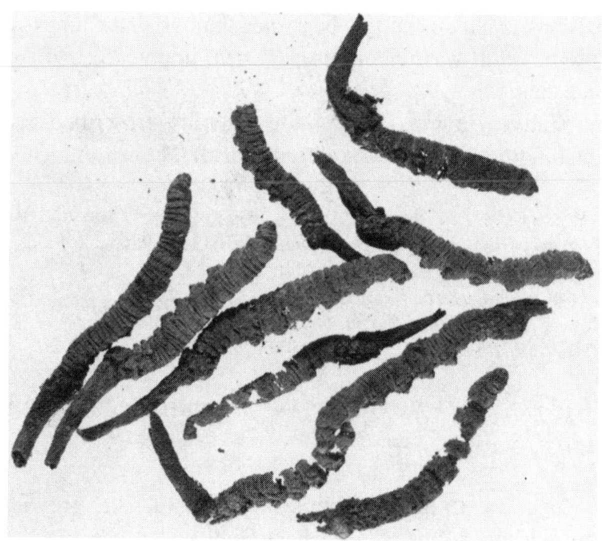

FIGURE 1. *Cordyceps sinensis.*

Chemistry

Between 25 and 32% of the herb consists of crude protein, cordycepin 3′-deoxyadenosine ($C_{10}H_{13}O_3N_6$), and *d*-mannitol.

NH_2 N N N N O CH_2OH OH

Cordycepin

Actions

The fungus has an antiasthmatic effect. It causes smooth muscle relaxation and can potentiate epinephrine effects. It is also an anticancer agent. Cordyceptin can inhibit the growth of Ehrlich ascites cells and has increased the survival time of animals inoculated with these cancer cells.

It has antibacterial properties and has been reputed to stimulate the reticuloendothelial and endocrine systems, with a stress-reducing effect. Huang et al.[2] fed the extract of this herb to a group of young male rats and found increases in the weight of the rats' testes and in the sperm counts. Lin et al.[3] reported that this herb can stimulate the immune system.

Chen et al.[1] treated 30 patients of chronic renal failure with *Dong Chong Xie Cao* and reported that a significant improvement of symptoms was observed, for example, an increase of creatinine clearance (C_{cr}), an increased hemoglobin, and stimulation of reticulocytes in the blood count, and also a decrease of BUN.

Tang et al.[4] also found that this herb can increase the monocyte phagocytosis, regulate the immunoactivity and increase the hepatic and spleen weight, and antagonize the anti-immunity effect induced by injection of cortisol in mice.

Therapeutic Uses

The Chinese considered this herb to be a tonic to nourish the lungs and kidneys, replenish sperm, and reinforce *qi*. It was widely used to treat impotence and spermatorrhea, neurasthenia, and backache.

The herb is relatively nontoxic. It is not generally prescribed as a drug or medicine, but as a nutritional additive to be cooked with meat. The recommended dose is 3 to 9 g.

References

1. **Chen, Y. P. et al.,** *Chin. Trad. Herbal Drug,* 17, 16–19, 1986.
2. **Huang, Y. C. et al.,** *Chin. Herbs Res.,* October, 24–25, 1987.
3. **Lin, S. M. et al.,** *Chin. Herbs Res.,* December, 22–23, 1987.
4. **Tang, R. J. et al.,** *Chin. Trad. Herbal Drug,* 17, 22–25, 1986.
5. **Zhang, S. M.,** *Bull. Pharmacol.,* 16, 13, 1981.

QIONG ZHI (琼脂), Agar-Agar — the mucous substance isolated from *Gelidium amansii* Lamx.

In *Ben Cao Kong Mu*, Li described this plant as coral-like, growing in the South China Sea. It is tasteful and nutritious. Placed in boiling water, the plant yields a substance in emulsion form that coagulates later.

Chemistry

Qiong Zhi contains two kinds of polysaccharides, the agarose (approximately 0.14%) and agaropectin (approximately 0.6%). It also has a small quantity of taurine, NH_2–$CH_2CH_2SO_3H$, *N,N*-dimethyltaurine, $(CH_3)_2NCH_2CH_2SO_3H$, choline, vitamin B_2 and 24-methylene cholesterol. Agarose $(C_{12}H_{18}O_9)_n$, has a molecular weight of 120,000. Water hydrolysis of agarose gives the product, agarobiose, $C_{12}H_{20}O_{10}$, enzyme hydrolysis yields the product neoagarobiose.

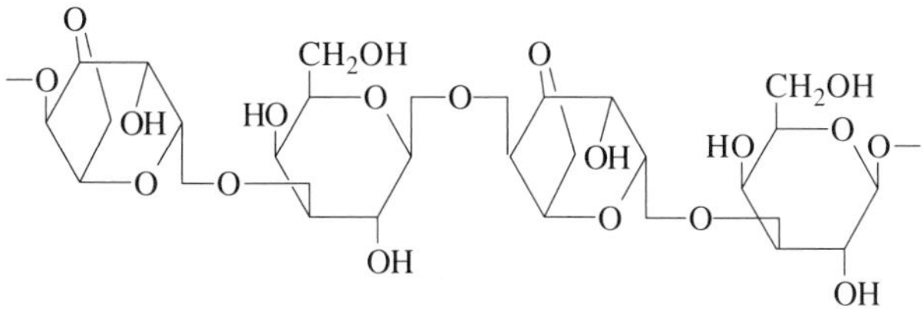

Partial structure of agarose.

Agaropectin has a molecular weight of 12,600. Its viscosity is slightly lower than that of agarose.

Therapeutic Uses

Agar-agar is mainly used as culture base for bacteria culture. It also has been used as mild laxative in the treatment of chronic constipation.

In Chinese food manual, agar-agar is served as a salad.

LU RONG (鹿茸) — hairy young horn of *Cervus nippon* Temminck or *C. elaphus* L.

Chemistry

The horn contains several amino acids (approximately 50% by weight), sugars, and vitamin A. In addition, *Lu Rong* contains the sex hormones estrone and estranediol, sphingomyeleine, ganglioside, and some prostaglandins.

Actions

The ''herb'' stimulates the growth of body tissues, especially reticuloendothelial cells and leucocytes. It enhances body activity and reduces fatigue by improving sleep and stimulating the appetite. It improves body metabolism and strengthens overall health, especially in elderly persons and young children. Additionally, it stimulates the sexual function in females, and can stimulate wound healing.

Therapeutic Uses

Lu Rong was used to reinvigorate the *yang* of the kidneys, replenish vital essences and blood, and strengthen muscles and bone.

It is prepared as a 20% tincture in wine, and administered in doses of 10 ml three times a day in the treatment of thrombocytopenia and leukopenia.

BAI SHAO (白芍) — dried peeled root of *Paeonia lactiflora* Pall.

Chemistry

The root contains several glucosides, including paeoniflorin ($C_{23}H_{23}O_{11}$), oxypaeoniflorin ($C_{23}H_{28}O_{12}$), albiflorin, and benzoylpaeoniflorin. Some essential oils are also present.

Actions

Bai Shao increases the leucocyte count, including lymphocytes. It has antispasmodic, sedative, analgesic, and antipyretic properties. It has been observed to cause coronary dilatation and peripheral vasodilatation.

The *Ben Cao Kong Mu* described it as having antibacterial and antidiarrheal activity. The water extract of the herb can inhibit the growth of dysentery bacilli and staphylococci. It also inhibits platelet aggregation.

Therapeutic Uses

In traditional Chinese medicine, this herb was used to check hyperactivity of the liver, to "soften" the liver and relieve pain, to act as an astringent on *yin*, and to nourish the blood.

It is commonly used in combination with other herbs to treat female diseases, stop excessive menstruation, and to relieve abdominal pain.

BA JI TIAN (巴戟天) — dried root of *Morinda officinalis*

Actions and Therapeutic Uses

The root can increase the leukocyte count and stimulate the endocrine system. It was used to invigorate the *yang* of the kidneys, and to strengthen muscles and bone. It is also prescribed as an antirheumatic agent.

XU DUAN (续断) — dried root of *Dipsacus asper*

Actions and Therapeutic Uses

The herb can increase the leukocyte count. It was used to tone the liver and kidneys, strengthen muscles and bones, prevent spontaneous abortion, and smooth blood circulation.

TU SIZI (菟丝子) — dried ripe seed of *Cuscuta chinensis*

Actions and Therapeutic Uses

This seed increases the rate of turnover of lymphatic tissues, improves body immunity, and increases blood sugar metabolism. It was traditionally used to improve the tone of the liver and kidneys, replenish vital essences, improve visual acuity, and to prevent spontaneous abortion.

SUO YANG (锁阳) — dried fresh stem of *Cynomorium songaricum*

Actions and Therapeutic Uses

Suo Yang has been reported to improve body immunity and to stimulate the endocrine system. It is prescribed by Chinese herbalists to improve the overall tone of the kidneys, to invigorate *yang*, and to lubricate the intestines and relieve constipation.

XIAN MAO (仙茅) — dried rhizome of *Curculigo orchioides*

Actions and Therapeutic Uses

The herb has been observed to improve body immunity and stimulate the endocrine system. It is generally used to "warm" the kidneys and invigorate the overall *yang* of patients, to dispel "coolness", and to remove "dampness".

BU GU ZHI (补骨脂) — dried ripe fruit of *Psoralea corylifolia* L.

The plant was imported from Persia and is now grown mainly in Szechuan, Kwangtung, and Yunnan provinces.

Chemistry

The major components isolated from this plant are: psoralen, isopsoralen, coryfolin, corylifolinin, bavachinin, and *d*-backuchiol.

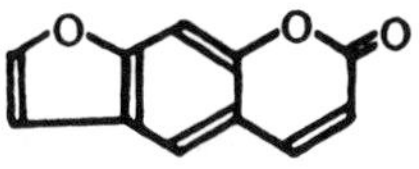

Psoralen 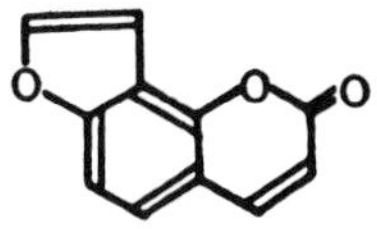 Angelicin 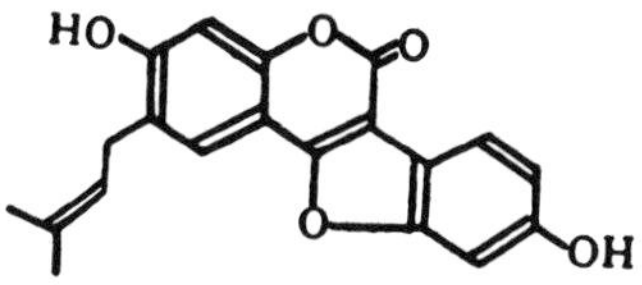 Psoralidin

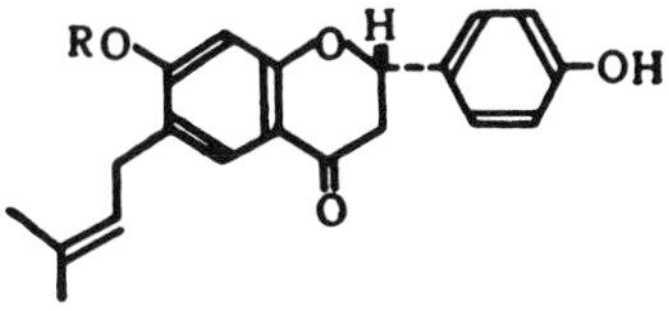

	R=
Coryfolin	H
Bavachinin	CH_3

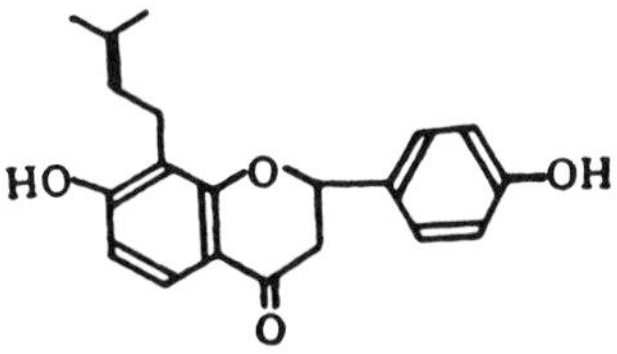

Isobavachin

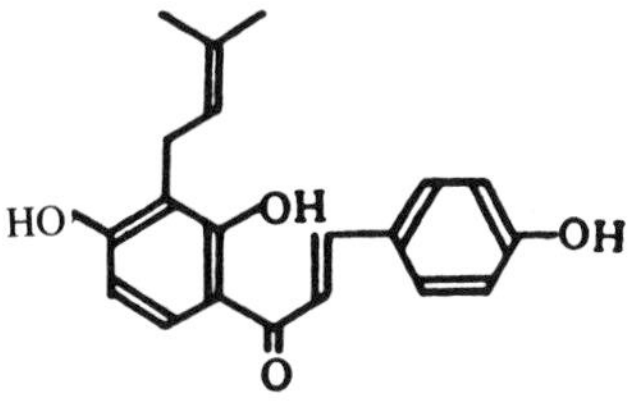

Corylifolinin

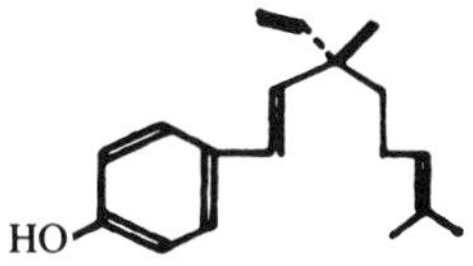

d-Backuchiol

Actions

At a concentration of 10^{-5} to 10^{-6}, coryfolinin produces a marked coronary vasodilatating effect and increases the myocardial contraction. It has both antibacterial and anticancer action. *In vitro* the herb can contract isolated smooth muscles preparations, but *in vivo* it can relax the bronchomuscles.

This herb can shorten the bleeding time and produces a photosensitive effect to stimulate pigmentation.

Therapeutic Uses

Folk medicine used it to invigorate the kidney-*yang* and to warm the spleen. The *Ben Cao Kong Mu* described this herb being used as abortifacient and to prevent early fertilization.

It was used in the treatment of neurosis, male impotence, and frequent nocturnal emission. Dose was 3 to 6 g made in a decoction.

GOU QI ZI (枸杞子) — dried fruit of *Lycium barbarum* L. or *L. chinense* Mill.

Chemistry

The fruit contains betaine (0.1% of content by weight), zeaxanthin, physalein, and several vitamins, including carotine, nicotinic acid, and vitamin C.

Actions and Therapeutic Uses

The herb's effects include an increase in leukocytes (including lymphocyte) count, an increase in nonspecific immunity, and stimulation of tissue development. Chinese physicians prescribe it to strengthen muscles and bone, protect liver functions, replenish vital essences, and to improve visual acuity. It is also claimed that the water extract of the herb can lower blood pressure and stimulate the heart.

BAI HE (百合) — dried bulb of *Lilium lancifolium, L. browni viridulum* or *L. lumilum*

Actions and Therapeutic Uses

Traditional Chinese medicine recommended this herb to ''moisten'' the lungs and arrest coughing, ease anxiety, and improve digestion. It has been found to increase the leukocyte count.

MI MENG HUA (蜜蒙花) — dried flower bud and inflorescence of *Buddleja officinalis*

Chemistry

It contains the glycoside buddleoglycoside ($C_{25}H_{32}O_{14}{\cdot}H_2O$), which on hydrolysis will yield the genin acacetin and sugar.

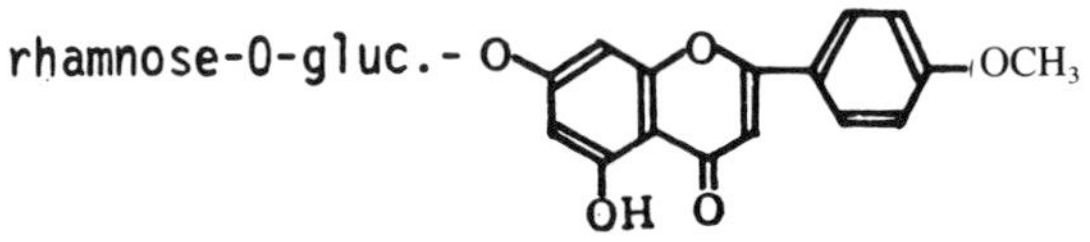

rhamnose-O-gluc.- O ... OCH_3

OH O

Buddleoglycoside

Therapeutic Uses

The herb was used to remove ''heat'' from the liver, improve visual acuity, and in the treatment of malaria.

Section VI
Respiratory System

THE RESPIRATORY SYSTEM

Coughing, sneezing, and breathing difficulties are symptoms frequently accompanying changes in climate. For thousands of years, herbs were widely tried as antitussives and expectorants. All over China, prepackaged herbs for treatment of coughing and the common cold are readily available in any apothecary, without need of a doctor's prescription; these are analogous to Western ''over-the-counter'' drugs.

In the 18th Century, after opium was introduced by force into China by Westerners, it was a common practice for the *Lau Bai Sin* (the common people) to walk into an opium den and pay a fee to take one or two inhalations from an opium pipe in order to treat coughing and reduce dyspnea. Certainly, such remedies were not mentioned in the medical textbooks or in the *Ban Cao Kong Mu*. Since opium was not originally grown in Chinese soil, Chinese apothecaries did not sell opium or its derivatives as antitussive drugs.

Chapter 23

ANTITUSSIVE HERBS

Coughing is a reflexive movement. The mechanism involves an irritation or stimulation of the tracheal or bronchial mucosa, initiating an impulse carried by afferent nerve fibers to the medullary cough center and then through the vagus nerve back to the respiratory muscles. Antitussive herbs exert their effect primarily by inhibiting the medullary cough center, though some may also act directly on the respiratory mucosa to reduce irritation.

In traditional Chinese medicine, ''coughing was considered to properly originate in the lung, but not the lung alone; the other 'Five Organs and Six Viscera' (五脏六腑) also participate . . . .'' Therefore, herbs which showed a supporting action on other organs or viscera were also considered beneficial in combating the cough.

XING REN (杏仁) — bitter almond, or the dried ripe kernel of *Prunus armeniaca* L. var. *ansa* Maxim., *P. sibirica* L., and *P. manshurica* Koehne

Chemistry

Bitter almond contains the glucoside amygdalin and the enzyme amygdalase.

$$C_6H_5\text{-}\underset{CN}{\overset{H}{C}}\text{-O-Gentiobiose} \xrightarrow[H_2O]{\text{Emulsin}} C_6H_5\text{-}\underset{CN}{\overset{H}{C}}\text{-O glucose} + \text{glucose} \longrightarrow C_6H_5\text{-}\underset{CN}{\overset{H}{C}}\text{-OH} + \text{glucose}$$

Amygdalin Prunasine Mandelonitrile

Actions

The enzyme amygdalase and pepsin of gastric juice can hydrolyze amygdalin to produce a small quantity of cyanic acid, HCN, which can stimulate the respiratory center reflexively and produce a tranquilizing effect. This produces the herb's antitussive and antiasthmatic effects.

Overdosage can cause cyanide intoxication, especially in children. The toxic symptoms usually appear between 0.5 and 5 h after ingestion. These include dizziness, nausea, and vomiting; in extreme cases, the patient may become comatose and die from tissue anoxia.

The bitter almond also contains a large amount of fat, which can lubricate the intestine and produce a laxative effect.

BAI BU (百部) — the dried root of *Stemona japonica* (Bl.) Miq., *S. sessilifolia* or *S. tuberosa*

Chemistry

The root contains the alkaloids stemonine, stemonidine, isostemonidine, protostemonine, and others.

Tuberostemonine Stenine Isostemonamine

Stemonine Protostemonine

Actions and Therapeutic Uses

The herb and its alkaloids can suppress excitation of the respiratory center and inhibit the cough reflex. It also exerts an antituberculous, antibacterial, and antifungal effect. It has a pesticidic effect, especially to kill *Pediculus capitus*. It was used to treat cold, cough, ascarids, tinea, and lice.

Generally the dose is 6 to 18 g.

The alcoholic extract of the herb is used as an insecticide and as a household spraying agent.

BEI MU (貝母), ZHI BEI MU (浙貝母), or CHUAN BEI MU (川貝母) — the dried bulbs of *Fritillaria verticillata* Willd. var *thunbergii* Mig., or *F. cirrhosa* D. Don

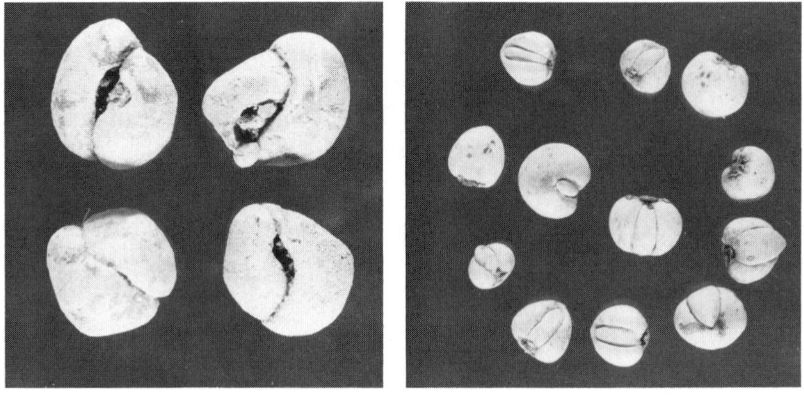

A B

FIGURE 1. (A) *Zhi Bei Mu*, (B) *Chuan Bei Mu*.

TABLE 1
Compounds Isolated from Different *Bei Mu*

Fritillaria	Habitat	Compound isolated	Melting point (°C)
F. verticillata	Japan	Fritilline, $C_{25}H_{41}O_3N{\cdot}H_2O$	214 Crystal form
		Fritillarine, $C_{19}H_{33}O_2N$	130–131 Amorphous
		Verticine, $C_{18}H_{33}O_2N$	224–224.5
		Verticinine, $C_{19}H_{33}O_2N$	130
F. verticillata var. *thunbergii*	Zhejiang	Peimine, $C_{26}H_{43}O_3N$	—
		Peiminine, $C_{26}H_{41}O_3N$	—
		Peimisine, $C_{27}H_{43}O_4N$	270
		Peimiphine, $C_{27}H_{42}O_3N$	127
		Peimidine, $C_{27}H_{45}O_2N$	222
		Peimilidine, $C_{27}H_{47}O_3N$	188
	Szechuan	Fritimine $C_{38}H_{62}O_3N_2$	167
		Fritiminine, beilupeimine, chinpeimine, sonpeimine (also a small quantity of verticine, verticinine, fritillarine)	
		Propeimin ($C_{27}H_{46}O_3$)	

Chemistry

Whether this herb was called *Zhi Bei Mu* or *Chuan Bei Mu* depended upon its habitat; *Zhi Bei Mu* grows in Zhejiang province on the east coast of China, while *Chuan Bei Mu* grows in Szechuan province in southwestern China.

Several alkaloids have been isolated from the plant. They differ in structure, depending upon whether they are derived from *Zhi Bei Mu* or *Chuan Bei Mu*. Table 1 provides a summary of these alkaloids.

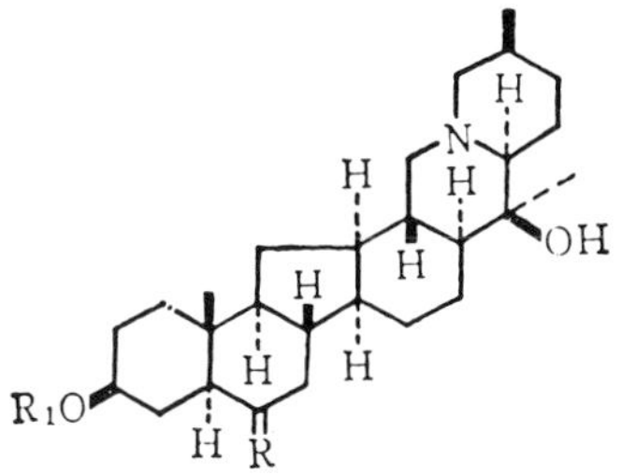

	R=	R_1=
Peimine (Verticine)	H,α-OH	H
Peiminine (Verticinone)	O	H
Peiminoside	H,α-OH	β-D-glucose

Actions

This herb causes bronchodilatation and inhibition of mucosal secretions. It also inhibits salivary secretion, but is less potent than atropine. It is a good antitussive agent. In laboratory experiments, coughing was induced in mice by administration of ammonia water. When the animals were treated with *Bei Mu* alkaloids, at 0.25 to 0.5 mg per mouse, a definite relief of coughing was observed.

The herb can stimulate uterine contraction, especially in pregnant women, and can stimulate intestinal contraction. It can cause pupillary dilatation. Verticine and verticinine have been observed to produce a prolonged hypotensive effect in cats.

Adverse effects include pupillary dilatation and respiratory depression. The MLD is 10 to 12 mg/kg in rabbits (intravenous) and 8 to 10 mg/kg in cats.

Therapeutic Uses

This herb is traditionally used to relieve coughing and dyspnea in cases of chronic bronchitis or upper respiratory infection. It is used to treat pulmonary tuberculosis and whooping cough. In addition, it can be used to treat gastric and duodenal ulcer, due to its antacid and analgesic effects.

KUAN DONG HUA (款冬花) — dried flower bud of *Tussilago farfara* L. (Compositae)

Chemistry

The flower bud contains faradiol, rutin, hyperin, taraxanthin, tannin, some essential oils, and saponins.

Actions

This herb is an effective antitussive, expectorant, and antiasthmatic agent. It has been observed to stimulate the medullary center and slowly raise blood pressure.

Therapeutic Uses

Therapeutically, it was used to treat chronic bronchitis, tuberculosis, and upper respiratory infections. Chinese medicine recommended it to ''moisten'' the lungs, arrest coughing, and reduce phlegm.

HAN CAI (焊菜) — the whole plant of *Rorippa indica* (L.) Hiern. or *R. montana* (Wall.) Small

Chemistry

Two major substances containing radicals of $-CN$ or $-CONH_2$ have been isolated from the plant: rorifone and rorifamide.

Structure of Rorifone and Rorifamide

$$CH_3{-}SO_2{-}CH_2(CH_2)_7{-}CH_2{-}R$$

$R = -CN$	$R = -CONH_2$
Rorifone	Rorifamide

Some plants also contain gluconasturtiin ($C_{15}H_{20}O_9NS_2K$) and α-phenylethylisothiocyanate.

$$C_6H_5{-}CH_2CH_2C(=NOSO_3K){-}S{-}glucose \xrightarrow{H_2O} C_6H_5{-}CH_2CH_2{-}N{=}C{=}S$$

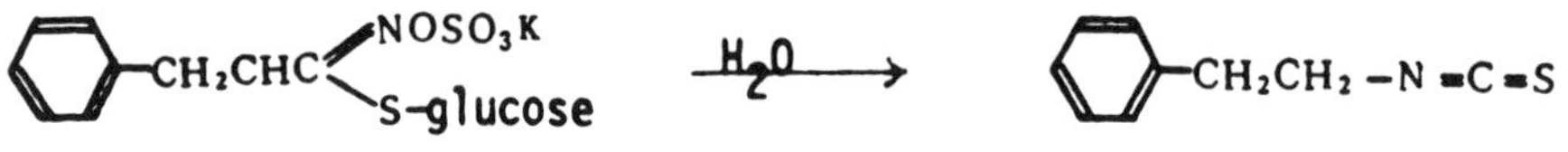

Gluconasturtiin α-Phenylethylisothiocyanate

Therapeutic Uses

Han Cai is used as an antitussive, expectorant, diuretic, and detoxicant.

HU TAO REN (胡桃仁) or HE TAO (核桃) — dried ripe seed of *Juglans regia* L.

Chemistry

The seed contains α-hydrojuglone-4-β-D-glucoside ($C_{15}H_{16}O_9$), jugone ($C_{10}H_6O_3$), and juglanin ($C_{20}H_{15}O_{10}$).

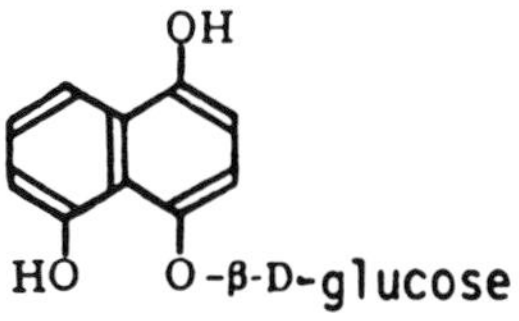

α-Hydrojuglone-4-β-D-glucoside

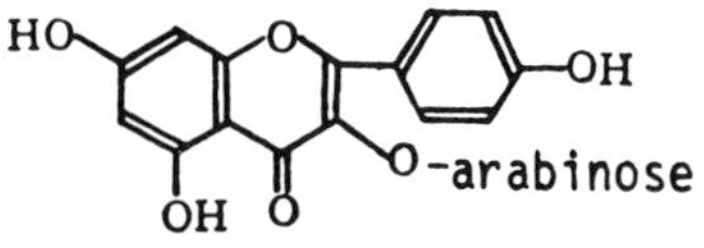

Juglanin

Therapeutic Uses

Traditional medicine prescribed the herb to nourish and invigorate the lungs and kidneys. It is commonly used as a supporting agent in the treatment of coughing and bronchitis.

MIAN HUA GAN (棉花根) — dried root of *Gossypium hirsutum* L.

Chemistry

The root contains gossypol ($C_{30}H_{30}O_6$), hemigossypol ($C_{15}H_{10}O_4$), 6,6′-dimethoxylgossypol, methoxylhemigossipol ($C_{16}H_{15}O_4$), and acetovanillone ($C_9H_{10}O_3$). The seed of the plant contains the glycoside aflatoxin B, while the leaf contains hirsutrin.

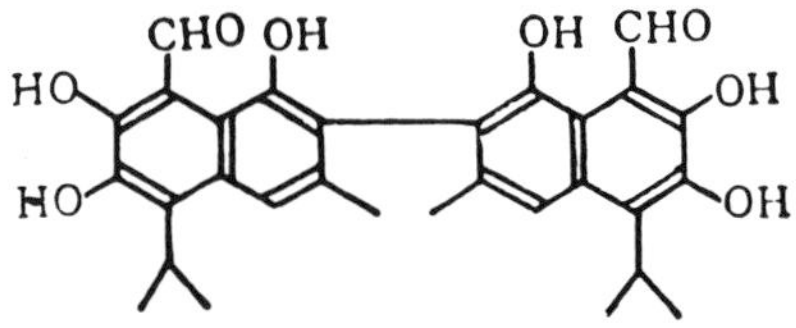

Gossypol

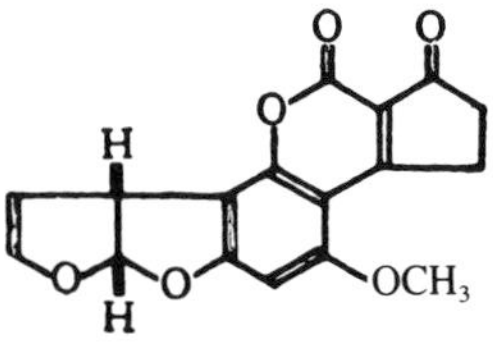

Aflatoxin B_1

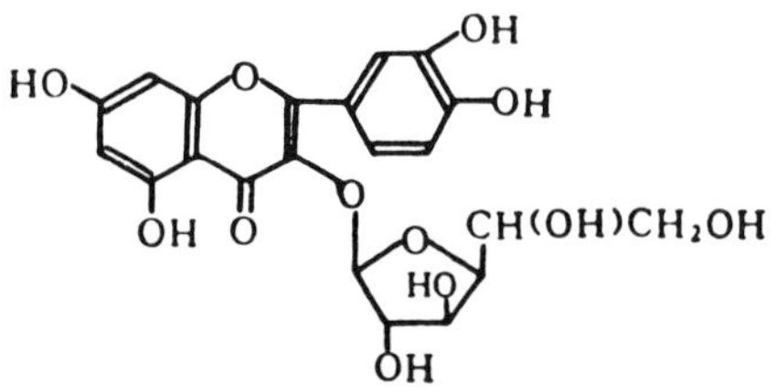

Hirsutrin

Therapeutic Uses

This herb is used as an antitussive in the treatment of bronchitis.

YU GAN ZI (余甘子) — dried fruit of *Phyllanthus emblica* L.

Chemistry

The fruit contains chebulinic acid ($C_{44}H_{32}O_{27}$) and chebulic acid ($C_{14}H_{14}O_{11}$); phyllemblic acid ($C_{16}H_{28}O_{17}\{COOH\}_8$), emblicol ($C_{20}H_{30}O_{19}\{OCH_3\}_6$), and mucic acid ($C_6H_{10}O_8$); and a glycoside whose hydrolyzed genin is α-leucodelphinidin.

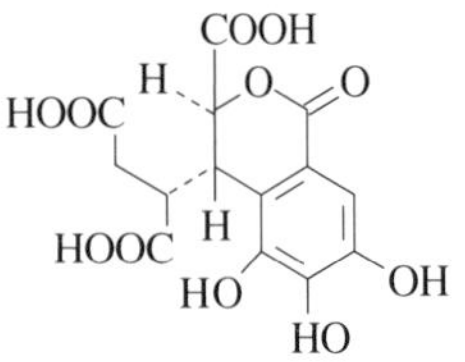

Chebulinic acid

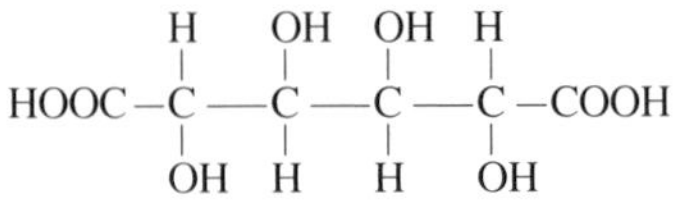

Mucic acid

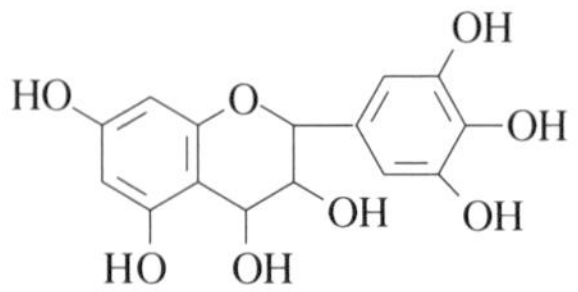

α-Leucodelphinidin

YE XIA ZHU (叶下珠) — derived from the same family as *Phyllanthus urinaria* L. or from *P. nirui* L.

Chemistry

This herb contains the alkaloids phyllanthine ($C_{14}H_{17}O_2N$) and phyllantidine ($C_{13}H_{15}O_3N$). Its leaf contains phyllanthin ($C_{24}H_{34}O_6$), hypophyllanthin ($C_{24}H_{30}O_7$), niranthin ($C_{24}H_{32}O_7$), nirtetralin, and phylteralin ($C_{24}H_{34}O_6$).

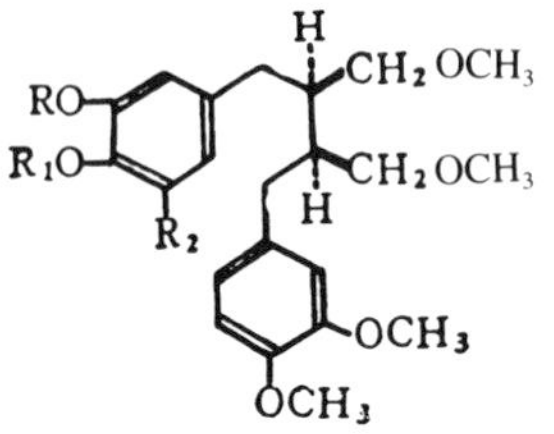

	R=	R_1=	R_2=
Phyllanthin	CH_3	CH_3	H
Niranthin		$-CH_2-$	OCH_3

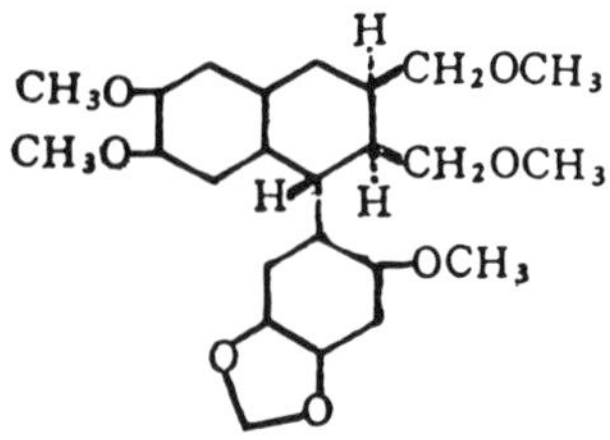

Hypophyllanthin

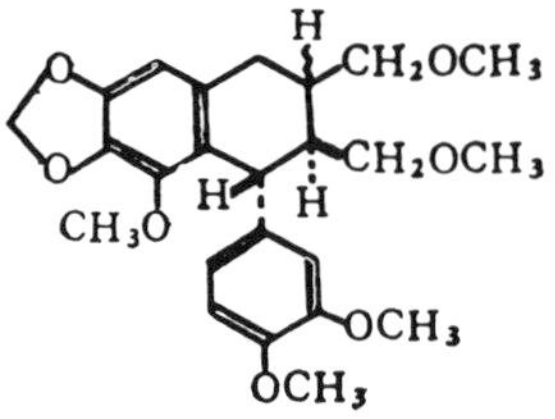

Nirtetralin

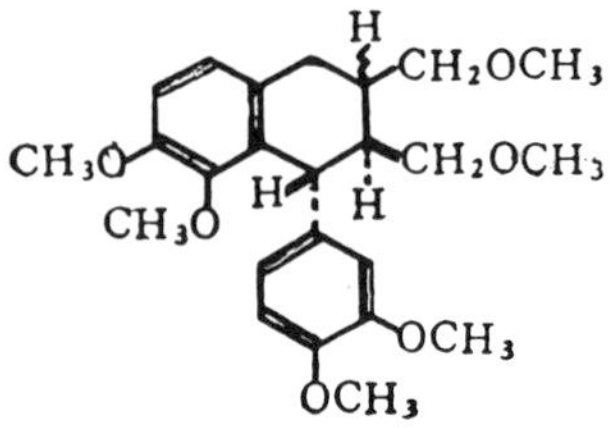

Phylteralin

Therapeutic Uses

Both herbs are traditionally used to arrest coughing, remove "heat" from blood, and promote digestion and secretion.

QIAN HU (前胡) — dried root of *Peucedanum praeruptorum* Dunn or *P. decursivum* (Miq.) Maxim.

Chemistry

The root contains essential oils, nodakenetin ($C_{14}H_{14}O_4$) and the glycoside nodakenin, decursidin, umbelliferone, and pencordin.

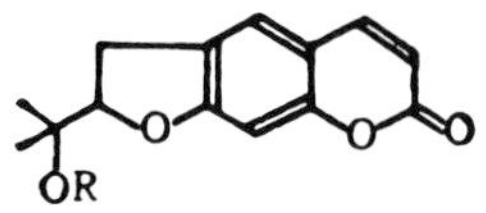

	R =
Nodakenetin	H
Nodakenin	–glucose

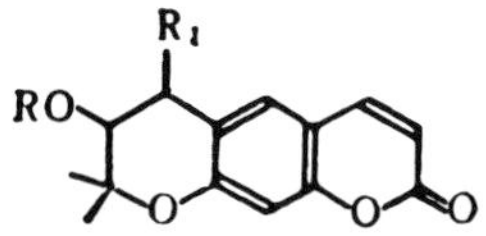

	R =	R_1 =
Decursin	CO–CH=C(CH_3)(CH_3)	H
Decursidin	–O–C(=O)–CH=C(CH_3)$_2$	–O–C(=O)–CH=C(CH_3)$_2$

Actions

Qian Hu is an effective antitussive and expectorant. Administration of the decoction to cats increases secretions in the respiratory tract. It has a long duration of action. The herb has antibacterial effects, and is also active against the influenza virus.

The herb has been observed to cause coronary dilatation. Pencordin, at a concentration of 2×10^{-5} g/ml, causes a marked increase in coronary flow in an isolated heart preparation. In an anesthetized pig, pencordin administered intravenously at 10 mg/kg results in an 82% increase in coronary flow and a decrease in oxygen consumption.

Therapeutic Uses

This herb was used in folk medicine as an antitussive and expectorant, to dispel ''wind'' and ''heat''.

TABLE 2
Other Antitussive Herbs

Name	Botanical source	Part	Major contents	Actions and uses
Man Shan Hong (满山红)	*Rhododendron dahuricum* L.	Dried leaf	Essential oils, germacrone and flavonoid, farrerol ($C_{17}H_{16}O_5$), foriol, quercetin, myricetin, andromedotoxin, rhodotoxin	Antitussive, expectorant, antiasthma; *Man Shan Hong* pills (0.05, 0.1 g) 1 pill t.i.d., syrup, 10 ml t.i.d., tablet (0.08 g) 3–4 tablet t.i.d.
Hong Tuan Yao (红管药)	*Aster ageratoides* Turcz.	Whole plant	Quercetin, kaempferol	Antitussive, expectorant, antiasthmatic; stimulates adrenal cortex; uses in chronic bronchitis; tablet (0.5 g) 5 tablets t.i.d.
Jin Gu Cao (筋骨草)	*Ajuga decumbens* Thunb.	Whole plant	Flavon glucoside; luteolin, alkaloids	Antitussive, antipyretic and expectorant, antiphlogistic, antibacterial; use in treating chronic bronchitis
Ai Di Cha (矮地茶)	*Ardisia japonica* (Thunb.)	Whole plant	Bergenin (0.1–0.2%) glucoside, essential oil	Antitussive, expectorant, antiphlegm, promotes blood circulation, hemostatic; use in chronic bronchitis, tuberculosis; dose, 30–60 g in decoction
Song Ta (松塔)	*Pinus bungeana* Zucc., *P. tabulaeformis*	Dried cone	Essential oil (0.2–1%), limonene, pinitol	Antitussive, antiasthmatic expectorant, antibacterial; use in chronic bronchitis
Quan Ye Qing Lan (全叶青兰)	*Dracocephalum integrifolium*	Dried aerial plant	Essential oil, flavon glucoside	Antitussive, expectorant, antiasthmatic; use to treat chronic bronchitis
Ye Kuan Mun (夜関門)	*Lespedeza cuneata* G.	Root or whole plant	β-Sitosterol, pinitol, flavonoid	Antitussive, expectorant, antiasthmatic, antiphlogistic antibacterial; use to treat chronic bronchitis, indigestion, and gastritis

Chapter 24

EXPECTORANTS

Most of the herbs described in this chapter have active principles that are essential oils or saponins. They tend to have an irritative effect on the gastric mucosa after oral administration, and reflexively cause an increase in bronchial secretions, resulting in a dilution of the sputum and an expectorant effect. Some of these herbs also possess antitussive and antiasthmatic effects.

JIE GENG (桔梗) — dried root of *Platycodon grandiflorum*

Chemistry

The major component of the root is the saponin platycodin. On hydrolysis, the saponin gives platycodigenin and polygalacid acid.

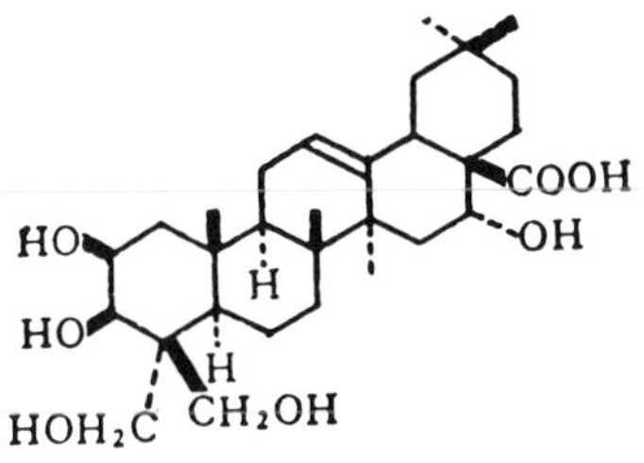

Platycodigenin

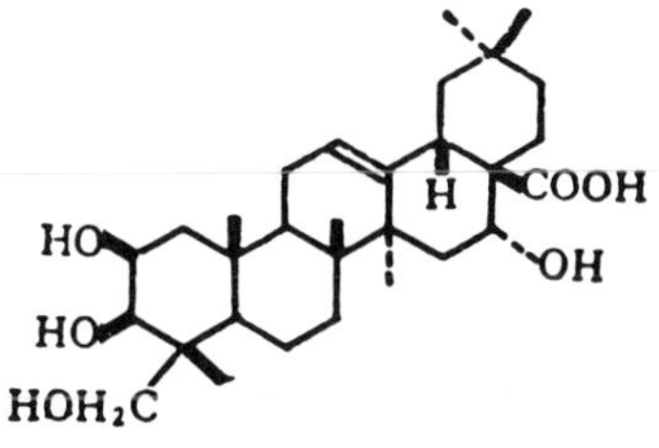

Polygalacid acid

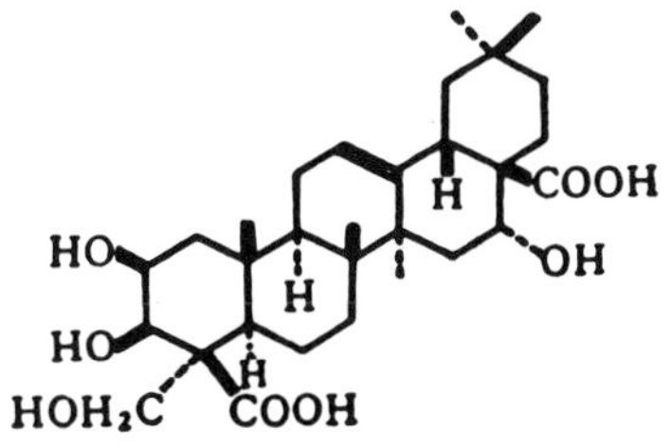

Platycogenic acid A

Actions

Platycodin exerts an expectorant action by irritating the gastric mucosa and causing an increase of bronchial secretion. It lowers blood cholesterol by stimulating the hepatic and biliary excretion of cholesterol and bile acids. The herb also lowers blood sugar and acts as an antibacterial agent. In addition, it has been observed to have both tranquilizing and antipyretic effects.

In clinical trials, the major adverse effects observed were nausea and vomiting. It must be administered cautiously to patients suffering from peptic ulcers and is contraindicated for parenteral administration, as the saponin itself is a potent hemolytic agent. The LD_{50} of the herb in mice is 420 mg/kg (oral) or 22.3 mg/kg (i.p.).

Therapeutic Uses

The herb is prescribed by traditional Chinese physicians to dispel phlegm and "ventilate" the lungs. It is also used to relieve sore throats and to promote pus discharge.

It is available as a *Je Gen* complex tablet, comprising 0.08 g of powdered *Je Gen*, 0.03 g of opium, and 0.18 g K_2SO_4. The standard dose in this form is one to two tablets t.i.d. A syrup is also available which is administered in doses of 4 to 8 ml, three to four times each day.

YUAN ZHI (远志) — dried root or bark of *Polygala tenuifolia* or *P. sibirica* (Polygalaceae)

Chemistry

The active principle is a saponin, which constitutes approximately 0.7% of the root. The hydrolysis products of the saponin are tenuigenin A1 and B1. Another substance found is tenuifolin or prosenegenin ($C_{36}H_{50}O_{12}$), which gives genin presenegenin ($C_{30}H_{40}O_7$) after further hydrolysis. the root also contains the alkaloid tenuidine ($C_{21}H_{31}O_5N_3$) and polygalitol.

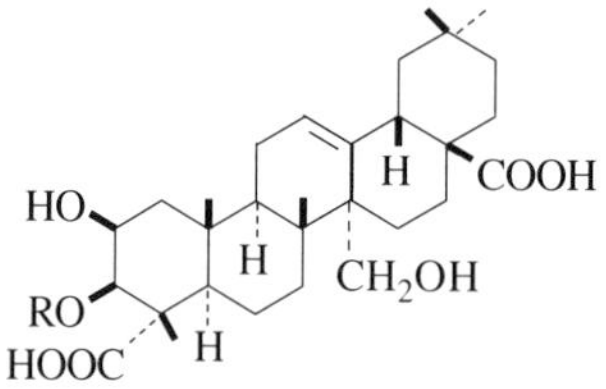

	R =
Tenuifolin	-glucose
Prosenegenin	H

Actions

The expectorant effect of the herb occurs when the saponins irritate the gastric mucosa, reflexively stimulating bronchial secretions. *Yuan Zhi* has also been observed to have a sedative and tranquilizing effect, and to stimulate uterine contraction. It exhibits some antibacterial properties.

Therapeutic Uses

This herb is traditionally used as a sedative, expectorant, and resuscitating agent. It can be prepared in several forms. The water extract of the herb is administered in doses of 0.2 to 2 ml t.i.d. In syrup form, it is given in doses of 2 to 8 ml t.i.d. A 2% tincture is also available, taken in doses of 2 to 5 ml t.i.d.

SHA SENG (沙参) or SOUTHERN SHA SENG — the root of *Adenophora tetraphylla* (Thunb.) Fisch. or *A. stricta* Miq.

Chemistry

The root contains saponins which produce expectorant effects lasting for more than 4 h.

Actions

It is highly hemolytic, and even at a 1:40 dilution, can hemolyze the blood cells quickly. It also stimulates myocardial contraction and has an antibacterial effect.

Therapeutic Uses

The Chinese pharmacopoeia recommends *Sha Seng* as an expectorant in the treatment of chronic bronchitis and whooping cough.

JU HONG (橘红) — red-colored external layer of the pericarp of *Citrus reticulata* or *C. grandis* (L.) Osbeck var. *tomentosa*

Chemistry

The herb contains citral ($C_{16}H_{16}O$), geraniol ($C_{16}H_{18}O$), linalool ($C_{10}H_{18}O$), methylanthranilate ($C_8H_9O_2N$), stachydrine ($C_7H_{13}O_2N$), putrescine ($C_{11}H_{12}O_2$), and apyrocatechol ($C_6H_6O_2$). Several glucosides are also present, including naringin ($C_{27}H_{32}O_{14}$), poncirin ($C_{28}H_{34}O_{14}$), hesperidin, neohespiridin ($C_{22}H_{34}O_{15}$), and nobiletin.

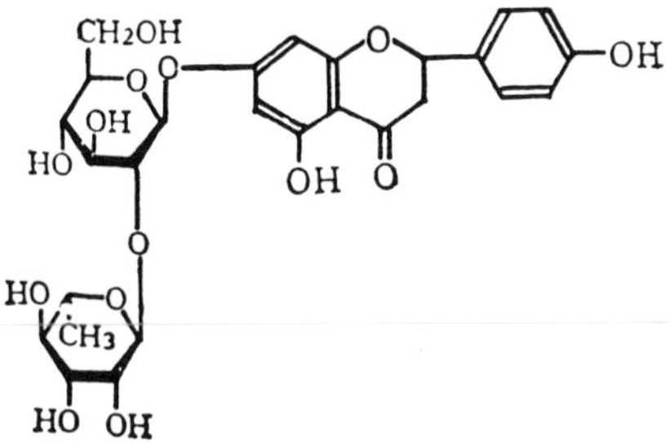

Naringin

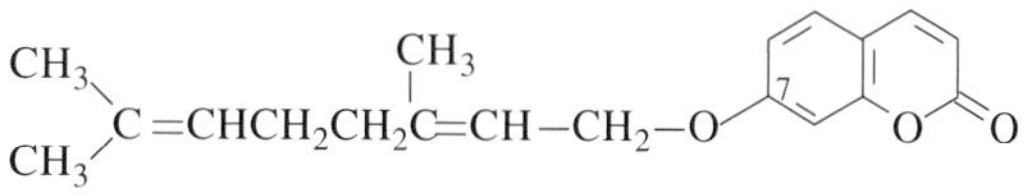

Aurapten

Therapeutic Uses

The herb is an effective expectorant and antitussive. Chinese medical books describe it as being capable of "warming" the lungs and reducing phlegm, normalizing the flow of *qi*, and removing "dampness". It is also prescribed in the treatment of indigestion and as an antiemetic agent.

Ju Hong is administered in a standard dose of 3 to 6 g, prepared in a decoction.

ER CHA (儿茶) — dried extract from the peeled branch and stem of *Acacia catechu* (L.) Willd.

Chemistry

The major active principles of the herb are *d*-catechin ($C_{15}H_{14}O_6$) and epicatechin. The herb also contains gambir-fluorescein, gambirine, mitraphylline, and roxburghine D.

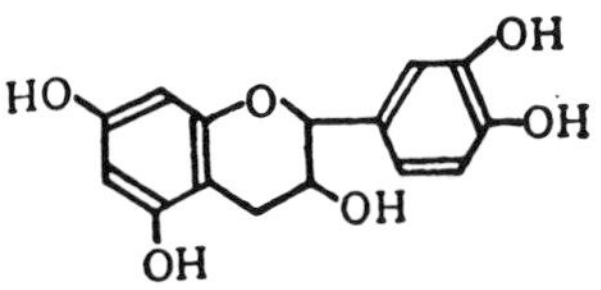

Epicatechin

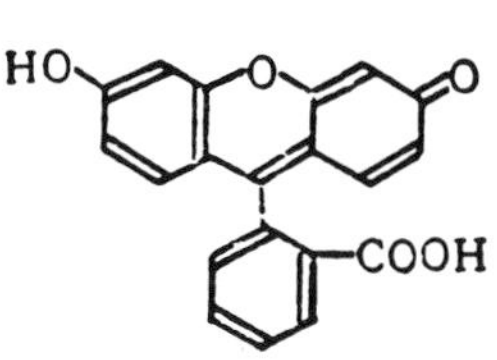

Gambir-fluorescein

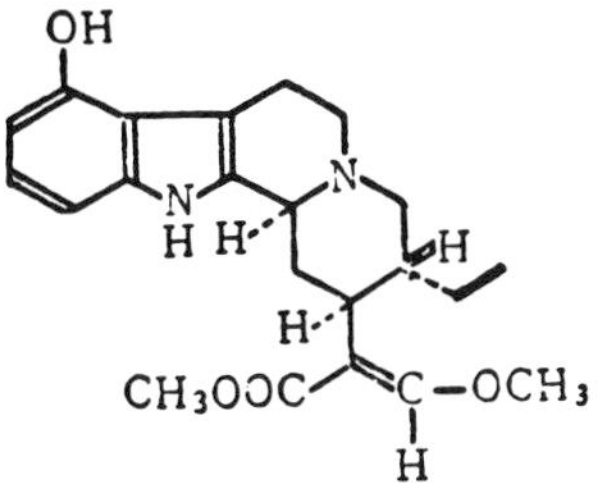

Gambirine

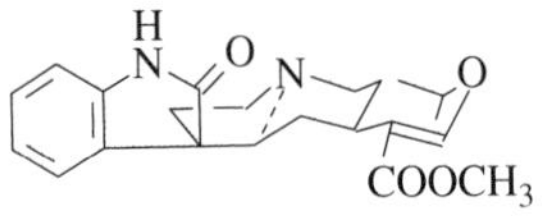

Mitraphylline

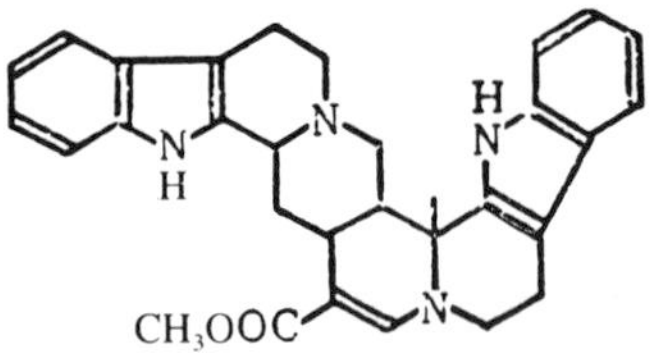

Roxburghine D

Therapeutic Uses

This herb is used to promote salivation, resolve phlegm, arrest bleeding, and to treat pyogenic infections.

LUO HAN GUO (罗汉果) — the dried ripe fruit of *Momordica grosvenori* Swingle (Cucurbitaceae)

FIGURE 1. *Luo Han Guo.*

Chemistry

The fruit contains the glucoside esgoside (1% of total content), whose hydrolysis product is chiefly fructose (approximately 14%) and is almost 300 times sweeter than cane sugar.

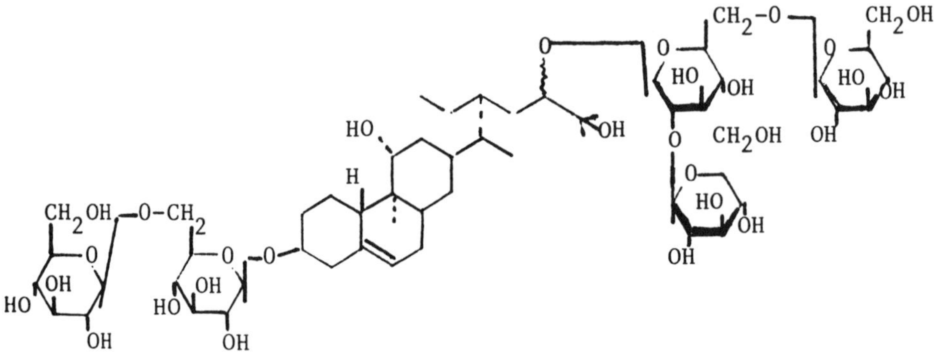

Esgoside

Actions

The herb and its glucoside have expectorant effects. It has been suggested that after administration, it is excreted into the bronchial tract, where it exerts a hypertonic function to moisten the mucosal surface and to reduce irritability. It can also control coughing.

Therapeutic Uses

This herb is not listed in traditional Chinese pharmacopoeia, and is generally limited to regional use in southern China; Chinese emigrants have also made it popular in southeast Asia and in the U.S. It is sold in markets, but is never served as a traditional fruit; it appears at the daily table only when someone in the family develops cold symptoms.

It is used to treat whooping cough and serves as an expectorant in cases of the common cold. The fruit is cut in half and hot water is poured over it to form a beverage. For export use, cubes of the powdered fruit are available, to be made into a tea.

Reference

Oriental Materia Medica, Oriental Healing Arts Institute, Long Beach, CA, 1986, 812–813.

DENG TAI YE (灯台叶) — dried leaf of *Alstonia scholaris*

Chemistry

The herb contains several alkaloids, including picrinine ($C_{20}H_{22}O_3N_2$) and picralinal ($C_{21}H_{22}O_4N$). In addition, the alkaloids echitamine ($C_{22}H_{28}O_4N_2$) and echitamidine ($C_{20}H_{26}O_3N_2$) are present.

Therapeutic Uses

This herb is used as an expectorant and as an antiphlogistic agent.

TABLE 1
Other Expectorant Herbs

Name	Source	Part	Active principles	Actions and uses
Mu Jing (牡荆)	*Vitex jeguado* L. var. *cannabifolia*	Dried leaf	17 Different kinds of essential oils, such as β-caryophyllene, caryophyllene oxide	Expectorant, antitussive, antiasthmatic, and antibacterial
Lie Xiang Du Juan (烈香杜鹃)	*Rhododendron anthopogonoides* Maxim.	Leaf	Essential oils and saponins; quercetin, gossypetin	Expectorant, antitussive, antiasthmatic; tablet (contains saponin, 80 mg/tablet), give 2 tablets t.i.d.
Qian Ri Hong (千日红)	*Gomphrena globosa* L.	Dried flower	Saponins; 7 kinds of betacyamines; gomphrenin I and II, amaranthin, isoamaranthin	Expectorant; used to treat chronic bronchitis, whooping cough; tablet (equivalent 3 g/tablet), 2 tablets t.i.d.
Shang Lu (商陆)	*Phytolacca acinosa* or *P. americana* L.	Root	Phytolacine, phytolaccatoxin, oxyristic acid, jaligonic acid, and saponins	Expectorant, antitussive, diuretic, antibacterial, anti-inflammatory; pill (equivalent to 4 g of raw material/pill), 1 pill t.i.d.

Chapter 25

ANTIASTHMATIC HERBS

The herbs discussed in this Chapter are used to relieve convulsive bronchitis and bronchial asthma. These illnesses are due primarily to a continuous contraction of bronchial smooth muscles, often accompanied by mucosal edema and increased secretions. Our present medical knowledge has shown that most of the antiasthmatic herbs act by stimulating β-adrenergic receptors, causing relaxation of the bronchial smooth muscles. Some may also act to desensitize histamine receptors (H_1 receptors).

The classic example of this group of herbs is *Ma Huang*, which has been used in China as an antiasthmatic, an anticoughing agent, to smooth the lungs, and to stimulate the heart and mind. The late Dr. K. K. Chen explored the pharmacology of *Ma Huang* while he was an instructor at Peking Medical Union College in the late 1920s. The structure of its active principle, ephedrine, was established, and it became one of the classic sympathomimetic agents described in medical pharmacology textbooks.

MA HUANG (麻黄) — the dried plant stem of *Ephedra sinica* Staph., *E. equisetina* Bge., or *E. intermedia* Schrenk et C. A. Meyer

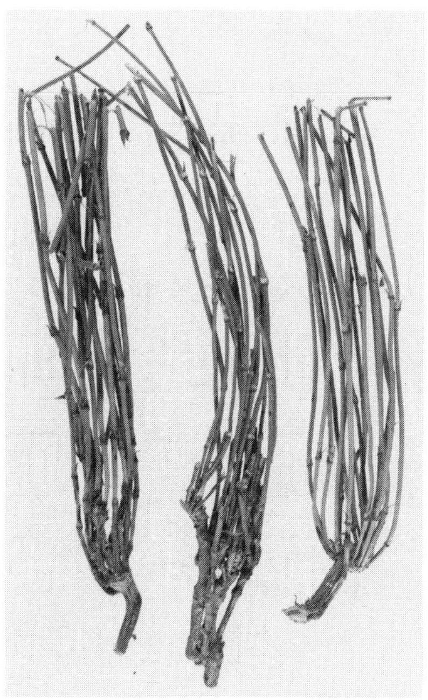

FIGURE 1. *Ma Huang.*

Chemistry

The active principles are alkaloids, which constitute about 1.32% of the herb by weight. Among the alkaloids, *l*-ephedrine accounts for 80 to 90%; the others include *d*-pseudoephedrine, methylephedrine, *d*-*N*-methylpseudoephedrine, and *l*-norephedrine.

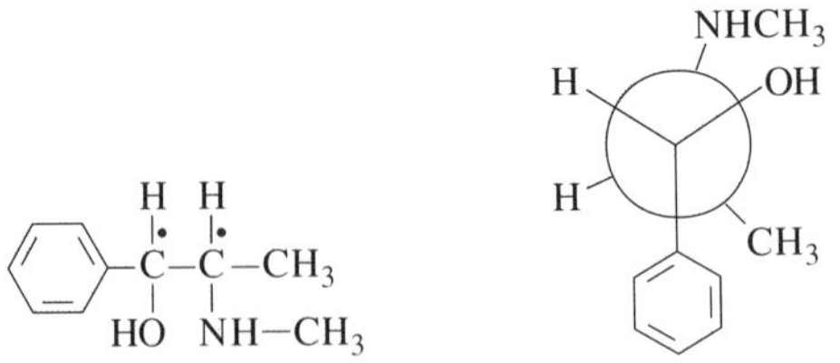

l-Ephedrine

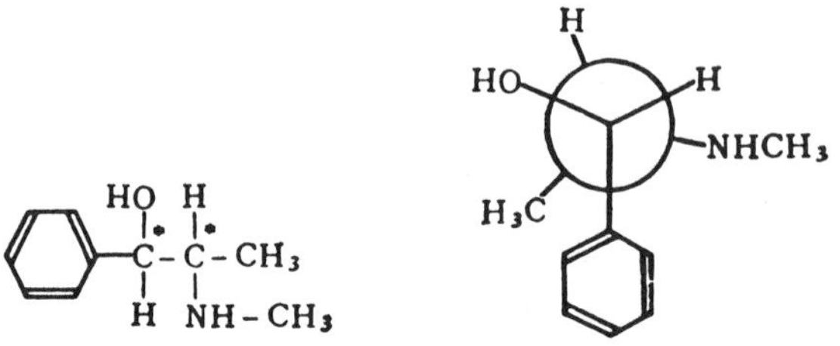

d-Pseudoephedrine

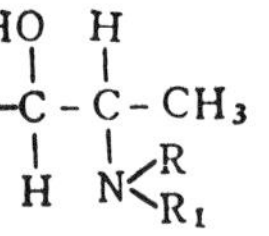

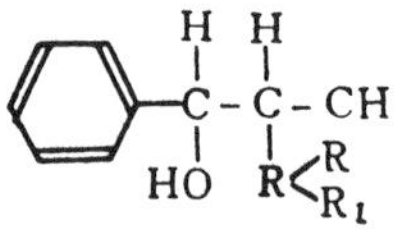

	R=	R_1=
d-N-Methylpseudoephedrine	CH_3	CH_3
d-Norpseudoephedrine	H	H

	R=	R_1=
l-N-Methylephedrine	CH_3	CH_3
l-Norephedrine	H	H

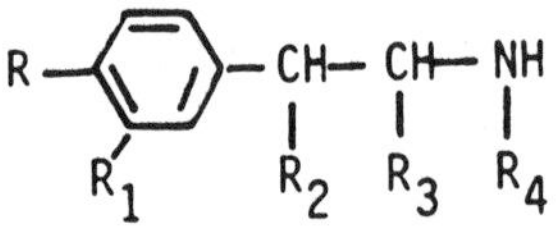

	R=	R_1=	R_2=	R_3=	R_4=
Phenylephrine	H	OH	OH	H	CH_3
Epinephrine	OH	OH	OH	H	CH_3
Ephedrine	H	H	OH	CH_3	CH_3

Actions

In the late 1920s, K. K. Chen of Peking Union Medical College, China, was the first to demonstrate the herb's sympathomimetic effect on dogs. Subsequently, he and his colleagues isolated and identified the structure of the active principle, ephedrine, which has a structure similar to epinephrine but without the two hydroxyl radicals on the benzene ring. Since then, the pharmacological actions and medical uses of ephedrine and its synthetic analogues have received great attention. In 1930, Chen and Schmidt published the monograph ''Ephedrine and Related Substances'', which contained references to more than 600 publications on this subject.

TABLE 1
Minimal Lethal Dose (MLD) of Three *Ma Huang* Compounds

Animal	Route of administration	MLD (mg/kg) *l*-Ephedrine	*dl*-Ephedrine	Pseudoephedrine
Frog	Lymph sac	640	630	770
Rat	I.p.	330	310	310
Rabbit				
Gray	Subcutaneous	230	360	400
Gray	I.v.	80	90	100
White	I.v.	50	70	130
Dog	I.v.	70	100	130

Experimental data have substantiated the finding that ephedrine is an adrenergic agonist, acting on both α- and β-adrenergic receptors. In contrast to epinephrine, ephedrine is completely absorbed from the intestine and has a much longer duration of action.

It differs structurally from epinephrine by the lack of two hydroxy groups on the benzene ring. Because of this lack, ephedrine is much more lipid soluble and can penetrate the blood-brain barrier, rapidly entering nerve tissue. This results in a stimulating effect on the brain stem, causing insomnia, tremors, and stimulation of the respiratory and vasomotor centers of the medulla.

l-Ephedrine is approximately 10 to 50 times more potent than the *d*-isomer and pseudoephedrine. By its β-adrenergic agonistic action, it can relax bronchial muscles to produce an effective antiasthmatic action. It produces myocardial stimulation by its β_1 agonistic effect. Ephedrine also constricts peripheral blood vessels by its α-agonistic effect, causing an increase in blood pressure and heart rate.

Acting centrally, the essential oils of *Ephedra* can lower body temperature. The herb can also reduce body temperature by increasing perspiration. In addition, it exhibits diuretic and antiviral effects.

When ephedrine or *Ma Huang* is administered at frequent intervals, its pharmacological effect is gradually reduced. A residue of the drug persists to bind the receptor sides longer, and thus lessens the number of free receptors available to react with newly administered drugs.

Toxicity

l-Ephedrine is slightly more toxic than *dl*-ephedrine and pseudoephedrine. Table 1 compares the MLD of these three principal compounds of *Ma Huang*.

In human subjects, overdosage often produces nervousness, insomnia and central stimulation, increased blood pressure, and heart palpitations. The herb is contraindicated in persons with hypertension, arteriosclerosis, hyperthyroidism, or diabetes mellitus.

Therapeutic Uses

Ma Huang has been used in Chinese medicine for thousands of years. Listed in *The Herbal Classic of the Divine Plowman* among the "middle class" herbs, it was used widely to induce perspiration and as an antiallergy agent.

Today, *Ma Huang* is the major ingredient of a cold formula used to relieve headaches, body aches, and coughing. This formula is also used to lower fever by increasing perspiration, via the actions of ephedrine. This formula is a decoction containing *Ma Huang, Gui Zhi* (cinnamon twig), *Gan Cao* (licorice root), and almond.

In the treatment of bronchial asthma, ephedrine is administered in tablet form (15, 25, or 30 mg) three times a day. Alternatively, an ephedrine HCl solution (30 mg/ampule) can be administered by subcutaneous or intramuscular injection.

To treat rhinitis and nasal congestion, ephedrine is also available in the form of a 1% nasal spray.

References

Chen, K. K. and Schmidt, C. F., *Medicine (Baltimore)*, 9, 1–117, 1930.

DI LONG (地龙) — dried body of *Pheretima aspergillum* (Perrier) or *Allobophora aliginosa trapezoides*

Chemistry

This ''herb'' contains hypoxanthine, lumbrofebrin, and lumbritin.

Actions

The N-containing principle isolated from this herb has a clear antiasthmatic effect; in experiments with animals, it has been shown to protect against the effects of inhaled histamine.

The herb has antipyretic, sedative, and anticonvulsant actions. It lowers blood pressure through a central action; the action is slow but long lasting.

Therapeutic Uses

In Chinese medicine, this herb is used as an antipyretic, anticonvulsant, and antiasthmatic agent. It is also used as a diuretic and an agent to activate the circulation collaterals.

In the treatment of chronic bronchitis and bronchial asthma, it is generally administered in powder form, in doses of 3 to 4 g t.i.d. A 40% tincture is available for the treatment of hypertension; the standard dose is 10 ml t.i.d.

The herb is also manufactured in an injection solution, with each milliliter equivalent to 1.5 g of raw material; this solution is administered intramuscularly. It must be cautioned that the injection may cause a hypersensitization reaction, probably due to impurities in the preparation.

YUN XIANG CAO (芸香草) — dried aerial parts of *Cymbopogon distans* (Nees) A. Camus

Chemistry

Between 0.7 and 1% of the herb consists of essential oils. Piperitone accounts for 40 to 50% of the oils.

Actions

The active principle of the herb has a smooth muscle relaxation effect. It can antagonize smooth muscle contraction induced by $BaCl_2$. It also has antitussive and antibacterial effects.

Therapeutic Uses

Traditional Chinese medicine recommended this herb to dispel ''cold'' and remove ''dampness'', relieve bronchial asthma, and arrest coughing. It was also used to promote the circulation of *qi* and relieve depression.

The preparations available include tablets and a piperitone aerosol.

HUA SHAN SENG (华山参) — the dried root of *Physochlaina infundibularis* Kuang

Chemistry

It contains about 0.26% alkaloids, which include hyocyamine, scopolamine, scopoletin, and scoplin.

Actions

These alkaloids are cholinergic blocking agents which have a relaxing effect on bronchial muscles and a depressant effect on the central nervous system.

The adverse effects and toxicity of this herb and its alkaloids are similar to those of atropine and scopolamine. The LD_{50} in mice is 43 g/kg (intraperitoneal). Dryness of mouth, dizziness, and pupil dilatation are the common complaints in patients taking this herb. It is contraindicated in patients suffering from glaucoma.

AI YE YU (艾叶油) — the essential oil extracted from dried *Artemisia argyi* Levl. et Vant.

Chemistry

This is a mixture of seven different oils: terpinenol-4, β-caryophyllene, artemisia alcohol, linalool, camphorae, borneol, and cineol or eucalyptol.

Actions

The actions of these oils include antiasthmatic, antitussive, and expectorant effects. In experiments, *Ai Ye Yu* at a concentration of 0.5 μg/ml in an isolated muscle bath showed a relaxing effect equivalent to 0.125 μg/ml of isoproterenol. It can antagonize the bronchial constriction induced by acetylcholine or histamine; the duration of action is relatively long. The herb also has an antibacterial effect.

Therapeutic Uses

It is used to treat chronic bronchitis, hypersensitivity, and oral infections. The herb is given in the form of a 0.075-ml capsule taken three times per day. An alternative form is an aerosol to spray into the larynx, with an approximate dose of 0.1 ml.

TONG GUAN TENG (通关藤) — dried stem of *Marsdenia tenacissima* (Roxb.) Wight. et Arn.

Chemistry

The stem contains several saponins. Upon hydrolysis, a product called sarcostin is found.

Actions

Sarcostin is the active principal responsible for the herb's antiasthmatic effect. It also has hypotensive and antibacterial effects.

Therapeutic Uses

This herb is used in the treatment of bronchitis and upper respiratory infections.

SHAN CONG ZI Oil (山苍子油) — the essential oils extracted from the fruit of *Litsea cubeba* (Lour.)

Chemistry

The active principles of the oil are citral and linalool.

Actions

Citral and linalool can relax bronchial smooth muscles and antagonize histamine- or acetylcholine-induced constriction. They also effectively protect against hypersensitization shock, prolong barbiturate-induced sleeping time, and exert an expectorant effect.

Therapeutic Uses

This herb is used in traditional Chinese medicine in the treatment of chronic bronchitis and bronchial asthma. Preparations include a 0.05-ml capsule (standard dose, 2 capsules t.i.d.) and a nasal spray.

TABLE 2
Other Herbs Claimed to Have Antitussive, Expectorant, and Antiasthmatic Effects

Name	Source	Part	Effects: Anti-tussive	Effects: Expectorant	Effects: Anti-asthmatic	Other
Bai Jie Zi (白芥子)	*Brassica alba* Boiss.; contains sinalbin	Seed	Yes		Yes	Analgesic
Yiang Jin Hua (洋金花)	*Datura tatula*	Flower		Yes		Central depression (see Chapter 8)
Ma Don Ling (马兜铃)	*Aristolochia debilis* or *A. contorta*	Aerial part		Yes		Antibacteria, bronchodilatation
Hu Tin Yie (胡颓叶)	*Elaeagnus pungens* T.; contains harmane, elaeagnine	Leaf	Yes			
Qian Hu (前胡)	*Pseudedanum praeruptorum, P. decursivum*	Root		Yes		Stimulate respiratory secretion
Kuan Bu (昆布)	*Laminaria japonica* Hrech.	Whole plant	Yes		Yes	Treat goiter (see Chapter 30)
Shu Qu Cao (鼠曲草)	*Gnaphalium affine*	Whole plant	Yes	Yes	Yes	Antirheumatic
Lao Jiun Xiu (老君须)	*Usnea diffracta* Vain., contains *d*-usnic acid	Whole plant	Yes	Yes		Antituberculous
Xao Yie Pi Pa (小叶枇杷)	*Rhododendron anthopogonoides* Maxim.	Leaf	Yes	Yes		
Pin Di Mu (平地木)	*Ardisia japonica* Bleme.		Yes			

Section VII
Genitourinary System

Chapter 26

DIURETIC HERBS

Although many herbs were described in *The Herbal Classic of the Divine Plowman* or other Chinese pharmacopoeias as having the ability to increase urinary flow, they were not categorized as ''diuretics'', and the mechanisms by which they stimulated urine flow were not understood. Only a few herbs described in this Chapter — not more than two dozen — were commonly used to ''unobstruct water flow and dissipate wetness . . . '' or to produce a net increase in urinary flow and remove edema fluid and toxic wastes from the body. Such herbs were mainly used in the treatment of liver, heart, and kidney diseases. They were also used in the treatment of toxemia of pregnancy.

MU TONG (木通) — the dried stem of *Aristolochia manshuriensis* Kom, or *Clematis armandii* or *C. montana* Buch.-Ham., or *Akebia quinata* (Thunb.) Dcne.

Chemistry

Regardless of which of the four species the herb is derived from, its active chemical principles remain the same. It contains aristolochic acid and the saponin akebin, which on hydrolysis yields the products akebigin, hederagonin, and oleanolic acid.

Actions

This herb has a diuretic effect, a myocardial-stimulating effect, and an antibacterial effect.

Therapeutic Uses

The Herbal Classic of the Divine Plowman described the herb as an agent to combat intensive ''heat'' and to promote milk secretion. It was also used in the treatment of acute urethritis, nephrotic edema, and mammary gland obstruction.

It is generally administered as a decoction, in doses of 9 g. Additionally, it is one of the main ingredients in a popular Chinese formula containing *Mu Tong, Dan Zhu Ye*, and licorice root.

ZHU LING (豬苓) — the dried fungus of *Polyporus umbellatus* (Pers.) Fr.

Chemistry

The major components isolated from this fungus are ergosterol, biotin, and protein.

Actions

This herb has strong diuretic action. Subjects who took the fungus orally in doses of 8 g had an increase in urinary output and electrolyte excretion approximately 62 and 45%, respectively, above normal.

This diuretic effect is thought to be due to inhibition of renal reabsorption of the electrolytes Na^+, K^+, and Cl^-. Experiments with animals showed that the intravenous injection of a *Zhu Ling* decoction solution can produce a remarkable diuresis; this effect could be reduced by pretreatment with deoxycorticosterone.

The herb can also stimulate the immune system and has anticancer activity.

Therapeutic Uses

The *Ben Cao Kong Mu* described this fungus as an agent to smooth the ''balance'' of the body, treat edema, increase vaginal secretion, and promote urination. It is used in the treatment of edema, including ascites; the standard dose in this treatment is 6 to 15 g, made into a decoction.

The herb is also one of the major ingredients in a Chinese formula called *Wu Ling* or ''decoction of the five *Ling* herbs''. this included *Zhu Ling, Fu Ling* (the *Foria* fungus), cinnamon, *Ze Shai*, and *Pai Shu*. Clinical trials with this decoction in cases of cirrhotic ascites showed a 92% improvement rate.

A new preparation (No. 757), which is an extract of this herb, has been tried in the treatment of lung and esophageal cancers; it is reported to produce good results.

BIAN XU (萹蓄) — the dried aerial part of *Polygonum aviculare*

Chemistry

The major components of this herb are the glycoside avicularin ($C_{20}H_{18}O_{11}$), caffeic acid, chlorogenic acid, and vitamin E.

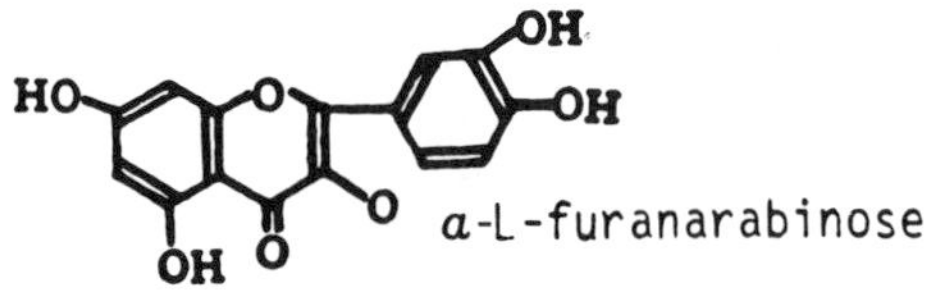

HO, OH, OH, OH, O, O, O, α-L-furanarabinose

Avicularin

Actions

This herb has diuretic, antibacterial, and anthelmintic effects.

Therapeutic Uses

Clinically, the herb is used to treat urethritis, lithiasis, and chyluria. It is effective against dysentery and parotitis and can also be used as an antiascardiasis agent. It is generally administered as a decoction, in doses of 30 g. It is also available as a 100% syrup, in doses of 50 ml taken two to three times a day.

QU MAI (瞿麦) — the dried aerial part of *Dianthus superbus* or *D. chinensis*

Chemistry

It contains the dianthus-saponin; the flower contains some essential oils. The major active component of the oil is eugenol.

Actions and Therapeutic Uses

The oils of this herb have antipyretic and diuretic actions. It is used to treat urinary tract infections and to relieve strangury.

The herb also displays antibacterial and anticancer activity. Decocted in combination with $CaCO_3$, it is reputed to be effective in the treatment of esophageal and rectal cancers.

JIN QIAN CAO (金錢草) — the dried whole plant of *Lysimachia christianae* Hance, *Desmodium styracifolium* Merr., *Dichondra repens* Forst., *Hydrocotyle sibthorpiodes* Lam., or *Glechoma longituba* (Nadoi)

Chemistry

In different regions of China, this herb is derived from different botanical species. All five, however, exhibit essentially the same properties and are thus referred to interchangeably as *Jin Qian Cao*.

The stem contains some essential oils; the major ones are *l*-pinocamphone, *l*-methone, *l*-pinene, limonene, 1,8-cineol, and P-cymene.

Actions

Jin Qian Cao is an effective diuretic and can acidify the urine. It is a choleretic agent and can relax the bile duct and increase bile secretion. It has been observed to increase coronary flow, lower blood pressure, and reduce myocardial oxygen consumption. In addition, it has antibacterial effects.

Therapeutic Uses

Chinese herbalists prescribe this herb to induce diuresis and remove calculi, reduce "heat" and swelling, remove "dampness", and treat jaundice. It is generally administered in combination with other herbs, decocted in doses of 30 to 60 g, to be drunk daily.

BAN BIAN LIAN (半边莲) — the whole plant of *Lobelia chinensis* Lour.

Chemistry

The active principle is lobeline. The herb also contains other alkaloids, including lobelanine, lobelanidine, and isolobelanine.

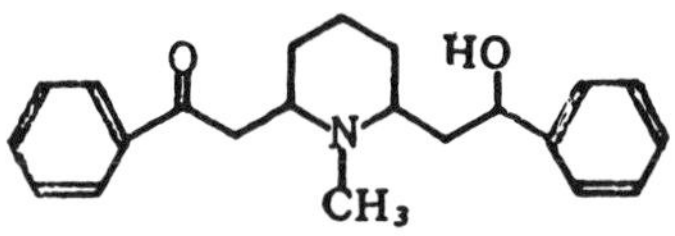

1-Lobeline

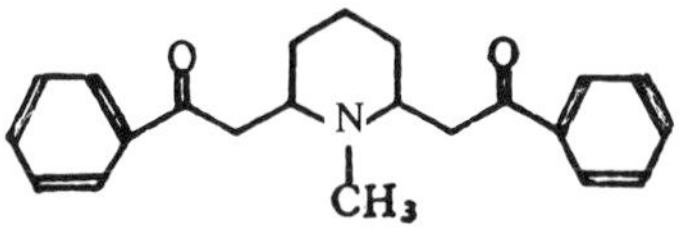

Lobelanine

Actions

Like nicotine, lobelin acts mainly on ganglia, the adrenal medulla, and medullary centers. Its potency is about $^1/_5$ to $^1/_{20}$ that of nicotine. The other alkaloids of this herb have similar pharmacological actions, but are less potent.

The herb is an effective diuretic. Intravenous administration of *Ban Bian Lian* extract or lobeline to dogs can produce an increase in urine output and Na^+ excretion. The duration of action is relatively long.

The herb can increase respiration via stimulation of carotid chemoreceptors. Water extracts of the plant can lower blood pressure; the intensity of this effect is related to the size of the dose. In addition, the herb is a choleretic agent and displays antibacterial action against staphylococci and *Escherichia coli*. Lobeline, but not the other alkaloids, has a marked emetic effect.

Toxicity

Clinical trials with extracts of the herb revealed no special adverse effects in patients, even when taken for a substantial period. If administered parenterally, excessive perspiration and dizziness may occur.

The LD_{50} of the total alkaloids in mice is 18.7 ± 2.0 mg/kg (intravenous). The LD_{50} of *Ban Bian Lian* extract in rats is 75.1 ± 13.1 g/kg (stomach).

Therapeutic Uses

The herb is used as a diuretic to remove edema fluid and ascites, especially in the later periods of schistosomiasis infestation. It has also been used to reduce swelling in cases of nephritis and snake bite.

SHEN JIN CAO (伸筋草) — dried whole plant of *Lycopodium clavatum*

Chemistry

The herb contains lycopodine ($C_{16}H_{25}ON$), clavatine ($C_{16}H_{25}O_2N$), clavoloninine, fawcettine, fawcettinine, and deacetylfawcettine.

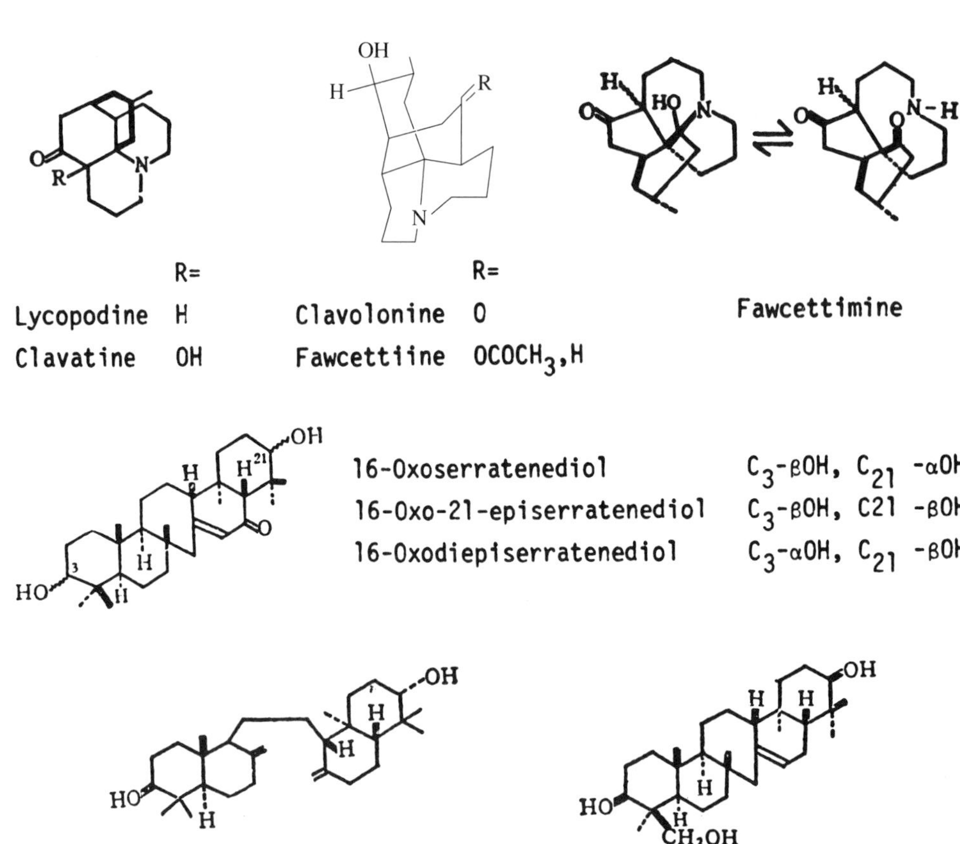

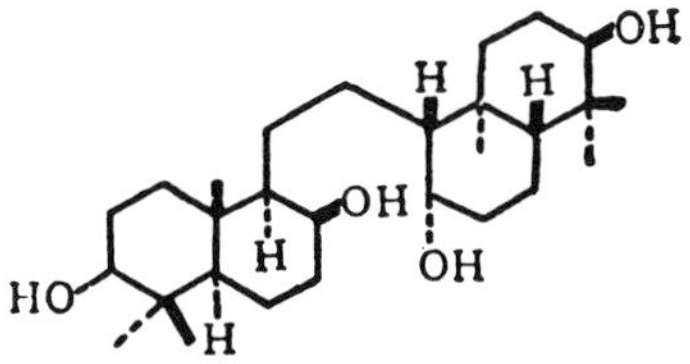

Lycoclavenol 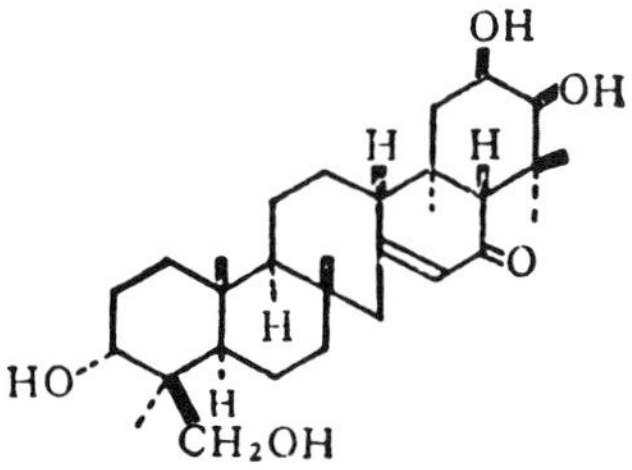 Clavatol

Therapeutic Uses

Shen Jin Cao was described in the Chinese pharmacopoeia as being able to dispel "wind" and remove "dampness" and to relieve rigidity of muscles and joints. It is also used to treat arthritis and dysmenorrhea.

ZHI WEI (石韦) — dried leaf of *Pyrrosia petiolosa* (Chris.) Ching, *P. sheareri*, or *P. lingna*

Chemistry

The herb contains isomangiferin ($C_{19}H_{18}O_{11}\cdot H_2O$) and diplotene ($C_{20}H_{50}$).

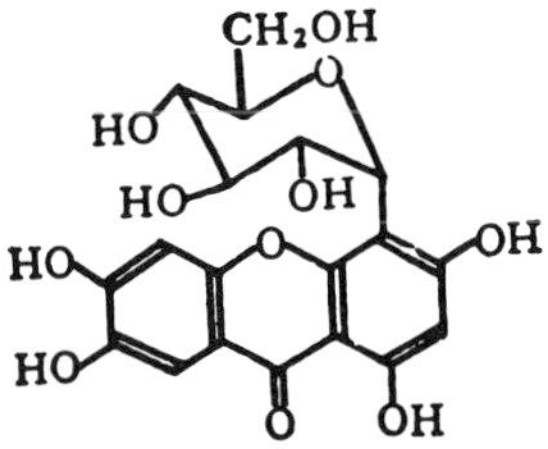

Isomangiferin

Therapeutic Uses

It is used as a diuretic in the treatment of urinary tract infections and urolithiasis. In addition, it can be used to arrest bleeding.

WU JIU (乌桕) — dried root bark of *Sapium sebiferum* (L.) Roxb.

Chemistry

The root bark contains xanthoxylin ($C_{10}H_{12}O_4$), corilagin, and sebiferic acid ($C_{30}H_{48}O_2$).

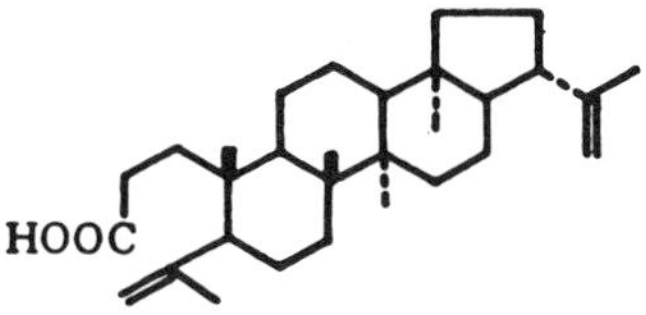

Sebiferic acid

Therapeutic Uses

The herb is used as a hydragogue, a diuretic to relieve edema.

QIAN JIN ZI (千金子) — dried ripe seed of *Euphorbia lathyris* L.

Chemistry

The seed contains the following steroid-like compounds: euphorbiasteroid ($C_{32}H_{40}O_8$), 7-hydroxylathyrol ($C_{20}H_{30}O_5$), lathyrol diacetate benzoate, lathyrol diacetate nicotinate ($C_{30}H_{37}O_7N$), euphol ($C_{30}H_{50}O$), euphorbol ($C_{31}H_{52}O$), euphorbetin ($C_{18}H_{10}O_8$), isoeuphorbetin, esculetin ($C_9H_6O_4$), and daphnetin.

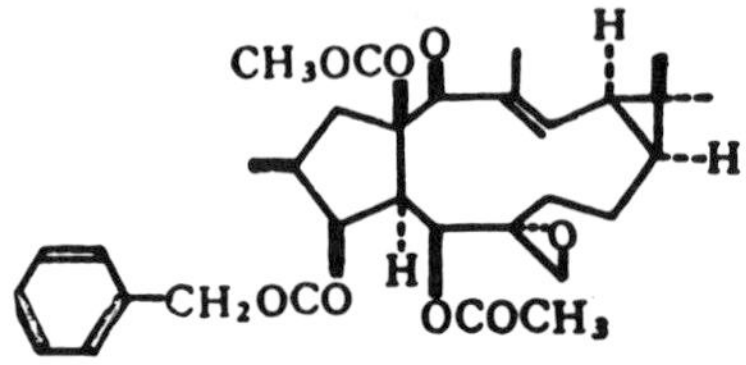

Euphorbiasteroid

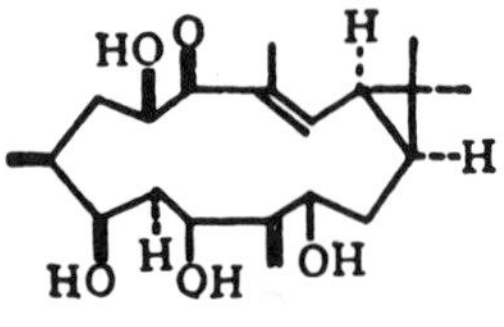

7-Hydroxylathyrol

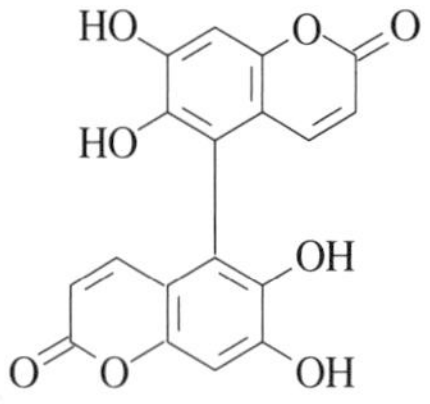

Euphorbetin

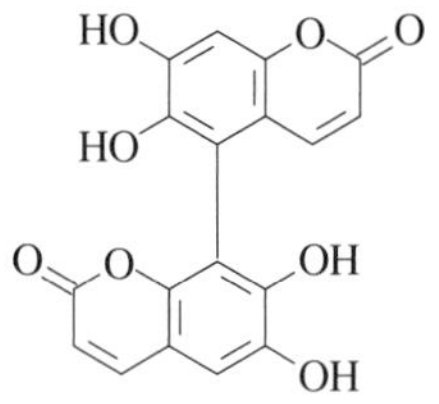

Isoeuphorbetin

Therapeutic Uses

This herb is used as a diuretic to remove edema, to eliminate blood stasis, and to resolve masses.

Three other herbs in the genus *Euphorbia* are also used in Chinese herbal medicine as diuretics:

GAN SUEI (甘遂) — dried tuberous root of *Euphorbia kansui* Lion.

Chemistry

The root contains α-euphol ($C_{30}H_{50}O$), tirucallol, α-euphorbol ($C_{31}H_{52}O$), kansuinine A ($C_{37}H_{40}O_{15}$), and kansuinine B ($C_{35}H_{42}O_{14}$). The latter two are both steroid derivatives and are very toxic.

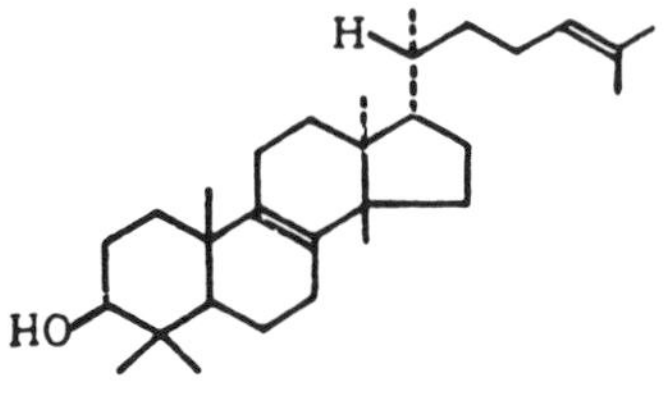

γ-Euphorbol

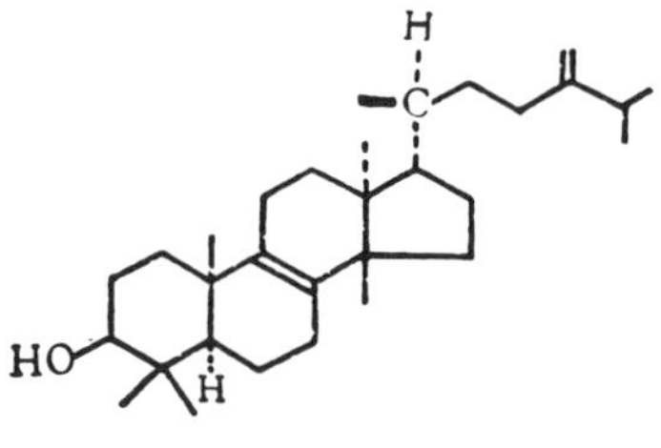

α-Euphorbol

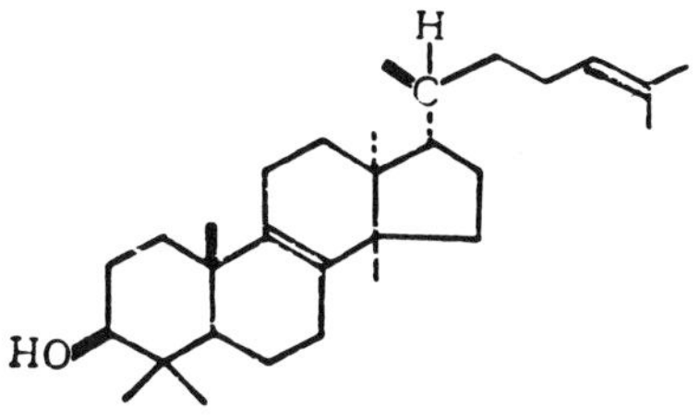

Tirucallol

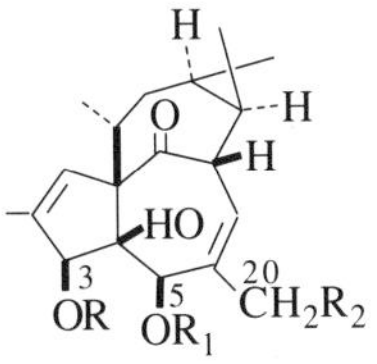

	R=	R_1=	R_2=
20-Deoxyingenol	H	H	H
Derivative I	COC_6H_5	H	H
Derivative II	H	COC_6H_5	H
Ingenol	H	H	OH
Derivative III	$CO(CH{=}CH)_2(CH_2)_4CH_3$	H	$OCOCH_3$

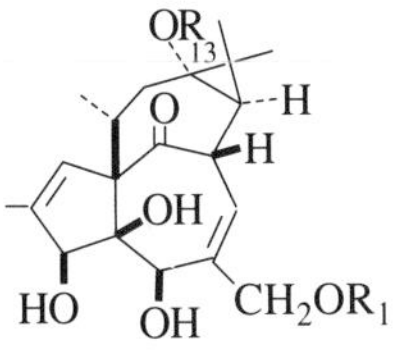

13-Oxyingenol R= R_1 = H

Derivative IV

R = $CO(CH_2)_{10}CH_3$

R_1= $CO(CH_2)_4CH_3$

ZE QI (泽漆) — the plant of *Euphorbia helioscopia* L.

Chemistry

The herb contains phasin, tithymalin, and heliscopiol ($C_{21}H_{44}O$).

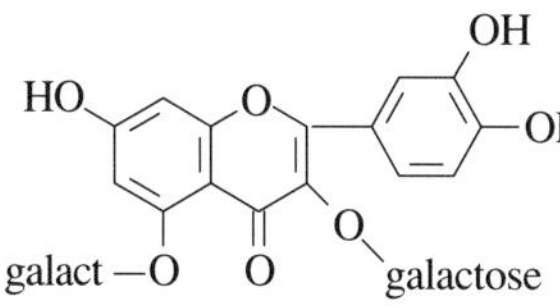

Tithymalin

HUO YU JIN (火殃勒) — the plant of *Euphorbia antiquorum* L.

Chemistry

This plant contains friedelan-3α-*ol* ($C_{30}H_{52}O$), α-taraxerol ($C_{30}H_{50}O$), β-amyrin ($C_{30}H_{55}O$), cycloarternol ($C_{30}H_{50}O$), euphol, and α-euphorbol.

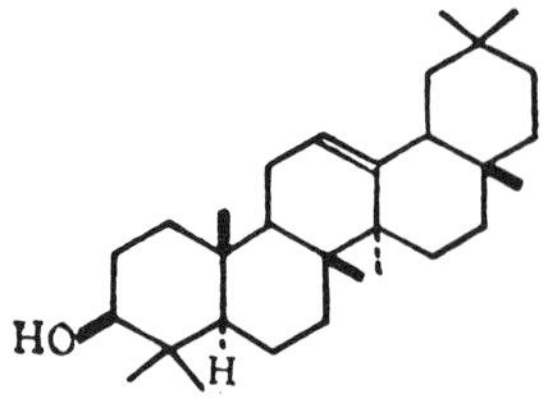

β-Amyrin

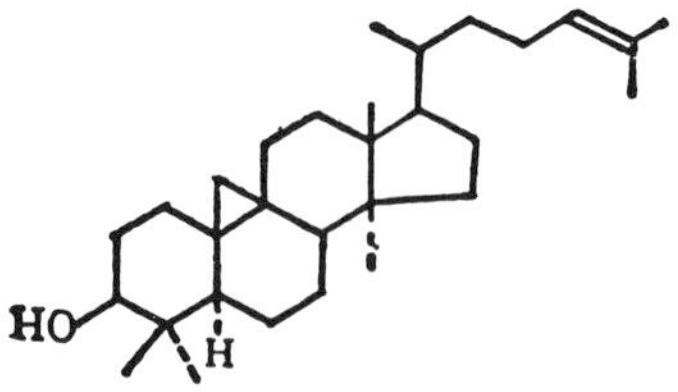

Cycloartenol

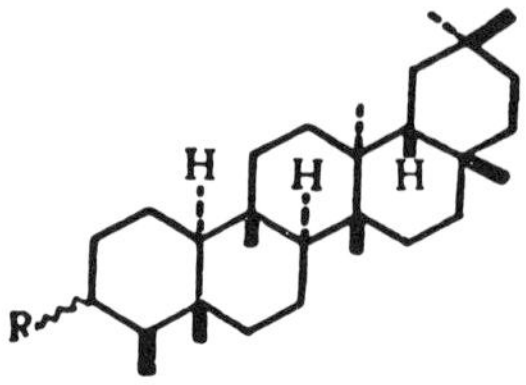

	R =
Friedelan-3α-ol	α-OH
Epi-friedelinol	β-OH

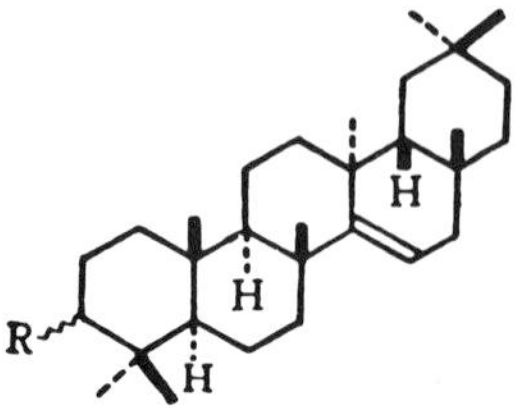

	R =
α-Taraxerol	α-OH
β-Taraxerol	β-OH

SHAN ZHU YU (山茱萸) — dried sarcocarp of *Cornus officinalis* Sieb. et Zucc.

Chemistry

The herb contains the following glycosides: morroniside ($C_{17}H_{26}O_{11}$), 7-*O*-methylmorroniside, ($C_{18}H_{28}O_{11}$), sworoside ($C_{16}H_{22}O_9$), and loganin ($C_{17}H_{26}O_{11}$). The leaf of *Cornus* contains longiceroside ($C_{17}H_{24}O_{16}$).

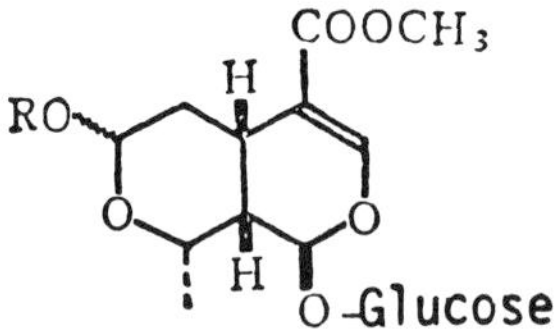

	R=
Morroniside	H
7-O-Methylmorroniside	CH_3

Actions

Shan Zhu Yu has been observed to increase urine flow. It has additional effects in lowering blood pressure.

Therapeutic Uses

This herb is prescribed as a diuretic, to treat dysmenorrhea, and to arrest spontaneous emission and perspiration.

BAI MAO GEN (白茅根) — dried root or rhizome of *Imperata cylindrica* (L.) Beauv. var. *major* or *I. cylindrica* Beauv. var. *koenigii* Durand

Chemistry

There are several substances that have been isolated from the root. They are as follows.

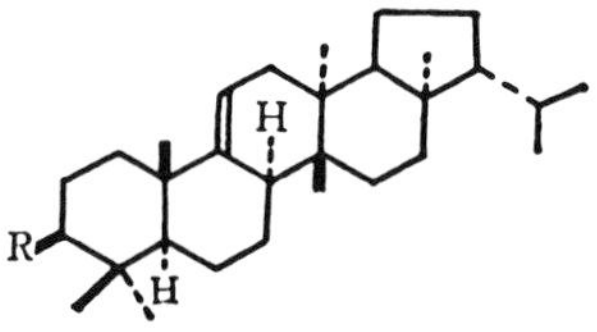

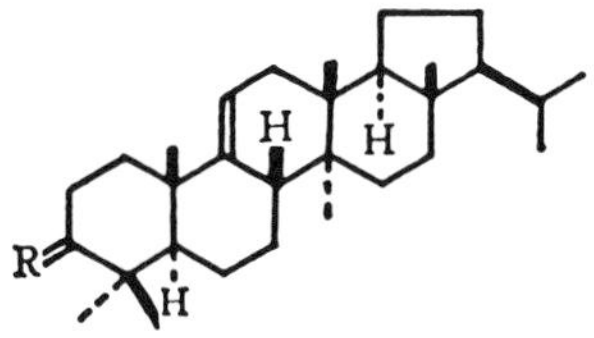

	R =
Fernenol	OH
Arundoin	OCH_3

	R =
Arborinol	β-H, α-OH
Arborino methyl ester	β-H, α-OCH_3
Arborinone	O
Glutinol	β-OH, α-H

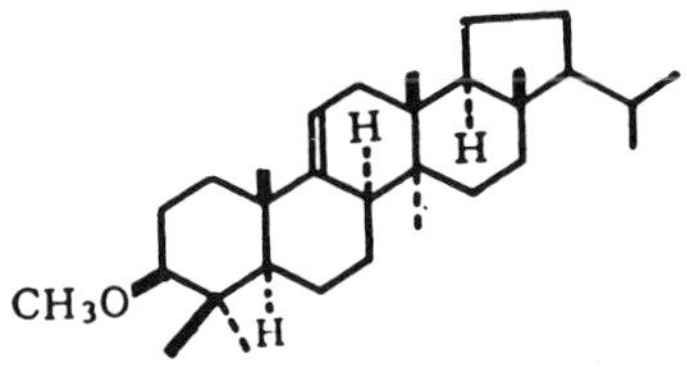

Cylindrin

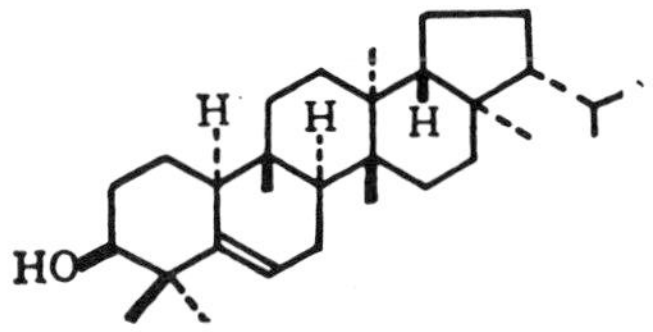

Simiarenol

Actions

The herb has a diuretic, hemostatic effect. It is also antibacterial.

Therapeutic Uses:

It was used to remove heat and to promote secretion and urination in acute and chronic glomerulonephritis. It was also used to treat acute toxic hepatitis and measles. It was given as a 250 to 500 g decoction.

TABLE 1
Other Diuretic Herbs

Name	Source	Part	Contains	Action and uses
Yu Min Zu (玉米須)	*Zea mays* L.	Dried style, and stigma	Glucoside (1.15%), saponin (3.18%), fat (2.5%), and essential oil (0.12%), cryptoxanthin, vitamin C, and vitamin K	Diuretic and antihypertensive, choleretic; used in chronic nephritis; removes edema; treats hypertension, jaundice, and chronic cholecystitis with gallstone
Fu Ping (浮萍)	*Spirodela polyrrhiza*	Whole plant	K^+ acetate, KCl, I_2, some flavoids	Diuretic, induces diaphoresis, promotes eruption, dispels wind; used to treat nephritis and measles

Chapter 27

HERBS AFFECTING THE UTERUS

The herbs that were generally used in chinese traditional medicine to treat obstetric and gynecological problems were reputed to make stagnant blood come to life, relax the uterus, stabilize pregnancy, regulate the menstrual cycle, and enrich the circulation of blood to the uterus. Each of the herbs described here may have different mechanisms of action and different uses in various female ailments.

DANG GUI (当归) — the dried root of *Angelica sinensis* (Oliv.) Diels.

FIGURE 1. *Dang Gui* root.

Chemistry

This is the most popular of Chinese herbs, and is widely used by both the Chinese and the Japanese. It contains a significant quantity of vitamin B_{12} and vitamin E. It also contains ferulic acid, succinic acid, nicotinic acid, uracil, adenine, butylidenephalide, ligustilide, folinic acid, and biotin.

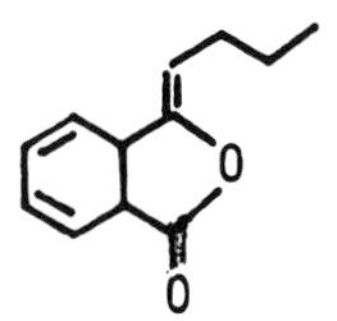

n-Butylidene phthalide

$CO(CH_2)_3CH_3$

COOH

n-Valerophonone-O-carboxylic acid

Actions

There are two components of the herb which exert either a stimulating or an inhibiting effect on the uterus. The water-soluble and nonvolatile component causes stimulation, while the alcohol soluble component, which is an essential oil with a high boiling point, exerts the pharmacological action of inhibiting or relaxing the uterus. This component increases DNA synthesis in the uterine tissue and increases its growth.

Dang Gui can prolong the refractory period of the myocardium and lower blood pressure. It dilates the coronary vessels and increases coronary flow. The herb also reduces the respiratory rate.

Because of its high content of vitamin B_{12} (0.25 to 0.4 μg/100 g dried root) as well as its content of folinic acid and biotin, the herb exerts a remarkable stimulating effect on hematopoiesis in the bone marrow. It also exhibits an antiplatelet action, inhibiting the release of 5-hydroxytryptamine (5HT) from platelets.

Other effects include lowering blood cholesterol. In experiments with animals, 5% *Dang Gui* mixed in animal food has reduced atherosclerosis formation. In addition, the herb has anti-inflammatory, analgesic, and antibacterial properties.

Toxicity

The MLD of *Dang Gui* extract in mice is 0.3 to 0.9 g/10 g body weight. Intravenous administration of the essential oil of *Dang Gui* in doses of 1 ml/kg in animals can cause a fall in blood pressure and depression of respiration. The alcoholic extract of the herb is much more toxic; at a dose of 0.06 to 0.2 ml/kg, it can cause death in animals due to respiratory arrest.

The adverse effects include excessive bleeding and, occasionally fever. Both of these effects probably involve hypersensitivity to the herb.

Therapeutic Uses

This herb is very popular in the treatment of women's diseases. In cases of irregular menstruation, it regulates bleeding, reduces pain, and relaxes the uterus. Furthermore, it has a hematopoietic effect useful in the treatment of anemia, which is commonly observed in cases of dysmenorrhea.

It is used in the treatment of thrombophlebitis, neuralgia, and arthritis. A 25% *Dang Gui* extract has been injected intravenously, directly into joints, or as a nerve block, with reportedly successful results.

It has been used to treat chronic nephritis, constrictive aortitis, and anemia; especially pernicious anemia and folic acid deficiency anemia. It is reported to be effective in treating skin diseases, including eczematous dermatitis, neurodermic dermatitis, and psoriasis. Other infectious diseases, including dysentery and hepatitis, have also been treated with this herb.

Many formulas containing *Dang Gui* are recommended by Chinese pharmacopoeias in the treatment of female ailments. Tablets made from *Dang Gui* extract (0.5 g) are administered in doses of two tablets twice a day, in the treatment of dysmenorrhea. The herb is also available in injection form (10 or 25%), at 20 ml/ampule; this is administered intramuscularly or intravenously.

References

1. **Wang, Y. S., Ed.,** *Pharmacology of Chinese Herbs and Their Uses* (in Chinese), People's Health Publication, Beijing, 1983, 424–438.
2. **Lue, F. H. et al.,** *Chin. Med. J.* (in Chinese), 40, 670, 1954.
3. **Wu, P. J., Ed.,** *Pharmacology of Chinese Herbs,* People's Health Publication, Beijing, 1983, 191–193.

YI MU CAO (益母草) — dried aerial parts of *Leonurus heterophyllus*

Chemistry

This herb contains several alkaloids, including leonurine (0.01% of the herb by weight), stachydrine, leonaridine, and leonurinine. In addition, vitamin A and fatty oils are present.

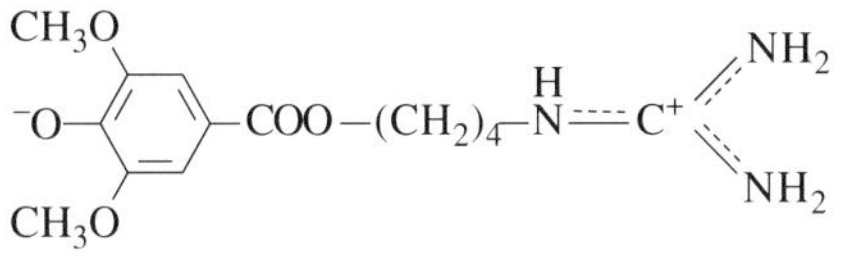

Leonurine

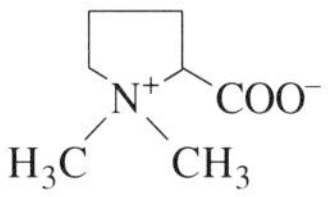

Stachydrine

Actions

The herb stimulates uterine contraction. This action is similar to that of oxytocin, but weaker. Small doses of *Yi Mu Chao* can stimulate the heart, while large doses will inhibit it. It is a vasodilatating agent and can lower blood pressure. It can hemolyze red blood cells, even at a concentration of 0.5 to 1 mg/ml.

The herb stimulates the respiratory center directly; both the rate and amplitude increase. Vagoectomy does not affect the stimulatory effect.

Leonurine causes a stimulation of the skeletal muscles, followed by a depression of the neuromuscular junction, producing a curare-like effect. In addition, leonurine is an effective diuretic, and can increase urine output by more than 200 to 300%.

Last, the herb exhibits antibacterial properties.

Therapeutic Uses

This herb was used to activate blood flow and remove blood stasis. It is used to treat irregular menstruation and postpartum bleeding, and to bring the uterus back to normal size after delivery.

It is administered in the form of an extract, which is taken in doses of 5 to 10 ml t.i.d., or in the form of pills, which are taken in doses of one to two pills b.i.d. In the treatment of acute nephritis, the herb is decocted in doses of 30 g, drunk daily to produce a diuretic effect and to remove edema.

HONG HUA (红花) — the dried flower of *Carthamus tinctorius* L.

Chemistry

Approximately 0.3% of the flower petal consists of the glycoside carthamin. This is a yellow substance; enzymes inside the petal can change the color by producing carthamone and carthemidin. Other substances isolated from the flower include neocarthamin and saflor yellow, which is actually made up by four different yellow substances.

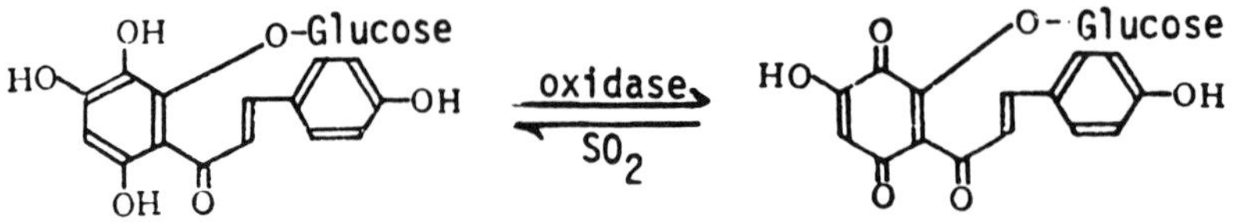

oxidase / SO_2

Carthamin (yellow color) Carthamone (red color)

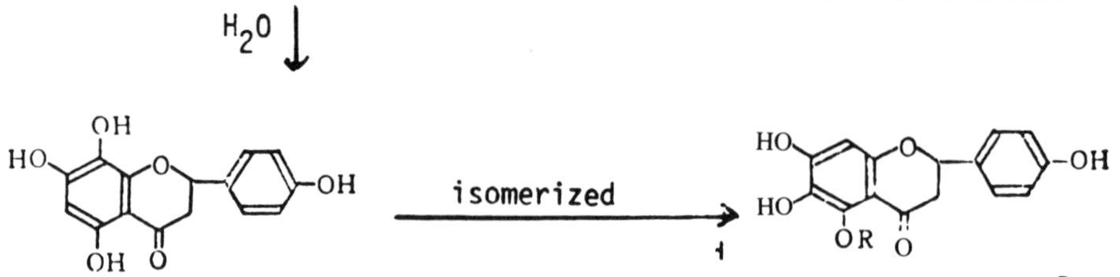

H_2O

isomerized

Carthamidin (colorless)

	R=
Isocarthamidin (colorless)	H
Neo-carthamin (colorless)	-glucose

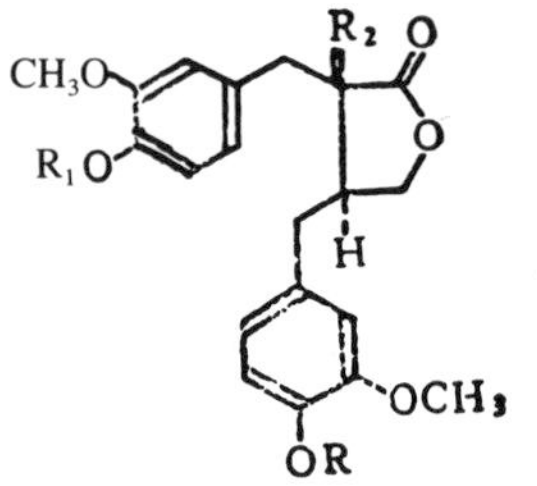

	R=	R_1=	R_2=
Matalresinol monoglucoside	-glucose	H	H
2-Hydroxyarctiin	CH_3	-glucose	OH

Actions

In small doses, this herb causes a rhythmic contraction of the uterus. When the dose increases, the tone of contraction increases, until the uterus finally reaches a convulsive condition.

The herb stimulates the heart in small doses and inhibits it when the dose is increased. The water extract of the herb causes coronary dilatation, increases tolerance to oxygen deprivation, and prolongs survival times. The herb can also lower blood pressure; its duration of action is relatively long.

Hong Hua has been observed to prolong blood coagulation time and inhibit platelet aggregation. It lowers plasma cholesterol and triglyceride levels. In addition, it causes bronchial contraction.

Therapeutic Uses

This herb is reputed to promote blood circulation, remove blood stastis, and restore normal menstruation. It is claimed to be especially effective in the treatment of dysmenorrhea and menopause. It was also used in treating angina pectoris to increase coronary circulation.

In the treatment of cerebral thrombosis, this herb is given in the form of a 50% injection solution, diluted with a 10% glucose solution for intravenous infusion. *Hong Hua* injection ampules (5%) are also available for intramuscular administration; the standard dose is 2 ml, given in the treatment of neuralgic dermatitis or to relieve pain due to hematoma swelling.

XIANG FU (香附) — dried tuber of *Cyperus rotundus* L.

Chemistry

Approximately 1% of the tuber's weight consists of an essential oil. The major components of the oil are α- and β-cyperene (30 to 40%), α- and β-cyperol (40 to 49%), and cyperoone (0.3%).

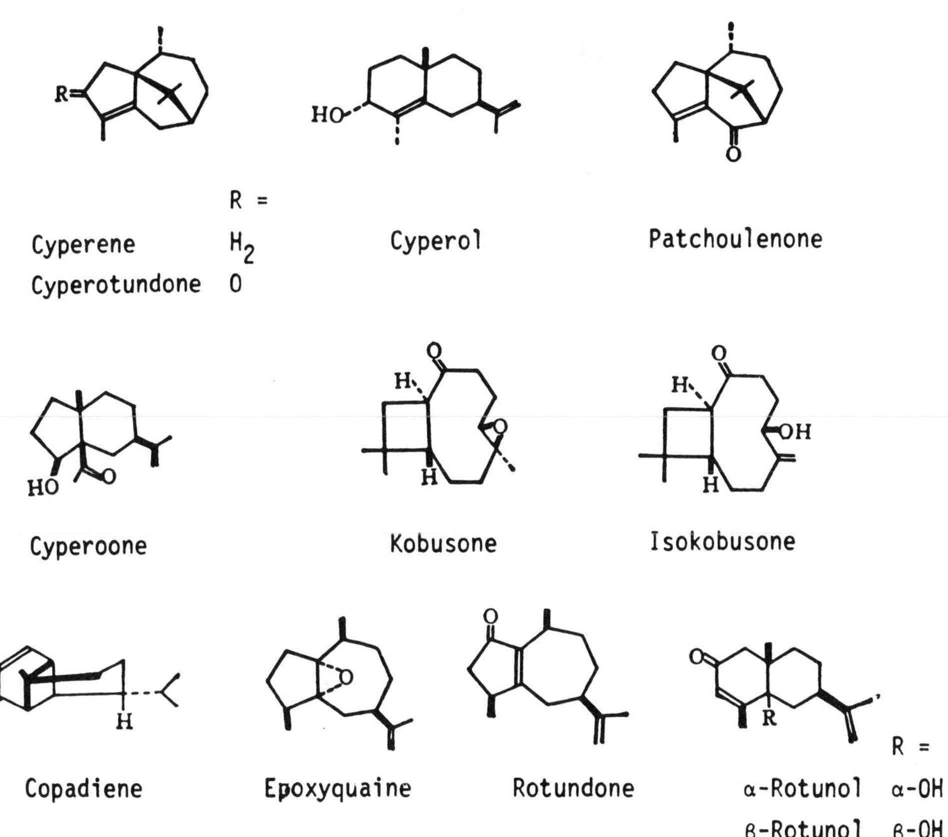

Actions

The water extract of the herb has an inhibitory effect on the uterus, causing uterine relaxation in both pregnant and nonpregnant women and relieving pain. The herb can stimulate gastric and salivary secretion, and increase the motility of the stomach. In addition, it has an antibacterial effect.

Therapeutic Uses

Therapeutically, this herb is used to treat dysmenorrhea and other menstrual irregularities. In the Chinese pharmacopoeia, it was described as an agent to "smooth" the liver, regulate the circulation of *qi*, normalize menstruation, and relieve pain.

A popular remedy for female ailments is a decoction of 12 g *Xiang Fu*, 12 g *Yi Mu Cao*, 15 g *Tan Seng*, and 9 g *Bai Shao* (peony root).

TIEN HUA FEN (天花粉) — dried root of *Trichosanthes kirilowii* Maxim. or *T. japonica* Regel.

FIGURE 2. The root of *Tien Hua Fen*.

Chemistry

This root grows wild in southern parts of China. It contains the protein trichosanthin, some polysaccharides, amino acids, and saponin.

Actions

The herb has a stimulating effect on the uterus similar to that of oxytocin. Intravenous administration of the extract is very effective in removing chorioma, hydatid mole, and ectopic pregnancies. It was reported that in 100 cases of midterm gestation, trichosanthin (0.48 mg) injected into the amniotic sac caused abortion in 99 cases by the 4th day after injection.

The mechanism of this action occurs when trichosanthin causes structural damage to the trophoblastic tissue, thus impairing the function of the placenta. In addition, the serum hCG level decreased to half of its initial level within 24 h after administration. PGE and $PGF_{2\alpha}$ levels in the amniotic fluid increased to twice their initial levels. Necrosis and tissue destruction resulted in a contraction of the uterus.

Toxicity

This herb is a potent immunosuppressant and irritative agent. Overdosage can cause a rise in body temperature, extreme malaise, sore throat, and severe headache. Hypersensitivity to the herb produces the symptoms of itching, rash, local swelling, and irritation, especially after parenteral injection.

The herb is eliminated chiefly by renal excretion and partially by hepatic metabolism. Overdosage may cause renal and hepatic damage.

Therapeutic Uses

The *Ben Cao Kong Mu* stated that this plant "can promote menstruation, facilitate the expulsion of the placenta, and has been used for hundreds of years to induce abortion."

The herb is generally prepared in injection form. For deep muscular injection or intra-amniotic injection, it is prepared at a strength of 5 mg/ampule and administered in doses of 5 to 10 mg, dissolved in 1 to 2 ml of saline. For intravenous infusion, 5 to 10 mg of the *Tian Hua Fen* injection solution is dissolved in 500 ml of physiological saline.

Purified trichosanthin is available in powder form. This is mixed with saline at 2 mg per ampule, for intramuscular or intraamniotic injection.

References

Jin, Y. G., Clinical study of trichosanthin, in *Advances in Chinese Medicinal Material Research,* Chang, H. M. et al., Eds., World Scientific Publishing Co., Singapore, 1985, 319–326.

YUAN HUA (芫花) — dried flower bud and bark of *Daphne genkwa* Sieb. et Zucc.

Chemistry

The flower bud contains several flavone glycosides. The hydrolyzed products of the glucosides include genkwanin, apigenin, hydroxygenkwanin, and yuanhuacin I. The herb also contains some oily substances which are quite toxic.

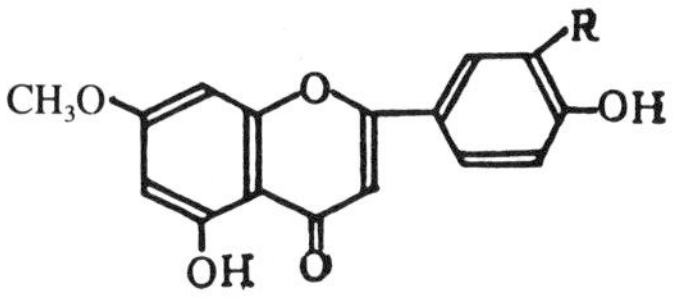

	R=
Genkwanin	H
Hydroxygenkwanin	OH

Actions

The herb increases uterine contraction. If taken in early pregnancy, it can poison the fetus and induce abortion.

In addition, the herb exhibits antitussive and expectorant effects. It is an antibacterial agent and parasiticide, especially in cases of ascaris. It exhibits diuretic, laxative, analgesic, and anticonvulsive properties.

The LD_{50} of this herb in rats is 9.25 g/kg (i.p.). Death occurs from convulsions and respiratory arrest.

Therapeutic Uses

This herb has been used to induce abortion by direct injection into the vagina or amniotic fluid. It is also used as an expectorant in the treatment of chronic bronchitis. In addition, it is used to treat malaria, as an anthelmintic against ascaris, and in the treatment of cutaneous infections. It has also been used to relieve toothache.

The herb is available as an injection preparation of 80 μg/2 ml, for intrauterine or intraamniotic administration.

MAI JIAO (麦角) or ERGOT — dried sclerotium of *Claviceps purpurea*, a parasite found in the ovary of *Secale cereale*

Chemistry

Several alkaloids have been isolated from this herb. These include six pairs of isomeric compounds: ergotoxine, ergothioneine, ergotamine, ergometrine, ergocryptine, and ergocristine.

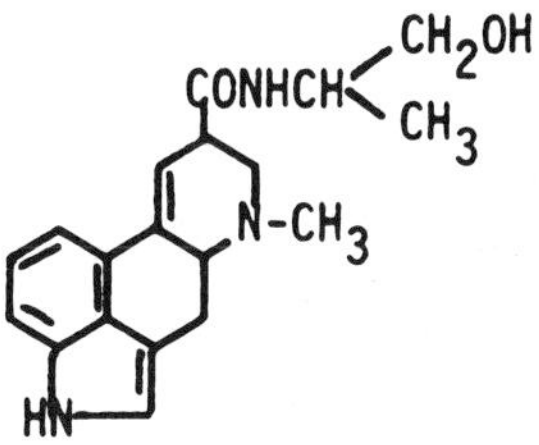

Ergometrine

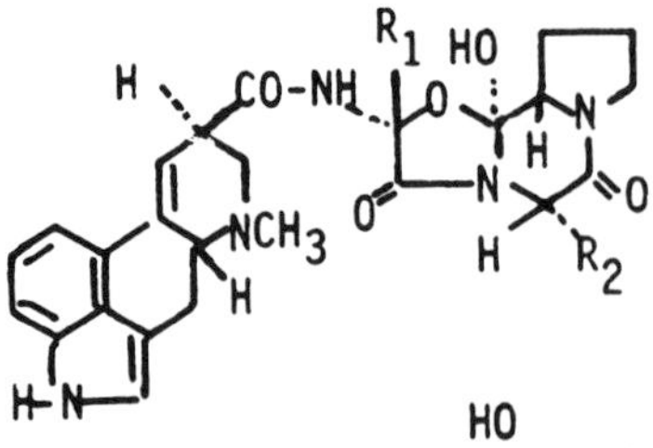

	R_1=	R_2=
Ergotamine	CH_3	$CH_2C_6H_5$
Ergosine	CH_3	$CH_2CH(CH_3)_2$
Ergocornine	$CH(CH_3)_2$	$CH(CH_3)_2$
Ergocryptine	$CH(CH_3)_2$	$CH_2CH(CH_3)_2$
Ergocristine	$CH(CH_3)_2$	$CH_2C_6H_5$

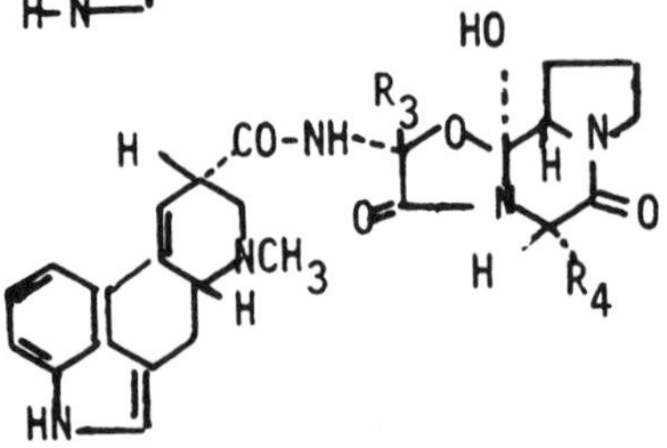

	R_3=	R_4=
Ergotaminine	CH_3	$CH_2C_6H_5$
Ergosinine	CH_3	$CH_2CH(CH_3)_2$
Ergocorninine	$CH(CH_3)_2$	$CH(CH_3)_2$
Ergocryptinine	$CH(CH_3)_2$	$CH_2CH(CH_3)_2$
Ergocristinine	$CH(CH_3)_2$	$CH_2C_6H_5$

Actions

The alkaloids have many pharmacodynamic actions. Besides their α-adrenergic blocking action, they are potent uterine contracting agents. They constrict the blood vessels and stop bleeding.

The major toxicities are induction of early miscarriage or early delivery, and gangrene of the extremities.

Therapeutic Uses

For hundreds of years, the herb was used in folk medicine to stop postpartum uterine bleeding and to contract the enlarged uterus.

APPENDIX I: Herbs Promoting Milk Secretion

WANG BU LIU XING (王不留行) — the dried seed of *Vaccaria segetalis* (Neck.) and *V. pyramidata* Medic.

Chemistry

The seed contains vacsegoside ($C_{75}H_{118}O_{40}$), vaccaroside ($C_{36}H_{54}O_{10}$), gypsogenin ($C_{30}H_{46}O_4$), and vaccarin ($C_{32}H_{38}O_{19}$).

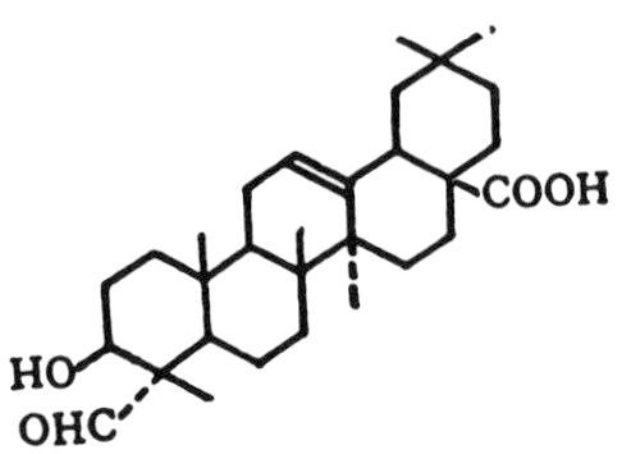

Gypsogenin

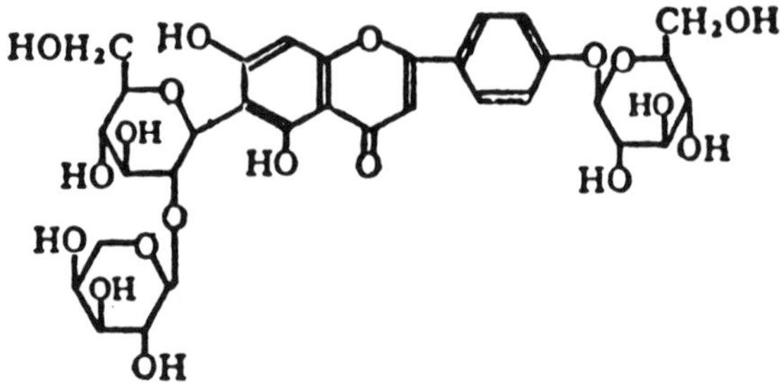

Vaccarin

Therapeutic Uses

It is used to activate blood flow and promote milk secretion. In addition, it is commonly used to treat amenorrhea and breast infections.

APPENDIX II: Herbs Affecting Reproduction

GOSSYPOL (棉子素) — compound derived from the seed, stem, or root of the cotton plant *Gossypium*

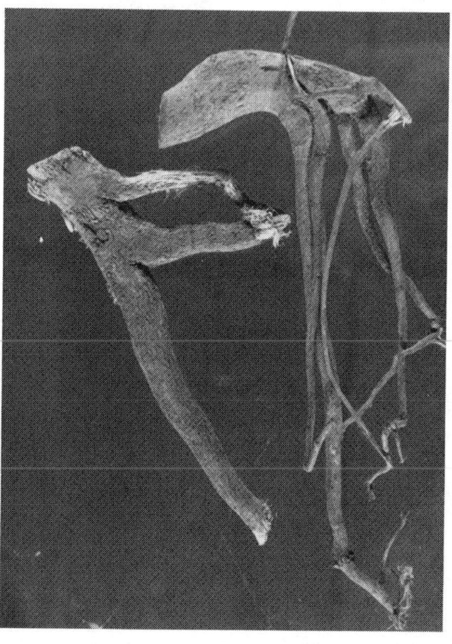

FIGURE 3. *Mian Hua Gan*, the root of *Gossypium herbaceum*.

Chemistry

This is a yellowish binaphthol compound isolated from the seed, stem, or root of the cotton plant ($C_{30}H_{30}O_8$).

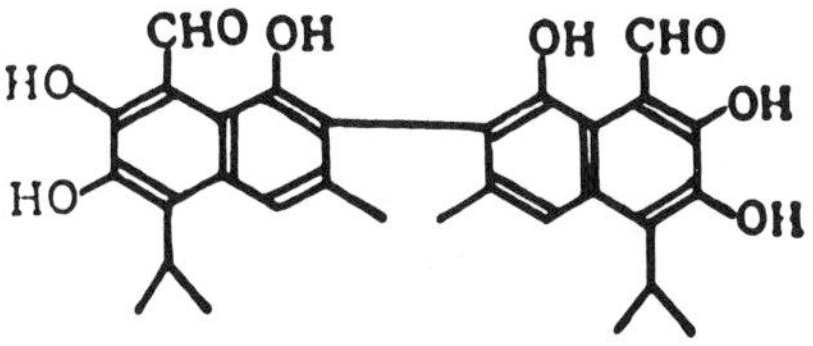

Gossypol

Actions

This compound suppresses spermatogenesis very effectively. In China, it has been widely used as an antifertility agent in men. It was recommended that it be given in the form of gossypol acetic acid or gossypol formic acid, in doses of 20 mg/d for 75 d. After treatment, sperm count decreases tremendously.

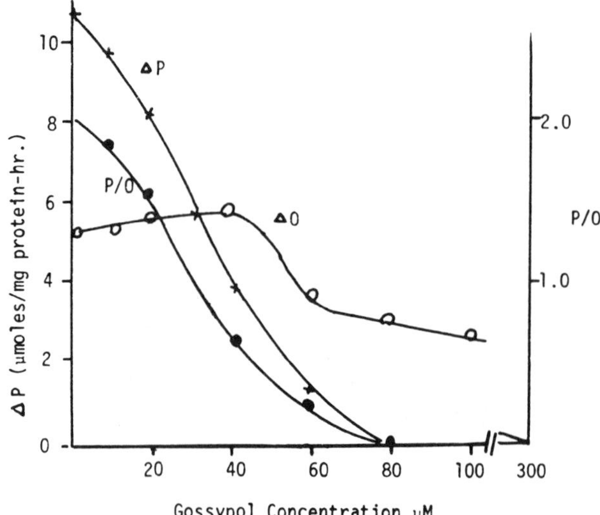

FIGURE 4. The effect of gossypol on O_2, phosphorylation, and P/O ratio of sperm cells. (From Xue, S. P. et al., in *Advances in Chinese Medicinal Materials Research,* Chang, H. M. et al., Eds., 1985, Figure 7. With permission.)

This agent causes damage to the testicular mitochondrial ultrastructure, and uncouples mitochondrial respiration and oxidative phosphorylation. It especially diminishes the mitochondrial energy-linked enzymes, ion regulatory enzymes, and the spermatogenetic enzyme, LDH-X. All of these actions suggest that testicular mitochondria might be the target organelles for gossypol.

It decreases glucose and fructose metabolism in human spermatozoa, indicating an action on energy-related enzymes.

It blocks neuromuscular transmission without affecting nerve conduction; neostigmine does not block this action. This suggests that gossypol is primarily a presynaptic blocking agent. It has been speculated that this effect is due to a decrease in Ca^{2+} uptake by the mitochondria at the presynaptic endings.

Gossypol inhibits Na-K ATPase activity by an noncompetitive mechanism; the mechanism of action is unclear. In addition, it selectively inactivates the isozymes of lactate dehydrogenase (IC_{50} = 50 to 75 μM).

At the dose of 20 to 30 mg/kg in rats, gossypol has no effect on LH, FSH, and PRL serum concentrations. It also has no effect on serum testosterone level, causing at most a slight increase.

Gossypol also has effect on females. At high doses, it can interrupt early pregnancy in female rats. This is due to its antiprogesterone and anti-corpus luteum effect. Peng et al.[2] reported that the endometrium became atrophic when animals received a dose of 30 mg/kg/d for 8 consecutive weeks. Similar results were reported for human subjects.

Toxicity

This compound has been used in China as a contraceptive, with the government's approval. It is reported that the herb does not affect blood pressure, hemoglobin, body weight, or plasma K^+ levels. However, some general complaints have been made by people taking this compound. These include fatigue (13%), gastrointestinal disturbances (7%), and a decrease in libido (9%).

The most difficult problem associated with the herb is that spermatogenic depression may become permanent, or the recovery less rapid than expected. Twelve months after stopping administration, more than 50% of patients still showed a zero sperm count. This is quite discouraging.

References

1. **Ling, G. Z.,** in *Advances in Chinese Medicinal Materials Research,* Chang, H. M. et al., Eds., World Scientific Publishing Co., Singapore, 1985, 605–612.
2. **Peng, L. H. et al.,** in *Advances in Chinese Medicinal Materials Research,* Chang, H. M. et al., Eds., 1985, 639–645.
3. **Xue, S. P. et al.,** in *Advances in Chinese Medicinal Materials Research,* Chang, H. M. et al., Eds., 1985, 625–638.
4. **Ye, Y. X. et al.,** *Acta Pharmacol. Sin.,* 7, 265–268, 1986.
5. **Yu, Z. H.,** *Acta Pharmacol. Sin.,* 3, 260–263, 1982.
6. **Zhang, L. Z. et al.,** *Acta Pharmacol. Sin.,* 7, 563–567, 1986.

Section VIII
Hematopoietic System

Chapter 28

HERBS PROMOTING BLOOD FORMATION

Chinese herbal practitioners use the term ''hollowness of the blood'' to refer to a number of conditions, including anemia, hemorrhage, and many disorders involving the blood or originating from the heart, liver, spleen, kidneys, or bone marrow. Such ''hollowness'' or weakness is often seen in gynecological problems such as menstrual irregularity and in infectious diseases. The signs or symptoms described as ''hollowness'' are the loss of facial color, lack of freshness and brightness, dizziness, and heart palpitations.

The herbs described in this chapter are used to combat such illnesses, to produce an antianemic effect, and to promote leukocyte formation.

E JIAO (阿胶) — gelatin from the skin of an ass, *Equus asinus* (Equidae)

Chemistry

The major components are the hydrolysis products of gelatin, amino acids, and the Ca^{2+} ion.

Actions

Recent clinical trials in China reported that this herb exhibits several actions. The primary action is hematopoiesis. It stimulates the formation of red blood cells and hemoglobin in the bone marrow. It is effective in the treatment of hemorrhagic anemia. In addition, it has a hemostatic effect, but does not change clotting times. It improves the muscle tissue regeneration.

The herb also has an effect on Ca^{2+} metabolism. It increases the intestinal absorption of Ca^{2+} and decreases the urinary excretion of Ca^{2+}.

Therapeutic Uses

According to the *Ben Cao Kong Mu, E Jiao* can arrest bleeding in cases of hematuria, hematemesis, bloody stools, dysmenorrhea, and postpartum bleeding. It was claimed that the herb could treat all kinds of *Feng* or ''wind''-related illnesses, which were believed to be caused by the climate or allergies, and which occurred in both sexes. It is also used to treat joint pain, edema, asthma, and septicemia.

The herb is given in doses of 6 to 9 g, dissolved in wine or water and drunk daily. This was recommended to improve blood tone, arrest bleeding, replenish vital essences, and ''moisten'' the lungs.

JI XUE TENG (鸡血藤) — dried stem of *Spatholobus suberectus* Dunn (Leguminosae)

Chemistry

The major components of the herb are friedelin, taraxerone, and other alcoholic derivatives.

Actions

This herb can slow the heart rate and lower blood pressure. The water decoction of *Ji Xue Teng* can stimulate rhythmic contraction of the uterus. Large doses can cause convulsive contractions. It is antibacterial, especially against staphylococci.

Therapeutic Uses

Traditional Chinese medicine prescribed the herb to improve blood tone and activate blood flow and to relieve rigidity of muscles and joints. It is used in the treatment of leukopenia, and malnutritional or hemorrhagic anemia.

The standard dose of this herb is 60 to 120 g made into decoction, drunk daily. It is also prepared in the form of a syrup (2 g/ml), administered in doses of 10 to 20 ml t.i.d., or in the form of 1-g tablets, taken in doses of four to eight tablets t.i.d.

NU ZHEN ZI (女貞子) — glossy privet fruit, from *Ligustrum lucidum* Ait.

Chemistry

The fruit contains the glycoside nuzhenide, oleanolic acid, and ursolic acid.

Actions

The herb has been observed to increase the leukocyte count. It is a cardiac tonic and diuretic. In addition, it exhibits antibacterial properties.

Therapeutic Uses

Nu Zhen Zi is used in the treatment of leukopenia, chronic bronchitis, and acute dysentery. The Chinese pharmacopoeia also recommended it to improve the tone of the kidneys, replenish vital essences, nourish the liver, and improve visual acuity.

The herb is generally administered as a decoction of either 60 g of the bark or 90 g of the leaf; this is drunk three times a day for ten consecutive days.

BEI YAO ZI (白药子) — dried tuberous root of *Stephania cepharentura* Hayata

Chemistry

The root contains several alkaloids, including cepharanthine, cycleanine, isotetrandrine, berbamine, and cepharmine.

Actions

The herb has been observed to stimulate leukocyte formation in the bone marrow, and to promote regeneration of red blood cells and hemoglobin. It has antibacterial effects, especially against the tubercular bacillus. It is an antitoxin effective against tetanus and diphtheria toxins. In addition, it promotes muscle relaxation.

Therapeutic Uses

The herb is used to treat leukopenia, acute hepatitis, bacterial dysentery, epidemic parotiditis, and neurodermatitis. It is also prescribed in cases of bleeding and blood stasis.

The standard dose is 40 mg of cepharanthine, taken either orally or injected intravenously.

Chapter 29

HEMOSTATIC AND ANTISTASIC HERBS

Hemostatic herbs which promoted blood coagulation were used by Chinese herbalists to treat all types of hemorrhagic diseases. Traditional Chinese medicine considered bleeding to result from an overflowing of blood from the circulatory system. It was speculated that there were three possible causes of bleeding, and three corresponding treatments:

1. An internal ''fire devil'' which created either a negative or positive fire which damaged blood vessels. The remedy for this condition was any herb with the ability to clear ''fire'' of the liver, remove lung ''heat'', cool the blood, and stop the bleeding.
2. Feebleness of *qi*, which results in stasis of circulation, leading to bleeding. The remedy for this type of ''heat'' was any herb capable of restoring the ''tone'' of *qi*, thereby removing the stasis.
3. Stagnant blood in the vessels, resulting in an intravascular blood obstruction which interferes with circulation and causes bleeding. In these cases, the remedy was any herb which could resolve the blood stasis.

Our present knowledge of blood coagulation mechanisms has helped us to understand that the liver is the key organ which manufactures certain factors important for the coagulation process. Deficiency of one of these factors interferes with blood coagulation. Although Chinese traditional medicine did not have an explicit description of how blood coagulates, it did make a strong correlation between liver function and coagulation process.

Most of the herbs described in this chapter act as astringents, precipitating blood proteins, thereby producing clotting.

CE BAI YE (侧柏叶) — dried leafy twig of *Biota orientalis* (L.) Endl.

Chemistry

The twig contains the glycoside quercitrin and others such as pinipicrin and thuzone, plus some essential oils.

Actions

The herb has a hemostatic effect, shortening blood clotting time. It is an antitussive and expectorant and causes bronchodilatation via a partial anticholinergic effect. It has antibacterial properties and can potentiate the sleeping effect induced by barbiturates and relax the smooth muscles.

The common adverse effects produced by this herb include dizziness, nausea and vomiting, and loss of appetite. In rare cases, it may cause edema and hypersensitivity.

The LD_{50} in mice is 15.2 g/kg (i.p.).

Therapeutic Uses

This herb is commonly used to stop bleeding in cases such as uterine bleeding, duodenal bleeding, and bloody stools. It is available in tablet form (0.5 g, equivalent to 3.4 g of raw material), an injection solution (equivalent to 2.5 g/ml), and a tincture or alcoholic extract.

Ce Bai Ye has also been used in the treatment of chronic bronchitis and whooping cough; it is also reputed to be effective in treating tuberculosis. For these treatments, 30 g of the fresh leaf is boiled with one bowl of water, reduced to 100 ml and mixed with

10 ml of honey. This concoction is given to the patient in doses of 15 to 20 ml, three times a day.

The glycoside quercitrin also has an anti-inflammatory action. It can inhibit the edematous swelling of the extremities induced by protein injection. Because of this, the herb is also recommended for external application on burn wounds or swelling. For external application, 9 to 15 g of the crude herb is decocted, and then applied to reduce swelling.

MA HAN LIAN (墨旱蓮) — dried aerial part of *Eclipta prostrata* (Compositae)

Chemistry

The plant contains essential oils, tannic acid, saponin, wedolactone, demethyl wedolactone, α-tertiary methanol, nicotine, and ediptine.

Therapeutic Uses

According to the Chinese pharmacopoeia, the herb's primary effect is to "cool" the blood and stop bleeding. It is generally administered in doses of 30 g, made into a decoction; this formula is prescribed in the treatment of hematuria, blood coughing, duodenal bleeding, and uterine bleeding. The herb can also be applied externally in the treatment of eczema or bleeding of the skin.

HUAI HUA (槐花) or HUAI MI (槐米) — dried flower or bud of *Sophora japonica* (Leguminosae)

Chemistry

Between 10 and 28% of the herb by weight consists of the glycoside, rutin ($C_{27}H_{30}O_{10}$), which yields the genin quercetin after water hydrolysis — one molecule of glucose and one of rhamnose. The plant also contains saponins which produce the hydrolysis products betulin and sophoradiol ($C_{30}H_{50}O_2$), sophorin A ($C_{20}H_{36}O_{17}$), sophorin B ($C_{27}H_{45}O_{10}$), and sophorin C ($C_{27}H_{48}O_{10}$). These have been isolated primarily from the seed of the plant.

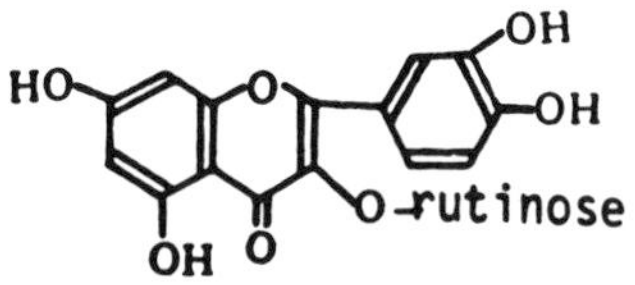

Rutin

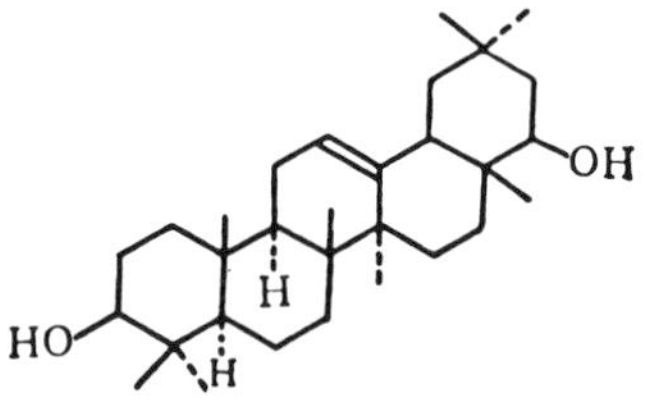

Sophoradiol

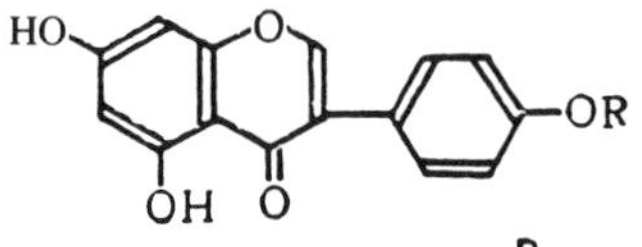

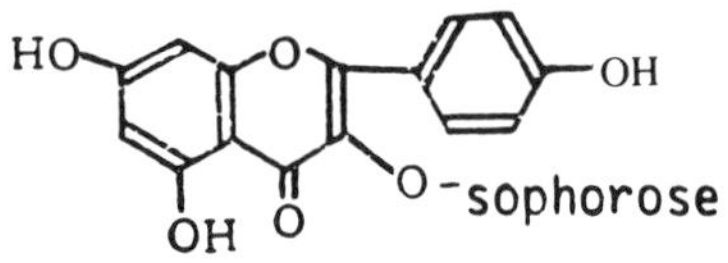

	R =
Genistein	H
Sophoricoside	β-glucose
Sophorabioside	-gluc.-O-rhamn.

Sophoraflavonoloside

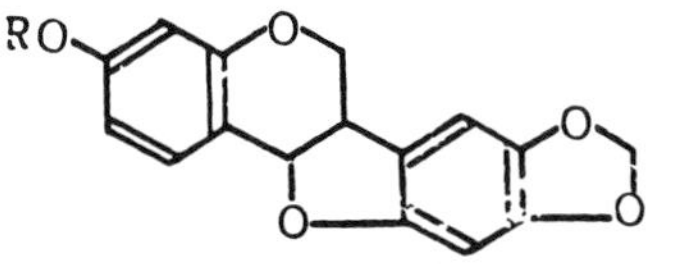

	R =
dl-Maackiain	H
Sophojaponicin	-glucose

Some other compounds that have been isolated from the *Sophora* plant include genistein ($C_{16}H_{16}O_5$), sophoricoside ($C_{21}H_{20}O_{10}$), sophorabioside ($C_{27}H_{36}O_{14}$), sophoraflavonoloside ($C_{27}H_{30}O_{16}$), *d*-maackiain-β-D-glucoside ($C_{22}H_{22}O_{10}$), and *dl*-maackiain ($C_{16}H_{12}O_6$).

Actions

Rutin and its genin quercetin have the properties of increasing capillary resistance while decreasing capillary fragility and permeability. It has a prophylactic effect which prevents hypertension and stops bleeding. In addition, quercetin can lower the heart rate, dilate coronary vessels, and increase systolic volume.

The rutin glycoside can also relax smooth muscles and has an anticonvulsant effect. In experiments with animals, it has been observed to lower hepatic and blood cholesterol levels.

The maackiain glucoside has anti-inflammatory, hemostatic, and anti-X-irradiation effects.

Therapeutic Uses

Therapeutically, this herb is used as a hemostatic agent in all kinds of bleeding. It can also be applied externally to reduce swelling.

In the treatment of hematuria, hematemesis, and hematochezia, the standard dose is 9 to 15 g made into a decoction. The glycosides of the herb are also marketed in tablet form (20 mg), with a standard dose of one to two tablets taken three times a day.

TIE SHU (鉄树) — dried leaf of *Cycas revoluta*

Chemistry

The leaf contains sotetsuflavone and hinokiflavone.

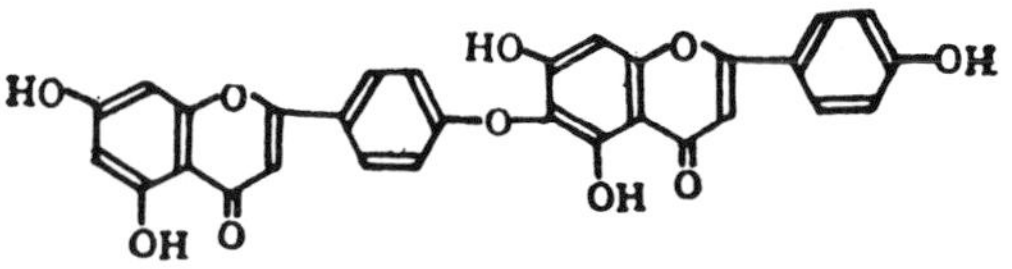

Hinokiflavone

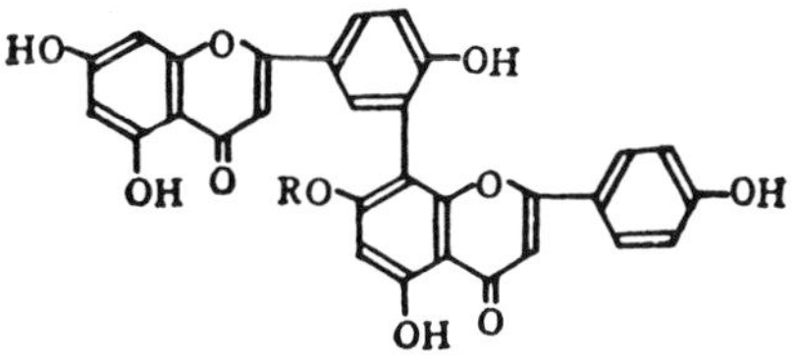

	R=
Sotetsuflavone	CH_3
Amentoflavone	H

Therapeutic Uses

Traditional medicine claimed that this herb would promote the circulation of *qi* and blood. It is used chiefly as a hemostatic agent, but is also reputed to be effective in the treatment of diarrhea and back pain.

JUAN BAI (卷柏) — dried whole plant of *Selaginella tamarisina*

Chemistry

The herb contains sotetsuflavone ($C_{31}H_{20}O_{10}$), amentoflavone ($C_{36}H_{18}O_{10}$), and apigenin ($C_{15}H_{10}O_6$).

Therapeutic Uses

Chinese herbalists prescribe *Juan Bai* in the treatment of hematuria and dysmenorrhea and to stop postpartum bleeding.

JING TIN SAN QI (景天三七) — dried whole plant of *Sedum aizoon* L.

Chemistry

This herb contains sedoflorin ($C_{23}H_{24}O_{13}$), sedocaulin, and sedocitrin ($C_{28}H_{24}O_{19}$).

Therapeutic Uses

The herb is used as a hemostatic and for removing blood stasis.

SU MU (苏木) — heartwood of *Caesalpinia sappan* L.

Chemistry

The wood contains brasilin ($C_{16}H_{14}O_5$), which forms brasilein ($C_{16}H_{12}O_5$) on oxidation.

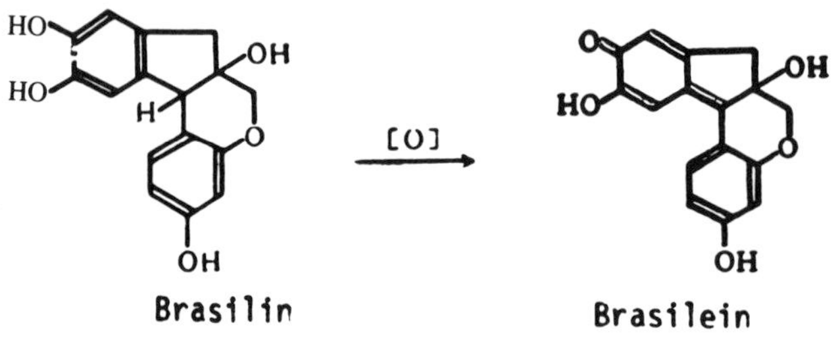

HO HO OH H O OH [O] O HO OH O OH

Brasilin Brasilein

Therapeutic Uses

This herb was used to activate blood flow, remove blood stasis, reduce swelling, and relieve pain.

GUI HUA (桂花) — dried flower of *Osmanthus fragrans* Lour.

Chemistry

The herb contains β-phellandrene; osmane ($C_{10}H_{20}$), and nerol; methyllaurate, methylmyristate, and methypalmitate; and uvaol ($C_{30}H_{50}O_2$).

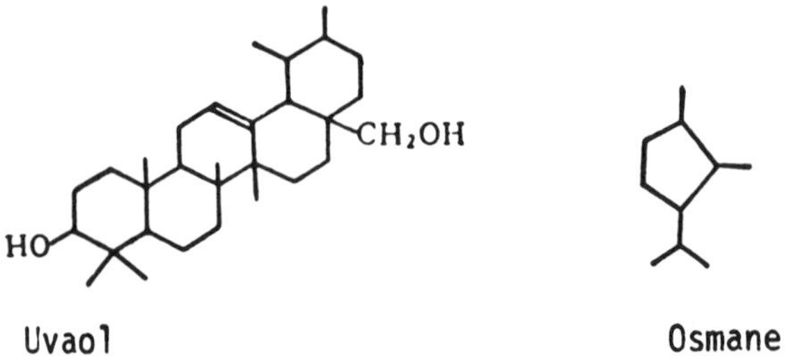

CH₂OH HO

Uvaol Osmane

Therapeutic Uses

Traditional Chinese medicine recommended this herb to reduce phlegm and remove blood stasis.

TABLE 1
Other Hemostatic Herbs

Name	Source	Content	Action and uses	Dose
Ji Cai (荠菜)	Dried whole plant of *Capsella bursa-pastoris* Medic.	Bursic acid, alkaloids, vitamin A, choline, citric acid, K^+ bursilate	Hemostatic, antihypertensive, chyluria; nephritis, edema, and hematuria	15–60 g, decoction
Xiao Ji (小薊)	Dried aerial part of *Cephalanoplos segetum* or *C. setasum* Kitam	Alkaloids, choline, saponins	Hemostatic by increasing the platelet aggregation; cardiac stimulation; antibacterial; used to treat dysentery	9–18 g, decoction
Qian Cao (茜草)	Dried root or rhizome of *Rubia cordifolia*	Rubierythrinic acid, alizarin, purpurin, pseudopurpurin	Hemostatic, shorten the clotting time; antibacterial, antitussive, expectorant, stimulates uterine contraction	3–9 g, decoction; 15 g, as external application
Duan Xue Liu (断血流)	Dried aerial part of *Clinopodium polycephalum* or *C. chinense* (Benth.)	Glucosides, tannin	Hemostatic; uterine contraction; antibacterial	3–5 Tablets t.i.d. (equal to 0.3 g raw drug/tablet)
Shu Liang (薯良)	Dried tuber of *Dioscorea cirrhosa*	Tannin, mucus	Hemostatic, increase platelet aggregation; increase uterine contraction; antibacterial used in dysentery; healing burn wound	20% Decoction, given 20 ml t.i.d.; 20% ointment for external application
Tu Tai Huang (土大黄)	Dried root of *Rumex patientia* L.	Chrysophanol, emodin, physcion, small amount of aloe-emodin; anthranol and its glucoside	Hemostatic used to treat thrombopenia and uterus; laxative, skin diseases	15 g oral as decoction or 0.5 g of extract t.i.d.
Zi Ju Cao (紫珠草)	Dried leaf or root of *Callicarpa macrophylla* or *C. pedunculata* R. Brown, or *C. dichotoma* K. Koch	Tannin, flavon, resin; Mg^{2+}, Ca^{2+}, and Fe^{2+} salts	Hemostatic, acts by constricting the blood vessels and increasing platelet aggregation; antibacterial; mainly used to treat tuberculosis bleeding	4% Extract, give 10 ml t.i.d. or tablet 0.3 g t.i.d. or injection ampule (40 mg/ml) for i.m. injection
Bai Ji (白芨)	Dried tuber of *Bletilla striata* (Thunb.)	Gelatin and essential oil	Hemostatic, promotes leukocyte and platelet aggregation, used to treat hematuria, blood splitting; antibacterial and antiviral; used in all kinds of bleeding	Powder, 3–9 g or 2% extract as plasma substitute, or tablet 0.3 g

Section IX
Endocrine System

Chapter 30

HERBS AFFECTING THE THYROID GLANDS

Goiter, an enlargement of the thyroid gland, is commonly caused by iodine deficiency. It occurs more frequently in females than in males and is also more common in certain geographical regions. Chinese history has recorded a high incidence of goiter in a county of Shensi Province, west of Beijing.

Traditional Chinese medicine theorized that goiter is caused by an abnormality in hepatic function or an accumulated stasis of sputum in the trachea. This results in an interference with and stagnation of blood or *qi*. Therefore, it was advocated that any remedy should clear the sputum, improve the tone of the blood, and circulate *qi*; this would soften the goiter.

For centuries, Chinese medicine used algae and seaweed to effectively treat goiter. Now it is understood that this treatment was effective because it replenished patients' iodine, a substance which is present in large quantities in algae and seaweed.

HAI DAI (海带) or KUN BU (昆布) — dried thallus of *Laminaria japonica* Aresch. or *Ecklonia kurome*

Chemistry

The herb contains iodine (approximately 0.34% of total weight), K^+, Ca^{2+}, amino acids, laminarin, laminine, and algin, which is the sodium salt of alginic acid.

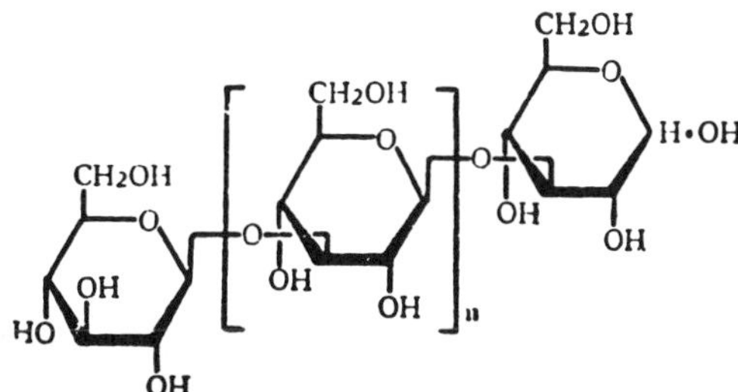

Laminarin, Laminariose

Actions

The herb readjusts thyroid functions and corrects the malignant vicious-cycle effect of iodine deficiency. It can lower blood pressure, and is hemostatic. The herb exhibits anti-X-irradiation effects.

Laminarin sulfate has a hypolipemic effect. It also has a heparin-like anticoagulant effect; the principle responsible for this effect is heat resistance.

Adverse effects include headache and heart palpitations. Because of its high iodine content, this herb should be used cautiously in cases of acute tuberculosis and chronic bronchitis.

Therapeutic Uses

This herb is a common food product which can be purchased in grocery stores; it is very popular in both China and Japan. A snack of rice wrapped with this herb is considered a delicacy.

Hai Dai used in the treatment of goiter disease in certain regions, with a daily dose of 9 to 21 g. It is also used as an antihypertensive agent; the standard dose in this treatment is 6 to 12 g daily.

Other treatments include the external use of alginic acid solution soaked into gauze to arrest bleeding. Additionally, when used as eye drops, this herb is reputed to be effective in preventing cataracts.

HAI ZAO (海藻) — dried seaweed of *Sargassum pallidum* or *S. fusiforme* (Harv.) Setch.

Chemistry

The seaweed contains 0.03% of iodine, alginic acid, algin, and Fe^{2+} and K^{+} salts.

Therapeutic Uses

It was used as an antigoiter agent and anticoagulant. It was generally administered in doses of 4.5 to 30 g in the treatment of goiter disease.

HUANG YAO ZI (黄药子) — dried rhizome of *Dioscorea bulbifera* L.

Chemistry

This herb contains a small amount of saponins, including dioscorecin, dioscoretoxin, and tannin. Some bitter substances are present, as are diosbulbine A, B, and C, and iodine.

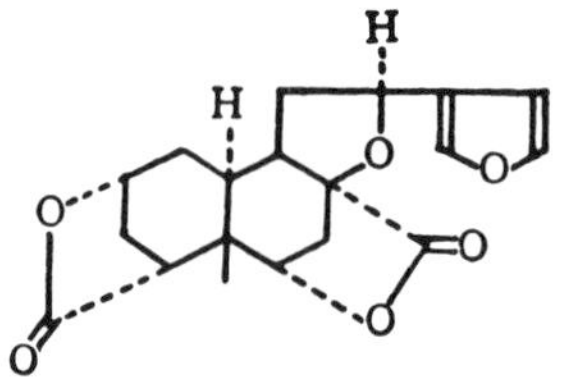

Diosbulbine A

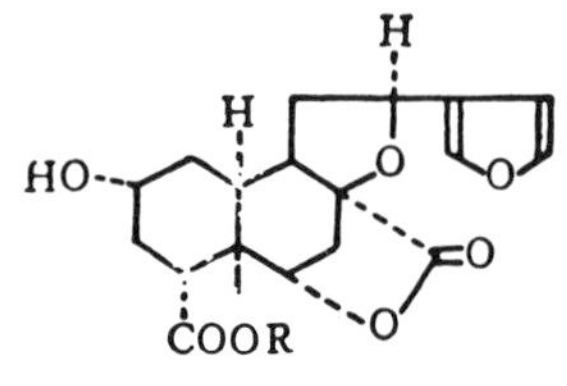

	R=
Diosbulbine B	CH_3
Diosbulbine C	H

Therapeutic Uses

The Chinese pharmacopoeia recommended this herb to remove ''heat'' from the blood, to remove toxic substances, and to treat cancer and goiter. It is marketed in the form of a tincture (standard dose of 100 ml/d), 500-mg tablets (standard dose of 15 tablets/d) and as a 25% decoction fluid (standard dose of 5 ml t.i.d.).

LIU YE (柳叶) — the leaf and fibrous root of *Salix babylonica* L.

Chemistry

This plant contains saligenin glucoside, iodine, pyrocaledol, and some saponins.

Actions

The herb has antigoiter and expectorant actions. It has some antibacterial properties and is especially effective against tubercule bacilli.

Therapeutic Uses

Liu Ye is primarily prescribed in the treatment of simple and edemic goiter. By placing one or two leaves of the herb in the family drinking jar, it is also used as a preventive measure against goiter.

It has also been used in the treatment of tuberculosis. Externally, it is applied to the skin to treat urticaria, and on the wounds to protect against infections.

Preparations include 500-mg tablets, administered in doses of eight to ten tablets a day. The herb may be prepared as a decoction, with 30 g decocted and drunk per day. It should be administered continuously for 2 months.

Chapter 31

HERBS AFFECTING THE ADRENAL CORTEX

GAN CAO (甘草), or Licorice root — the dried root and rhizome of *Glycyrrhiza uralensis, G. inflata*, or *G. glabra*

FIGURE 1. *Gan Cao*, the licorice root.

Chemistry

Between 6 and 14% of the herb by weight consists of glycyrrhizin, which is the Ca^{2+} or K^+ salt of glycyrrhinic acid. Glycyrrhizin is approximately 50 times sweeter than cane sugar. After water hydrolysis, it gives one molecule of glycyrrhetic acid and two molecules of glycuronic acid.

The plant also contains small amounts of three glycosides called liquiritin, isoliquiritin, and neoliquiritin. Recently, an antiulcerative substance called FM 100, licorione, and an immunosuppressant substance called LX were isolated from the plant.

Glycyrrhizic acid
R = glycyrrhetic acid

18 β-Glycyrrhetic acid

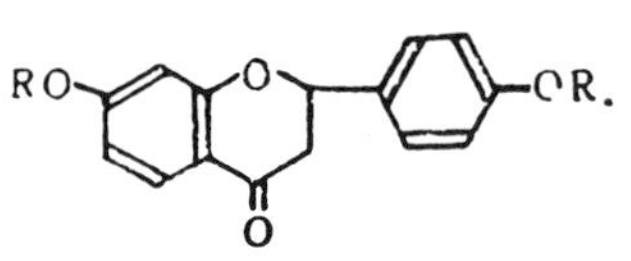

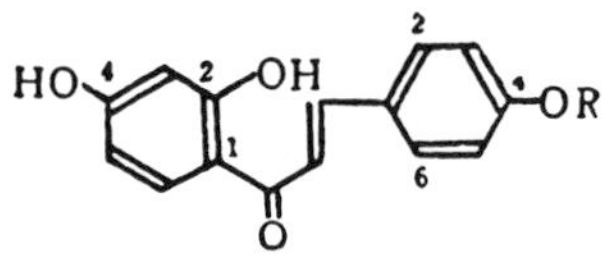

	R=	R_1=
Liquiritigenin	H	H
Liquiritin	H	-glucose
Neo-liquiritin	-glu.	H

	R=
Isoliquiritigenin	H
Isoliquiritin	-glucose

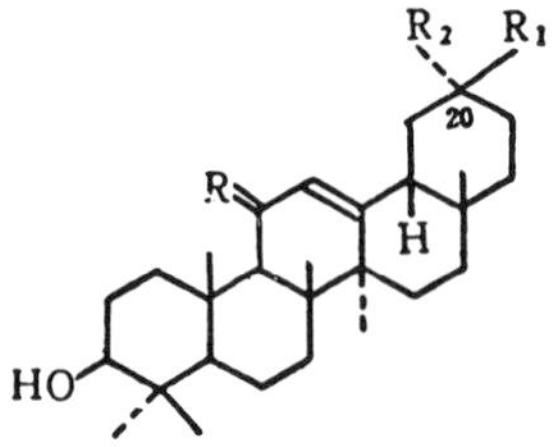

	R=	R_1=	R_2=
Acid I	H_2	COOH	CH_3
Liquiritic acid	O	CH_3	COOH
Glycyrrhetol	O	CH_2OH	CH_3

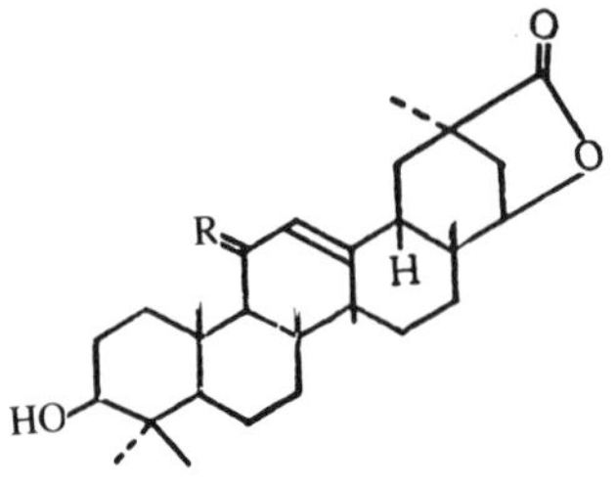

	R=
Glabrolide	O
Deoxyglabrolide	H_2

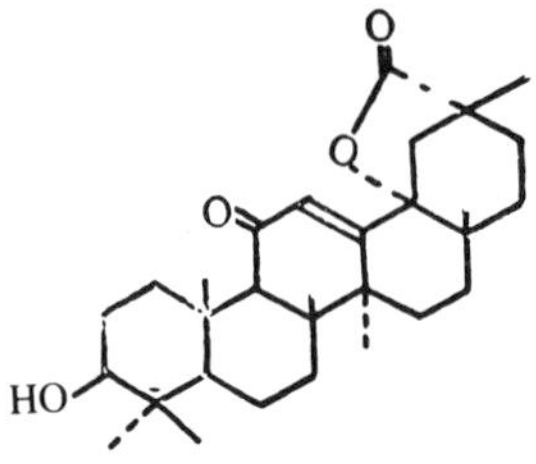

Isoglabrolide

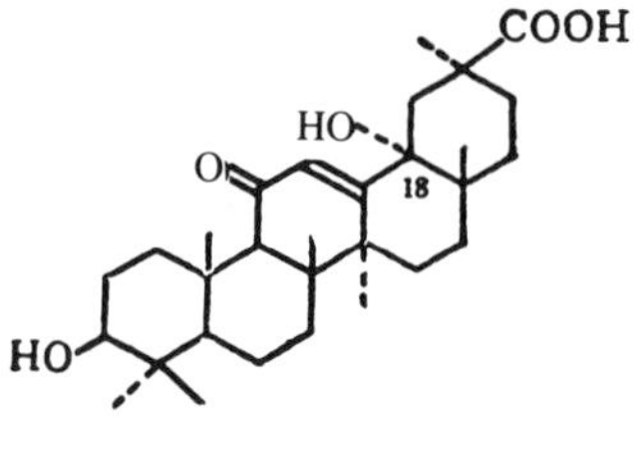

18α-Hydroxyglycyrrhetic acid

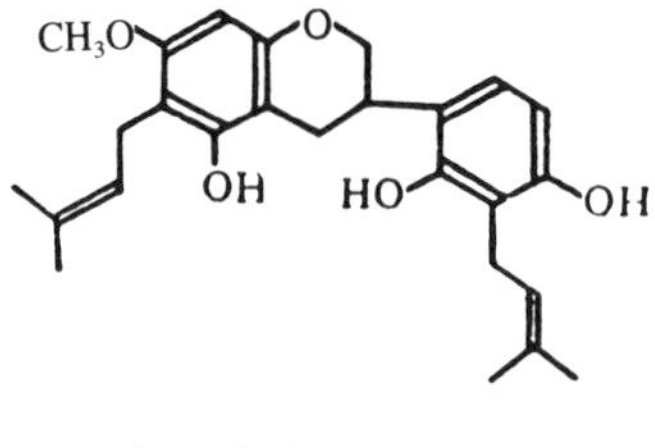

Licoricidin

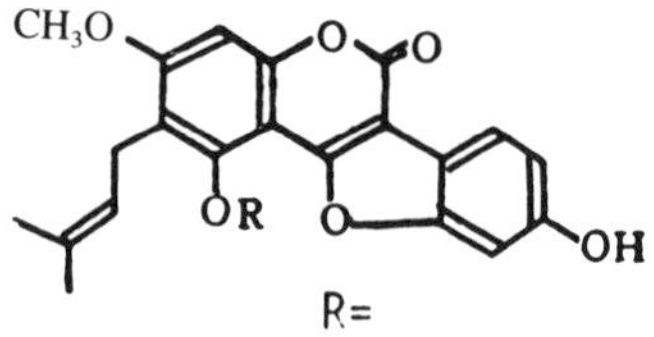

	R=
Glycyrol	H
5-O-Methyl-glycyrol	CH_3

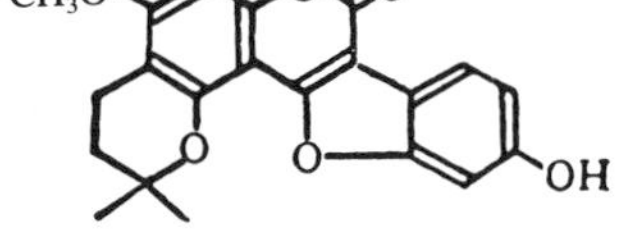

Isoglycyrol

Actions

The herb stimulates the adrenal cortex, increasing mineralocorticoid secretion, which leads to a decrease in urinary Na^+ excretion and an increase in urinary K^+. Plasma Na^+ levels increase and water is retained. Blood Ca^{2+} decreases. Such effects are not observed if the adrenal cortex is removed.

Gan Cao can potentiate and prolong the action of cortisol and increase the urinary excretion of 17-ketosterone. It inhibits the release of the melanin-stimulating hormone from the pituitary. It causes a fall in vitamin C levels in the adrenal gland and an increase in adrenal weight. It also causes a decrease in eosinophil and leukocyte counts.

The herb has anti-inflammatory effects. It can reduce hypersensitivity reactions and capillary permeability. In addition, it can prolong the survival time of transplanted tissue and inhibit the production of antibodies. The active principle responsible for this effect belongs to the heat-stable LX immunosuppressant. It has been suggested that this mechanism of action involves blocking the transfer of the immune signal from phagocytes.

Gan Cao also inhibits gastric secretion and ulcer formation. This activity is due to FM 100 and licorione, particularly the former, which can lower gastric acidity, reduce pepsin activity, and inhibit gastric secretion. The mechanism of action is not yet clear.

The herb is a potent antitoxin. Chinese medical classics state that "*Gan Cao* can detoxify hundreds of toxic substances . . . " Experiments have shown that glycyrrhinic acid can lower the toxicity of strychnine, histamine, chloral hydrate, arsenate, snake venom, diphtheria toxin, and tetanus toxin, to mention just a few. In isolated perfused heart experiments, *Gan Cao* has antagonized the actions of physostigmine and acetylcholine.

In small doses, the herb can lower the plasma levels of cholesterol and triglycerol in hypertensive patients. It can have a preventive effect on arteriosclerosis formation.

Gan Cao is an effective antitussive and expectorant. Oral administration can reduce inflammation of the laryngeal mucosa and has a protective action to reduce irritation. The 18-β-glycyrrhetic acid has a definite antitussive effect, acting both locally and centrally.

The herb has been observed to protect hepatic function. It has analgesic and anticonvulsive effects. In addition, it is claimed that the herb, through its glycyrrhinic acid and water hydrolysis products, can transform several toxins in the liver into the insoluble products.

Continuous use of *Gan Cao* can result in sodium retention and water accumulation, leading to edema and hypertension. The herb also has a tendency to lower the basal metabolic rate and decrease thyroid function.

Therapeutic Uses

Chinese medical texts describe *Gan Cao* as an agent to "improve the tone of the 'middle *Jiao*' (the digestive system) and replenish *qi*, to remove 'heat' and toxic substances, to moisturize the lungs and arrest coughing, and to relieve spasms and pain".

It is effective in the treatment of mild or moderate cases of hypocorticosteroidism, or Addison's disease. It can be used alone or in combination with cortisol to produce a synergistic effect.

The herb is also used to treat bronchitis, tuberculosis, and peptic ulcers. It is administered in doses of 30 g, decocted and taken twice a day in the treatment of thrombocytopenia purpurea. In addition, it is used as an adjuvant to other herbs, to smooth their taste and reduce their side effects.

There are many formulas in which *Gan Cao* serves as the principal adjuvant, or supporting herb. Standard preparations include a *Gan Cao* extract, which is administered in doses of 5 to 15 ml, t.i.d. In tablet form, the herb is administered in doses of three to four

TABLE 1
Effect of β-Escin on Plasma Level of ACTH, Corticosterone, and Glucose in the Rat

		Plasma concentration of		
Treatment	**Dose**	**ACTH (pg/ml)**	**Corticosterol (μg/dl)**	**Glucose (mg/dl)**
Saline	5 ml/kg	103 ± 4	2.3 ± 2.0	149 ± 4
Escin	5 mg/kg	1167 ± 71[a]	46 ± 1[a]	241 ± 16[b]

[a] $p < 0.001$
[b] $p < 0.01$

From Hiai, S., in *Advances in Chinese Medicinal Material Research*, Chang, H. M. et al., Eds., World Scientific Publishing Co., Singapore, 1985, 49–59. With permission.

tablets, t.i.d. The drug biogastrone contains glycyrrhetic acid (50 mg per tablet); it is taken in doses of two tablets t.i.d.

SUO HUO ZI (沙罗子), also called "Seven Leaves Tree", "Monkey Chestnut" — the fruit and seed of the tree *Aesculus chinensis* Bge., *A. hippocastanum* L., or *A. wilsonii* Rehd.

Chemistry

The seed contains large quantity of fatty acid, ranging from 20 to 31%. *A. hippocastanum* grown in Jiangsi province contains a mixture of saponins, escin, after enzyme hydrolysis, which would produce protoescigenin, ($C_{30}H_{50}O_6$), barringtogenol C ($C_{30}H_{50}O_5$), and their derivatives.

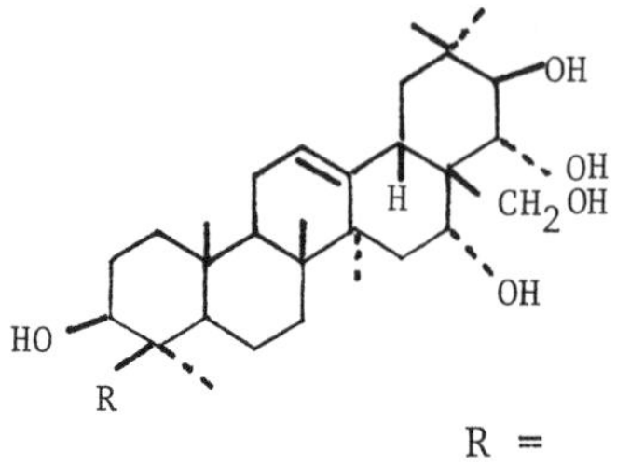

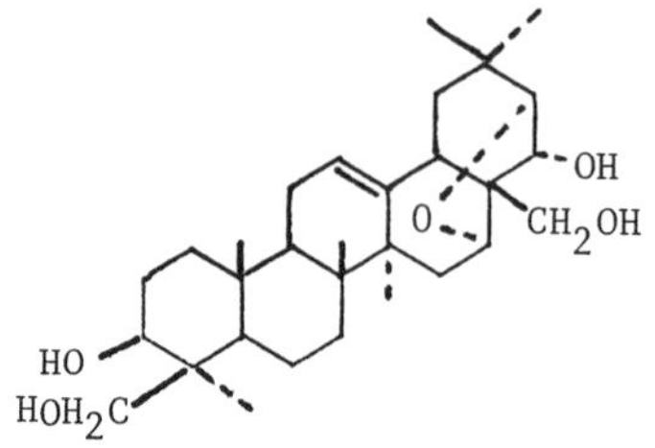

	R =
Protoescigenin	CH_2OH
Barringtogenol C	CH_3

Escigenin

Actions

Experimentally it has been shown that β-escin can cause a 10-fold increase of plasma adrenocorticotropic hormone (ACTH) in rats and a 20-fold increase of plasma corticosterol level (see Table 1).

Escin also produces an anti-inflammatory action. Such an effect is not shown in adrenectomized or hypophysectomized animals.

GINSENG (人参) — the dried root of *Panax ginseng* C. A. Meyer

Actions

This root contains several ginsenosides which possess a remarkable stimulatory effect on corticosteroid secretion from the adrenal cortex. Its actions are described in Chapter 1.

Chapter 32

ANTIDIABETIC HERBS

Diabetes mellitus was common among the Chinese and was referred to in Chinese medicine as a "thirst disease" (消渴病). This illness is due chiefly to insulin deficiency. Traditional Chinese medicine, however, considered it to be a sign of weakness of the *yang* of the kidney and excessive "heat" or "fire" in the stomach. Patients suffering from diabetes often complain of backache, dizziness, and polyuria, which are diagnosed as the symptoms of weak *yang*. These patients eat more but still feel hungry. They also experience dryness of mouth and increased thirst, dry skin, and constipation. These are the symptoms of "heat" in the stomach and excessive *qi* in the spleen.

The proposed remedy for this disease was to strengthen the *yang*, drain the excessive "heat", and smooth the *qi*. The Chinese pharmacopoeia recommends many herbs to serve these purposes, including *Dong Seng* (党参), *Wu Mei Zi* (五味子), *Ge Gen* (葛根), *Tian Hua Fen* (天花粉), *Gan Cao* (甘草), *Di Gu Pi* (地骨皮), and *Zhi Mu* (知母). The first five herbs were described in earlier chapters.

DI GU PI (地骨皮) — the dried root bark of *Lycium chinense* Mill. or *L. barbarum* L.

Chemistry

The root bark contains cinnamic acid, betaine, and other organic acids.

Actions

The herb has been observed to lower blood sugar levels. The peak action is observed 3 to 4 h after oral administration; the duration of action is 7 to 8 h. The herb is less effective if given by subcutaneous injection.

Other effects include a lowering of blood pressure, although the duration of this effect is short. It can lower blood cholesterol, but is less effective in lowering triglyceride levels.

The herb has antipyretic properties. It can stimulate uterine contraction. It has antibacterial and antihypersensitivity actions.

Adverse effects may include nausea and vomiting. The LD_{50} in mice is 12.83 ± 1.9 g/kg (i.p.); in dogs, the LD_{50} is 120 g/kg (by stomach) or 30 g/kg (i.p.).

Therapeutic Uses

This herb is commonly prescribed in doses of 6 to 12 g, decocted for treatment of diabetes mellitus. A decoction extract of 30 g/100 ml is administered daily in the treatment of hypertension. The herb is also prescribed in doses of 30 g, brewed as a tea for treatment of malaria.

ZHI MU (知母) or DI SENG (地参) — the dried rhizome of *Anemarrhena asphodeloides* Bunge

Chemistry

The rhizome contains several saponin and mucus substances. The aerial part of this plant also contains the glucosides mangiferin and isomangiferin.

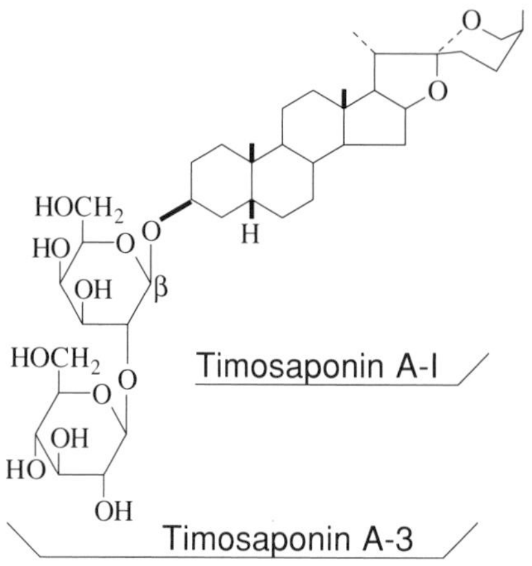

Timosaponin A-I

Timosaponin A-3

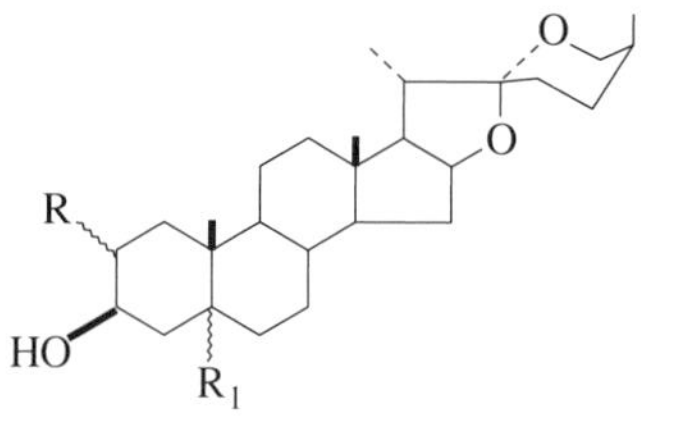

	R =	R_1 =
Markogenin	β-OH	β-H
Neogitogenin	α-OH	α-H

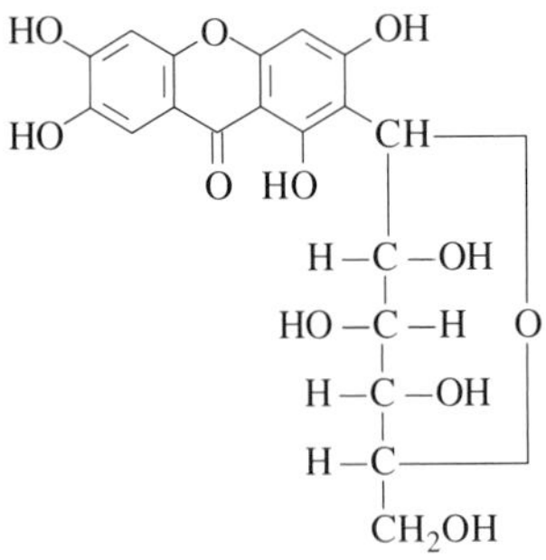

Mangiferin

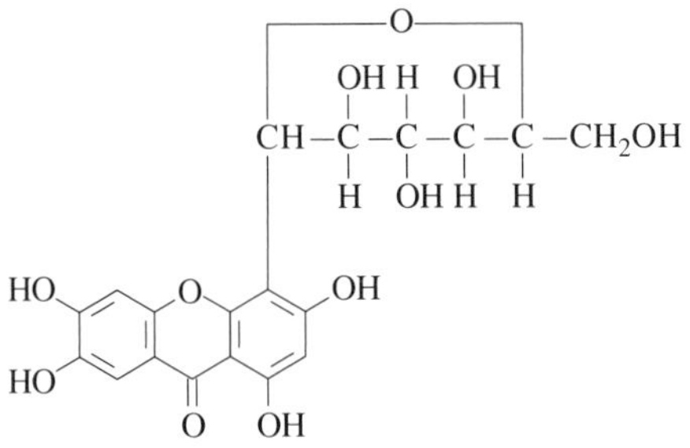

Isomangiferin

Actions

Modern Chinese medical texts explain that the herb can lower blood sugar by increasing the metabolism of glucose in the body and increasing glycogen synthesis in the liver. The herb also has antibacterial properties and has been observed to lower the body temperature of animals inoculated with *Escherichia coli*.

Therapeutic Uses

The Herbal Classic of the Divine Plowman indicated that *Zhi Mu* should be used mainly to treat dryness of mouth and ''thirst'', removing evil *qi* and edema fluid and remedying insufficiency of *qi*. The herb is generally prescribed in doses of 12 g, in combination with 4.5 g of *Huang Lian* (黄連), prepared as a decoction to treat diabetes mellitus and acute infectious diseases.

DI HUANG (地黄) — the root of *Rehmannia glutinosa* Libosch.

Chemistry

The major components isolated from this plant are sterol, campesterol, catalpol, rehmannin, and some alkaloids.

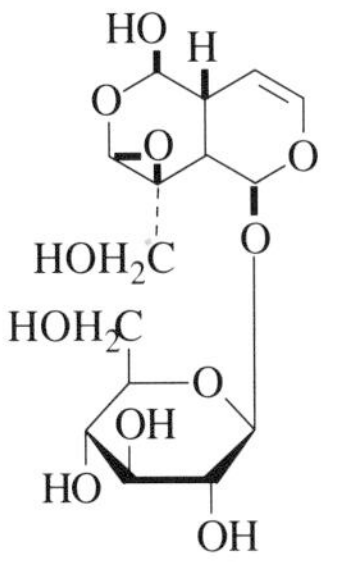

Catalpol

Actions

This root lowers blood sugar. It is the major herb used in Chinese medicine to treat "thirst disease". It is a hemostatic agent, promoting blood coagulation. It is a cardiac tonic and diuretic; these actions are probably effected by producing renal vasodilatation. In addition, the herb has antibacterial and anti-inflammatory properties.

Therapeutic Uses

Chinese apothecaries use either the fresh or the dried root as an herb. If used fresh, the herb is known as "raw" *Di Huang*; if used dried, it is known as "cooked" *Di Huang*.

It is used in the treatment of diabetes mellitus. It can also be administered to stop bleeding, or to treat dermatitis, rheumatitis, diphtheria, and acute tonsillitis.

XUAN SENG (玄参) — the dried root of *Scrophularia ningpoensis* Hemsl. and *S. buergeriana* Miq.

Chemistry

These plants are in the family Scrophulariaceae, to which *Digitalis purpurea*, *D. lanata*, and *Rehmannia glutinosa* also belong.

The root contains scrophularin and iridoid glycosides. The latter are the active principles; 20 to 30% of these glycosides consist of 8-(*O*-methyl-*p*-coumaroyl)-harpagide, while 70 to 80% is harpagoside. In addition, small amounts of essential oils, alkaloids, flavonoids, and *p*-methoxycinnamic acid are present.

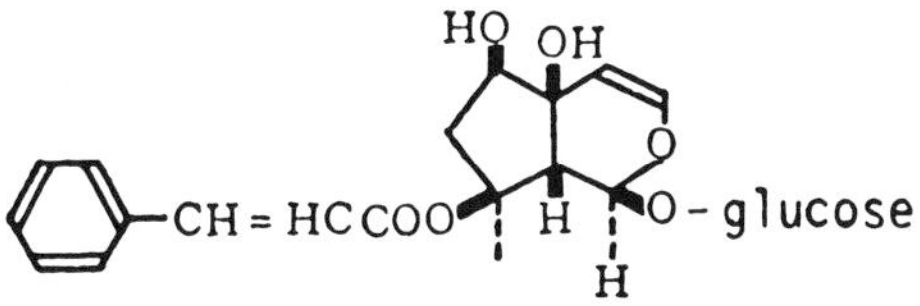

Harpagoside

Actions

The water extract of *Xuan Seng* can lower blood pressure in both anesthetized and conscious animals. Infusion of a 5 to 10% extract into an extremity vessel of cats causes a vasodilatory effect. In cats and dogs, the extract can slow the heart rate, prolong the PQ interval, and increase myocardial contractility.

Xuan Seng can lower blood sugar levels. Although this effect is less marked than that of *Di Huang*, the duration of action is long. Rabbits receiving a dose of extract equivalent to 5 g/kg showed a 16% drop in blood sugar levels that lasted for 5 h.

The herb has a tranquilizing and anticonvulsant effect. It can prolong the sleeping time of barbiturate-treated animals. It has antibacterial and antitoxicant effects. In addition, the herb has a choleretic effect and can reduce capillary permeability. The saponin component of the herb can cause hemolysis of red blood cells and local irritation.

Therapeutic Uses

This herb is commonly used in combination with other herbs as a nutrient, a health strengthening agent, and to remove ''heat'' and replenish vital essences. It is also decocted with licorice root and peppermint to clean the lungs and to treat sore throat. In some Chinese textbooks, this herb is listed as an antibacterial agent (see Chapter 33).

The herb is generally administered in doses of 9 to 15 g. That dose can be increased to 30 g with little toxicity.

ZHANG SHU (蒼术) — the dried root of *Atractylis orata* Th.

Chemistry

About 1.5% of the root consists of essential oils, of which 20% is atractylone ($C_{14}H_{28}O$), which has a distinctive aroma. The other component of the essential oil is atractylol ($C_{15}H_{24}O$).

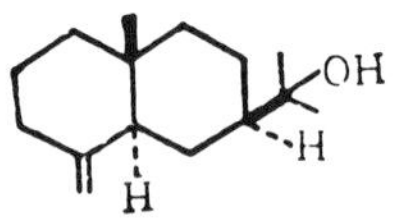

β-Eudesmol

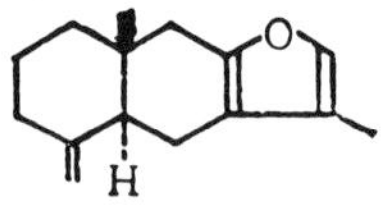

Atractylone

Hinesol

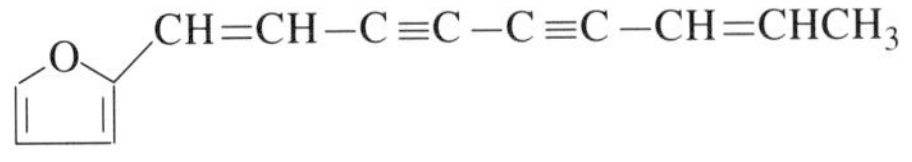

$CH{=}CH{-}C{\equiv}C{-}C{\equiv}C{-}CH{=}CHCH_3$ (on furan ring)

Atractylodin

$CH_3CH{=}CH{-}(C{\equiv}C)_2{-}CH{=}CH{-}CH_2{-}CH(O{-}CO{-}CH_3){-}CH_2{-}CH_2{-}O{-}CO{-}CH_3$

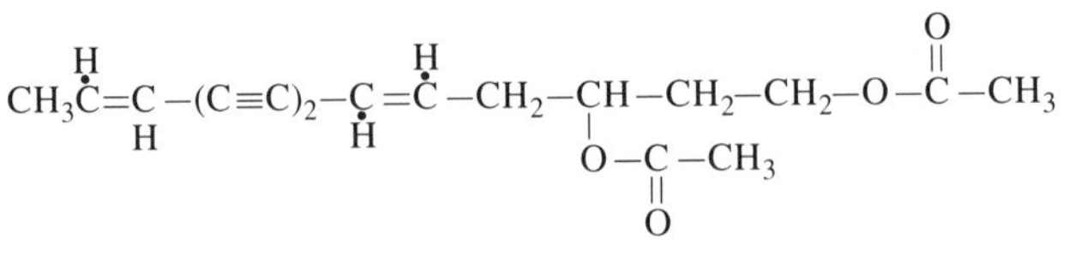

Diacetyl-atractylodiol

Actions

In rabbits, the herb can lower blood sugar to 40% of the original level after subcutaneous injection of the water extract at a dose of 6 g/kg.

The herb can also produce a sedative effect. Overdosage may cause death due to respiratory paralysis.

ZE XIE (泽泻) — the whole plant of *Alisma plantago-aquatica* J.

Actions

This herb has a weak hypoglycemic effect. Li Shih Chen's *Ben Cao Kong Mu* recommended this herb as an agent to relieve thirst and produce diuresis.

GINSENG (人参) — the dried root of *Panax ginseng* C. A. Meyer

Actions

In addition to its multiple other uses, this herb is well described and accepted in China as a thirst-relieving agent (see Chapter 1). It was not, however, commonly used to treat diabetes mellitus. In 1927, Japanese investigators demonstrated that the ginseng glucoside can effectively lower blood sugar levels.

Section X
Chemotherapy

Chapter 33

ANTIBACTERIAL, ANTIVIRAL, AND ANTIFUNGUS HERBS

The herbs described in this Chapter are listed in the Chinese pharmacopoeia as ''heat removing'' agents, because they are effective in lowering fever caused either by bacterial infections or by noninfectious diseases. Chinese medicine recognized several types of heat or fever, describing them as ''surface heat'', ''internal heat'', ''solid heat'', and ''asthentic heat''.

''Surface heat'' was a perception of fever which was thought to be caused by an invasion of evil substances from the outside. It could be arrested effectively by the use of diaphoresis-producing herbs. This condition could also be treated effectively with antipyretics such as *Ban Lan Jen* (板藍根) and *Ta Chin Yi* (大青叶).

''Internal heat'' referred to symptoms of high temperature, tremors, oliguria, constipation, and sometimes coma. These were thought to result from evil hiding within the body, rather than external causes. Today, we know that these fevers are caused mainly by microorganisms. Antimicrobial herbs serve a therapeutic use to combat these diseases.

HUANG LIAN (黃連) — dried rhizome of *Coptis chinensis* Franch., *C. deltoidea*, or *C. teetoides* C. Y. Cheng

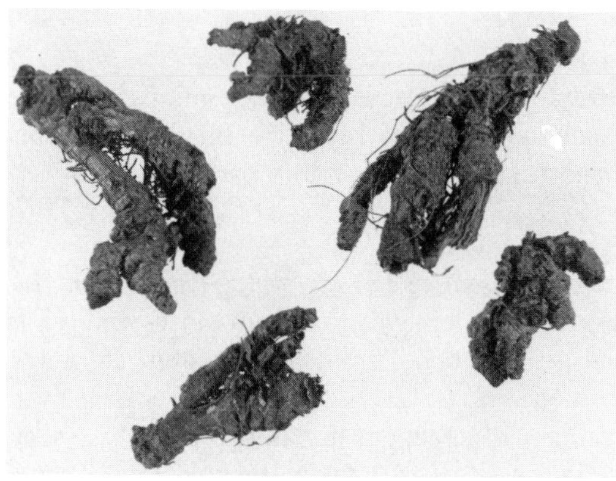

FIGURE 1. *Huang Lian.*

Chemistry

By weight, approximately 7 to 9% of the rhizome consists of berberine, which is the active principle of the herb. In addition, the plant contains small amounts of coptisine, urbenine, worenine, palmatine, jatrorrhizine, and columbamine.

Berberine is a yellow, nonbasic, quaternary ammonium alkaloid. It is not only found in *Coptis chinensis*, but also in many species of Berberidaceae, Anonaceae, Menispermaceae, Papaveraceae, and Rutaceae. It is now manufactured by chemical synthesis. On reduction, this compound will produce the colorless substance tetrahydroberberine, which has different pharmacological properties.

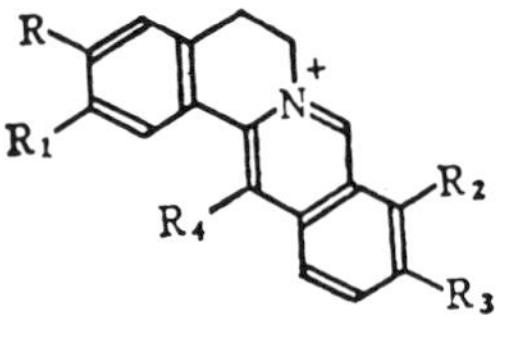

	R=	R_1=	R_2=	R_3=	R_4=
Berberine	-O-CH_2-O-		OCH_3	OCH_3	H
Worenine	-O-CH_2-O-		-O-CH_2-O-		CH_3
Coptisine	-O-CH_2-O-		-O-CH_2-O-		H
Palmatine	OCH_3	OCH_3	OCH_3	OCH_3	H
Jatrorrhizine	OH	OCH_3	OCH_3	OCH_3	H
Columbamine	OCH_3	OH	OCH_3	OCH_3	H

Actions

The herb is an antibacterial agent. It is claimed that berberine can inhibit the carbohydrate metabolism of bacteria. Such an inhibitory effect can be antagonized by vitamin B_6, vitamin PP, and histidine. Berberine can also inhibit the protein synthesis of bacteria. It can complex with nucleic acid to interfere with bacterial growth. Reports indicate that berberine can be used as a chloramphenicol substitute in the treatment of cholera.

The herb also has antiviral and antiprotozoal effects. It improves the immune system and phagocytic process. It is an antitoxin and anti-inflammatory agent. It can cause a fall in blood pressure, due to a muscarinic stimulatory effect and inhibition of cholinesterase activity. As well as having antiarrhythmic activity (see Chapter 3) it is a choleretic.

Small doses of berberine can stimulate the heart, but large doses are inhibitory. It also lowers blood pressure.

Toxicity

In overdosage, *Huang Lian* can cause nausea, vomiting, dyspnea, and convulsions. In animals, intravenous administration of berberine may cause respiratory depression and bronchial contraction and death from respiratory paralysis.

Therapeutic Uses

The Herbal Classic of the Divine Plowman stated that ''*Huang Lian* regulates the 'heat' and 'cold' of the five viscerals, removes toxic substances and wastes in the blood, stops thirst, drains water and benefits bone, smoothes the stomach, helps the alimentary tract, and promotes bile secretion . . . ''

It is used in the treatment of infectious diseases, especially dysentery, whooping cough, diphtheria, respiratory infection, scarlet fever, typhoid fever, acute conjunctivitis, otitis media, surgical wound infections, septicemia, hepatitis, trichomonal vaginitis, and eruptive dermatitis. It is also used as an antihypertensive agent.

Standard preparations include berberine tablets (50 or 100 mg), given in doses of one to three tablets t.i.d. A berberine injection solution (HCl salt) is available in strengths of 2 mg/2 ml or 50 mg/2 ml for intramuscular use. In the treatment of trichomonal vaginitis, a 20% *Huang Lian* extract is soaked into gauze and directly applied to the vaginal mucosal membrane; the reported efficiency rate is 96%.

Reference

Mitscher, L. A., Plant-derived antibiotics, *Antibiotics,* Vol. 15, Weinstein, M. J. and Wagman, G. H., Eds., Plenum Press, New York, 1978, 363–477.

MA WEI LIAN (马尾連) or MA WEI HUANG LIAN (马尾黄連) — dried root and rhizome of *Thalictrum glandulissimum*, *T. culturatum*, or *T. foliolosum* DC.

The *Ben Cao Kong Mu* described this herb being grown mainly in Yunnan Province. The dried herb is silk-like.

There are 70 kinds of *Thalictrum* plants in China; among them, 26 were used as medical herbs. The one grown in Yunnan is also called *Cao Huang Lian*, one grown in Szechuan Province is called *Golden Silk Huang Lian*.

Chemistry

This herb contains berberine, palmatine, jatrorhizine, thalictrine, thalidasine ($C_{39}H_{44}O_7N_2$), thalicarpine ($C_{41}H_{48}O_8N_2$) and saponaretin ($C_{21}H_{20}O_{10}$).

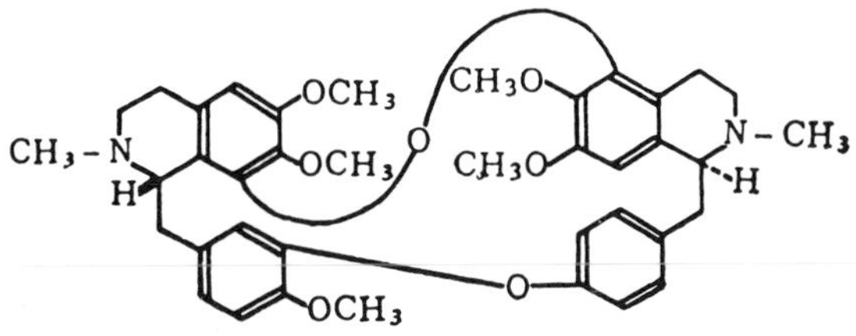

Thalidasine

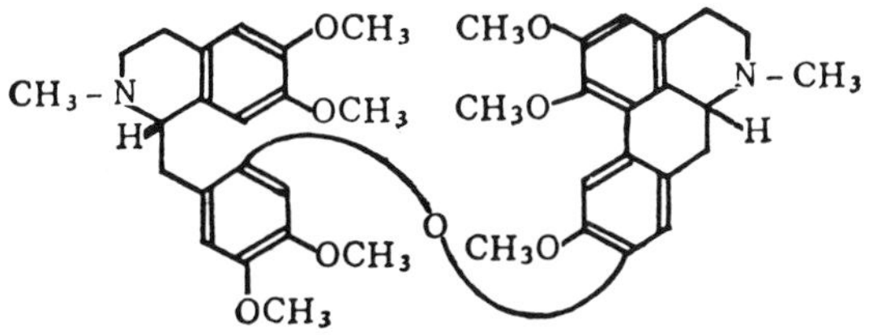

Thalicarpine

Actions

The pharmacological actions of this herb are similar to those of *Huang Lian*. The active principle is probably berberine or palmatine, which serve as antibacterial agents. Traditional Chinese medicine claims that the herb removes "heat" and "dampness".

Therapeutic Uses

It was used to combat influenza, pediatric fever, measles, and malaria. The usual dose is 2 to 6 g in decoction.

HUANG BAI (黄柏) — dried bark of *Phellodendron chinense* or *P. amurense*

Chemistry

The bark contains many alkaloids; the major ones are berberine (0.6 to 2.5% of total content), palmatine, and phellodendrine.

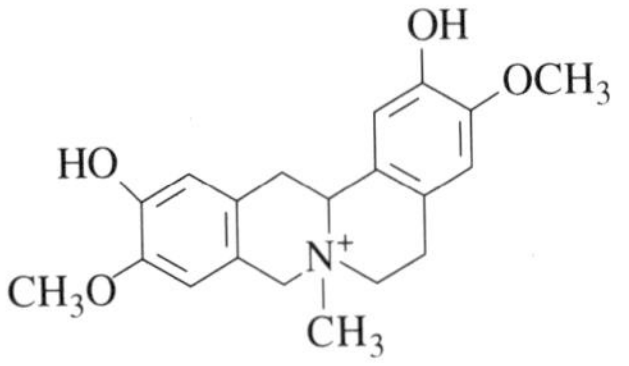

Phellodendrine

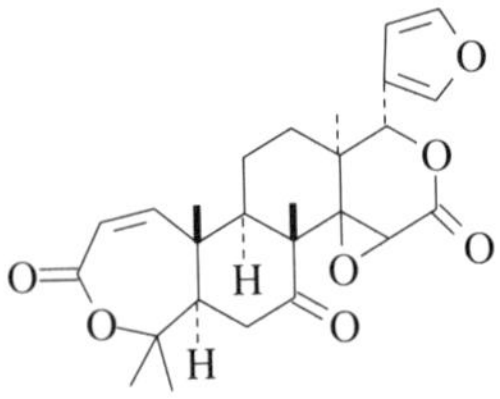

Obacunone

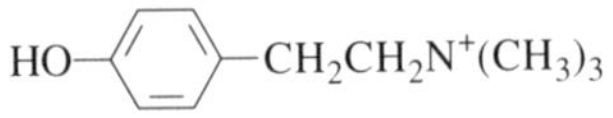

Candicine

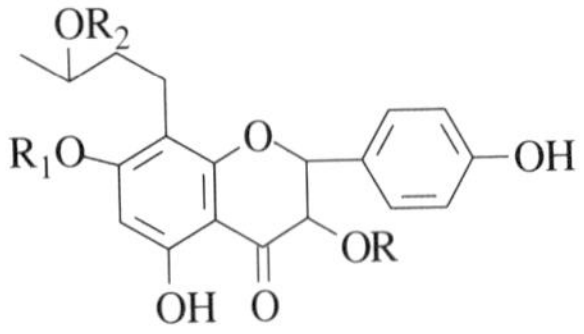

	R=	R_1=	R_2=	C_2 C_3
Phellodendroside	-gluc.	H	H	-
Phellamurin	H	-gluc	H	-
Dihydrophelloside	H	-gluc	-gluc	-
Phelloside	H	-gluc	-gluc	double bond
Nor-icariside	H	H	-gluc	double bond

Actions

Like *Huang Lian, Huang Bai* exerts an antibacterial effect, and is especially effective against diphtheria bacilli, streptococci, and dysentery bacilli. In addition, it can stimulate the phagocytic activity of leukocytes. It also has a vasodilatory effect, increasing coronary flow and lowering the blood pressure.

Therapeutic Uses

The Chinese pharamacopoeia recommended the herb to purge intensive "heat", and to remove dampness and toxic substances. Therapeutically, it is effective against dysentery and used in the treatment of jaundice, tuberculosis, epidemic meningitis, acute conjunctivitis, and trachoma.

The herb is available as an extract, equivalent to 1 g of raw material/ml, and as a 0.2% injection solution, for intramuscular administration.

HUANG QIN (黄芩) — the dried root of *Scutellaria baicalensis* Georgi, *S. viscidula* Bge., *S. amoena* C. H. Wright, and *S. ikoninkovii* Ju

Chemistry

Several active substances have been isolated from the root. The major ones are baicalein, baicalin, wogonin, and β-sitosterol.

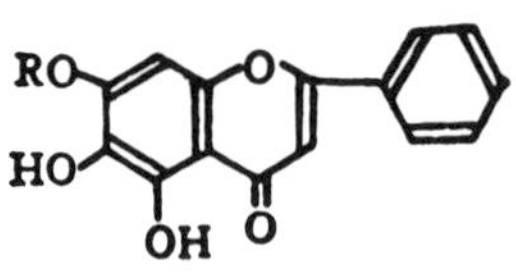

	R=
Baicalein	H
Baicalin	-glucose
7-Methoxy-baicalein	CH_3

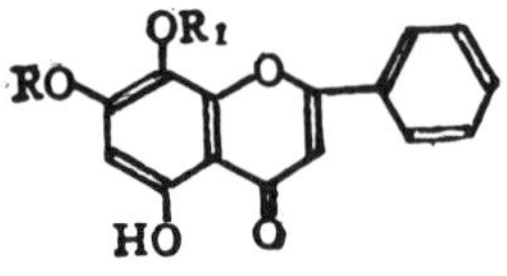

	R=	R_1=
Wogonin	H	H
Wogonoside	-gluc	CH_3
7-Methoxynorwogonin	CH_3	H

	R=	R_1=
Skullcapflavone I	H	H
Skullcapflavone II	OCH_3	OCH_3

Actions

The herb has antibacterial activity against staphylococci, cholera, typhoid, paratyphoid, dysentery, diphtheria, hemolytic streptococci, *Escherichia coli*, pneumococci, and spirochaeta. It is an antiviral agent effective against influenza virus.

Baicalein has antipyretic properties, though its effects are slightly weaker than those of aspirin. It is also an antitoxin. At a dose of 10 mg, baicalein can increase the LD_{50} of strychnine in mice by a factor of two and one-half. Also, it can increase the liver glycogen content in CCl_4-intoxicated mice.

Other actions of the herb include diuretic, sedative, antihypertensive, anti-inflammatory, and antihypersensitivity effects. It has choleretic properties; baicalein is more effective in this area than baicalin. In addition, the herb has been observed to directly relax smooth muscles.

Huang Qin is relatively nontoxic when given orally. Clinical trials showed that when 150 mg of baicalin is administered by intramuscular injection, it can cause fever and muscle aches. Intravenous administration of 27 mg of baicalin produces adverse effects including fever and a sudden drop of leukocyte count.

Therapeutic Uses

This herb and its active principle, baicalin, are used in the treatment of upper respiratory infections, such as acute tonsillitis and acute laryngopharyngitis. They have also been tried in cases of scarlet fever, viral hepatitis, nephritis, and pelvitis.

Baicalin is available in 250-mg tablets; in the treatment of viral hepatitis, the standard dose is two tablets t.i.d. A complex tablet comprising 50 mg of baicalin and 100 mg of *Shung Hua* (双花) is available for treatment of upper respiratory infections; the standard dose is two to three tablets, taken four to six times per day.

SAN KE ZHEN (三顆針) or XIAO YEH GEN (小蘗根) — root of *Berberis soulieana, B. wilsonae, B. poiretii,* or *B. vernae*

FIGURE 2. *San Ke Zhen.*

Chemistry

The major ingredient of the root is berberine. In addition, it contains berbamine ($C_{37}H_{40}O_6N_2$), palmatine, jatrorrhizine, and oxycanthine ($C_{20}H_{17}O_5N$).

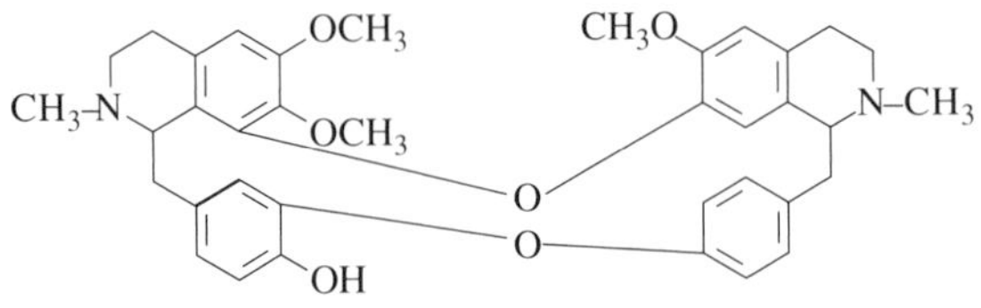

Berbamine

Actions

The herb is an antibacterial agent. It also promotes leukocytosis, and is an effective choleretic.

Therapeutic Uses

Traditional Chinese medicine considers this herb to be useful to remove "heat" and "dampness", to purge intense "heat", and to counteract toxins. It is effective against bacterial dysentery, upper respiratory infections, urethral infections, and cholcystitis.

JIN YIN HUA (金銀花) — dried flower buds of *Lonicera japonica, L. hypoglauca, L. confusa*, or *L. dasystyla*

Chemistry

The flower bud contains luteolin, inositol, saponins, and chlorogenic acid. The latter is the main antibacterial principle of the herb.

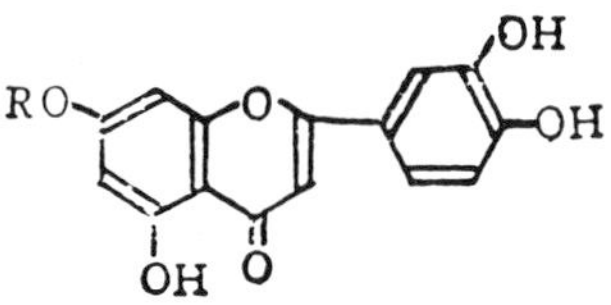

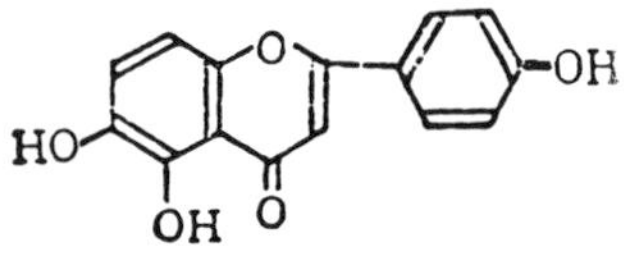

	R=
Luteolin	H
Lonicein	neohesperose

Loniceraflavone

Actions

The herb has antibacterial effects against *Staphylococcus aureus*, streptococci, pneumococci, *Bacillus dysenterii, B. typhoid*, and paratyphoid. It is an antiviral agent. It has antilipemic actions, interfering with lipid absorption from the gut.

Therapeutic Uses

The herb is used mainly in the treatment of upper respiratory infections, such as tonsillitis and acute laryngitis. It is also used in the treatment of skin suppurations, such as carbuncles, and to treat viral conjunctivitis, influenza, pneumonia, and mastitis.

The herb has also been administered in cases of acute appendicitis; 62 to 93 g of the herb is added to 9 to 10 g of licorice root to make a decoction.

Preparations include tablets containing 100 mg of *Jin Yin Hua* extract; the standard dose is two to three tablets, taken every 4 to 6 h. An injection solution consisting of 50 mg of extract and 40 mg of wogonin per 2 ml of solution is available.

LIAN QIAO (连翘) — dried fruit of *Forsythia suspensa* (Thunb.) Vahl

Chemistry

The fruit contains forsythol ($C_{15}H_{18}O_7$), phillyroside ($C_{27}H_{34}O_{11}$), oleanolic acid, and rutoside.

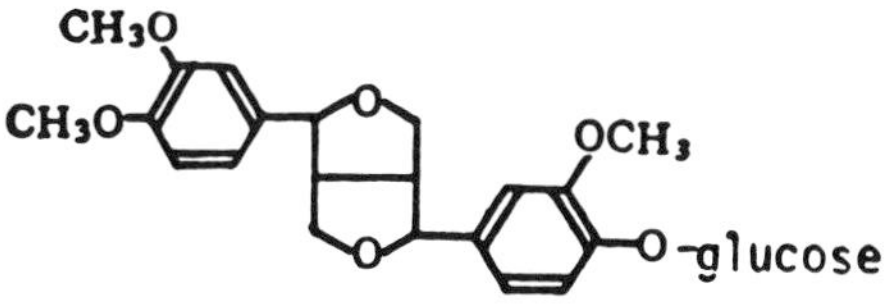

Phillyrin or Phillyroside

Actions

The herb has antibacterial actions against *Salmonella typhi*, cholera, *E. coli*, diphtheria, plague, tuberculosis, staphylococci, and pneumococci. It has anti-inflammatory properties and can lower body temperature. It can increase body immunity. It protects hepatic function and is an effective choleretic. On the cardiovascular system, it can produce vasodilation and a hypotensive effect. It is a diuretic and an antiemetic; the latter effect occurs via inhibition of the chemoreceptor trigger zone (CTZ).

Therapeutic Uses

In the treatment of early influenza and cold, the herb is administered in doses of 9 to 16 g, prepared in a decoction. It is also used in the treatment of encephalitis, hepatitis, carbuncle, and tuberculosis.

CHUAN XIN LIAN (穿心莲) — dried aerial parts of *Andrographis paniculata*

Chemistry

The major ingredients found in this plant include deoxyandrographolide, andrographolide, neoandrographolide, and dehydroandrographolide.

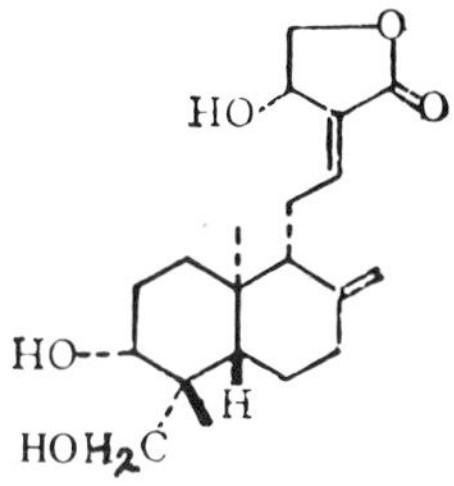

Andrographolide 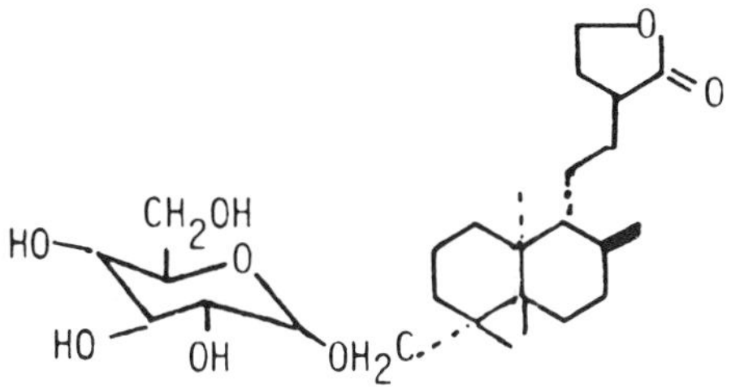 Neoandrographolide 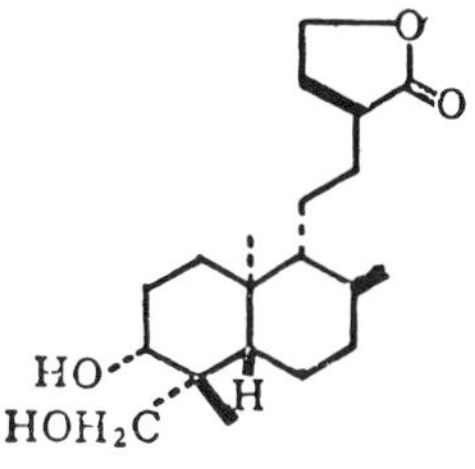 14-Deoxyandrographolide

Actions

The herb is an effective antibacterial agent against staphylococci, pneumococci, hemolytic streptococci, dysentery, and typhoid bacilli. It has been observed to increase body defense mechanisms, and to act as an antipyretic. It has anti-inflammatory properties, probably due to stimulation of adrenocortical hormone release and ACTH secretion.

Other effects include sedative action. The herb can prolong the sleeping time induced by barbiturates and shorten the time in onset of action. In addition, *Chuan Xin Lian* has reportedly been used to terminate pregnancies.

Adverse effects can include dizziness and palpitations of the heart. The LD_{50} of this herb in mice is 2.7 g/kg (intravenous).

Therapeutic Uses

The herb is effective in the treatment of bacterial dysentery, arresting diarrhea and lowering fever quickly. It is also effective in the treatment of upper respiratory infections, tonsillitis, pneumonia, tuberculosis, and renal pelvitis. It has been used to treat snake bite, but the mechanism of action is unknown.

Standard preparations include 50-mg andrographolide tablets, taken in doses of two tablets t.i.d. An injection solution of 50 mg/2 ml is used for intravenous infusion or intramuscular injection.

PU GONG YING (蒲公英) — dried aerial parts of *Taraxacum mongolicum* Hand-Mazz., *T. sinicum*, or *T. hetrolepis*

Chemistry

The plant contains taraxasterol, taraxerol, taraxacerin, taraxacin, and vitamins A, B, and D.

Actions

The herb has antibacterial action against *Staphylococcus aureus, Streptococcus hemolyticus*, typhoid bacilli, dysentery bacilli, tuberculosis, and most Gram-positive bacteria. It has antispirochetic and antiviral effects. In addition, it has been observed to protect the liver function and act as a choleretic agent.

Therapeutic Uses

In the treatment of epidemic paroditis, the herb is ground with water to make a paste, which is applied externally on the inflamed glands. In cases of mastitis, 31 g of the herb is decocted with water or mixed with wine and taken orally. It is also used in the treatment of hepatitis and upper respiratory infection, such as tonsillitis and laryngitis.

Other standard preparations include tablets containing 0.3 g of extract; the normal dose is three to five tablets, taken four times a day. *Pu Gong Ying* is also available in injection form, equivalent to 10 g of raw material/2 ml, for intramuscular use.

QIAN JI GUANG (千里光) — dried aerial part of *Senecio scandens*

Chemistry

The herb contains lavoxanthin, chrysanthemaxanthin, and some alkaloids.

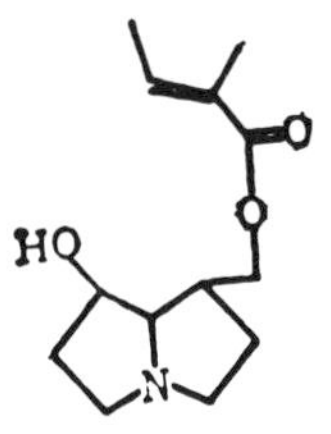

Macrophylline 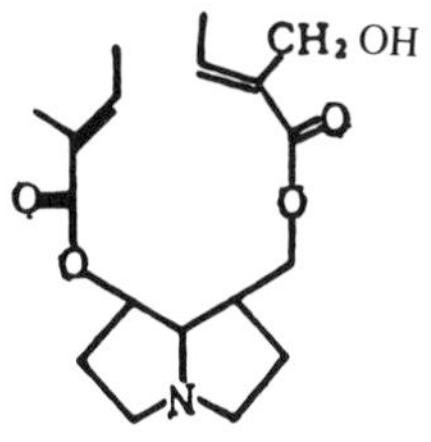   Sarracine

Actions and Therapeutic Uses

It is claimed that this herb has antibacterial and antiplasmodial activity. It is generally used in the treatment of acute bacterial dysentery, appendicitis, pneumonia, and bronchitis. It is also used in cases of conjunctivitis and trichomonas vaginitis.

The toxicity of the herb is minimal. A similar plant, however, *Senecio nemrensis*, is quite toxic; it is grown near the peak of the Ting-Mo mountain in China. Domestic animals that have eaten this plant have died of hepatic necrosis.

Sugar-coated tablets made from an extract of the herb, equivalent to 2 g of raw material, are administered in doses of three to five tablets t.i.d. An injection solution equivalent to 1 to 3 g raw material/ml is available, administered intramuscularly in doses of 2 to 4 ml. Also available are *Qian Ji Guang* eye drops (50%).

HU ZHANG (虎杖) — dried rhizome of *Polygonum cuspidatum* Sieb.

Chemistry

This herb contains the glycosides polygonin, glucofranglin, polydatin, and emodin, as well as a large quantity of tannin substances. The following glycosides have been isolated from the leaf of this plant.

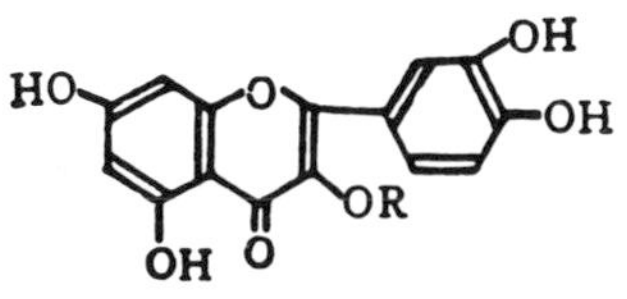

	R =
Reynoutrin	D-xylcse
Avicularin	α-L-furanarabinose
Hyperin	D-galactose

Actions

The herb has antibacterial actions, effective against *Staphylococcus aureus*, streptococci, *E. coil*, spirochetes, and tuberculosis bacilli. It has antiviral activity, and is both a laxative and antitussive. It can lower blood sugar and blood cholesterol levels.

The herb may cause nausea, vomiting, diarrhea, and indigestion. In rare cases, it may cause hepatic damage and respiratory depression.

Therapeutic Uses

In cases of burn, this herb is applied externally to prevent infection. A 5% decoction fluid is prepared for external use, or a 20 to 30% ointment.

It is also used to treat toxic jaundice, hepatitis, appendicitis, and newborn jaundice. In the treatment of candida vaginitis, a decoction of the herb is used to wash the infected vagina daily. Tablets made from an extract of the herb can be given orally to treat chronic bronchitis.

YU XING CAO (魚腥草) — dried aerial parts of *Houttuynia cordata* Thunb.

Chemistry

Approximately 0.005% of the herb consists of an essential oil. The major component of this oil is decanoylacetaldehyde, which is unstable and disintegrates into a complex. The synthetic sodium derivative of this compound is called houttuynium. The other component of this oil is methyl-nonylketone lauric aldehyde.

The leaf of the plant contains the glycoside quercitrin; the flower and fruit of the plant contain isoquercitrin.

Decanoylacetaldehyde	Methyl-nonylketone
$CH_3(CH_2)_8COCH_2CHO$	$CH_3(CH_2)_8COCH_3$

Actions

The herb has antibacterial actions. At a concentration of 1:40,000 it can inhibit staphylococci, typhoid bacilli, pneumococci, *E. coli*, dysentery bacilli, and leptospira. In addition, it has antiviral activity and can increase immunity and the phagocytic activity of leuokcytes. Other actions include analgesic, hemostatic, antitussive, and promotion of tissue regeneration.

Quercitrin has diuretic effects, causing renal vasodilatation and increasing renal blood flow.

Pharmacokinetic studies with ^{14}C-houttuynium show that this agent is distributed mainly in the lung and is metabolized rapidly in the liver.

Therapeutic Uses

The herb is used in the treatment of lung abscess, chronic bronchitis, and pneumonia; for the latter, it is given in a dose of 30 g, decocted with 15 g of *Tie Geng* (桔梗), divided into six portions, and taken every 8 h.

Other uses include the treatment of chronic uterine cervical infection, leptospirosis, otitis media, cystitis, urinary infection, dysentery, and mastitis.

Houttuynium is available in 20-mg tablets, which are administered in doses of four to five tablets t.i.d. In injection form (4 mg/2 ml), houttuynium is used intramuscularly.

YA ZHI CAO (鴨跖草) — dried aerial parts of *Commelina communis* L.

Chemistry

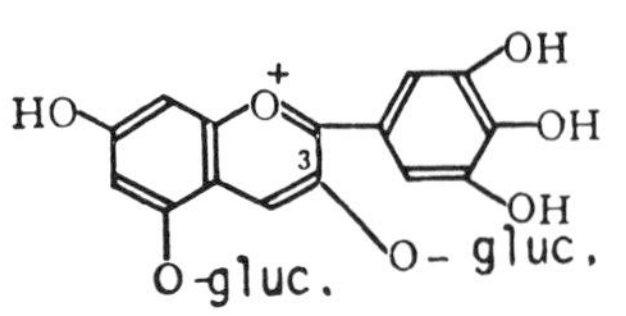

Awobanin

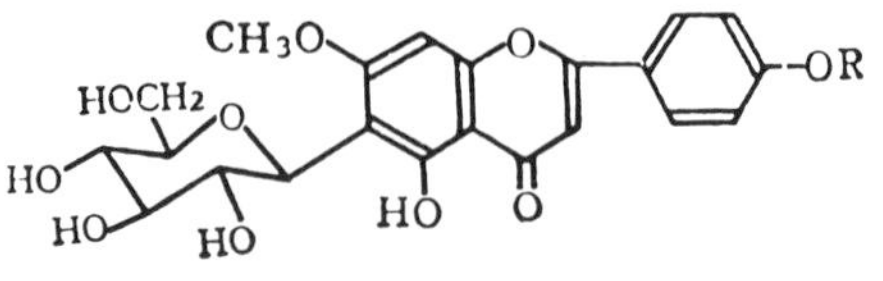

	R=
Flavocommelitin	H
Flavocommelin	-glucose

Actions

The herb has antibacterial actions. It is an antipyretic agent, and possesses diuretic and antiedematic properties.

Therapeutic Uses

The herb is used to treat the common cold and influenza. It is given in doses of 60 g in a decoction, taken for 3 to 5 d. It has also been used as an antipyretic for noninfectious fever, and to treat ascites, edema, and hordeolum.

DAN ZHI (胆汁) — concentrated animal bile

Chemistry

There are more than 20 kinds of animal bile listed in the *Ben Cao Kong Mu*. The most common types sold in Chinese apothecaries come from pigs, cows, sheep, and chickens. Snake bile is classified into a different category and is considered a precious herb.

The main components of bile are bile acids, biliverdin, cholesterol, and lecithin.

Actions

Bile is antibacterial, able to inhibit most Gram-positive bacteria. Experimental data show that the antibacterial activity of bile on whooping cough bacilli is better than that of garlic, but less potent than that of *Huang Lian* (黃連). Deoxybile acid has a much more potent antibacterial activity than bile acid.

It has antiascaris effects and is an effective antitussive and expectorant. It can emulsify fatty acids in the alimentary tract, and stimulates the intestinal absorption of fat soluble vitamins. It inhibits smooth muscle contraction and has a choleretic effect. It was reported by some Chinese investigators that bile has sedative and antipyretic effects.

Bile can cause hemolysis of red blood cells if it is injected into the blood stream directly; it can also cause a sudden fall in blood pressure.

The LD_{50} of bile in mice is 1 ± 0.08 g/kg (oral).

Therapeutic Uses

Chinese medicine uses bile in the treatment of chronic bronchitis, whooping cough, tuberculosis, jaundice, hepatitis, and trachoma (10% solution used as eye drops). Preparations include a concentrate of 5:1 strength, an injection solution, and 0.35-g tablets; the latter are used as an antitussive and to treat whooping cough, taken in doses of one to three tablets t.i.d. A complex tablet comprising 100 mg of bile, 0.05 mg of atropine, and 3.25 mg of sinolbin is used to treat chronic bronchitis; the dose is two to four tablets t.i.d.

DA SUAN (大蒜) or Garlic — the bulb of *Allium sativum* L.

Garlic has been universally touted for thousands of years as a magical medicinal herb with effects ranging from headache healing to aphrodisiac. Chinese folk medicine used *Da Suan* in cooking, as a stomachic, and for prevention and treatment of the common cold and other infectious diseases. Classified by the Chinese as vegetables, the name *Chong Suan* (葱蒜) encompasses many species in the genus *Allium:*

1. *Da Suan*, from *A. sativum* L.
2. *Hu* (葫), from *A. scorodoprasm* L.
3. *Chong* (葱), from *A. fistulosum* L.
4. Foreign *Chong* (洋葱), from *A. cepa* L.

5. *Jiu* (韭), from *A. odorum* L.
6. *Chau Min Jiu* (孝文韭), from *A. ledebourianum* Schult.
7. *Xie* (薤), from *A. bakeri* Regel.

Chemistry

The active principle isolated from the bulb is allicin, which is a form of essential oil and comprises approximately 1.5% of the plant. It is a yellowish liquid with a strong odor and is very unstable under heat. The density of the oil is 1.046 to 1.057.

Allicin exists in the fresh plant as a prodrug called alliin, which is colorless and odorless. Under catalysis reaction by the enzyme alliinase, it is hydrolyzed to form one molecule of allicin and two molecules of diallyl disulfide. The latter has an antilipemic effect.

```
        CH2
        ||
        CH
        |
        CH2                       [ CH2 ]         CH2           CH2
        |                         [ ||  ]         ||            |
        S         Alliinase  2 of [ CH  ]  --->   CH      +     CH     --->
        |        ---------->      [ |   ]         |             |
2 of    CH2                       [ CH2 ]         CH2           CH2
        |                         [ |   ]         |             |
  H2N—CH                          [H-S→O]      O←S—OH          SH
        |
        COOH                                  CH2  CH2                 CH2  CH2
                                              ||   ||                  ||   ||
                                              CH   CH      steam       CH   CH
                                              |    |    ---------->    |    |
   Alliin                                     CH2  CH2   distilled     CH2  CH2
   (Pro-drug)                                 |    |                   |    |
                                           O←S —— S                    S —— SO2

                                              Allicin                  Diallyl
                                                                       thiosulfonate
```

Da Suan also contains other substances, including allistatin and glucominol. From the essential oils of the plant, another antibacterial principle, neoallicin, has been isolated. This is a diallylthiosulfonate, which is now prepared synthetically for marketing.

Actions

The herb has antibacterial actions against staphylococci, *E. coli*, typhoid and paratyphoid, dysentery, cholera, diphtheria bacilli, and pneumococci. It is also effective against tubercule bacilli. It has antiplasmodial effects, especially against ameba.

It is reputed to have anticancer properties. A survey of 4000 people living in regions of Italy and China showed that those whose daily diets in the last 20 years contained the most Allium vegetables had the lowest incidence of stomach cancer, by as much as 60%.

Da Suan lowers plasma cholesterol and low-density lipoproteins, decreases plasma fibrinogen, prolongs blood coagulation time, and prevents atherosclerosis. In India, 432 heart-attack survivors were given fresh garlic each day for 3 years. The garlic eaters had 32% fewer recurrent heart attacks and 45% fewer deaths from heart attack than the patients with unsupplemented diets.

It has an antithrombotic effect. Block reported that a product especially extracted from *Da Suan* called ajoene ($C_9H_{14}S_3O$) prevents platelet aggregation, although dehydrated garlic powder does not have such an effect. He proposed that ajoene acts by inhibiting exposure of fibrogen receptors on platelet membrane.

The herb has been observed to lower blood sugar and stimulate phagocytic activity of leukocytes. It is a stomachic which stimulates gastric secretion and gastric motility. It has anti-inflammatory properties.

The LD_{50} of allium oil in mice is 50 to 78 mg/kg (intravenous); the LD_{50} of neoallicin is 70 mg/kg (intravenous) and 600 mg/kg (oral).

Therapeutic Uses

Garlic has been used to treat many microbial infectious diseases, such as bacterial and amebial dysentery. Rectal infusion of 5 to 10% garlic extract, in combination with oral administration, gives a good curative result in both acute and chronic cases. *Da Suan* is also used in Chinese medicine in the treatment of whooping cough, tuberculosis, epidemic encephalitis, cholera, oxyuris, and taenia infections.

In the treatment of trichomoniasis, garlic juice is placed in bath water or the vagina is washed with garlic solution for 7 to 10 d. An intravenous infusion of *Da Suan* solution (containing 1.5% allium oil) has been used in the treatment of laryngeal cancer.

Externally, garlic is used to dissolve clavus, and to treat furuncles and carbuncles. Gauze is soaked with 10% *Da Suan* solution and applied to the suppurative wounds.

References

1. **Block, E.,** in *Folk Medicine,* Steiner, R. P., Ed., American Chemical Society, Washington, D.C., 1986, 125–137.
2. **Stolzenburg, W.,** *Sci. News,* September 8, 1990, p. 153.

TIE XIAN CAI (鉄莧菜) — dried aerial parts of *Acalypha australis*

Chemistry

The plant contains the alkaloid acalyphine, tannic acid, and gallic acid.

Actions

This herb has an antibacterial action against staphlococci, dysentery, typhoid bacilli, and *Pseudomonas pyocyanea*. It has an astringent effect which protects mucosal membranes. In addition, it has been observed to have antiasthmatic effects.

Therapeutic Uses

In Chinese folk medicine, the herb is used as an antipyretic, detoxicant, antidysenteric, and hemostatic agent. It is given in tablet form (0.4 g/tablet) in doses of four tablets, four times per day.

DI JI CAO (地錦草) — the whole plant of *Euphorbia humifusa* or *E. supina*

Chemistry

The major components are tannin and gallic acid.

Actions

It has antibacterial activity against diphteria, staphylococci, streptococci, *Pseudomonas pyocyanea*, typhoid, paratyphoid, dysentery, *E. coli*, and whooping-cough bacilli. The herb has been shown to have a detoxicant action, effective against diphtheria toxin.

Therapeutic Uses

Decocted in combination with *Tie Xian Cai* (12 g each), the herb is an effective remedy for the treatment of dysentery. The herb is used to treat other kinds of gastrointestinal infections, as an antipyretic and to stop noninfectious diarrhea. It is also used as a hemostatic agent, and to arrest hematuria, hematemesis, and hemoptysis.

LIAO GE WANG (了哥王) — dried root or root bark of *Wikstroemia indica* C. A. Mey.

Chemistry

The root and root bark contain wikstroemin, hydroxygenkwanin, daphnetin, and acidic resin.

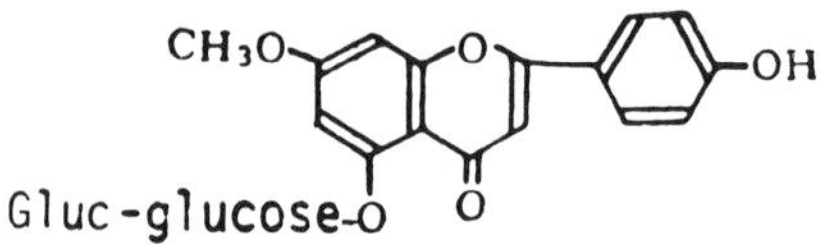

CH_3O OH

Gluc-glucose-O O

Wikstroemin

Actions

The herb has an antibacterial action, especially against staphylococci, streptococci, and pneumococci.

Therapeutic Uses

In Chinese folk medicine, it was used to remove ''heat'' and toxic substances, to activate blood flow and dissipate blood stasis, and to cause diuresis and reduce edema. Therapeutic uses include rheumatic arthritis, leprosy, bronchial asthma, whooping cough, amebic dysentery, and abscesses. It is also used to induce abortion, by inserting a fresh root or root bark into the cervix.

Standard preparations include tablets containing 0.22 g of extract, taken in doses of three tablets t.i.d. An injection solution equivalent to 3 g/2 ml is also available for intramuscular use.

MA CHI XIAN (马齿苋) — dried aerial parts of *Portulaca oleracea* L.

Chemistry

The herb contains large quantities of potassium salts (7.5% of total weight) and several catecholamines, such as norepinephrine, dopamine, and dopa. Also present are vitamins A, B_1, B_2, PP, and C; of these, the vitamin A content is the highest, at 40 U/g.

Actions

It is an antibacterial agent effective against *Escherichia coli, Proteus*, dysentery, typhoid, and paratyphoid. It causes vasoconstriction, stimulates uterine and intestinal smooth muscle contraction, and is a diuretic. It promotes wound healing.

Therapeutic Uses

Chinese herbal medicine used this herb to remove ''heat'', ''dampness'', and toxic substances, and to stop bleeding. Therapeutically, it is used in the treatment of acute bacterial dysentery, whooping cough, enteritis, appendicitis, ascaris, and hookworm infestation. It has also been used to stop postpartum bleeding.

WEI LING CAI (委陵菜) — dried whole plant of *Potentilla chinensis* Ser. and *P. discolor* Bge.

Chemistry

The major components of the plant are tannin, protein, vitamin C, and Ca^{2+} salts.

Actions

The herb has antibacterial, antiamebial, and antitrichomonas actions. It also causes smooth muscle relaxation, especially of the bronchial and intestinal muscles.

Therapeutic Uses

This herb is used mainly in the treatment of amebic dysentery, trichomonas vaginitis, and tuberculosis involving the lymph nodes of the neck. Internally, it is given in a dose of 45 to 60 g extracted with wine or water, drunk daily. Externally, the decoction of this herb is used to wash the infected vagina in cases of trichomonas vaginitis.

JIN DENG LONG (錦灯笼) — dried persistent calyx with fruit or the dried calyx only of *Physalis alkekengi franchetti*

Chemistry

The herb contains small amounts of alkaloids, vitamin C, physalien, physalin A, B, and C, and hystonin.

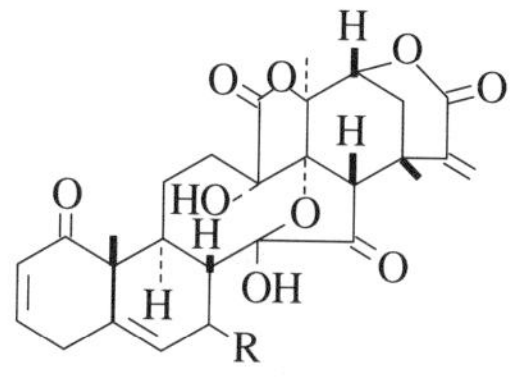

	R=
Physalin A	α-OH
Physalin C	H

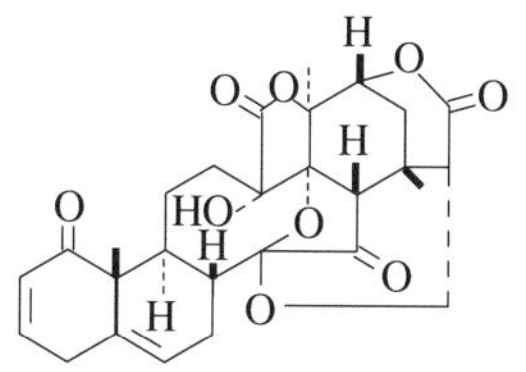

Physalin B

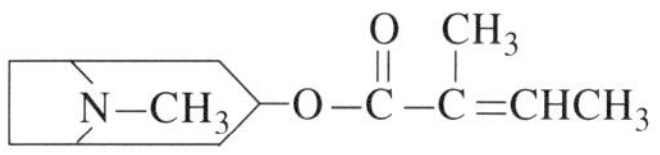

3α-Tigloidine

Actions

The extract of the herb has antibacterial effects, inhibiting dysentery bacilli and staphylococci. The herb stimulates myocardial contraction and causes vasoconstriction, resulting in a rise in the blood pressure. Hystonin can cause uterine contraction; it has a faster onset of action but shorter duration than oxytocin.

Therapeutic Uses

This herb is used to treat acute tonsillitis, sore throat, and laryngeal infections in children. It is marketed in solution form equivalent to 1 g of raw material per milliliter, used for intramuscular injection.

Chinese traditional medicine also prescribed the fruit to induce labor.

GUANG DOU GEN (廣豆根) or SHAN DOU GEN (山豆根) — dried root and rhizome of *Sophora subprostrata* Chun et T. Chen

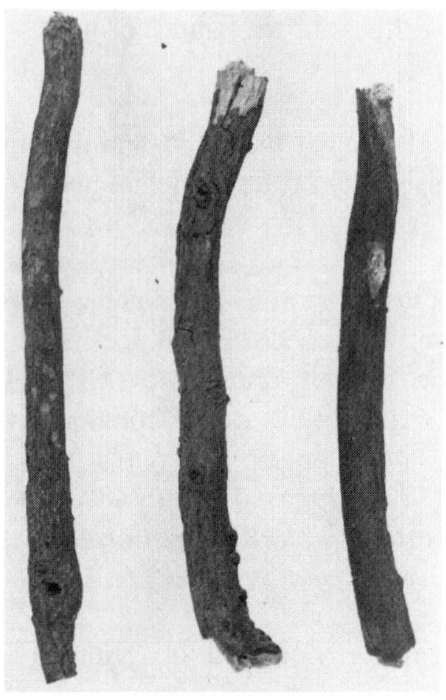

FIGURE 3. The root of *Guang Dou Gen*.

Chemistry

Approximately 0.93% of the root consists of alkaloids, including matrine, oxymatrine, anagyrine, and methylcytisine. The root contains other nonalkaloid substances, including sophoranone, sophoranochromene, sophoradin, daidzein ($C_{15}H_{10}O_4$). Also present is *l*-trifolirhizin ($C_{22}H_{22}O_{10}$), a glucoside whose hydrolyzed product is *l*-maackinin ($C_{16}H_{12}O_5$).

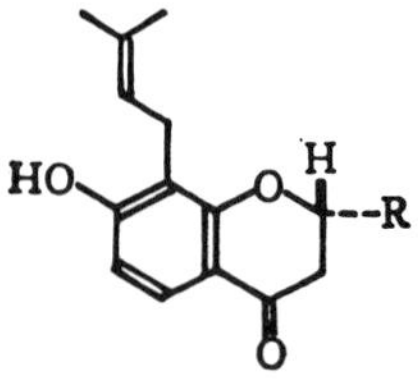

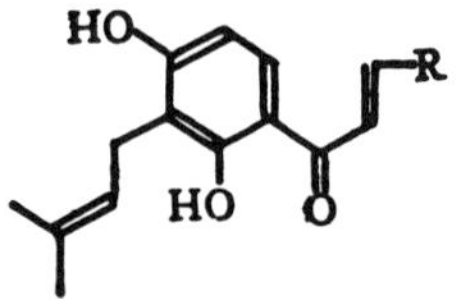

Sophoranone R =

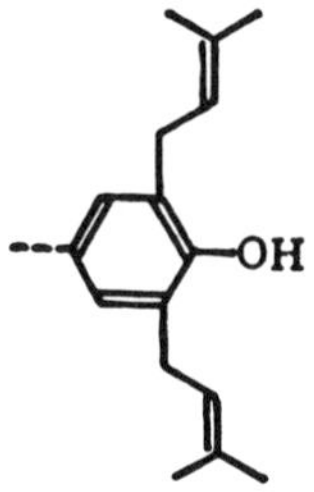

Sophoradin R =

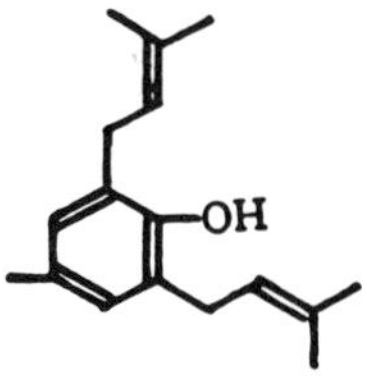

Sophoranochromene R =

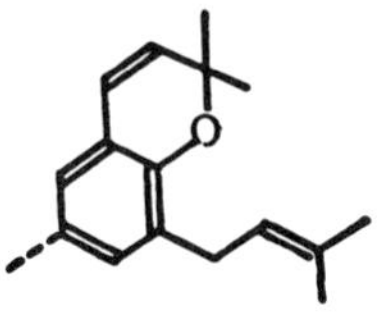

Sophoradochromene R =

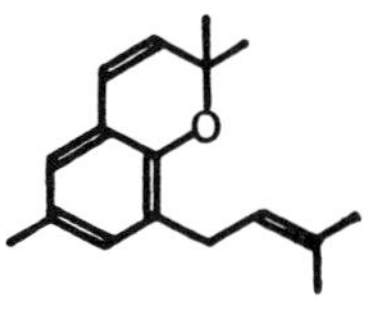

Other substances isolated from *Guang Dou Gen* include pterocarpin ($C_{17}H_{14}O_6$) and *l*-maachlain.

As mentioned in Chapter 8, the name *Shan Dou Gen* is used for two very different herbs. *Guang Dou Gen*, which is mainly grown in the Chinese province of Kwangsi, shares that name with *Menispermum dauricum*, to which the *Zhong Cao Yao Xue* also gives the alternate name *Ye Dou Gen*, or northern *Dou Gen*.

Actions

The herb is highly effective against tubercule bacilli, staphylococci, epidermophyton, and *Candida albicans*. It has an antiarrhythmic effect, due to a central action. It has antiasthmatic properties, and is especially effective against histamine-induced asthma.

The herb is an anticancer agent. It can inhibit the growth of mouse S180 sarcoma cells. Kojama et al.[1] reported that the chemotherapeutic index of oxymatrine against S180 cells is approximately seven to eight times greater than that of mitomycin C. However, the herb has no effect on Ehrlich ascites cells. Mori et al.[2] reported that this herb can enhance tumor immunity in animals and increase humoral immunity-stimulating activity.

It promotes leukocytosis. Rabbits that receive a dose of the total alkaloid — 30 mg/kg intravenously or 60 mg/kg intramuscularly — show an antagonizing effect on X-irradiation-induced leukopenia. In mice receiving a dose of oxymatrine every other day for five doses, and then 300 rads of total body X-irradiation, there is no significant fall in leukocyte count.

Oxymatrine is metabolized in the intestine, if given orally. When it is injected either intramuscularly or intravenousy, it is distributed uniformly in all tissues and excreted unchanged in the urine; approximately 88% of the administered dose is excreted within 24 h.

Toxicity

Clinical trials showed that nausea, vomiting, dizziness, and headache are common side effects. In severe cases, the herb may cause sweating, heart palpitations, dyspnea, and convulsions.

The water extract of this agent is relatively nontoxic, but oxymatrine is toxic. The LD_{50} in mice is 0.57 ± 0.49 g/kg (intraperitoneal) and 0.95 ± 0.116 g/kg (subcutaneous). These animals died of respiratory depression.

Therapeutic Uses

In the treatment of leukopenia (below 4000/mm^3), 200 to 400 mg of total alkaloids is given intramuscularly each day. An 84% improvement has been observed. Daily injection of the total alkaloid to patients receiving ^{60}Co-radiation therapy gives a 93% effective rate in preventing a fall in leukocyte count.

In the treatment of uterine cervical ulceration or chronic cervicitis, the powder of total alkaloids is used as a spray applied directly to the ulcerated surface for 10 d. To treat bacterial dysentery and enteritis, total alkaloids are given in a dose of two tablets (0.3 g per tablet) t.i.d.

This agent is also used in the treatment of hydatid moles and chorioblastoma; asthma and chronic bronchitis; and acute tonsillitis and laryngitis.

References

1. **Kojama, R. et al.,** *Chem. Pharm. Bull.*, 18, 2555, 1970.
2. **Mori, H. et al.,** *Jpn. J. Pharmacol.*, 49, 423–432, 1989.

DA QING YE (大青叶) — dried leaf of *Isatis indigotica* Fort., or *I. tinctoria* L., *Baphicacanthus cusia* (Nees) Bremek., *Clerodendron cyrtophyllum* Turcx., and *Polygonum tinctorium* Lous.

Chinese herbists also called this herb *Qing Dai* (青黛).

Chemistry

The leaf contains the glycosides indican and isatan B (approximately 1% of total content). On hydrolysis, indican gives indoxyl, which in turn further oxidizes to give the compound indigo.

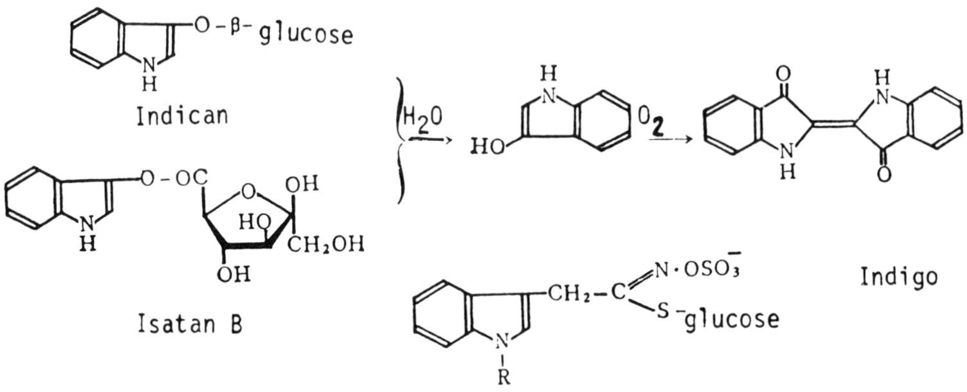

Indican

Isatan B

Indigo

	R=
Glucobrassicin	H
Neoglucobrassicin	OCH_3
1-Sulpho-3-indolylmethyl glucosinolate	SO_3^-

Actions

The herb has antibacterial actions effective against staphylococci, pneumococci, and meningococci. It is an antiviral agent, effective against influenza virus. It has antipyretic, anti-inflammatory, and choleretic properties. It has been observed to increase phagocytic activity of leukocytes. It can relax intestinal smooth muscles, but contracts uterine muscle.

Pharmacokinetic studies showed that oral administration of indican to rabbits produced a maximum plasma concentration after 3 h. The agent was distributed into the liver, kidney, muscle, and alimentary tract. Of the ingested dose, 90% was excreted in the urine.

Adverse effects included nausea and vomiting. In some cases, it may cause hematuria after intramuscular injection.

Therapeutic Uses

This herb is used to treat acute parotitis, upper respiratory infection, encephalitis, hepatitis, lung abscess, dysentery, and acute gastroenteritis.

Recently it was tried to treat HIV viral infection, but no data were reported.

BAN LAN GEN (板藍根) — the dried root of *Isatis tinctoria* L., or *I. indigotica* Fort. or *Baphicacanthus cusia* (Nees) Bremek.

Chemistry

This herb is listed in both the old and recent Chinese pharmacopoeias, but it is unclear whether it is considered to be identical to *Da Qing Ye*, which is derived from the same plant. In *Zhong Cao Yao Xue* it was listed in a different category than *Da Qing Ye*.

In Zhou's *Pharmacology of Chinese Herbs*, these two herbs are categorized in the chapter on fever-reducing agents, as being similar to the group of *Huang Lian* and its alkaloid berberine. He stated that *Da Qing Ye* has antibacterial and antipyretic actions, while *Ben Lan Gen* has antibacterial and antioxicant effects; *Ban Lan Gen* was not listed as an antipyretic agent. Wu's *Pharmacology of Chinese Herbs* gives more explicit descriptions of the actions and uses of *Ban Lan Gen*.

Actions

Its actions include antibacterial and antiviral effects. It can increase the blood flow, improve microcirculation and lower blood pressure, reduce capillary permeability, and reduce the oxygen consumption of heart muscle.

Therapeutic Uses

This herb is used in the treatment of epidemic encephalitis, measles, paroditis, influenza, trachoma, and verruca plana. It is available in injection form, equivalent to 1 g/2 ml, for intramuscular administration. Tablets equivalent to 2 g of raw material are also available, taken in doses of five to six tablets, four to six times each day. In syrup form, the herb is available in concentrations of 1 g/ml.

KONG XIN LIAN ZI CAO (空心蓮子草) — fresh aerial parts of *Alternanthera philoxeroides* Griseb.

Chemistry

The herb contains saponin, coumarin, tannin, and some flavins.

Therapeutic Uses

This herb was recently found to be effective against viral infections. It was claimed to have a remarkable inhibitory effect on certain viruses, such as influenza, encephalitis, and rabies.

In the treatment of measles, oral administration of this plant can shorten the disease process, reducing the fever period and skin eruptions. It can minimize the complications often observed during the recovery period.

If the herb is given in the early stages of encephalitis, it can shorten the duration of the disease, lower mortality, and improve recovery rates. At a dose of 20 to 50 g/kg/d in a group of encephalitis patients, the mortality rate was reduced from 50 to 13%.

Other uses include the treatment of hemorrhagic fever, toxic hepatitis, and icteric hepatitis. In the treatment of epidemic hemorrhagic conjunctivitis, a 1:1 extract is used as eyedrops and applied daily for a couple of days; this reportedly gives good results.

QIN PI (秦皮) — dried bark of *Fraxinus rhynchopylla* Hance, *F. chinensis* or *F. stylosa, F. bungeana* DC. or *F. paxiana* Lingelsh.

Chemistry

The active principles are the glycosides fraxin ($C_{10}H_{18}O_{10}$) and aesculin. Their hydrolysis products are fraxetin ($C_{10}H_8O_5$) and aesculetin ($C_9H_6O_4$).

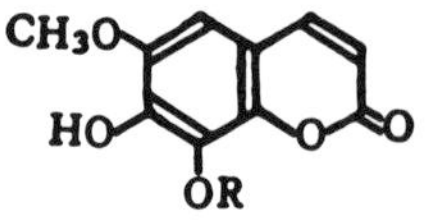

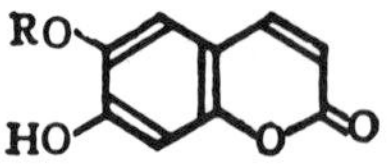

	R=			R=
Fraxetin	H		Aesculetin	H
Fraxin	-glucose		Aesculin	-glucose

Actions

The herb is an antibacterial agent effective against dysentery bacilli, staphylococci, and streptococci. It can exert hypnotic, analgesic, and anticonvulsant effects; its analgesic effect is greater than that of aspirin. It has anti-inflammatory effects, probably due to stimulation of the adrenal cortex. It is an antitussive, expectorant, and antiasthmatic.

Toxic signs are coma and respiratory depression. The LD_{50} of fraxetin in mice is 2.39 g/kg.

Therapeutic Uses

To treat dysentery, fraxetin is administered orally in daily doses of 50 to 100 mg/kg; the cure rate is reported at 80%. This agent can quickly lower body temperature and stop diarrhea. It is also used in cases of chronic bronchitis and rheumatic arthritis.

Preparations include tablets prepared from *Qin Pi* extract (0.3 g/tablet); to treat chronic bronchitis, two to four tablets t.i.d. are taken for 10 d. The herb can also be placed in steaming water and sprayed or inhaled into the respiratory tract for 30 min.

MU DAN PI (牡丹皮) — dried root bark of *Paeonia suffruticosa* Andr.

Chemistry

The bark contains essential oils and several glycosides, including paeonolide ($C_{20}H_{28}O_{12}$), paeonoside ($C_{15}H_{20}O_8$), paeonol ($C_9H_{10}O_3$), paeoniflorin, astragalin ($C_{21}H_{20}O_{11}$), paeonin ($C_{22}H_{33}O_{16}Cl{\cdot}5H_2O$), and pelargonin ($C_{21}H_{31}O_{15}Cl{\cdot}3\ 1/2\ H_2O$).

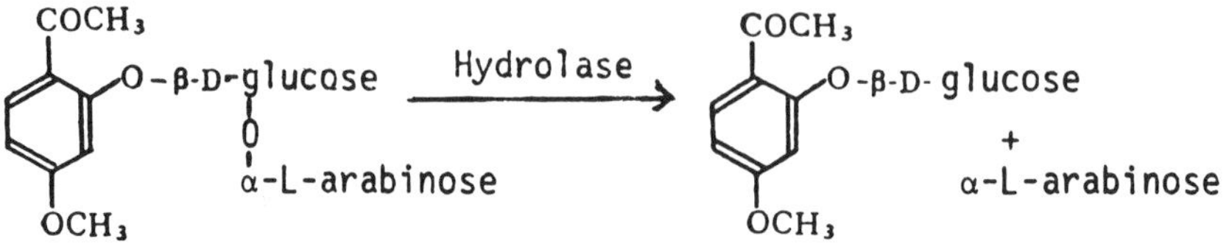

COCH₃ / O-β-D-glucose–O–α-L-arabinose / OCH₃ —Hydrolase→ COCH₃ / O-β-D-glucose + α-L-arabinose / OCH₃

Paeonolide Paeonoside

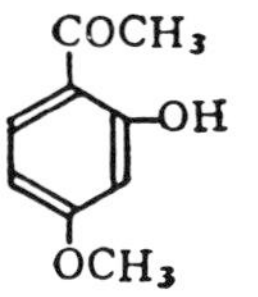

Paeonol | Astragalin | Paeonin R= OCH_3 | Pelargonin R= OH

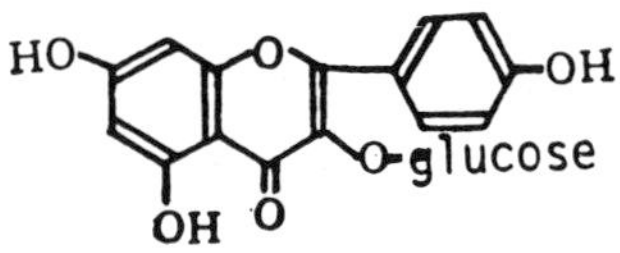

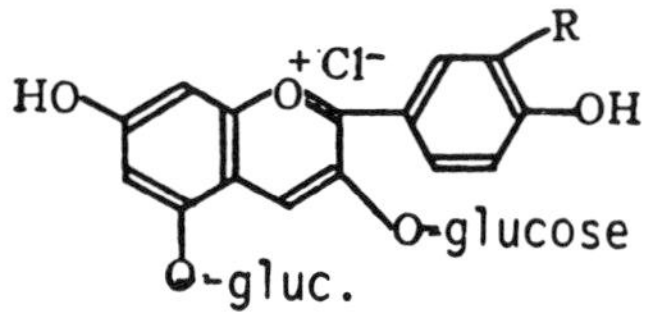

Actions

The herb has antimicrobial actions against *E. coli*, typhoid and paratyphoid bacilli, *Staphylococcus aureus, Streptococcus hemolyticus*, pneumococci, and cholera vibrio.

It is an anti-inflammatory agent. A 70% alcohol extract of *Mu Dan Pi* has a remarkable inhibitory effect on experimentally induced swelling of the rat's paw, reducing capillary permeability. The water extract of this herb can effectively reduce arthritic joint swelling.

It is antihypertensive. It has analgesic, sedative, and anticonvulsant effects. It has been observed to lower the body temperature of mice and exert an antipyretic action in typhoid-infected animals.

This herb is absorbed rapidly from the intestine. A maximal plasma concentration is reached 20 min after administration. After absorption, it is distributed mainly in the liver, followed by the kidneys and lungs. Within 24 h, 89% of the administered dose is excreted in the urine, partially in the form of metabolic products.

It has relatively low toxicity. The LD_{50} in mice is 3.43 g/kg (by stomach), 0.78 g/kg (intraperitoneal), and 0.196 g/kg (intravenous).

Therapeutic Uses

The herb is prescribed to treat appendicitis. It is effective in the treatment of dysentery, with a 50% decoction found to be curative in 28 out of 29 cases. In the treatment of hypertension, it is given in a dose of 15 to 18 g daily, made into a decoction; the dose can be increased to 50 g daily if no side effects are observed. In the treatment of allergic rhinitis, 50 ml of a 10% decoction is given in the evening for 10 consecutive days, producing excellent results.

SHUI HONG CAO (水红草) — dried plant of *Polygonum orientale* L.

Chemistry

This plant contains orientin, vitexin, isovitexin, isoorientin, and plastoquinone-9.

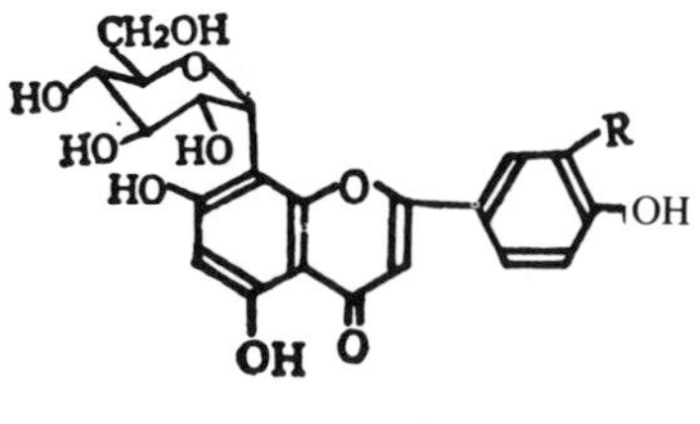

	R=
Vitexin	H
Orientin	OH

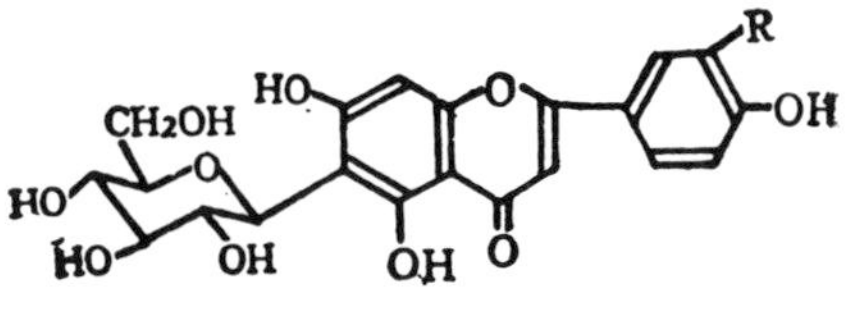

	R=
Isovitexin	H
Isoorientin	OH

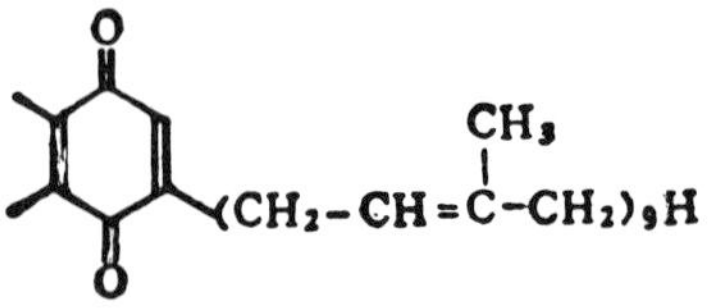

CH_3
$(CH_2-CH=C-CH_2)_9H$

Plastoquinone-9

Therapeutic Uses

The herb has an antibacterial effect. It was used in traditional Chinese medicine to remove ''heat'' and treat infection.

BAI XIAN PI (白鲜皮) — dried root bark of *Dictamnus dasycarpus* Turcz.

Chemistry

Several substances have been isolated from the bark. These include; the alkaloids dictamnine ($C_{12}H_9O_2N$), skimmianine ($C_{14}H_{13}O_4N$), γ-fragarine ($C_{13}H_{11}O_3N$), preskimmianine, and isomaculosindine; limonin ($C_{26}H_{30}O_8$) and obakinone; fraxinellone; psoralen ($C_{11}H_6O_3$), aurapten, and bergapten; and saponins and essential oils.

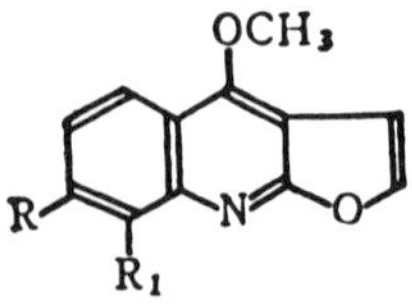

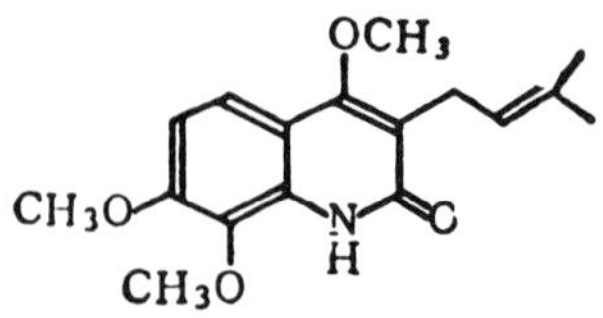

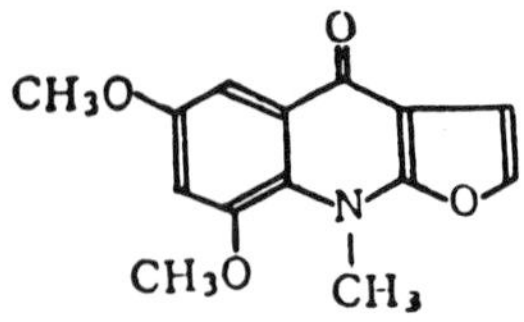

	R=	R_1=
Dictamnine	H	H
Skimmianine	OCH_3	OCH_3
γ-Fagarine	H	OCH_3

Preskimmianine

Isomaculosindine

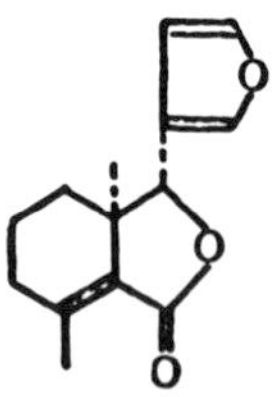

Fraxinellone

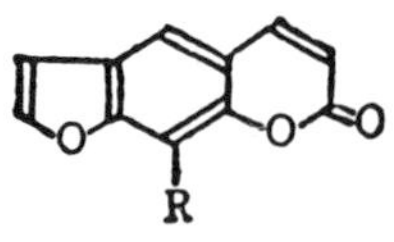

	R=
Psoralen	H
Xanthotoxin	OCH_3

Actions

The herb is noted for its antifungal activity. It also displays antipyretic effects.

Therapeutic Uses

This herb is used to treat dermatitis, psoriasis, and itching of the skin. It reportedly is also effective in the treatment of icteric hepatitis. The standard dose is 3 to 10 g.

LU XIAN CAO (鹿銜草) — dried whole plant of *Pyrola decorata* H. Andres, *P. rotundifolia chinensis*, or *P. rotundifolia* L.

Chemistry

The plant contains several glucosides, including arbutin ($C_{12}H_{16}O_7$), homoarbutin, and isohomoarbutin. Other substances present include chimapillin and monotropein.

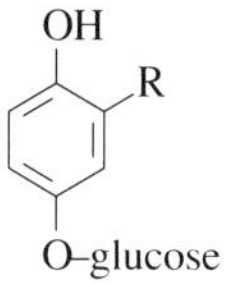

	R=
Arbutin	H
Homoarbutin	CH_3

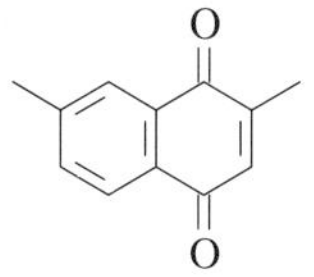

Chimapillin

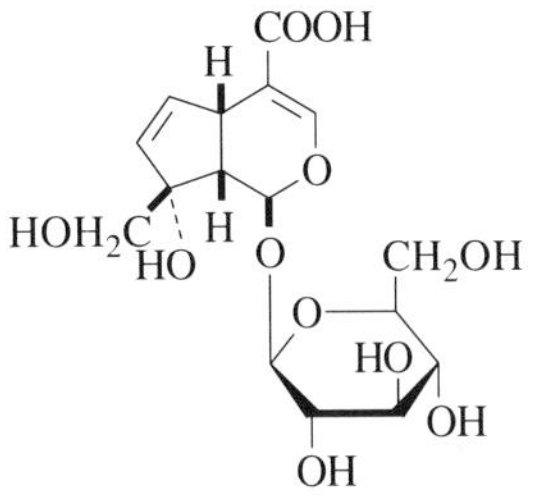

Monotropein

Actions

The herb has antibacterial effects against *Staphylococcus aureus*, dysentery bacilli, typhoid bacilli, and *Bacillus pyrogenes*. The minimal inhibitory concentration ranges from 12.5 to 50 μg/ml.

Extracts of the herb can increase myocardial contractility and exert an antiarrhythmic effect. It has a vasodilatory effect and can lower blood pressure. The herb also has hemostatic effects.

It has contraceptive actions. A 20% *Lu Xian Cao* extract was given to a group of mature female mice at a dose of 0.06 ml/10 g daily for 10 d; fertilization was completely prevented.

Therapeutic Uses

Extracts of the herbs can be given orally or by intravenous infusion for successful treatment of pulmonary infection. The herb is also administered in cases of intestinal and urinary infections, especially in pediatric diarrhea and acute dysentery. It is used in the treatment of liver infection and as an antirheumatic agent.

SHI CAO (蓍草) or Yarrow — the dried aerial parts of *Achillea alpina* L.

Chemistry

The plant contains several alkaloids, essential oils, and some flavonoids. Two other species belonging to the genus *Achillea* include *A. millifolium* and *A. sibirica* Ledeb. The composition of the former includes 0.6 to 0.85% essential oils and 0.05% of the active principles achillin, betonicine, or achilleine. The latter is composed of *d*-camphor and desacetylmatricarin.

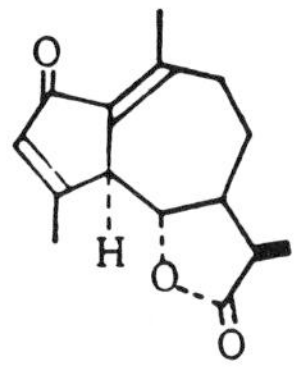

Achillin

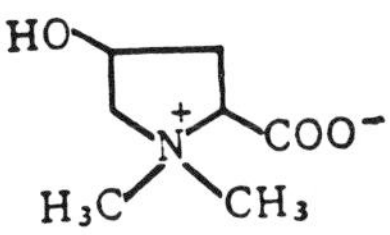

Betonicine (Achilleine)

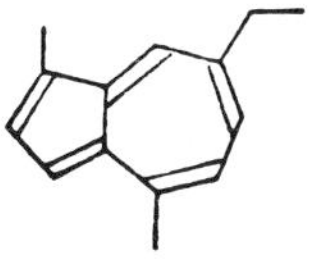

Chamazulene

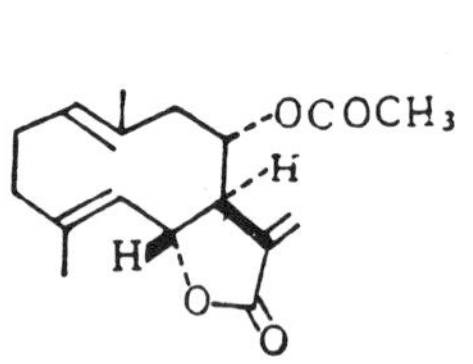

Acetylbalchanolide

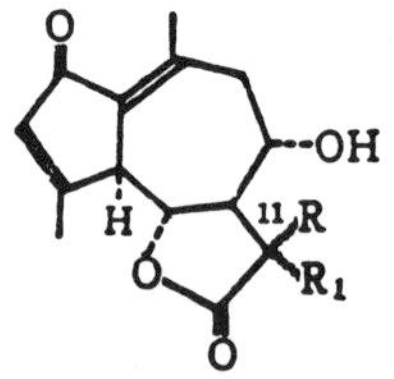

	R=	R_1=
Desacetylmatricarin	α-CH_3	β-H
Desacetyl-isomatricarin	β-CH_3	α-H

Actions and Therapeutic Uses

The herb is listed in *The Herbal Classic of the Divine Plowman* as an "upper class" herb. The *Ben Cao Kong Mu* describes it as "useful in treating abdominal fullness." It has antibacterial actions against *Staphylococcus aureus*, pneumococci, *E. coli* and *Bacillus dysenteria*. Folk medicine used it to treat menopause, abdominal pain, acute intestinitis, wound infections, and snake bite. It is also used to stop internal bleeding of the alimentary tract, hemorrhoids, and gastrointestinal ulcers. It is contraindicated during pregnancy.

It is administered in doses of 3 to 9 g, prepared as a decoction.

Reference

Kresanek, J., *Healing Plants*, Dorset Press, New York, 1989, 38.

LONG KUI (龙葵) — the dried whole plant of *Solanum nigrum* L.

Chemistry

Approximately 0.04% of the herb consists of saponin-like alkaloids. Six different solanigrines have been isolated: α, β, γ, δ, ϵ, and ζ solanigrine. Other saponins have also been found; the hydrolyze products or genin of these are diosgenin and tigogenin. The plant also contains vitamin C and resin.

Table 1 summarizes the structure and products derived from four of the solanigrines.

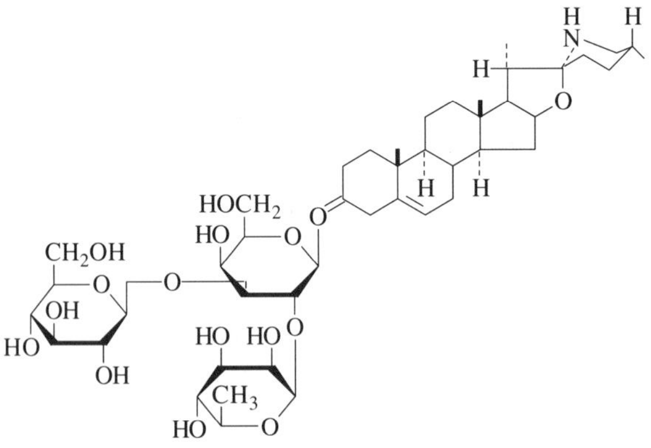

γ-Solanigrine

Actions

The herb has antibacterial and diuretic properties. It should be noted, however, that children who have eaten berries from the plant have complained of headache, vertigo, nausea, vomiting, and tenesmus.

TABLE 1
Solanigrines and Their Products

Saponin-like alkaloid	Other name	Genin	Sugars
β-Solanigrine	Solasodamine ($C_{51}H_{83}O_2N$)		1 mol glucose, 1 mol galactose, and 2 mol rhamnose
γ-Solanigrine	Solasonine ($C_{45}H_{73}O_{16}N$)	Solasodine + ($C_{27}H_{43}O_2N$)	–*O*–galactose–*O*–glucose | rhamnose
δ-Solanigrine	Solamargine ($C_{45}H_{75}O_{15}N$)		1 mol glucose and 2 mol rhamnose
ϵ-Solanigrine	($C_{39}H_{63}O_{11}N$)		1 mol glucose and 1 mol rhamnose

Therapeutic Uses

Long Kui is prescribed in the treatment of mastitis, cervitis, chronic bronchitis, dysentery, and skin infection. It is also used to treat leucorrhea and oliguria in children.

The normal preparation for the herb is a dose of 9 to 15 g in the form of a decoction. This decoction fluid or an ointment can be applied externally on the cervix to treat cervitis.

Reference

Millspaught, C. F., *American Medical Plants,* Dover Publications, New York, 1974, 486–490.

TABLE 2
Other Antimicrobial Herbs

Name	Source	Anti-bacterial	Anti-fungus	Anti-plasmodium	Other effects
Hong Teng (紅藤)	*Sargentodoxa cuneata*	+			Antipyretic, activates blood flow
Hai Jin Sha Teng (海金沙藤)	*Lygodium japonicum* (spore)	+			
Bai Hua She She Cao (白花蛇舌草)	*Hedyotis diffusa*	+			Antipyretic, detoxicant, diuretic, anticancer
Bai Jiang Cao (敗醬草)	*Patrinia scabiosaefolia*	+			
She Gan (射干)	*Belamcanda chinensis*		+		Antipyretic, analgesic, detoxicant
Bai Lian (白蘞)	*Ampelopsis japonica*		+		
Ku Seng (苦參)	*Sophora flavescens*	+	+	+	Antiarrhythmic (see Chapter 3)
Chi Shao (赤芍)	*Paeonia veitchii*	+			Antispasmodic Anticonvulsive
Zhi Zi (梔子)	*Gardenia jasminoides*	+			Antipyretic
Xia Ku Cao (夏枯草)	*Prunella vulgaris*	+			Cardiac tonic, diuretic, anticancer

From Mitscher, L. A., Plant-derived antibiotics, *Antibiotics*, Vol. 15, Weinstein, M. J. and Wagman, G. H., Eds., Plenum Press, New York, 1978, 463.

Chapter 34

ANTITUBERCULAR HERBS

Traditionally, Chinese medicine considered tuberculosis to be a sign of weakness of *yin* and excessive ''heat''. Therefore it was suggested that herbs should be used which would support *yin* and reduce ''heat''.

Our present medical knowledge has well established the causes of tuberculosis, rendering such reasoning meaningless and obsolete. The actions of herbs described in this chapter are based on the results of recent investigations by Chinese researchers, working either in the laboratory or clinic. Their inclusion in this text have nothing to do with their effects on *yin* or ''heat''.

PI JIU HUA (啤酒花) — the dried unripe fruit of *Humulus lupulus* L.

Chemistry

The active principles of the herb are stored in its resin. They include lupulone, humulone, isohumulone, and isovaleric acid.

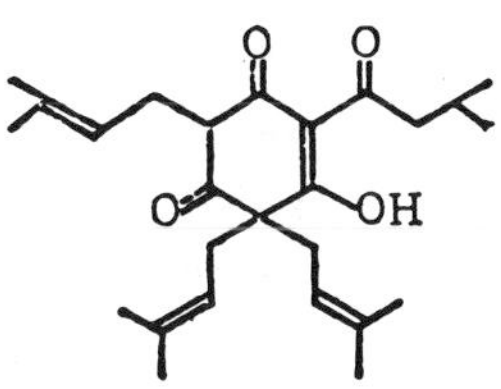

Humulone

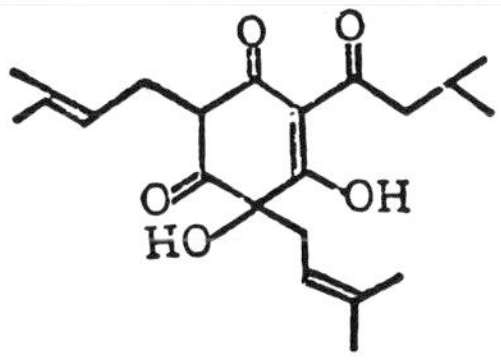

Lupulone

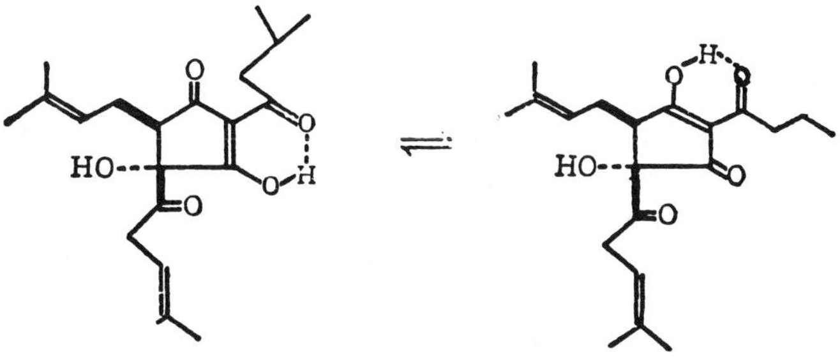

Isohumulone A (colorless crystal)

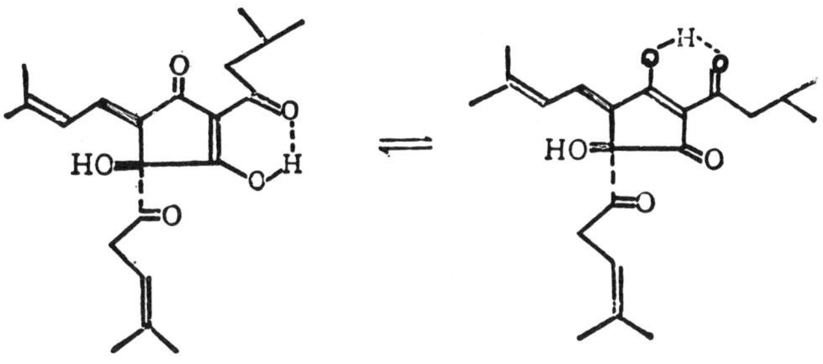

Isohumulone B (oily)

TABLE 1
Antibacterial Activity of Lupulone and Humulone

	Minimal inhibitory concentration (μg/ml)					
Compound	***M. tuberculosis***	**Staphylococci**	***C. diphtheriae***	**Pneumococci**	***B. anthracis***	***B. subtilis***
Lupulone	10.4	0.6 (use synthetic product)	14	3.3	3.3	1
Humulone	100	—	100	200	10	20

Actions

The herb markedly inhibits the growth of tubercule bacilli, at a concentration of 1 to 10 μg/ml of lupulone or 7.5 μg/ml of *Pi Jiu Hua* decoction fluid. The decoction can also inhibit leprosy bacilli, staphylococci, dysentery bacilli, *E. coli, B. pyocyaneus*, typhoid bacilli, and pneumococci.

Table 1 gives a summary of the antibacterial activity of humulone and lupulone.

The isovaleric acid of the herb has a tranquilizing effect, although large doses can cause paralysis. The herb has anticonvulsant properties. It reportedly has a sexual-stimulating effect on females.

Toxicity

Patients taking either lupulone or the combination formula *San Ho Shu* occasionally complain of nausea, vomiting, loose stools, and dizziness. Such adverse effects disappear when the dose is reduced. In rare cases, patients may show hypersensitivity to the herb, and hemoptysis.

The LD_{50} of *Pi Jiu Hua* extract in mice is 175 mg/kg (i.p.); that of lupulone is 600 mg/kg (i.m.) and 528 mg/kg (by stomach). Toxicity is manifested as excitation and death from convulsions and respiratory depression.

Therapeutic Uses

Clinical trials showed a good curative effect with a combination formula *San Ho Shu* (三合素); the lung cavities were closed or became smaller, exudative pleuritis resolved, and the symptoms improved. The herb is also effective in arresting tuberculosilicosis. Patients usually show an increase in body weight and pulmonary vital capacity after treatment. Other uses of the herb include treatment of leprosy, wound infections, mastitis, and dysentery.

The *San Ho Shu* formula is available in either capsule or ointment form. In capsule form (三合素丸), each capsule contains either 18 or 36 mg of *Pi Jiu Hua* resin; the standard dose is 12 to 30 capsules t.i.d. The ointment contains 8% resin and is used externally.

Humulone is available in tablet form, at 0.4 g per tablet; the standard dose is three tablets t.i.d.

LU CAO (葎草) — the aerial parts of *Humulus scandens* (Lour.) Merr. or *H. japonicus*

Chemistry

The plant contains humulone, lupulone, asparagine, choline, and luteolin.

Actions

Lu Cao effectively inhibits tubercule bacillus, as well as diphtheria, typhoid bacilli, and staphylococci. It also has antipyretic and diuretic effects.

Therapeutic Uses

Traditional Chinese medicine prescribed the herb in the treatment of tuberculosis, dysentery, and infections such as cystitis, mastitis, and tonsillitis.

SHI DIAO LAN (石吊兰) — the aerial parts of *Lysionotus pauciflorus* Maxim.

Chemistry

The plant contains some organic acids, flavons, and the genin lysionotinum.

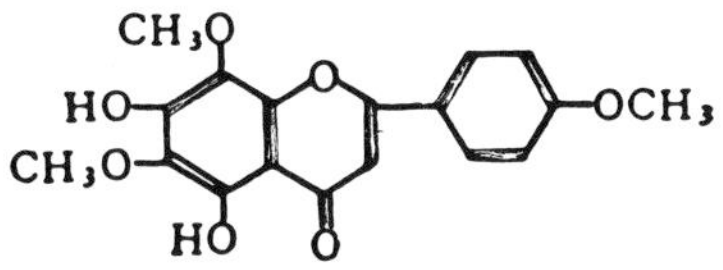

Nevadensin

Actions

The herb has antibacterial, antitussive, anti-inflammatory, and expectorant effects. It can lower the blood pressure for more than 2 h. Atropine can weaken its hypotensive effect.

Therapeutic Uses

This herb is traditionally used to treat tuberculosis, including cases of lymph node and bone involvement. It is also used to treat chronic bronchitis, furuncles, and many kinds of wound infection.

It is available in syrup form, equivalent to 1 g/ml; the standard dose is 10 ml t.i.d. Lysionotinum is also available in 50-mg tablets, administered in doses of one to two tablets t.i.d.

Chapter 35

ANTISEPTICS AND DISINFECTANTS

The herbs listed in this chapter are applied externally and are valued for their antibacterial, astringent, and hemostatic effects. They can diminish tissue swelling and exert an analgesic effect. They are used in the treatment of burns, to reduce the exudation of fluid and promote healing.

LUO HUA ZI ZHU (裸花紫珠) — dried leaf of *Callicarpa nudiflora* Hook. et Arn.

Chemistry

The major active ingredient of the leaf is a tannin substance.

Therapeutic Uses

The herb is commonly used to treat suppurative skin infections and burns. Gauze is soaked with a 3:1 decoction fluid and applied to the wound daily for 3 to 5 d. New tissue grows under the wound.

It is also used to disinfect the skin and mucosa. A 1:1 decoction fluid is used as a substitute for tincture of iodine for disinfection.

The concentrated decoction fluid may be irritating. Occasionally, it causes a hypersensitivity reaction.

ZI CAO (紫草) — dried root of *Arnebia euchroma* Johnst. or *Lithospermum erythrorhizon* Sieb.

Chemistry

The root contains shikonin ($C_{16}H_{16}O_5$), acetylshikonin β-β-dimethylacrylshikonin, and β-*OH*-isovalerylshikonin. Acetylshikonin is the active principle displaying anti-inflammatory and antiseptic activity. The general structure of these compounds is as follows:

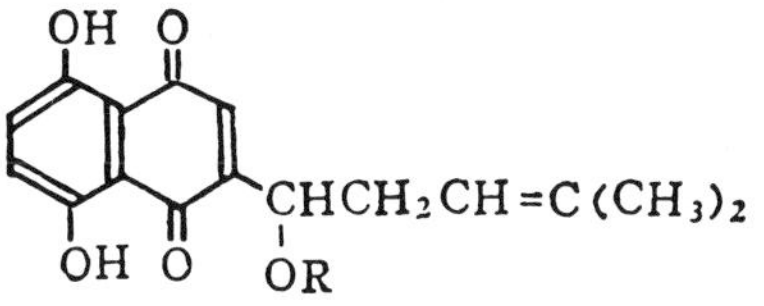

OH O

$CHCH_2CH{=}C(CH_3)_2$

OH O OR

	R =
Shikonin	H
Acetylshikonin	$COCH_3$
β,β-Dimethylacrylshikonin	$COCH{=}C(CH_3)_2$
Isobutylshikonin	$COCH(CH_3)_2$
β-Hydroxyisovaleryshikonin	$COCHC(CH_3)_2OH$
Teracrylshikonin	$COCH_2C(CH_3){=}C(CH_3)CH_3$
Isovalerylshikonin	$COCH_2CH(CH_3)_2$
α-Methyl-n-butylshikonin	$COCH(CH_3)CH_2CH_3$

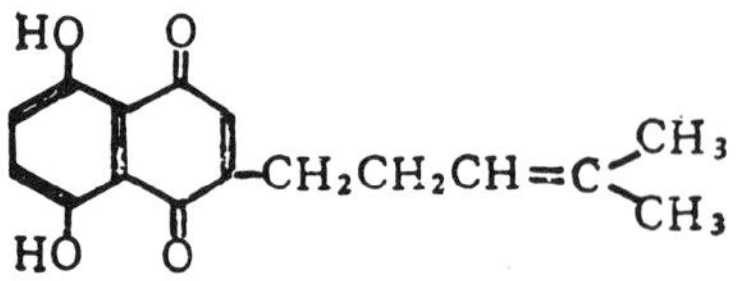

Deoxyshikonin

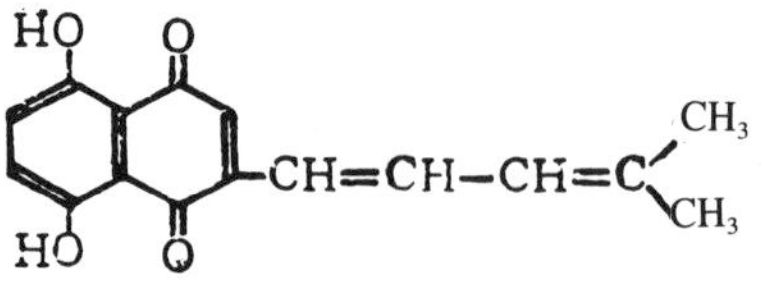

Anhydroalkannin

Actions

The herb has a definite antibacterial effect. A 2.5% solution can inhibit the growth of staphylococci, *E. coli*, *B. pyocyaneus*, and dysentery bacilli. It is an anti-inflammatory agent and can reduce capillary permeability. This effect is not related to the adrenocorticotropic-hormone-cortisol system.

Zi Cao has anti-fertilization effects. It is suggested that the herb has a central inhibitory effect which reduces luteinizing hormone (LH) secretion. Animals fed with this herb for a long period show decreases in the weights of their ovaries, uterus, and pituitary glands.

The water extract of *Zi Cao* has a myocardial-stimulating effect and can contract the uterus. The herb also inhibits HeLa cells by inhibiting the early mitotic stages. It effectively inhibits the growth of choriocarcinoma and hydatid mole.

This herb is toxic. Experimentally, in a group of mice fed with 30% extract for a week, most of the animals lost weight and died of a fatty degeneration of tissues. Rabbits fed with the root at a dose of 5 g/kg developed diarrhea and renal irritation.

Therapeutic Uses

The herb is used to treat burns. *Zi Cao* decoction or *Zi Cao* oil, applied externally to wound surfaces, can exert analgesic, anti-inflammatory, and astringents effect on the wound and promote the regeneration of the epidermis.

It is also effective in the treatment of vaginitis, vulvitis, and cervicitis. A 1:4 to 1:5 extract solution is used to wash the infected area. It is also effective in treating genital eczema.

Other reported uses of the herb include treatment of hepatitis, pityriasis rosea, and choriocarcinoma. In the treatment of the latter, 60 g of the root is made into a decoction and drunk daily.

DA YE AN (大叶桉) — the leaf of *Eucalyptus robusta* Smith

Chemistry

The herb contains essential oils whose active ingredients are cineols and thymol. In addition, the leaf contains gallic acid.

Actions

The water extract of the leaf possesses antibacterial activity against staphylococci, streptococci, and dysentery bacilli. The herb also has antimalarial activity. Used externally, it can rapidly kill the organism causing trichomonas vaginalis.

Therapeutic Uses

The herb is used as a skin disinfectant. A 20% water decoction or 0.25 to 0.5% ether-ethanol extraction of *Da Ye An* is used as a substitute for tincture of iodine. It is also prescribed in the treatment of infectious conjunctivitis; a 10% water decoction is applied externally on the infected region.

Other uses of the herb include treatment of renal pelvisitis, influenza, malaria, and amebic dysentery. For the latter, a 2% water decoction is administered rectally, in doses of 100 to 200 ml, twice a day for 7 to 10 d.

Preparations include tablets equivalent to 6 g or raw material; the standard dose is five tablets t.i.d. A syrup equivalent to 2 g/ml is an alternative preparation, administered in doses of 10 to 30 ml t.i.d. Eye drops in solution strength of 1.7% are also available.

YI GI JIAN (一支箭) — the whole plant of *Ophioglossum vulgatum* L.

Chemistry

The plant contains 3-*O*-methyl-quercetin and several glycosides.

Therapeutic Uses

It is used externally in the treatment of snake bite.

Chapter 36

ANTHELMINTICS

The most common intestinal parasitic diseases in China are ascariasis, hookworm, oxyuriasis, and taeniasis. They usually produce symptoms of malaise, loss of body weight even with frequent eating, yellowish skin, and abdominal distension. Chinese called these conditions *Gan Ji* (疳积), which can be translated as malnutrition, a symptom often seen in those patients. Several popular folk medicines were widely used in these cases, acting by purging the parasite(s) from the intestine.

KU LIAN PI (苦楝皮) or CHUAN LIAN PI (川楝皮) — dried stem or root bark of *Melia toosendan* or *M. azedarach* L.

Chemistry

The root bark and the fruit of *Melia toosendan* contain several compounds whose structures have been identified. Toosendanin is the active principle. Also present are kulinone ($C_{30}H_{48}O_2$), and methylkulonate ($C_{31}H_{48}O_4$), melianol ($C_{30}H_{48}O_4$), melianodiol, melianotriol, melialactone, azadiarachtin ($C_{35}H_{44}O_{16}$), nimbolin A and B, fraxinella ($C_{14}H_{16}O_3$), gedunin, and cycloencalenol.

Another species belonging to the genus *Melia, M. azedarach* L. *var. japonica* Mak., contains azaduarachtin plus sendanone A ($C_{30}H_{46}O_4$) and sendanone B, melianone, and nimbin ($C_{30}H_{30}O_3$).

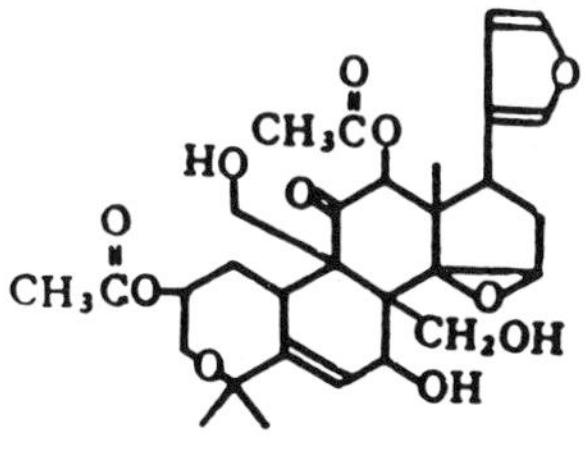

Toosendanin

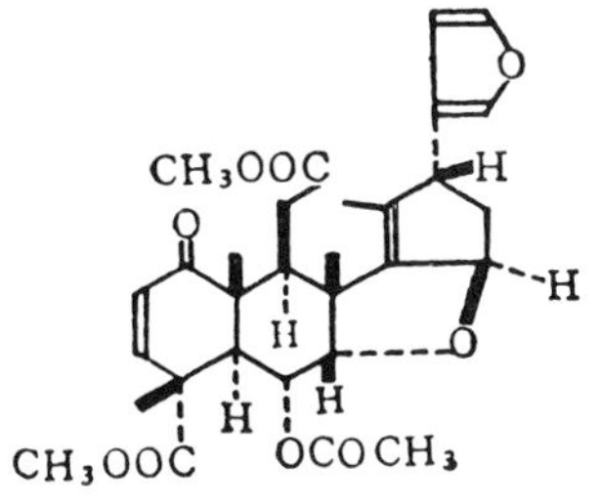

Nimbin

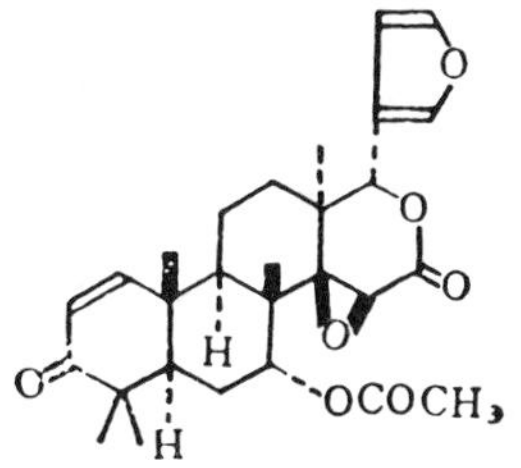

Gedunin

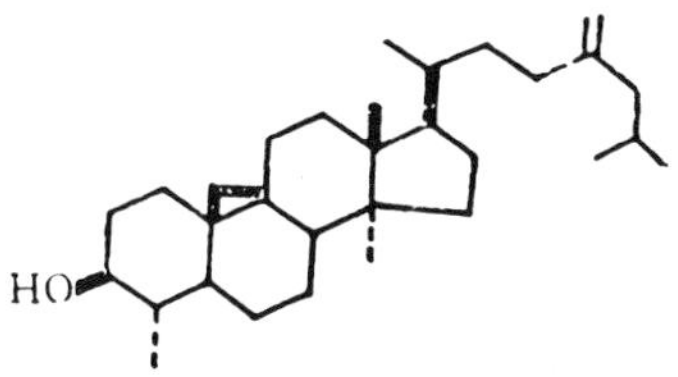

Cycloeucalenol

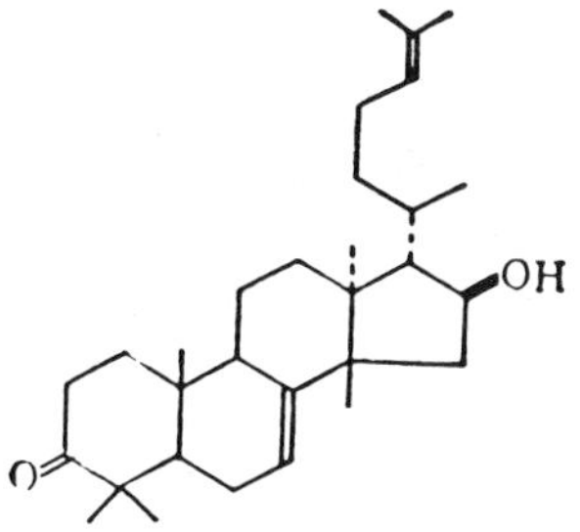

Kulinone

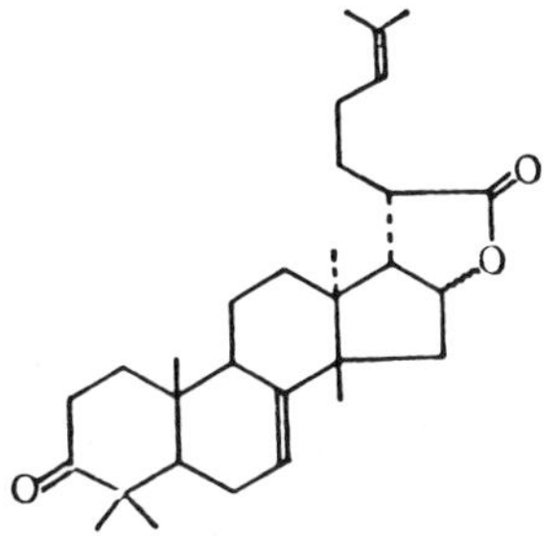

Kulactone

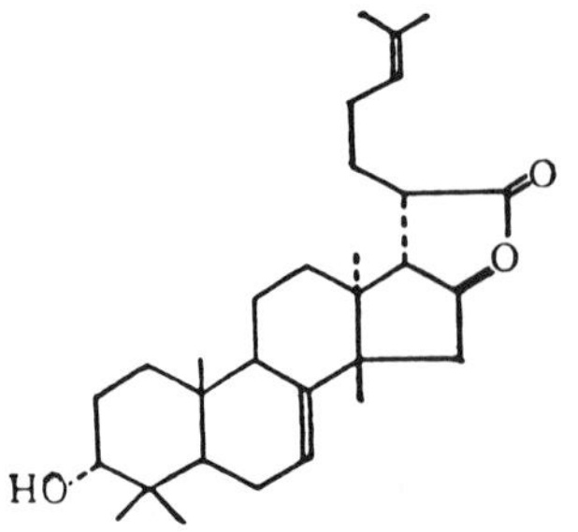

Kulolactone

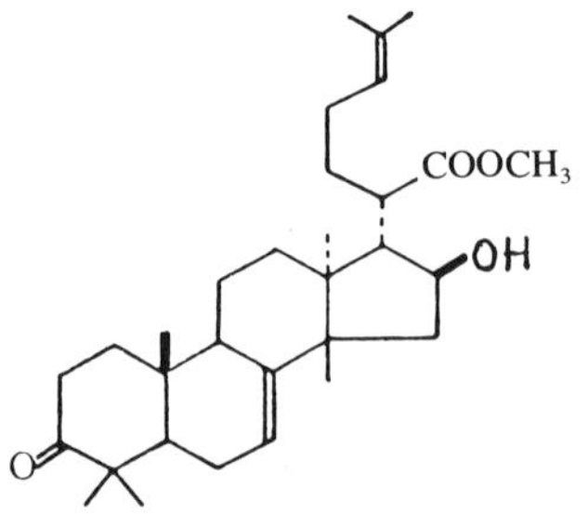

Methylkulonate

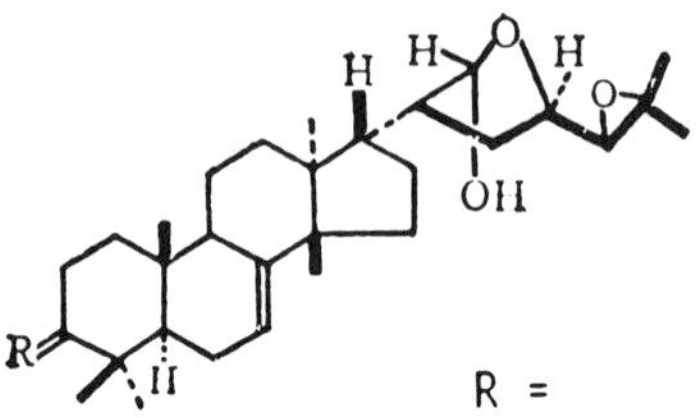

	R =
Melianone	O
Melianol	β-OH, α-H

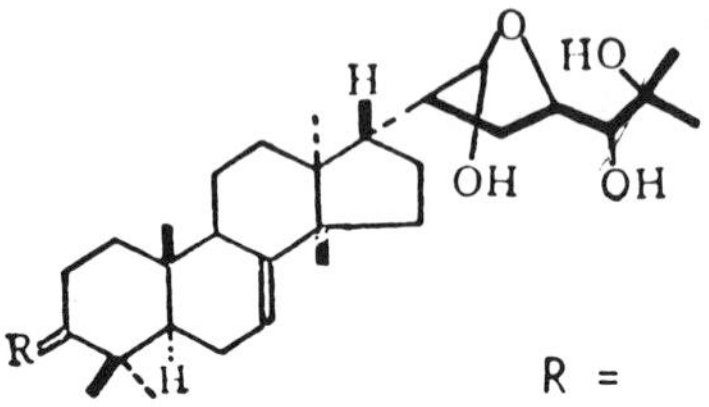

	R =
Melianotriol	H, OH
Melianodiol	O

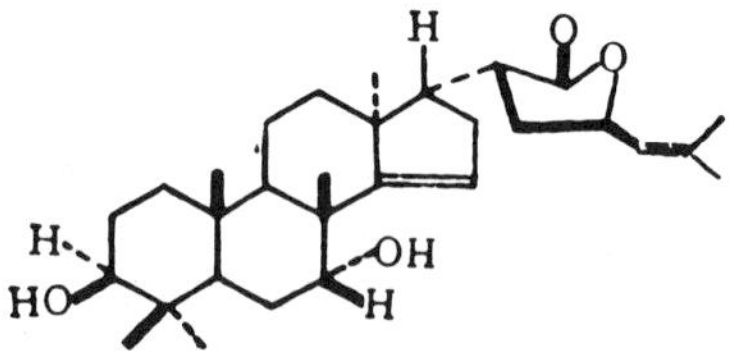

Melialactone

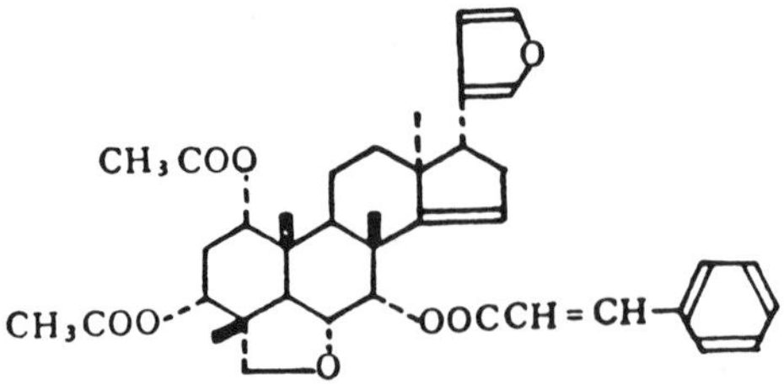

Nimbolin A

Actions

Toosendanin can cause parasites to contract, thereby loosening their attachment to the intestinal wall. It also increases the contractility of the host intestine. The net effect is to purge parasites from the intestine. In addition, the herb has antibacterial effects.

Toxicity

Patients taking the herb may develop gastrointestinal symptoms including nausea, vomiting, and abdominal pain. Occasionally, blurred vision and skin eruption are reported. Contraindications include heart diseases, acute tuberculosis, peptic ulcers, and liver diseases.

The LD_{50} of toosendanin in mice is 480 ± 63.4 mg/kg, and 120 ± 38.5 mg/kg in rats. Tissue damage, especially of the stomach and liver, is seen in these intoxicated animals.

Therapeutic Uses

These plants have been used for thousands of years as anthelmintics, and were mentioned in *The Herbal Classic of the Divine Plowman* as toxic agents.

It is used clinically to purge ascaris, oxyuris, trichuris, and hookworm. It was reported that a 90% curative rate can be obtained in the treatment of ascariasis. There are some advantages in using this herb as an anthelmintic. First, there is no need to use a cathartic after the herb, because the herb itself can increase bowel motility. Second, the herb is not fat soluble, so dietary oil or fat is not contraindicated when it is used.

This herb is also used externally to treat skin sarcoidosis and ringworm diseases. A water decoction is first used to clean the infected area, and then a powder or ointment of *Ku Lian Pi* is applied.

Standard preparations include a syrup equivalent to 2.5 g of raw material/ml; this is administered in doses of 35 to 40 ml in the morning, after overnight fasting. Also available are 25-mg toosendanin tablets; the adult dose is eight to ten tablets, taken in the morning.

SHI JIUN ZI (使君子) — dried fruit of *Quisqualis indica* L.

Chemistry

The active principles of the fruit are quisqualic acid ($C_5H_7O_5N_3$) and trigonelline ($C_7H_7O_2N$).

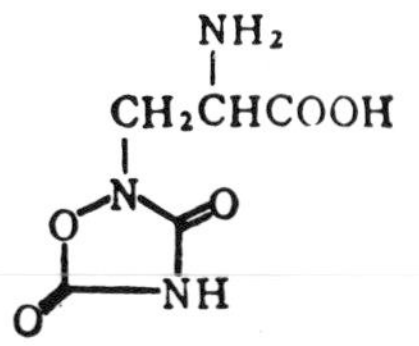

Quisqualic acid

Actions

The herb acts on parasites by producing an initial stimulation, followed by paralysis. In cases of ascariasis and oxyuriasis, a 70 to 80% curative rate has been obtained.

Therapeutic Uses

The herb has a mildly sweet taste, making it suitable for use in the treatment of children. In adults, however, its anthelmintic effect is relatively weak. It is therefore more commonly used in combination with other preparations. Potassium quisqualate is administered in doses of 0.025 to 0.125 g; the *Shi Jiun Zi* kernel is administered in doses of 1 g for children between the ages of 3 and 5, and 3.5 g for 6- to 8-year olds.

MAI SHE CAO (美舌藻) or JE KOO CAI (鹧鸪菜) — the whole plant of *Caloglossa epieurii* (Mont.) J. Ag.

Chemistry

It contains α-kainic acid, a digeneaside. The α-kainic acid can cause the ascaris to contract at first and then become paralyzed.

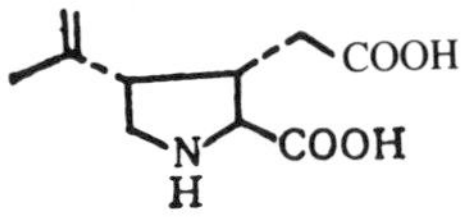

α-Kainic acid

Actions

The herb inhibits the myocardium and causes a fall in blood pressure. Large doses may cause damage to the kidney glomeruli and tubules. Side effects include nausea, abdominal pain, and dizziness.

Therapeutic Uses

Its major use is as an ascaricide; an 80% effective rate has been reported. In combination with 50 mg of santonin, 10 mg of α-kainic acid can produce an 83 to 100% curative rate as an ascaricide. Standard preparations include tablets equivalent to 7.5 g of raw material; the adult dose is eight tablets. Also available is a syrup equivalent to 2 g of raw material per ml; the adult dose is 20 to 30 ml.

The herb is also used as an insecticide for many house insects, including flies and crickets.

SHI LIU PI (石榴皮) — dried pericarp of *Punica granatum* L.

Chemistry

The herb contains five alkaloids: pelletierine, isopelletierine, methylpelletierine, methylisopelletierine, and pseudopelletierine. The first four are liquids. The major active principles are pelletierine and isopelletierine.

In addition to these alkaloids, the herb also contains a large amount of tannic acid. The skin of *Punica* also contains granatin and some mucous substances.

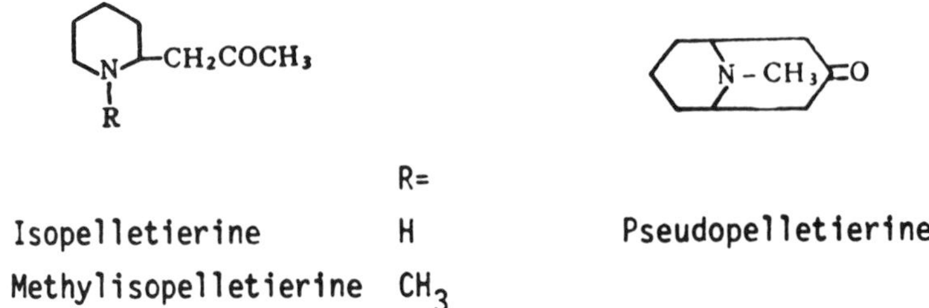

Actions

At a concentration of 1:10,000 pelletierine HCl can kill taenia within 5 to 10 min. Clinical trials found that the pelletierine tannate is much more effective *in vivo*, because the tannate salt is less soluble and less absorbable in the host intestine. Therefore, a higher concentration is maintained in the intestinal lumen, where killing of the parasites occurs.

The herb also has antibacterial actions against staphylococci, diphtheria, and dysentery bacilli.

Toxicity

The alkaloid has a central depressant effect, can inhibit the respiratory center, and has a curare-like effect on skeletal muscles. Intoxication with this herb can produce tremors, convulsions, and coma.

Adverse effects observed in patients taking this herb include signs of gastrointestinal irritation, such as nausea, vomiting, diarrhea, and abdominal pain. Patients may also experience headache, dizziness, and visual difficulty.

Therapeutic Uses

For more than a thousand years, this herb has been widely used to purge parasites from the intestine. The root bark and rhizome, in a dose of 30 to 60 g, is made into a decoction which is drunk in the morning; this is followed 0.5 h later by a dose of 25 to 35 g of $MgSO_4$, taken orally to purge the dead parasites from the lumen.

The herb has also been used in the treatment of dysentery and enteritis; 10 ml of a 50% decoction is taken four times a day. In the treatment of cervicitis, the powdered herb is applied externally to the infected region after thorough cleaning. Other uses include the treatment of chronic tonsillitis and burns.

It is available in tablet form, containing 0.5 g of extract; the standard dose is four tablets taken four times a day.

GUAN ZHONG (貫众) or MIAN MA GUAN ZHONG (绵马貫众) — dried rhizome of *Dryopteris crassirhizoma* Nakai

Chemistry

The name *Guan Zhong* covers more than 59 different kinds of plants grown in different regions of China, including *Cyrtomium falcatum* and *C. fortunei* J. Sm. The substances isolated from these plants vary. The major principles isolated from *Dryopteris* are filmarone, filixic acid, albaspididin, flavaspidin, dryocrassin, 9(11)-fernene, and diplotene. From the *Crytomium* plant, cyrtonin and cyrtopterine have been isolated.

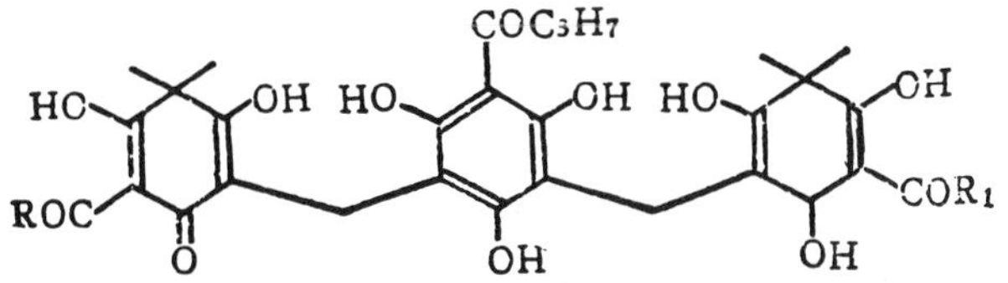

	R=	R_1=
Filicic acid BBB	C_3H_7	C_3H_7
Filicic acid PBB	C_2H_5	C_3H_7
Filicic acid PBP	C_2H_5	C_2H_5

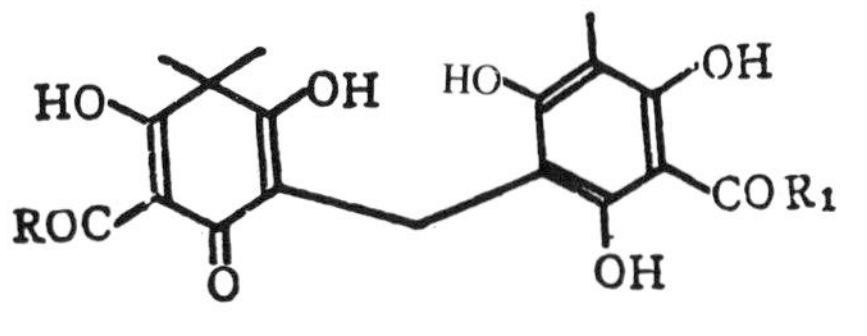

	R=	R_1=
Flavaspidic acid BB	C_3H_7	C_3H_7
Flavaspidic acid PB	C_2H_5	C_3H_7
Flavaspidic acid AB	CH_3	C_3H_7

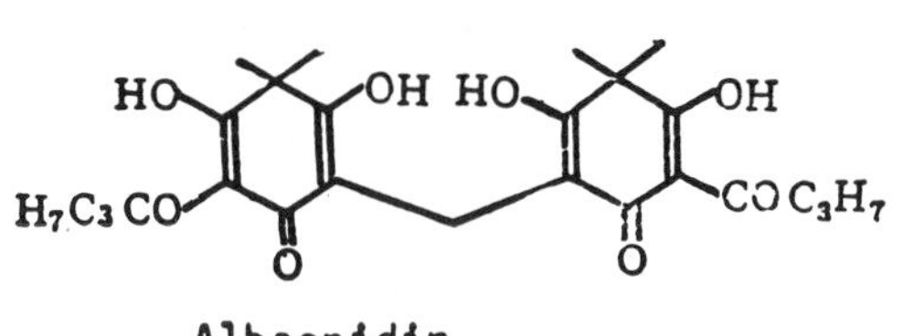

Albaspidin

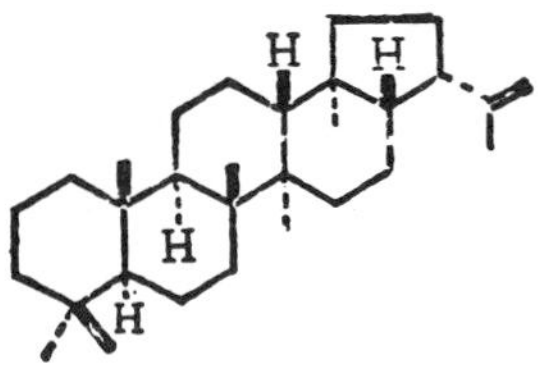

9(11)-Fernene

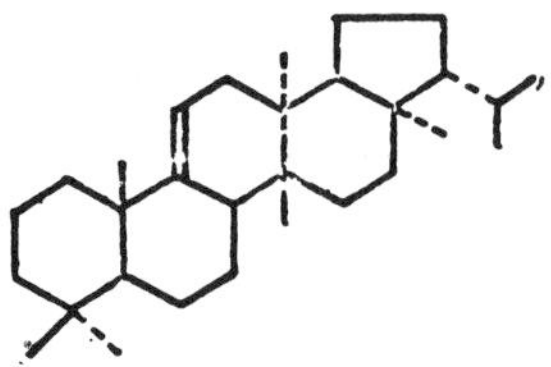

Diploptene

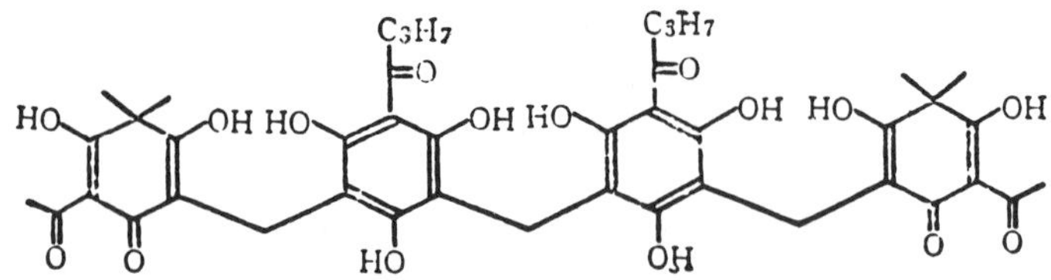

Dryocrassin

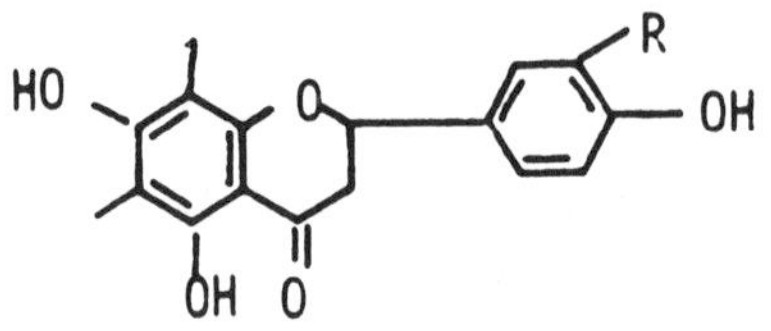

	R=
Cyrtominetin	OH
Cyrtopterinetin	H

Actions

The herb is an anthelmintic and insecticide. It can paralyze taenia and ascaria, which can then be removed from the intestine with a cathartic. It has antibacterial effects against meningococci, *Hemophilus influenzae*, and *Bacillus dysenteriae*. It is an antiviral and antifungal agent. It stimulates the uterus and female sex hormones, plus has abortifacient activity. It is hemostatic.

Toxicity

This herb and its active principle are toxic and irritative. They produce an irritant effect on the gastrointestinal tract when taken orally. Nausea, vomiting, abdominal pain, and bloody stools are common adverse effects. In pregnant women, they may cause abortion.

It is highly lipid soluble and rapidly absorbed from the intestine when some dietary fat is present. After absorption, it causes central depression. In mild cases, headache and dizziness are common. In severe cases, tremors, convulsions, paralysis, and constriction of retinal vessels may be observed.

This herb is contraindicated in children and pregnant women. Oily foods should not be eaten when taking the herb.

Therapeutic Uses

This herb is used to treat ascariasis, oxyuriasis, and hookworm infestation. It is commonly prescribed in doses of 15 g, combined with *Wu Mei* (乌梅) and *Tai Huang* (大黄) in a decoction, to drink twice a day.

It is also used prophylactically and therapeutically in treating influenza. In addition, it can be used to stop excessive menstruation and postpartum bleeding.

There are two herbs similar to *Guan Zhong* which carry the same Chinese name, but differ in their botanical families and chemical ingredients.

ZI KEE GUAN ZHONG (紫萁贯众) — dried plant of *Osmunda japonica* Thunb.

Chemistry

This plant has several steroid-like substances, including ponasterone A ($C_{27}H_{44}O_3$), ecdysterone, custeodysine ($C_{27}H_{44}O_7$), and ecdysone ($C_{27}H_{44}O_6$).

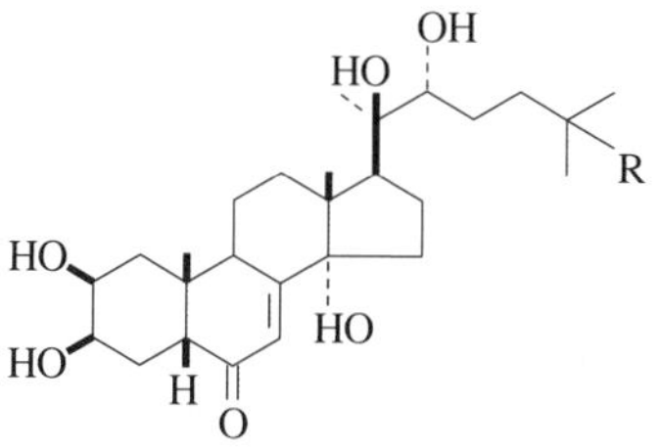

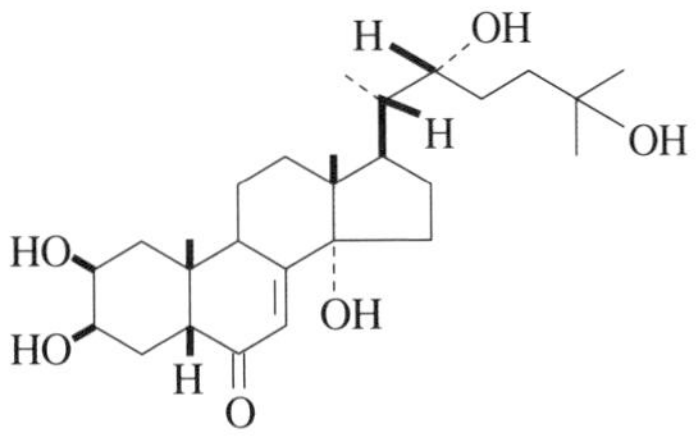

OH HO HO HO HO H O R

	R=
Ponasterone	H
Ecdysterone	OH

Ecdysone

Therapeutic Uses

It is used as an anthelmintic agent and to treat inflammation of salivary glands.

GOU GI GUAN ZHONG (狗脊贯众) — dried rhizome of *Woodwardia japonica* (Lif.) Sm.

Chemistry

The rhizome contains inokosterone and woodwardic acid.

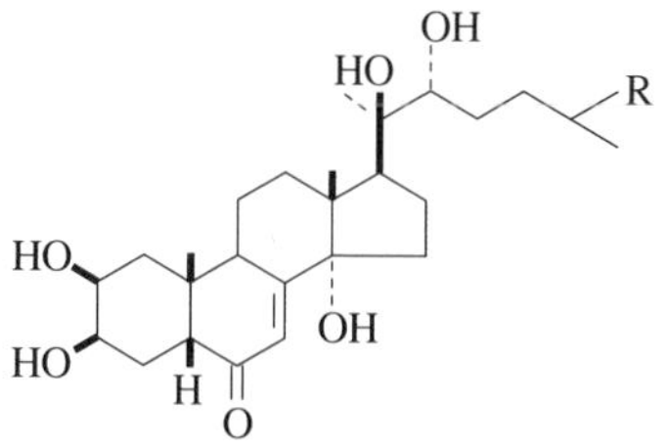

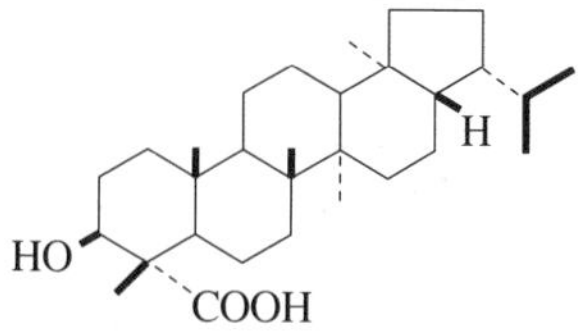

	R =
Inokosterone	CH_2OH
Ponasterone A	CH_3

Woodwardic acid

WU MEI (乌梅) — the fruit of *Prunus mume* Sieb. et Zucc.

Chemistry

The fruit contains the glucoside prudomenin, malic acid, and succinic acid.

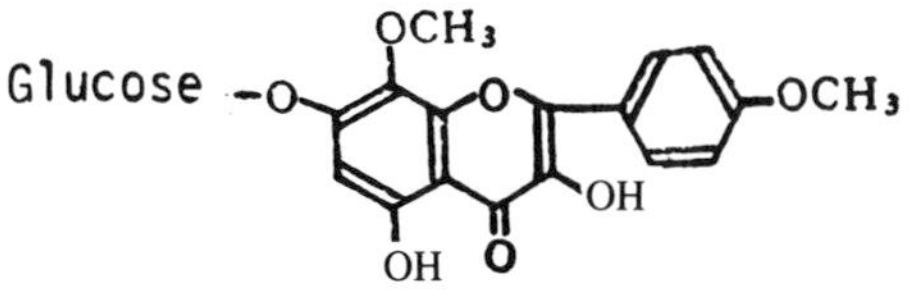

Glucose $-O$ OCH_3 OCH_3 OH OH O

Prudomenin

Actions

The herb can stimulate contraction of the muscles of intestinal parasites and also of the gall bladder, but causes relaxation of the bile duct. Thus, it is beneficial in purging ascaris from the bile duct and intestine.

It has antibacterial actions against *Escherichia coli*, dysentery bacilli, typhoid, and cholera. In addition, it is an antifungal agent.

Therapeutic Uses

This herb is commonly used to treat biliary ascariasis and hookworm. It is given in doses of 15 to 30 g made into a decoction, drunk twice a day; the reported effective rate is 70%. An alternative treatment consists of 45 g of *Wu Mei*, combined with *Huang Pei* and *Tai Huang* in a decoction.

It is also used in the treatment of cholecystitis and gallstone disease.

BING LANG (槟榔) — dried seed of *Areca catechu* L.

Chemistry

The seed contains several alkaloids (0.3 to 0.7% of total content). The major active principles are cholinergic agents: arecoline, arecolidine, guvacoline, and guvacine.

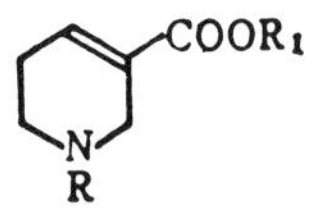

[Structure: tetrahydropyridine ring with $COOR_1$ substituent, N–R]

	R=	R_1=
Arecoline	CH_3	CH_3
Guvacoline	H	CH_3
Arecaidine	CH_3	H
Guvacine	H	H

Actions

Arecholine can paralyze taenia, especially *Taenia solium*. The head and segments of the parasite will be freed, unattached from the intestinal wall, and can then be purged by a cathartic agent. The herb also can paralyze ascaris, fluke, and oxyuris, but is slightly less potent against these organisms.

It has antiviral and antifungal properties. It is a cholinergic; the alkaloid is a muscarinic agonist and can increase intestinal motility and secretions, slow the heart rate, and lower blood pressure.

Therapeutic Uses

The herb is used to treat taeniasis. The adult dose is 80 to 100 g (50 to 60 g for children), made into a decoction, and taken on an empty stomach; 1.5 h later, a dose of $MgSO_4$ is administered to remove the paralyzed parasites from the intestine. It has a 90% effective rate in the treatment of pork taenia. For other taeniasis, a combination with *Nan Gua Zi* (南瓜子) is recommended.

NAN GUA ZI (南瓜子) — dried seed of *Cucurbita moschata* Duch.

Chemistry

The active principle of the seed is cucurbitine. The seed also contains a large amount of fat, protein, and urease. A synthetic cucurbitine is now on the market to replace the natural product for therapeutic uses.

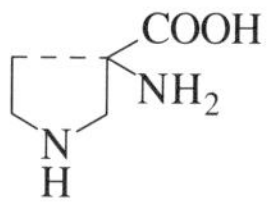

[Structure: pyrrolidine ring (N–H) bearing COOH and NH_2]

Cucurbitine

Actions

Cucurbitine markedly inhibits taenia, both *T. solium* and *T. saginata*. However, its anthelmintic action differs from that of *Bing Lang*. Cucurbitine acts mainly on the egg and segments of the taenia. Therefore, it is seldom used alone, but rather in combination with *Bing Lang*. The alcoholic extract of the combined herbs can kill taenia within 40 to 60 min., the water extract can kill ascaris and oxyuria at a concentration of 1:4000.

It has been reported that this herb can also inhibit and kill both the young and adult forms of schistosomia.

Adverse effects include nausea and vomiting.

Therapeutic Uses

In the treatment of taeniasis, 90 to 120 g of the herb is taken in the form of a powder, followed 1 to 2 h later by a dose of the cathartic $MgSO_4$. It is commonly given with *Bing Lang*, producing a curative rate reaching 95%.

References

1. **Fang, S. D. et al.,** *Acta Chim. Sin.*, 28, 244–251, 1962.
2. **Chen, Z. K. et al.,** *Acta Pharmacol. Sin.*, 1, 124–126, 1980.

XIAN HE CAO (仙鶴草) — dried aerial parts of *Agrimonia pilosa* Ledeb.

Chemistry

The active principle of this herb is agrimophol ($C_{26}H_{34}O_8$). The plant also contains agrimonine, agrimonolide ($C_{18}H_{18}O_5$), vitamins C and K, and large amounts of tannin.

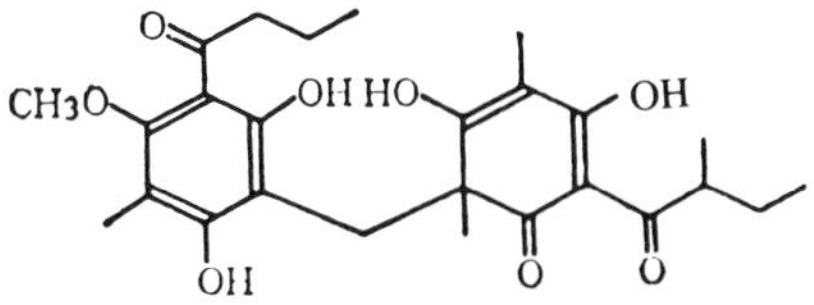

Agrimophol

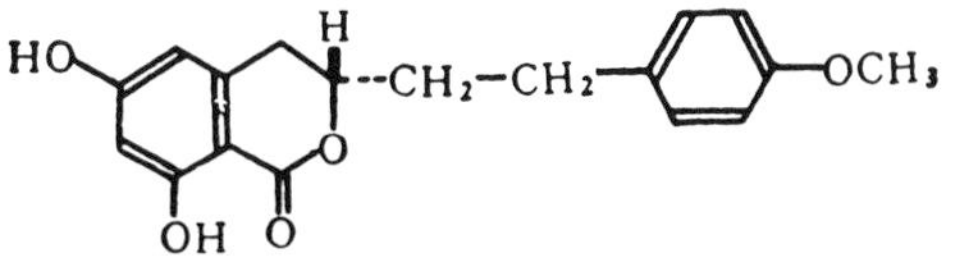

Agrimonolide

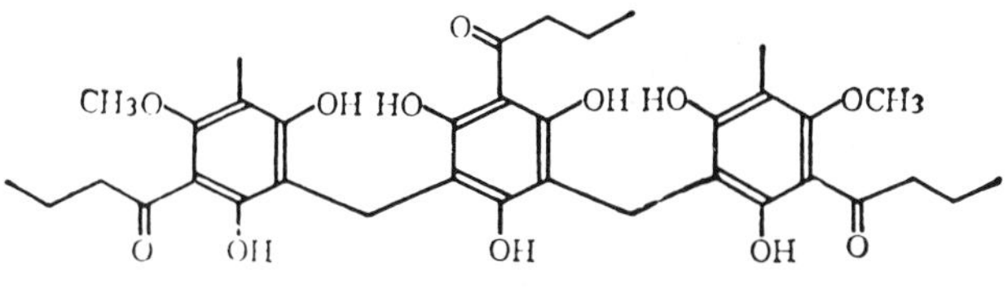

Agrimol C

Actions

In folk medicine, this herb was used to expel taenia. It is now known that the taeniacidal effect is due to the principle agrimophol, which is rapidly absorbed into the body of the parasite, where it directly inhibits the nervous system of the parasite. It also inhibits glycogenolysis in the parasite, by affecting both the aerobic and anaerobic metabolism. The herb is highly effective against *T. solium*.

It is trichomonasidal and can inhibit schistosomiasis and malaria plasmodium. It has hemostatic effects, though the mechanisms of action are not yet understood. In addition, it is an antibacterial agent against staphylococci, *E. coli, Bacillus pyocyaneus*, dysentery, and typhoid bacilli.

Pharmacokinetic Actions

The water extract of this herb is slowly absorbed from the intestine. Twelve hours after oral administration, about 58% remains in the intestinal lumen. After absorption, the active principle is distributed and metabolized mainly in the liver. The half-life, $t_{1/2}$, is 54 min.

Adverse effects include nausea, vomiting, dizziness, and sweating.

Therapeutic Uses

The herb is used in the treatment of taeniasis, for which a 95% effective rate has been reported. It is also used to stop bleeding, such as hemoptysis or hematemesis; 20-mg agrimonine tablets are used as a hemostatic, taken one to three tablets t.i.d. It can be used to treat wound infections and furuncles.

In the treatment of vaginal trichomoniasis, a 200% water decoction is used to wash the infected area. The decoction is also soaked into gauze to be placed in the vagina for at least 3 to 4 h/d for 7 d.

Standard preparations include 0.1-g agrimophol tablets. The adult dose is eight tablets, taken on an empty stomach, followed in 1 h by a dose of $MgSO_4$; the dose for children is 25 mg/kg. Also available is an agrimonine injection solution of 20 mg/2 ml, for intramuscular use.

HUA JIAO (花椒) — dried pericarp of *Zanthoxylum schinifolium* Sieb. et Zucc.

Chemistry

Several substances have been isolated from the pericarp, including estragol ($C_{10}H_{12}O$), citronellol ($C_{10}H_{20}O$), phellandrene ($C_{10}H_{16}$), and the essential oil zanthoxylene ($C_{10}H_{12}O_4$).

The bark of the plant also contains skimmianine ($C_{14}H_{13}O_4N^+$), magnoflorine ($C_{14}H_{13}O_4N^+$), and xanthoplanine ($C_{21}H_{26}O_4N$); γ-ragarine ($C_{13}H_{11}O_3N$) and dictamnine ($C_{12}H_{21}O_2N$); and the essential oil xanthoxylene.

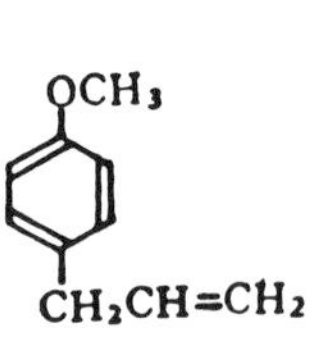

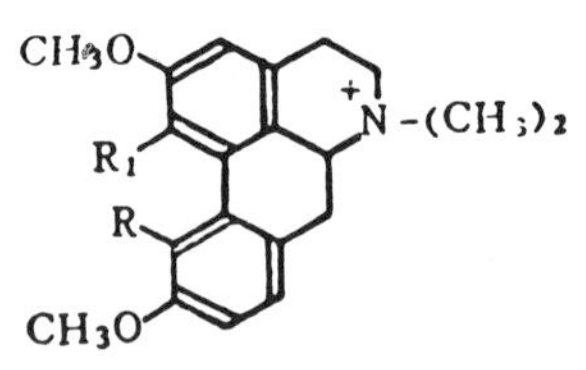

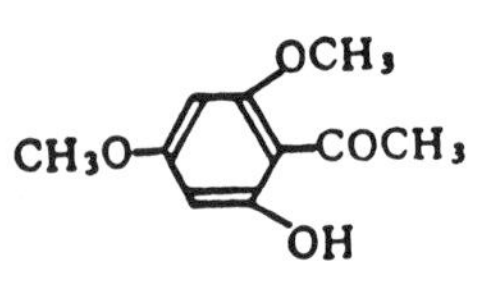

OCH₃ / CH₂CH=CH₂ — Estragol

CH₃O, R₁, R, CH₃O, N⁺-(CH₃)₂

	R=	R₁=
Magnoflorine	OH	OH
Xanthoplanine	H	OCH_3

OCH₃ / CH₃O / COCH₃ / OH — Xanthoxylin

Actions and Therapeutic Uses

The herb can kill ascaris and relieve abdominal pain due to ascaria or obstruction. Chinese folk medicine used it to "warm" the visceral organs, relieve pain, and remove "dampness".

DA FENG ZI (大风子) — dried seed of *Hydnocarpus anthelmintica* Pierre, *H. wightiana*

Chemistry

The seed contains hydnocarpus oil, hynocarpic acid ($C_{16}H_{28}O_2$), chaulmoogric acid ($C_{18}H_{52}O_2$), and gorlic acid ($C_{18}H_{30}O_2$).

$CH_2(CH_2)_{11}COOH$ $\qquad$ $CH_2(CH_2)_9COOH$ $\qquad$ $CH_2(CH_2)_5CH{=}CH(CH_2)_4COOH$

Chaulmoogric acid $\qquad$ Hydnocarpic acid $\qquad$ Gorlic acid

Therapeutic Uses

The herb is used as an anthelmintic. Chinese medicine also recommends it to dispel "wind", and remove "dampness" and toxic substances.

NAN HE SHI (南鹤虱) — dried ripe fruit of *Daucus carota* L.

Chemistry

The fruit contains asarone ($C_{12}H_{16}O_3$), asaraldehyde ($C_{10}H_{12}O_4$), bisakolene, daucol, and carotol. It is used as an anthelmintic.

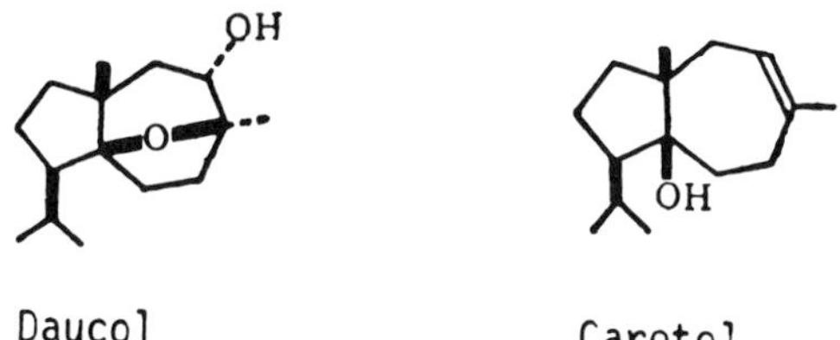

Daucol $\qquad$ Carotol

A WEI (阿魏) — gum resin obtained from *Ferula asafoetida* L., *F. sinkiangensis*, or *F. fukanensis*

Chemistry

The resin has a garlic-like fragrance and contains vanillin ($C_8H_8O_3$), asarensinotannol, and farnesiferol A, B, and C.

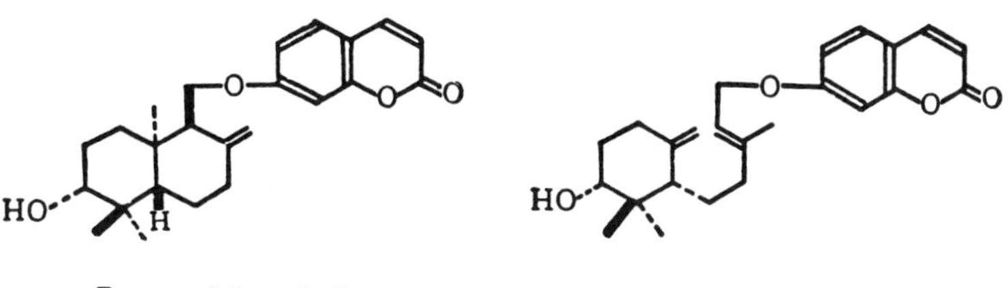

Farnesiferol A $\qquad$ Farnesiferol B

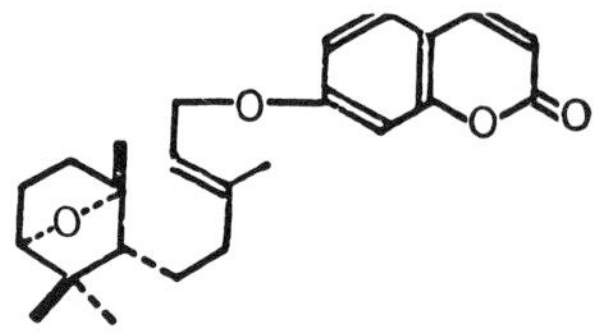

Farnesiferol C

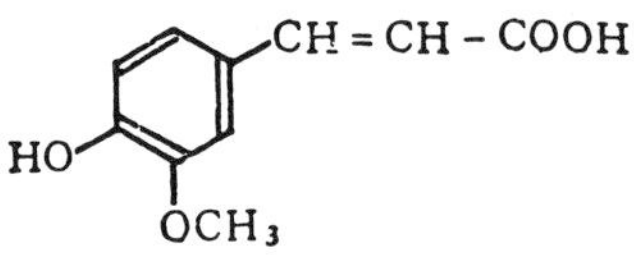

Ferulic acid

Therapeutic Uses

The herb is used as an anthelmintic. It has also been used to treat ascites, dysentery, and malaria.

ARTEMETER (蒿甲醚) — methyl dihydroartemisnin

Chemistry

This "herb" is a synthetic product, methyl dihydroartemisnin, manufactured by a Chinese pharmaceutical company.

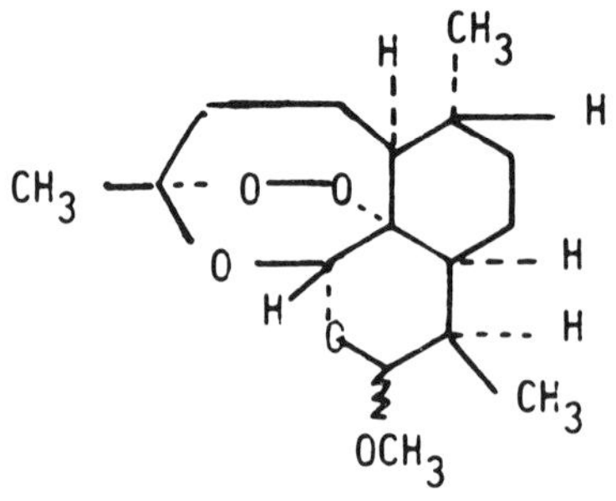

Artemeter

Actions

Artemeter was tested on a group of mice infected with *Schistosoma japonica* and found to be very effective in causing vacuolation and degeneration of the parasites. Figure 1 compares the effects of artemeter with two other antischistosomia agents.

Xuan et al. applied artemeter (1.56 mg/ml) on the skin of *Plasmodium berghes*-infected animals and obtained a 100% curative rate. They claim that this agent is very stable and is absorbed completely from the dermis.

References

1. **Yang, Y. Q. et al.,** *Acta Pharmacol. Sin.*, 7, 276–278, 1986.
2. **Yue, W. J.,** *Acta Pharmacol. Sin.*, 5, 60–63, 1984.
3. **Xuan, W. et al.,** *J. Trad. Chin. Med.*, 8, 282–284, 1988.

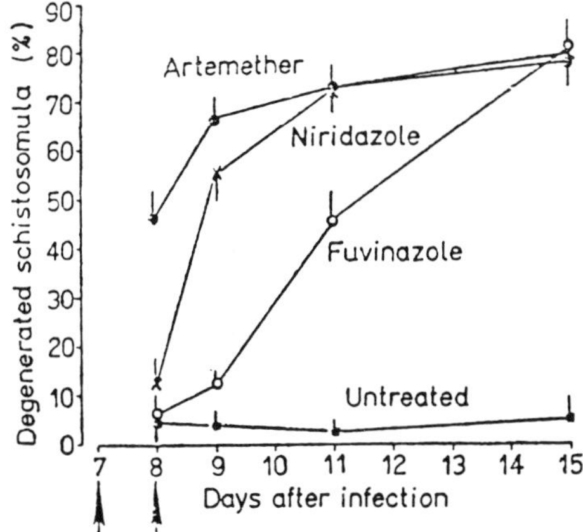

FIGURE 1. The antischistosoma effect of artemeter, niridazole, and fuvinazole. (From Yang, Y. Q. et al., *Acta Pharmacol. Sin.*, 7, 276–278, 1986. With permission.)

TABLE 1
Other Anthelmintic Herbs

Name	Source	Part of	Used
Lei Wan (雷丸)	*Omphalia lapidescens*	Dried sclerotium	Treats ascariasis, taeniasis, and ankylostomiasis; can cause uterus contraction
He Shi (鶴虱)	*Carpesium abrotamoides* L.	Whole plant or fruit	Ascariasis, enterobiasis, taeniasis

Chapter 37

ANTIAMEBIAL AND ANTITRICHOMONIAL HERBS

YA DAN ZI (鸦胆子) — dried fruit of *Brucea javanica* (L.) Merr.

Chemistry

Nine substances have been isolated from the fruit. Among them, seven are the alkaloids bruceine A, B, C, D, E, F, and G; the others are bruceolide and brusatol. All nine have antiamebial activity, while bruceine D and brusatol also have anticancer activity.

An oleic acid, *Ya Dan Zi* Oil, has been isolated from the fruit; it has anticancer activity. The glucoside yatanoside ($C_{20}H_{28}O_9$) is also present in the seed.

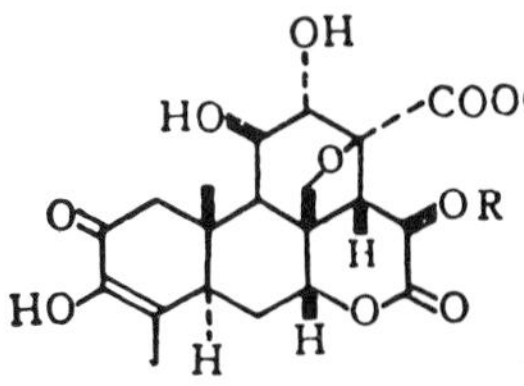

	R=
Bruceine A	$COCH_2CH(CH_3)_2$
Bruceine B	$COCH_3$
Bruceine C	$COCH=C(CH_3)-C(OH)(CH_3)_2$

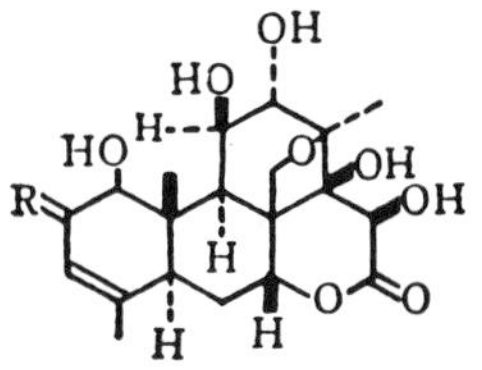

	R=
Bruceine D	O
Bruceine E	α-OH, β-H

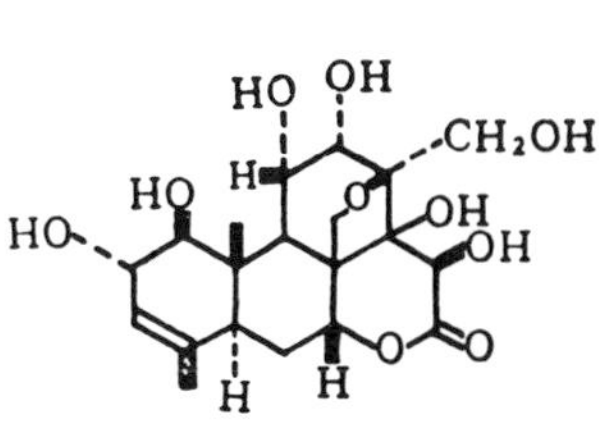

Bruceine F 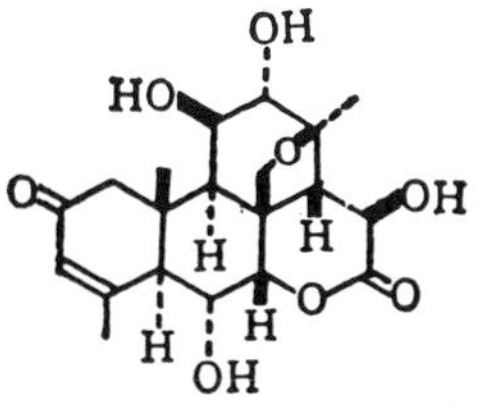   Bruceine G 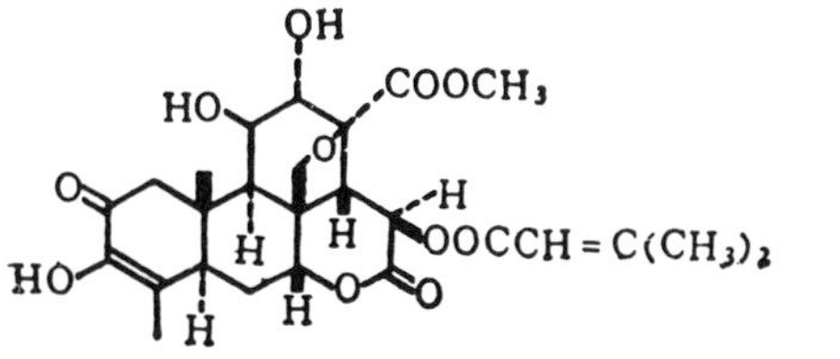   Brusatol

Actions

Recent investigations have shown that the bruceine alkaloids are potent antiprotozoan agents. In both *in vivo* and *in vitro* experiments, they have proven to be amebicides at low concentrations. Early Chinese publications indicate that yatanine, or the total alkaloids, at a concentration of 1:20,000 can kill amebas in test tubes; however, even at a concentration of 1:2,000, no antibacterial effects were observed.

Trials with *Ya Dan Zi* to treat amebic dysentery in dogs showed remarkable results. The amebas disappeared completely from the feces on the second day after drug administration.

The herb has antimalarial properties; its alkaloids can inhibit the growth and multiplication of the plasmodium. It is anthelmintic against trichnosis, ascariasis, and oxyuriasis. It has antiwart and anticancer actions; in experiments with animals, *Ya Dan Zi* oil had a destructive effect on Ehrlich ascites cells and inhibited the growth of S-37 and S-180 sarcoma.

Toxicity

This herb is toxic. Oral administration may cause gastrointestinal disturbances, such as nausea, vomiting, diarrhea, and abdominal pain. Rectal administration of the herb can cause a burning sensation of the anus and abdominal pain.

Intravenous administration of the water extract in animals causes obliguria, hematuria, and central depression. The LD_{50} of the water extract in mice is 2.16 g/kg (intravenous).

Therapeutic Uses

The *Ben Cao Kong Mu* described this herb as "an agent to treat cold dysentery and persistent diarrhea. One dose will stop the illness, while other herbs are ineffective".

In the treatment of amebial dysentery, both acute and chronic amebiasis, a capsule containing 10 to 15 kernels of *Ya Dan Zi* is administered orally three times a day for 7 to 10 consecutive days. The herb also can be administered by rectal perfusion; 20 kernels are ground with 100 to 200 ml of 1% $NaHCO_3$ solution for 2 h and perfused into the rectum either daily or every other day.

In the treatment of malaria, a capsule containing 10 kernels is swallowed three times a day. On the 3rd day, the dose is reduced in half, and therapy continues for another 2 d. The curative rate is equal to that of quinine, but side effects are frequent.

To treat warts and clavas, the kernels are ground with water into a paste, which is applied to the wart or corn. To treat cervical or rectal carcinoma, 5 to 10% *Ya Dan Zi* oil is injected directly into the cancer tissue, or an 10% emulsion preparation (2 ml per ampule) is used for intravenous injection. The dose depends on the size of the cancer.

References

1. **Jang, C. S.,** *Recent Research of Chinese Herbs,* 1956, 141–152.
2. **Lien, E. J. and Li, W. Y.,** in *Advances in Chinese Medicinal Material Research,* Chang, H. M. et al., Eds., World Scientific Publishing Co., Singapore, 1985, 433–452.
3. **Wang, C. Y. et al.,** *Chin. Med. J.,* 36, 409–474, 1940.

BAI TOU WENG (白头翁) — dried root of *Pulsatilla chinensis* (Bge.) Regel

Chemistry

The active principles of the herb are protoanemonin and anemonin. The herb also contains okinalin, okinalein, and some saponins.

Actions

The herb has antiamebial actions, and is an effective antibacterial agent against staphylococci, streptococci, *Bacillus pyocyaneus, B. diphtheria*, and *B. dysenteriae*. It has antitrichomonal effects; at a concentration of 2 mg/ml, proanemonin has a trichomonicidal effect. Okinalin and okinalein have a digitalis-like myocardial stimulating effect. In addition, the alcoholic extract of the root has sedative, analgesic, and anticonvulsant effects.

This herb is relatively nontoxic. Oral administration of proanemonin may cause an irritation of the mucosa.

Therapeutic Uses

Bai Tou Weng is used to treat amebic dysentery. Its effect is better than that of emetine and is less toxic. The standard dose is 15 to 30 g made into a decoction, taken daily for 5 to 7 d. Alternatively, extract tablets equivalent to 3 g of raw material can be taken in doses of four to six tablets t.i.d.

In the treatment of neuralgic dermatitis, the fresh leaf of the herb is rubbed on the affected area; the residue is left on the area and covered with gauze, which is changed daily. A 94% curative rate has been reported.

The herb is also used as an insecticide.

SHE CHUANG ZI (蛇床子) — dried fruit of *Cnidium monnieri* (L.) Cuss.

Chemistry

The fruit contains archangelicin, columbianetin, *O*-acetylcolumbianetin, *O*-isovaleryl columbianetin, cnidiadin, and cnidimine. Also present are the essential oils *l*-pinene and *l*-camphen.

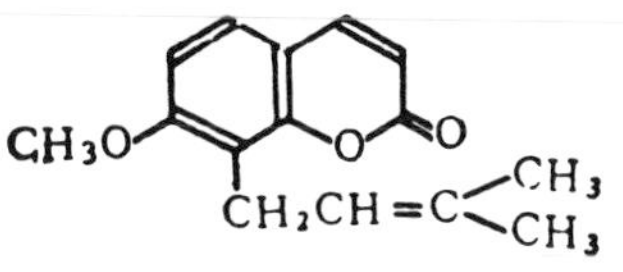

Osthol

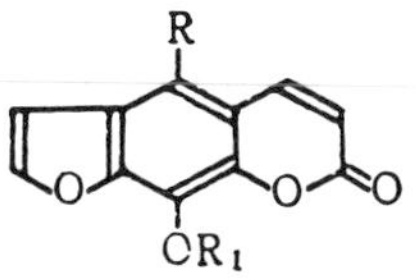

	R=	R_1=
Isopimpinellin	OCH_3	CH_3
Xanthotoxol	H	H

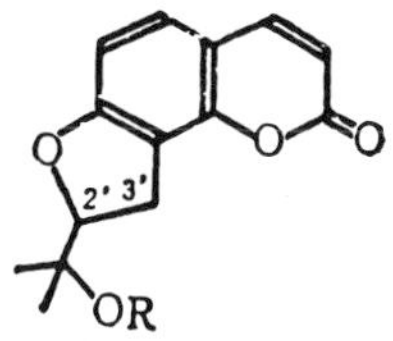

	R=
Columbianetin	H
Columbianadin	$COC(CH_3)\text{-}CH.CH_3$
O-Acetylcolumbianetin	$COCH_3$
O-Isovalerylcolumbianetin	$COCH_2CH(CH_3)_2$
Cinidiadin	$COCH(CH_3)_2$

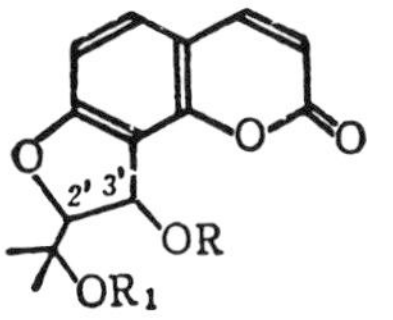

	R =	R_1=
Archangelicin	$COC(CH_3){=}CH(CH_3)$	$COC(CH_3){=}CH(CH_3)$
Edultin	$COC(CH_3){=}CH(CH_3)$	$COCH_3$

Actions

The herb is an effective trichomonicidal agent. It has antiascariac and antifungal actions. It has sexual-stimulating effects in castrated female animals and can increase the weight of the uterus and ovaries of these animals.

Therapeutic Uses

Chinese traditional medicine claimed that this herb could reinforce the *yang* of the kidney.

It is used in the treatment of vaginal trichomonasis. Three to 18 g in combination with *Ku Seng* (*Sophora flavescens*) is made into a decoction which is used to wash the vagina or as a vaginal suppository.

In the treatment of eczema, 30 g of *She Chuang Zi* powder is mixed with Vaseline, applied externally to the affected area. The herb has also been prescribed in the treatment of impotency.

BO LOR HUI (博落回) — dried root, rhizome, fruit, or whole plant of *Macleaya cordata* (Willd.) R. Br. (Papaveraceae)

Chemistry

More than ten alkaloids have been isolated from the plant; the fruit has the highest quantity. The alkaloids include sanguinarine, oxysanguinarine ($C_{20}H_{15}O_5N$), ethoxysanguinarine, protopine, α-allocryptopine ($C_{21}H_{23}O_5N$), chelerythrine ($C_{21}H_{18}O_4N$), coptisine, berberine, corysamine ($C_{20}H_{16}O_4N$), ethoxychelerythrine, chelilutine, chelirubine, bocconine, and bocconoline.

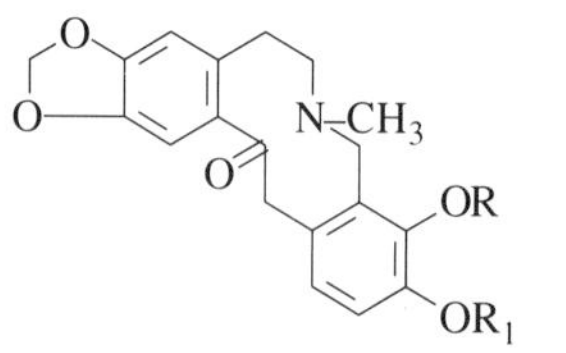

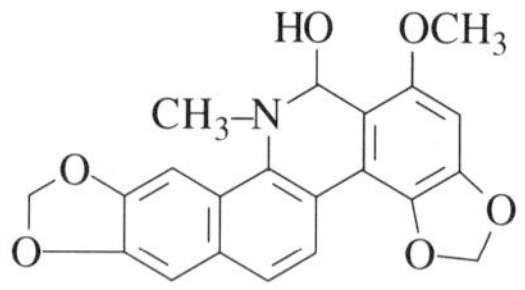

	R =	R_1 =
Protopine	$-CH_2-$	
α-Allocryptopine	CH_3	CH_3

Bocconine

Actions

The herb has antiplasmodial actions, and is very effective against vaginal trichomonas. It is an antibacterial agent, effective against Gram-positive and -negative bacteria, and spirochetes. Experimental results showed that the total alkaloids of this herb have a potent inhibitory effect on staphylococci, pneumococci, *Escherichia coli*, and dysentery bacilli. The minimum inhibitory concentration is about 2.5 to 50 μg/ml.

Sanguinarine has been found to inhibit the enzyme cholinesterase and cause an increase in intestinal contractions. It was also found that the alkaloid can stimulate myocardial contraction and produce a diuretic effect. The alkaloid also can produce a local anesthetic effect if given subcutaneously.

In addition, the herb is larvicidal, especially against fly larvae.

Toxicity

Clinical trials with this herb report some serious adverse effects, including irregularities of cardiac rhythm, occasional heart blockage, and central excitation. In large doses, it may cause coma. Atropine can be used as an antidote. Vaginal irrigation with the extract or decoction may cause local irritation.

Ethoxysanguinarine and ethoxychelerythrine are more toxic than sanguinarine. In mice, the first two alkaloids have an LD_{50} of 5 mg/kg (i.p.); in rabbits, the LD_{50} is 18 mg/kg (i.p.). Intravenous injection of the alkaloids may cause thrombophlebitis and cardiac arrhythmias. The LD_{50} of sanguinarine in mice is 19.4 mg/kg (intravenous).

Therapeutic Uses

This herb is primarily used to treat vaginal trichomoniasis and cervical ulceration. A decoction or extract (equivalent to 25 g/ml) is applied locally or irrigated into the vagina, or used as a suppository on the cervix, for 7 d. Reports indicate a 100% cure in cases of trichomoniasis and about 46% in cervical infections.

The herb is used to treat thyroid cancer and cervical carcinoma. It is prescribed in the treatment of some infectious diseases, including pneumonia, tonsillitis, and chronic bronchitis.

The herb is also used as a toilet disinfectant in Chinese rural areas. *Bo Lor Hui* leaf is chopped into small pieces and placed in the toilet bowl. This inhibits fly larvae growth; a single application was observed to be effective for 25 d.

Chapter 38

ANTIMALARIAL HERBS

Malaria is an endemic infectious disease in southwestern China, where mosquitos and the climate play important roles in transmitting the plasmodium from person to person. Chinese called it *Nue Ji* (瘧疾), or "the foul weather disease". There are at least three major plasmodia commonly found in the region which can cause malaria.

For more than a thousand years herbs were used to treat *Nue Ji*, but their selectivity against particular types of plasmodia was not specified. In the Chinese medical classics, many herbs were mentioned as antimalarial. Most, however, have not been proven experimentally to kill plasmodium; rather, they are antipyretic and provide only symptomatic relief of high fever.

CHANG SHAN (常山) — dried root of *Dichroa febrifuga* Lour.

Chemistry

Some alkaloids have been isolated from the root and have proven to be the herb's active principles. They are α-, β-, and γ-dichroine ($C_{16}H_{19}O_3N_3$), and dichroidine ($C_{18}H_{25}O_3N_3$). The root also contains 4-quinazolone.

β-Dichroine ⇌ α-Dichroine

Actions

In the early 1940s, Jang and his colleagues[1,2] identified the structure of dichroine and found it effective against chicken malaria. Table 1 presents the physical properties of various substances isolated from *Chang Shan*.

Clinical trials using the water extract of *Chang Shan* to treat patients suffering from *Plasmodium falciparum* or *P. vivax* showed a significant effective rate. The total alkaloids of *Chang Shan* are 26 to 50 times more potent than quinine in antimalarial activity. Among them, γ-dichroine is the most effective, being about 100 times more potent than quinine; β-dichroine is the next most potent. It has not yet been determined whether dichroidine has similar antimalarial activity.

The herb has antiamebial actions. Studies from both *in vitro* and *in vivo* experiments showed that it was as potent as emetine in its activity. Also, *Chang Shan* has antiarrhythmia properties; like quinidine, β-dichroine can slow the heart rate and conduction.

The herb is antipyretic. In addition, the alkaloids can lower the blood pressure, dilate blood vessels, and stimulate uterine contraction. All of these are quinine-like effects.

Toxicity

Chang Shan and its alkaloid, dichroine, are toxic. γ-Dichroine is approximately 150 times more toxic than quinine; the next most toxic is β-dichroine. Oral or parenteral administration of the alkaloids can cause nausea, vomiting, diarrhea, and hemorrhage of the intestinal mucosa.

TABLE 1
***Chang Shan* Alkaloids**

Principle	Chemical formula	Melting point (°C)	Properties
α-Dichroine	$C_{10}H_{19}O_3N_3$	136	Soluble in hot water, alcohol, and $CHCl_3$, not in ether
β-Dichroine	$C_{10}H_{19}O_3N_3$	146	Slightly soluble in $CHCl_3$ but not ether
γ-Dichroine	$C_{10}H_{19}O_3N_3$	161	Similar to β-dichroine
Dichroidine	$C_{18}H_{25}O_3N_3$	213	Soluble in hot water, alcohol, and $CHCl_3$; slightly in ether
4-Quinazolone	$C_{18}H_6ON_2$	212	Soluble in hot water, alcohol; slightly in $CHCl_3$ and ether
Dichrin A	$C_9H_6O_3$	228	Soluble in alcohol and $CHCl_3$
Dichrin B	$C_9H_6O_3$	180	Soluble in alcohol and ether

From Jang, C. S. *Recent Research on Chinese Herbs,* Chinese Science Library and Equipment Co., Shanghai, 1954, 136. With permission.

Patients taking this herb or its alkaloids may complain of nausea and vomiting. Overdosage may cause kidney injury and liver damage. It should be prescribed cautiously for elderly persons and is contraindicated in pregnant women.

The LD_{50} of the total alkaloids in mice is 7.8 mg ± 1.3 mg/kg (oral); the LD_{50} of β-dichroine is 6.6 ± 0.7 mg/kg.

Therapeutic Uses

Chang Shan is primarily used in the treatment of malaria. The herb or its alkaloid can be rapidly absorbed from the intestine, lowering body temperature and stopping malarial attacks.

As a preventative measure prior to malarial attacks, the herb is prescribed in doses of 15 g, in combination with 9 g of *Chang Bun Sha* (姜半夏) and 15 g of *Chai Hu* (柴胡), made into a water extract for oral administration. Dichroine may also be administered by intramuscular injection, prepared in a concentration of 8 mg/2 ml.

Chang Shan is also prescribed to treat flagellate disease. For this treatment, 9 g is decocted and taken daily for 7 d.

References

1. **Jang, C. S., Fu, F. Y., and Huang, K. C.,** *Nature,* 161, 400–401, 1948.
2. **Jang, C. S. et al.,** *Science,* 103, 59, 1946.
3. **Chou, T. C., Jang, C. S., and Huang, K. C.,** *Chin. Med. J.,* 65, 189, 1947.

MA BIAN CAO (马鞭草) — dried aerial parts of *Verbena officinalis* L.

Chemistry

The plant contains the glycosides verbenalin and verbenalol, adenosine, tannin, and some essential oils.

O-β-D→Glucose
H
O
O
H COOCH₃

Verbenalin

Actions

The herb exhibits antiplasmodial effects. It is an antibacterial agent, effective against staphylococci, dysentery, diphtheria, and spirochetes. It is an antitoxin against diphtheria toxin. It has analgesic and anti-inflammatory properties and has been observed to promote blood coagulation.

Gastrointestinal disturbances such as nausea, vomiting, and abdominal pain are common adverse effects observed in 70% of patients taking this herb.

Therapeutic Uses

In the treatment of malaria, this herb is decocted in a dose of 60 g, given three times a day for 3 to 7 d. Clinical trial report a curative rate of 94.8%. It should, however, be avoided in pregnancy. Note also that the herb is more effective against *P. vivax*.

Chinese medicine also recommends the herb in the treatment of diphtheria, schistosomiasis, and dysentery.

PAI CHIEN CAO (排錢草) — dried aerial parts of *Desmodium palchellum* (L.) Benth.

Chemistry

The plant contains the alkaloid bufotenine ($C_{12}H_{16}ON_2$) and its methyl ester, nigerin. Also present is donoxime.

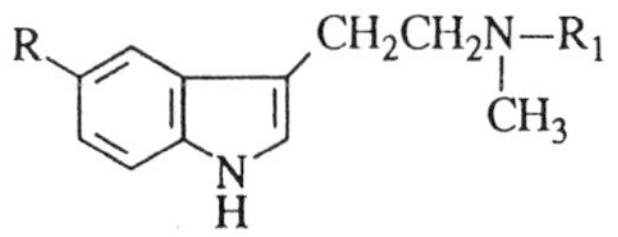

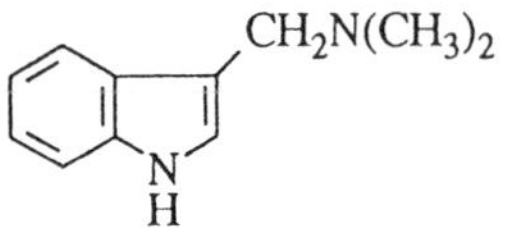

	P =	R_1 =
Bufotenine	OH	CH_3
Bufotenine-*O*-methyl ether	OCH_3	CH_3
Nigerin	H	CH_3
5-Methoxy-*N*-methyltryptamine	OCH_3	H

Gramine

Actions and Therapeutic Uses

The herb has antimalarial and antischistosomiasis activity. It is used in traditional Chinese medicine to treat the common cold and as an antipyretic agent.

ARTEMISININE (青蒿素) — an active antimalarial agent isolated from *Huang Hua Guo* (黄花蒿), *Artemisia annua* L., or *A. aspiacea* Hance, or *A. capillares*

Chemistry

In total, 25 derivatives have been synthesized from this substance. Some have the following structures:[1]

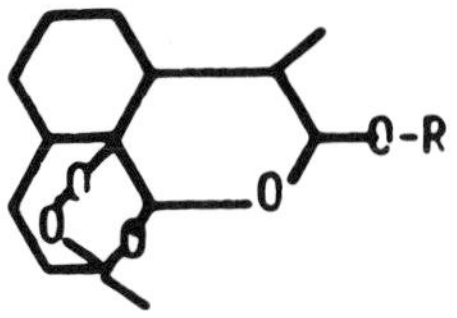

R =	Antimalarial Activity ED_{90} mg/Kg
CH_3	1.02
$CO\text{-}OCH_2CH_2CH_3$	0.50
CO-O-(phenyl) 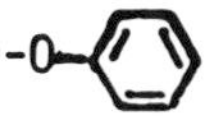	0.63

Actions

Chinese investigators have shown that artemisinine is effective against both *P. vivax* and *P. falciparum*, especially the asexual form. The ED_{50} for artemisinine in mice with malaria is 89.64 mg/kg (oral). On average, within 2 to 3 d after drug administration the plasmodia disappeared from the infected blood.

During clinical trials in patients, 1 g of artemisinine was given per day, producing a 95% effective rate (determined by the disappearance of the plasmodia) within 20 h.

Guan et al.[2] tested the antimalarial activity of three different compounds on human *P. falciparum* in culture and found that the ED_{50} at the end of 48 h after culturing was as follows:[2]

Artemisinine	1.99 mg/ml
Sodium artemisinate	0.14
Chloroquine	2.24

Artemisinine affects neither the exoerythrocyte form nor the sexual form of the plasmodium. Its mechanism of action is believed to involve a blockade of the plasmodium's utilization of the host's erythrocyte protein, thereby starving the plasmodium to death.

Artemisinine also effectively inhibits the schistosomia. The synthetic analog of artemisinine, artemeter, is found to be very effective in this area (see Chapter 36, Anthelmintic Agents). Mori et al.[4] studied the *in vitro* antitumor effect of *Artemisia capillaris* and reported that this herb has a direct tumoricidal action, but no immunosuppressive effect.

The intestinal absorption of artemisinine is incomplete. After absorption, this agent is distributed in the body water and crosses the blood-brain barrier. The half-life is approximately 4 h. It is mainly metabolized in the body and excreted in the urine.

Toxicity

Liang[3] and Wang and Liu[5] reported that oral administration of large doses of artemisinine to monkeys caused severe myocardial damage, decreasing the glycogen granules and producing swelling and vacuolar degeneration of the muscle cells.

Therapeutic Uses

The *Ben Cao Kong Mu* stated that "*Huang Hua Guo* cures hot and cold fevers." Artemisinine tablets are available in 100-mg strength. These tablets are administered in an initial dose of 1 g, followed 6 to 8 h later by 0.5 g. On the 2nd and 3rd days, 0.5 g is administered each day for a total dose of 2.5 g.

Also available is *ginghaosu* suspension in oil (100 mg/2 ml).[5] This is given intramuscularly in an initial dose of 200 mg, with another 100 mg given 6 to 8 h later; 100 mg is administered each day on the 2nd and 3rd days.

References

1. **Gu, H. M. et al.,** *Acta Pharmacol. Sin.*, 1, 48–50, 1980, 2, 138–144, 1981.
2. **Guan, W. B. et al.,** *Acta Pharmacol. Sin.*, 3, 139–141, 1982.
3. **Liang, X. T.,** in *Advances in Chinese Medicinal Material Research,* Chang, H. M. et al., Eds., World Scientific Publishing Co., Singapore, 1985, 427.
4. **Mori, H. et al.,** *Jpn. J. Pharmacol.*, 48, 37–46, 1988; *Jpn. J. Pharmacol.*, 49, 423–432, 1989.
5. **Wang, D. W. and Liu, X. I.,** *Acta Pharmacol. Sin.*, 4, 191–194, 1983.

Chapter 39

ANTICANCER HERBS

Cancer is an abnormal malignant growth of body tissues or cells. It was known and described in early Chinese medical literature. In the 7th century A.D., the *Tsin Shu* (晉書) recorded that "the Emperor had a tumor growth in his eye and sent for a surgeon to cut it." Many different kinds of cancer have been described in Chinese medical texts. Some even pointed out that "early removal of the cancer will save the life; delay will cause destruction of the internal viscerals and death." The treatment of cancer was also substantially documented in medical literature. Although surgical removal was often quoted, this treatment was used only for external tumors. For internal visceral cancers, herbs were administered to "help the righteousness and to repel the evil."

Traditional Chinese medicine did not discuss the etiology and pathogenesis of cancer growth, nor did medical practitioners know how herbal treatments worked. They evaluated the effectiveness of anticancer herbs based on whether patients taking the herbs showed an increase in appetite and body weight. Alternatively, if applied externally, herbs were deemed effective if they could reduce the size of the tumor.

In the last 30 years, the Chinese government and medical institutes have gradually paid more attention to the development of anticancer agents. Investigators using cancer-carrying animals or cancer cells in culture have performed screening tests to determine whether an herb can actually inhibit the growth of cancer cells or increase body immunity against cancer. In this chapter, only the classical herbs which have been mentioned in medical literature as being effective against cancer will be described. It should be noted, however, that many herbs that are listed in other chapters also have anticancer activity in addition to their principal pharmacodynamic actions. Evaluation of their effectiveness is still in the exploratory stages.

YE BAI HE (野百合) or NUNG GI LI (农吉利) — the dried whole plant of *Crotalaria sessiliflora* L. or *C. assamica* Benth.

Chemistry

The plant contains seven alkaloids; among these, monocrotaline A, B, and C are the major principles. Monocrotaline A is credited for the effective anticancer activity of the herb.

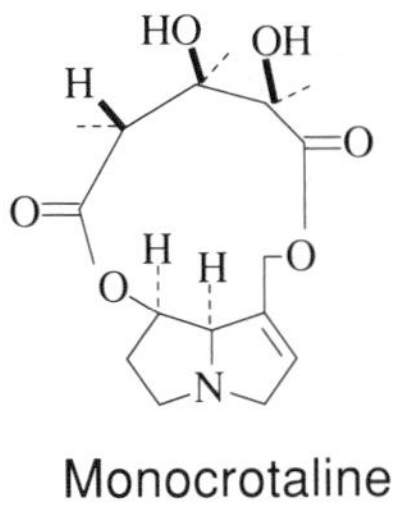

Monocrotaline

Actions

In experiments with animals, the alkaloids of *Ye Bai He* showed a remarkable inhibitory effect on mouse S-180, L-615 leukemia, and Ehrlich ascites cell tumors. Clinical trials showed that the herb is effective against skin carcinoma, cervical cancer, and rectal cancer.

Monocrotaline A is hydrolyzed in the liver. Its metabolite is strongly bound to RNA and inhibits DNA biosynthesis, resulting in a reduction of protein synthesis in the cancer cells. Experimental evidence indicates that monocrotaline A can be incorporated into cells of the uterus, kidney, liver, and skin. This may be involved in the anticancer effect of this agent.

The alkaloid is rapidly absorbed from the intestine or from the injection areas. A maximal plasma concentration, C_{max}, is quickly reached within an hour. It stays in the body for a long period. After absorption, it is primarily distributed into the liver, kidney, and lungs. Even 22 to 90 d after administration, a small amount of the alkaloid and its metabolite can be detected in the urine.

Monocrotaline A can also stimulate smooth muscle contraction, but inhibits the myocardium.

Toxicity

Because this alkaloid remains in the body for a long period, a cumulative toxicity is possible. It is claimed that vitamin C, glycyrrhizic acid, and glucuronic acid can protect the body from monocrotaline intoxication.

The LD_{50} of monocrotaline in mice is 296 mg/kg (i.p.), and in rats is 134 mg/kg (i.p.).

Therapeutic Uses

This herb is used to treat cancers of the uterine cervix, esophagus, lung, liver, and stomach. It is especially effective in the treatment of squanous carcinoma of the skin and cervical carcinoma.

Ye Bai He is available as a 50% injection solution (2 ml per ampule); the standard dose is 2 to 4 ml injected directly around the edges of the cancer daily. A monocrotaline HCl injection solution (50 mg/2 ml) is available for intramuscular use or direct injection into the cancer base daily for 15 d. Also available are a suppository and a 10% monocrotaline ointment.

E ZHU (莪术) — dried rhizome of *Curcuma zedoaria* Rosc., *C. aromatica* Salisk., or *C. kwangsiensis* A. Lee

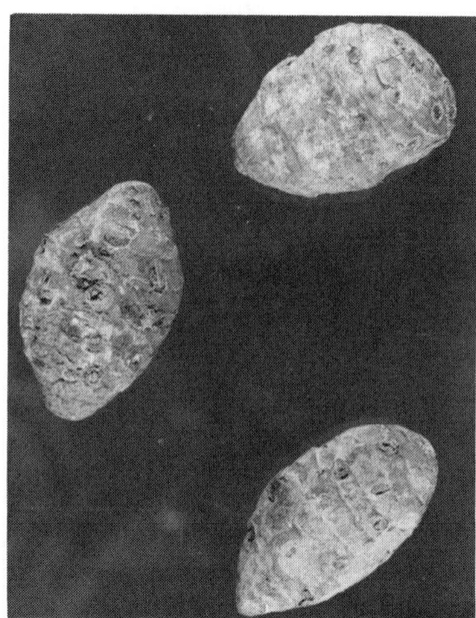

FIGURE 1. *E Zhu.*

Chemistry

Approximately 1 to 2.5% of the herb consists of an essential oil. Fifteen different substances have been isolated from the oil, including curzerenone, curcumenol, and curdione; the latter two are the active anticancer principles of the herb.

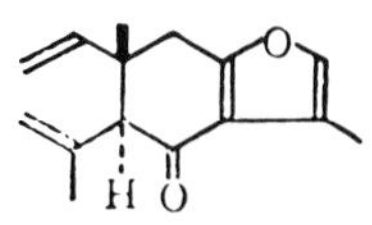

Curzerenone 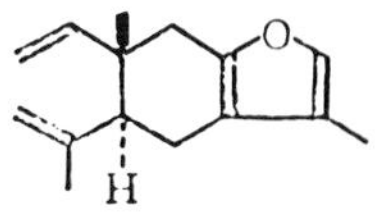 Curzenene 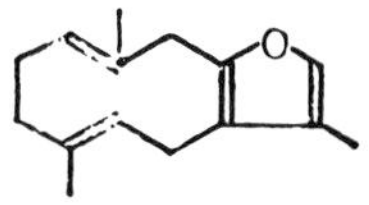 Furanodiene

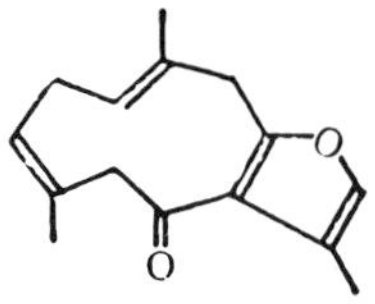

Furanodienone 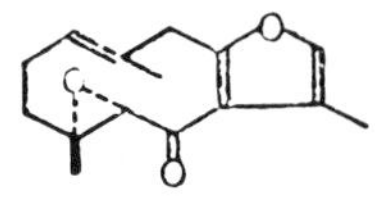 Zederone 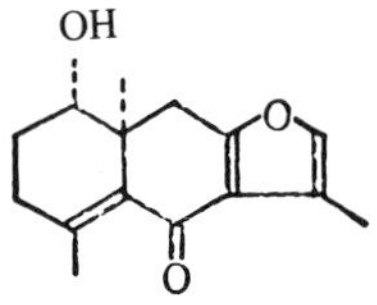 Curcolone

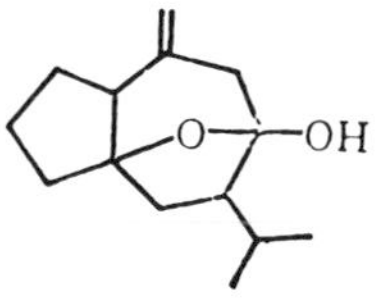

Curcumenol 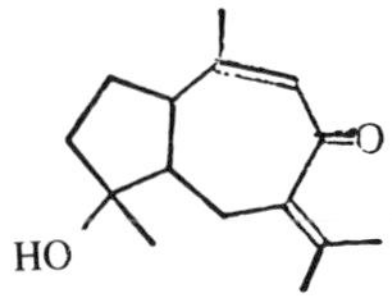 Procurcumenol 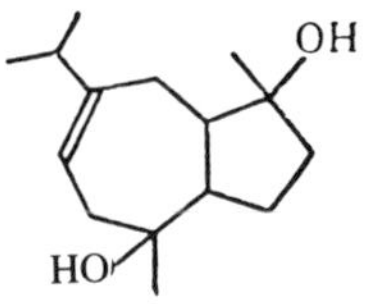 Curcumadiol 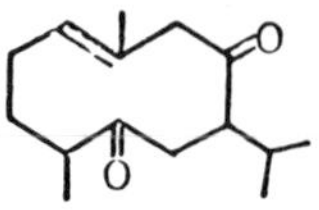 Curdione

Actions

E Zhu or curcumenol are broadly applied as anticancer remedies. Clinical trials in 209 cases of cervical carcinoma showed a 64.4% effective rate.

Curcumenol can inhibit the growth of mouse S-180, ascites cells, and mouse cervical U-14 cells. It was reported that this agent can improve patients' immunity against cancer and protect the leukocytes after X-irradiation. However, the herb itself does not exhibit any beneficial effect against leukopenia.

In experiments with animals, the water extract of *E Zhu* prevented the fertilization of female mice, causing early atrophy of the corpus luteum and inhibiting the secretory stage of the endometrium.

The essential oil of *E Zhu* is a stomachic, which can stimulate gastric acid secretion.

Curcumenol is rapidly absorbed from the intestine and after absorption is distributed throughout all body tissues, including the brain. It is mainly eliminated from the body in the urine and bile.

Toxicity

In patients, curcumenol injected intravenously or directly into the cancer tissue produces no damage to the kidney or liver. When the agent is taken orally, occasional dizziness, nausea, and vomiting; difficulty in breathing, burning sensations, palpitations, and fatigue may be reported. In severe cases, shock has been observed.

The LD_{50} of crude extract of *E Zhu* in mice is 16.75 g/kg, and 316.5 mg/kg (i.p.) for curcumenol.

Therapeutic Uses

Before cancerous conditions were fully recognized, Chinese medicine used *E Zhu* in the treatment of indigestion. A decoction of 3 to 9 g of the herb is drunk daily, to ease mild gastric disturbances.

The herb is used in the treatment of ovarian carcinoma, skin cancer, genital carcinoma, malignant lymphoma, primary hepatoma, thyroid cancer, gastric carcinoma, and lung carcinoma. If it is given in the early stages, the effective rate is much better than when used in the later stages. Administration of *E Zhu* shows a significant improvement of surrounding cancer tissue. For example, after injection of the agent around the cancer base, the size of the cancer becomes smaller, the cancer tissue shows signs of necrosis, and a clear line of demarcation develops between the normal and cancer tissues. The tissue on the edge is less inflammatory, secretion is reduced, and bad odors diminish.

The agent can potentiate the effect of radium therapy. It is therefore used as a substitute for deep X-irradiation, to produce a reduction in tumor size. For this treatment, the herb is used in the form of a 1.25% oil (10 mg/ml), injected locally around the cervix carcinoma in doses of 5 to 10 ml daily; for intravenous infusion, 20 ml of the oil is diluted with a 5% glucose solution. Alternatively, a 3% *E Zhu* oil emulsion can be directly applied to the cervix.

In the treatment of cervix ulceration, the *E Zhu* oil emulsion applied directly on the ulcer can give a 60% curative rate and a 98% effective rate. Gauze or cotton is used as a tampon, soaked with the emulsion, and applied to the ulcer once or twice a day.

When X-irradiation is used for deep tissue therapy, skin burns can be prevented or protected by rubbing the exposed area with a 4% *E Zhu* ointment. It takes 40 days to obtain maximal results.

TAIWAN PEI LAN (台灣佩蘭) — the plant of *Eupatorium formosanum* Hay.

Chemistry

Although primarily found in Taiwan, several types of *Eupatorium* also grow in the southwestern provinces of China. A number of sesquiterpine lactones have been isolated from this herb, including eupatolide, eupaformonin, eupaformosanin, michelenolide, costunolide, parthenolide, and santamarine.

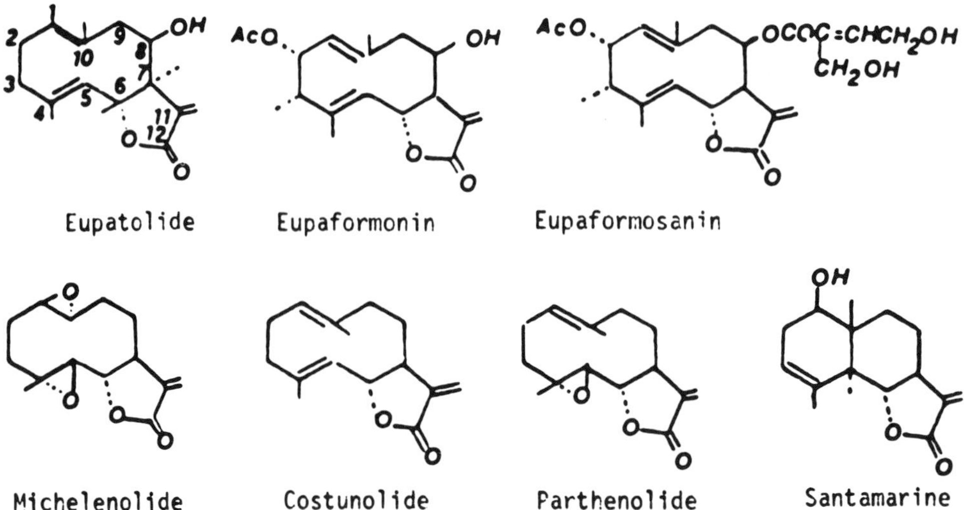

Eupatolide　Eupaformonin　Eupaformosanin

Michelenolide　Costunolide　Parthenolide　Santamarine

Actions

It has been shown that a number of sesquiterpene lactones containing the α-methylene-γ-lactone moiety are potent inhibitors of Walker-256 carcinoma and Ehrlich ascites tumor growth. *In vitro* experiments show that these compounds can inhibit phosphofructokinase and glycogen synthetase, due to their reaction with the essential SH groups of these enzymes.

ZHONG JIE FENG (肿节风) — the dried whole plant of *Sarcandra glabra* (Thunb.) Nakei

Chemistry

The plant contains glucosides, essential oils, fumaric acid, and succinic acid. Recently, several other substances were isolated from the plant and were named CI, CII, CIII, and CIV. Experimental data indicate that the total essential oil and the substance CII are the principles responsible for the herb's anticancer action.

Actions

Zhong Jie Feng extract can inhibit the growth of mouse L-415, ascites cells, and mouse S-180 cells *in vitro*. Trials in patients also showed that administration of the extract causes a decrease in tumor size and improves patients' sense of well being. For example, patients experienced increases in appetite and body weight. In some late stage cancers causing icteric symptoms, daily administration of the extract reduces jaundice of the skin and lowers the icteric index of the blood.

The herb also has antibacterial properties. It is especially effective against staphylococci, dysentery and typhoid bacilli, *Escherichia coli*, and pyogenic bacteria.

Therapeutic Uses

The herb is used in the treatment of malignant solid tumors, such as carcinoma of the pancreas, stomach, esophagus, rectum, bladder, colon, lungs, and thyroid. It may also be prescribed to treat infectious diseases such as tonsillitis, respiratory infections, urinary infections, appendicitis, and dysentery.

Zhong Jie Feng is available in tablet form, with 0.5 g extract per tablet (equivalent to 2.5 g of raw material); the standard dose for this preparation is three tablets t.i.d. An injection solution of the herb is also available (1 g extract per milliliter); the daily dose is 2 to 4 ml, administered intramuscularly or intravenously.

SAN JIN SHAN (三尖杉) — the branch of *Cephalotoxus fortunei* Hook., *C. qinensis* (Rehd. et Wils.) Li, *C. oliveri* Mast., or *C. haiuansis*

Chemistry

Twenty different alkaloids have been isolated from the *Cephalotoxus* plants. Between 0.2 to 0.7% of the leaf and branches of *C. fortunei* consists of alkaloids, including cephalotaxine, epicephalotaxine, demethylcephalotaxine, and cephalotaxinone. Other alkaloids have been isolated from the leaf of *C. wilsoniana*, including wilsonine and epiwilsonine.

From the root and rhizome of *C. harringtonia* K. Koch, two groups of alkaloids have been identified. The first group includes four active anticancer principles: harringtonine, homoharringtonine, isoharringtonine and deoxyharringtonine; the first two are present in highest quantity, comprising half of the total alkaloids. The second group includes the homoerythrina alkaloids.

FIGURE 2. *San Jin Shan.*

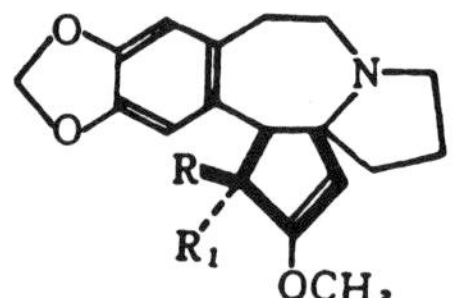

HO
H
O
N

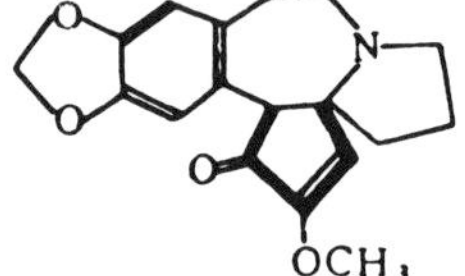

	R=	R_1=
Cephalotaxine	OH	H
Epicephalotaxine	H	OH
Acetylcephalotaxine	$OCOCH_3$	H

Demethylcephalotaxine

Cephalotaxinone

R
H
OCH_3
N

Cephalotaxine	R = OH
Harringtonine	$R = CH_3C(OH)_2(CH_2)_2-C(OH)(COO^-)-CH_2COOCH_3$
Isoharringtonine	$R = CH_3CH(CH_3)(CH_2)_2-C(OH)(COO^-)-CH(OH)COOCH_3$
Homoharringtonine	$R = CH_3C(OH)(CH_3)(CH_2)_3-C(OH)(COO^-)-CH_2COOCH_3$
Deoxyharringtonine	$R = CH_3CH(CH_3)(CH_2)_2-C(OH)(COO^-)-CH_2COOCH_3$

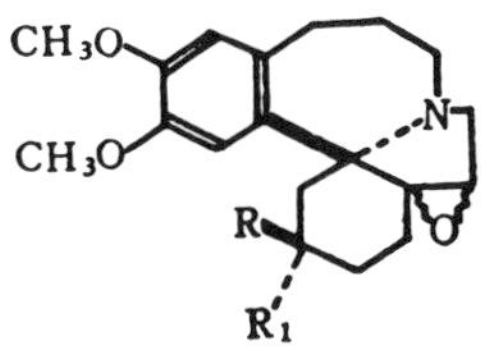

	R =	R_1=
Wilsonine	OCH_3	H
3-Epiwilsonine	H	OCH_3

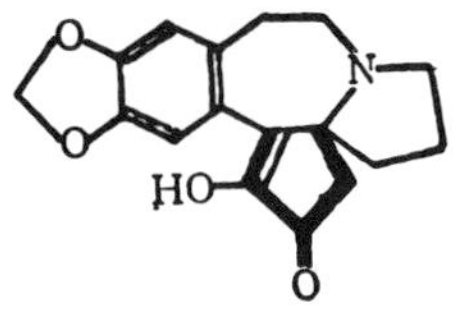

Demethylcephalotaxinone

Actions

In experiments with animals, *San Jin Shan* alkaloids markedly inhibited the growth of sarcoma cells and leukemia cells. The action was nonselective in that the herb universally inhibited the G1, G2, and S stages of cell division. Mitosis of the cancer cells was greatly reduced.

Recent reports from Chinese hospital physicians indicated that compounds of the harringtonine group can inhibit the hemopoietic system and reduce the leukocyte count. It is thus effective in the treatment of leukemia, especially acute granulocytic leukemia, chronic granulocytic leukemia, and mononuclear leukemia. The effective rate is 72.7% and a complete remission occurs in 20% of patients.

This group of alkaloids is effective against malignant lymphoma. Clinical trials showed a reduction in tumor size and improvement of patient's overall condition. Reports also indicate that the herb is effective in the treatment of polycythemia vera. The alkaloid is given by intravenous infusion in doses of 2.5 mg daily for 14 d, for a total dose of 55 mg.

Other actions of this herb include a central depressant effect and constriction of coronary vessels.

Toxicity

Clinically, observed adverse effects have included nausea, vomiting, abdominal pain, diarrhea, leukopenia, anemia, and thrombocytopenia. Some patients report palpitations and irregular heart rhythm. These symptoms are not severe and are usually resolved after stopping drug administration. The acute adverse effects are chiefly gastrointestinal disturbances, hemopoietic toxicity, and cardiovascular toxicity.

Harringtonine may damage the myocardium and reduce coronary circulation. Intravenous administration of this agent should be done with great caution, especially in patients with coronary artery disease. During the course of treatment with harringtonine, hair loss may occur, but regrowth occurs after terminating administration.

The LD_{50} of *San Jin Shan* total alkaloids in mice is 110 mg/kg (i.p.); the LD_{50} of harringtonine is 4.3 mg/kg.

Therapeutic Uses

In folk medicine, *San Jin Shan* was used to treat malignant tumors. Today, harringtonine is available as an injection solution of 1 mg/ml. The adult dose is 2 mg/d (1 mg/d for children), dissolved in 250 to 300 ml of 10% glucose solution, administered by slow intravenous infusion. A full course of treatment lasts 2 months. A homoharringtonine injection solution of 1 mg/ml is also available.

DONG LING CAO (冬凌草) — dried aerial parts of *Rabdosia rubescens* Hora.

Chemistry

The plant contains several terpines. One of the active principles is rubescensine B. Other anticancer principles isolated are oridonin and ponicidine. In addition, the herb also contains essential oils and tannic acid.

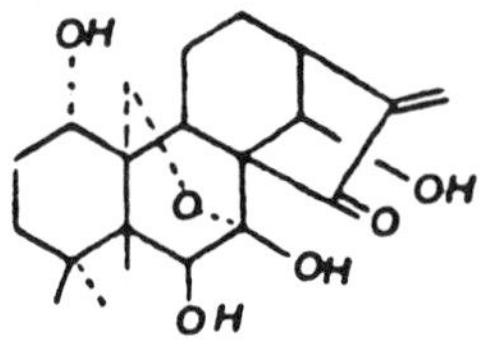

Oridonin 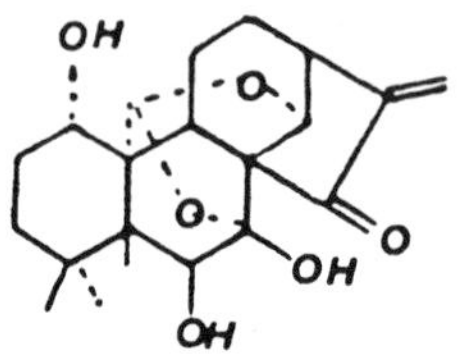 Ponicidine

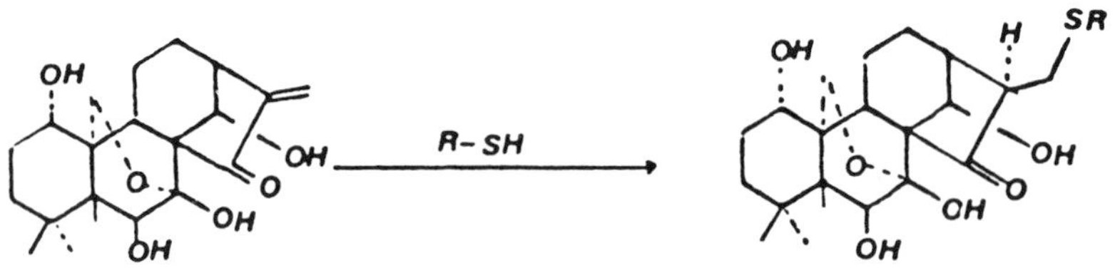

R‑SH

R = butyl, C_2H_5, C_3H_7, sec-Bu or $-CH_2-CH(NH_2)-COOCH_3$

Actions

This herb is widely used in the province of Hunan to treat esophageal cancer. It is claimed that DNA synthesis in Ehrlich ascites cells is inhibited by 74% after treatment with oridonin at a concentration of 10 μg/ml.

Experimental data also show that the alcohol extract of this herb is highly toxic to HeLa cells. Oral or parenteral administration of the extract in mice bearing Ehrlich ascites cells or S-180 sarcoma cells produces a remarkable inhibitory effect.

Rubescensine B possesses anticancer activity against hepatoma cells at a concentration of 4 μg/ml. It is also reported that this agent can increase the immunity of cancer-bearing animals.

The herb can relax smooth muscles and has antispasmodic, anti-inflammatory, and analgesic effects. The essential oil of the herb possesses antibacterial activity.

Toxicity

The LD_{50} of rubescensine B in mice is 55.9 ± 5.7 mg/kg (i.p.). In patients, it may cause nausea, vomiting, abdominal distension, and diarrhea.

Therapeutic Uses

The herb is used in the treatment of malignant cancer, especially in breast and esophageal cancers. After several doses of the agent, the patient can start to swallow food and shows an improvement in appetite, along with a reduction in pain.

The herb can be given in the form of tablets (4.3 g); the standard dose is five tablets t.i.d., with a full course of treatment lasting 1 to 1.5 months. An injection solution equivalent to 2 g/ml is also available; the standard dose is 4 ml/d, administered intramuscularly.

Chinese traditional medicine also prescribes the herb to treat acute tonsillitis. For this treatment, it is generally given in doses of 6 g, prepared as a decoction.

BAN SAO (斑蝥) — Chinese cantharides, or the dried body of *Mylabris phalerata* Pall. or *M. cichoorii* L.

Chemistry

Approximately 1 to 1.2% of the herb consists of cantharidin, which is the active anticancer ingredient.

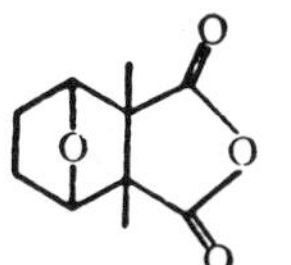

Cantharidin N-Hydroxy-canthatidinimide (synthetic)

Actions

Cantharidin is a very irritating substance, which can shrink the size of solid tumors and inhibit their growth. Clinical trials in primary hepatoma and cancers of the esophagus, anus, lungs, and breast showed a 72.6% effective rate and an increase in survival time.

Cantharidin has a stimulatory effect on the bone marrow and increases the leukocyte count in peripheral blood. Therefore, it is advantageous to combine this agent with other cancer chemotherapic agents to counteract their leukocyte depressant effect. It is also effective in suppressing hypersensitive rhinitis.

When given orally, it is rapidly absorbed from the intestine. A maximal plasma concentration is reached 1 h after administration. The agent stays in the body for a longer period and is mainly excreted through the kidneys and intestine.

Toxicity

Patients taking this herb orally usually complain of nausea, vomiting, and urinary irritation. They may experience difficulty in urination and hematuria. The agent is contraindicated in patients with a history of peptic ulcer or kidney diseases, and in pregnant women.

The LD_{50} of cantharidin in mice is 1.71 mg/kg.

Therapeutic Uses

The preparation consists of sugar-coated tablets (0.25 mg) at a dose of one to two tablets t.i.d.

XI ZHU (喜树) — the fruit of *Camptotheca acuminata* Decne.

Chemistry

The fruit contains camptothecine, hydroxyl-camptothecine, methoxylcamptothecine, and venoterpine.

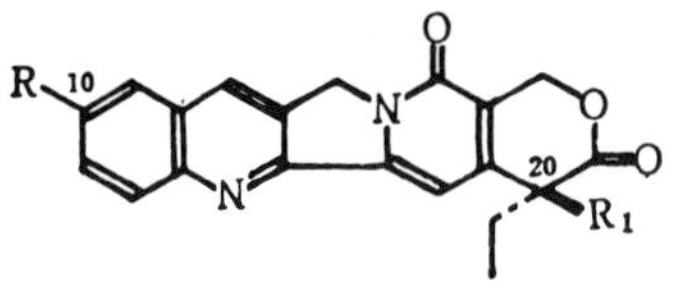

	R=	R_1=
Camptothecine	H	CH
Hydroxycamptothecine	OH	OH
Methoxycamptothecine	OCH_3	OH
Deoxycamptothecine	H	H

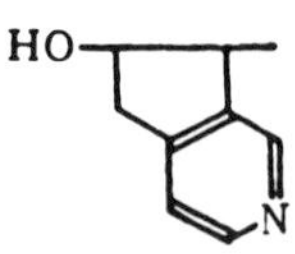

Venoterpine

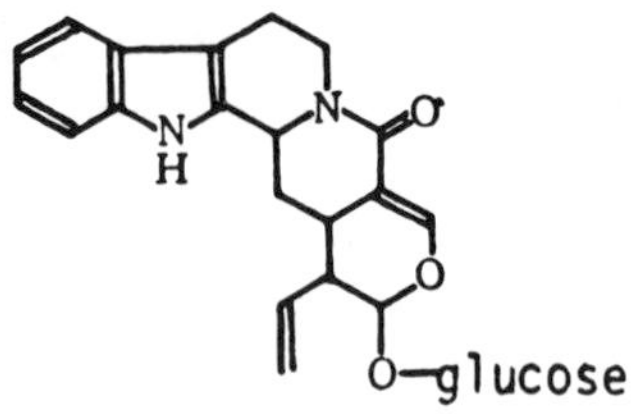

Camptoglucoside

Actions

Camptothecine and methoxycamptothecine have been found to inhibit the growth of L-1210, P-388, L-5178, and YK-1864 leukemia cells, Ehrlich ascites cells, and rat Walker sarcoma cells *in vitro*. Experiments with animals also showed that they can inhibit HeLa cells isolated from a patients' cervical carcinoma, L-16 cells from patients with hepatoma, and D-6 cells from lung cancer patients.

The mechanism of action involves inhibition of nucleic acid synthesis. These agents are crosstolerant to vincristine.

Toxicity

In patients, the toxic symptoms include bone marrow depression. Additionally, they may experience irritation of the gastrointestinal and urinary tracts, such as hematuria, which can be relieved with diuretic treatment.

The LD_{50} of camptothecine and methoxycamptothecine in mice is 68.4 mg/kg and 104 mg/kg (i.p.), respectively.

Therapeutic Uses

In Chinese folk medicine, this herb was used to treat carcinoma of the stomach, rectum, colon, and bladder, as well as treating chronic leukemia.

LEI GONG TENG (雷公藤) — dried root and leaf of *Tripterygium wilfordii* Hook.

Chemistry

It contains the alkaloid wilfordine and other substances, including as triptolide, A, B, and C ($C_{20}H_{24}O_6$), triptonide ($C_{20}H_{22}O_6$), and tripterin ($C_{29}H_{58}O_4$).

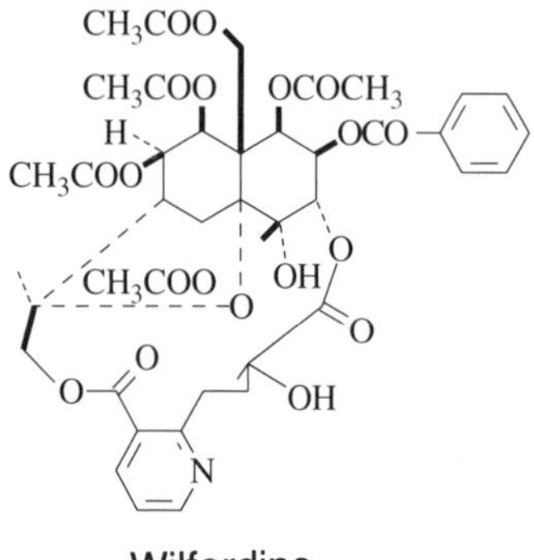

Wilfordine

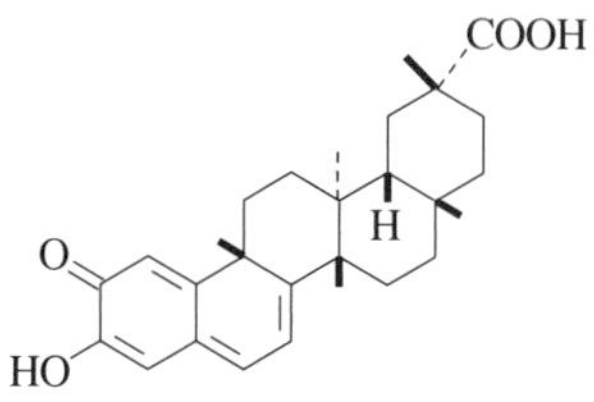

Tripterin

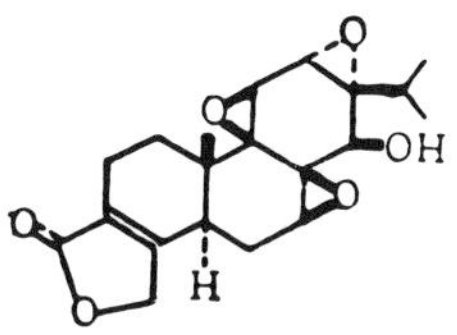

Triptclide Tripdiolide 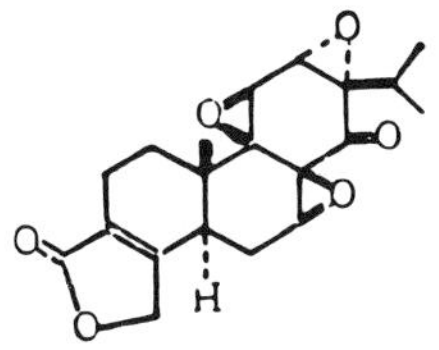 Triptonide

Actions

Experiments with animals show that triptolide A is the major active principle, with an ability to inhibit mouse L-615 leukemia cells and prolong the survival time of the animals. This agent is toxic; it can damage the cardiovascular system and the CNS.

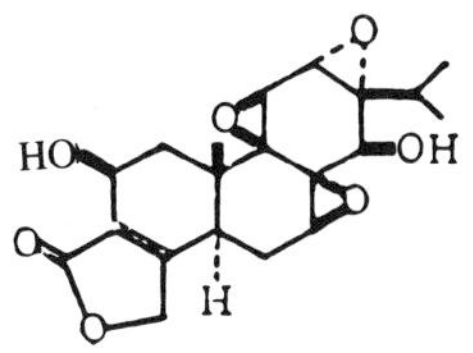

CHANG CHU HUA (长春花) — the dried whole plant of *Catharanthus roseus* (L.) G. Don (*Vinca rosea* L., *Lochnera rosa* Reichenb.)

FIGURE 3. *Chang Chu Hua.*

Chemistry

This plant is grown in the southern provinces of China. It contains more than 70 different alkaloids. The major ones are vinblastine (VLB), vincristine (VCR), vinrosidine, leurosine, leurosivine, rovidine, carosine, perivine, perividine, vindolinine, and pericalline.

Recently, a semisynthetic compound, vindensine (VDS), has been introduced for therapeutic use.

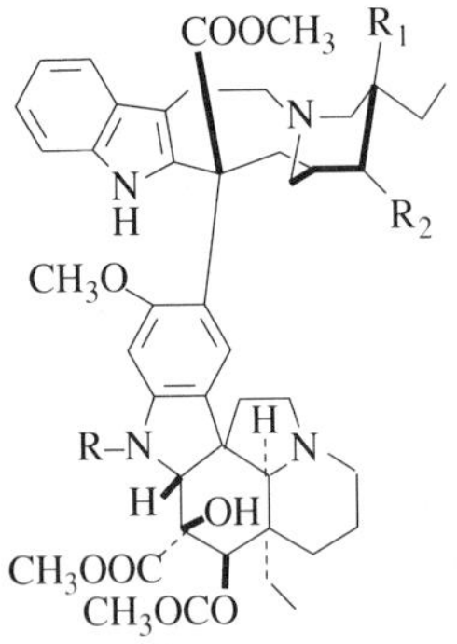

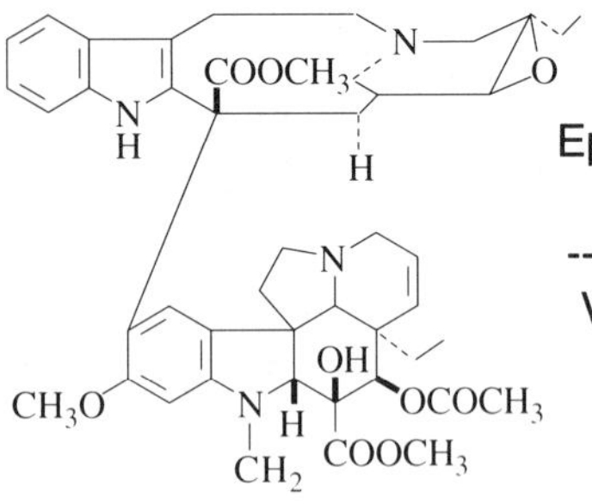

Epoxyvelbanamine moiety

Vindoline moiety

Leurosine

	R=	R_1=	R_2=
Vinblastine, VLB	CH_3	OH	H_2
Vincristine, VCR	CHO	OH	H_2
Vinrosicine, VRD	CH_3	H	α-OH

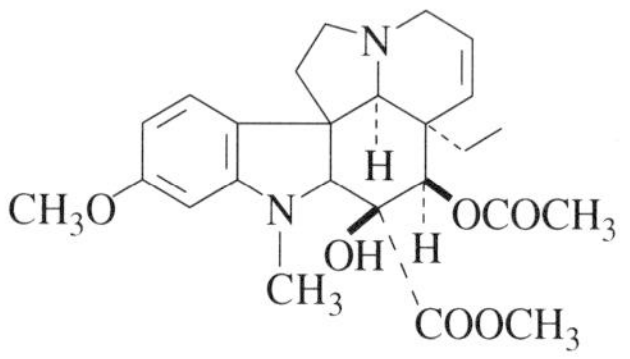

Vindoline

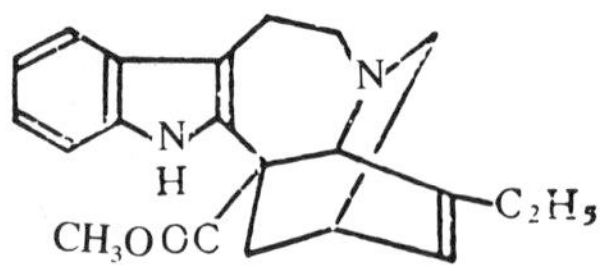

Catharanthine

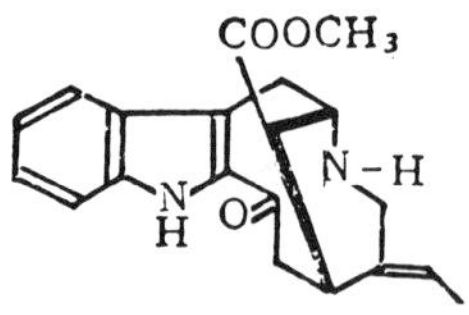

Perivine

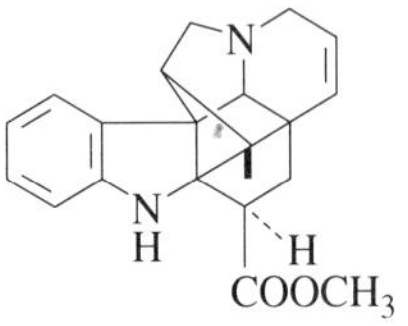

Vindolinine

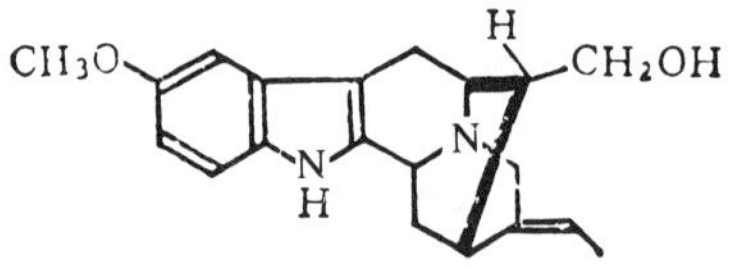

Lochnerine

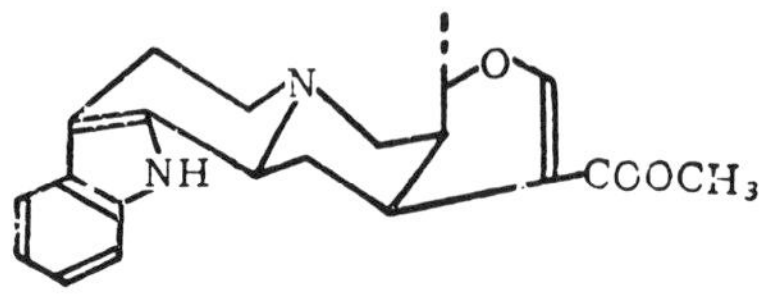

Tetrahydroalstonine

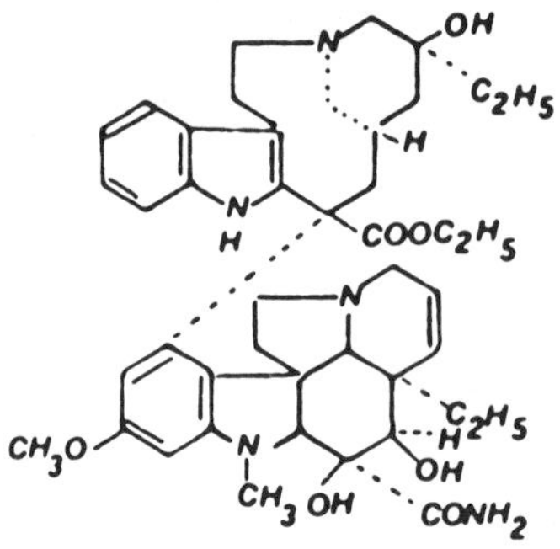

Vindensine

Actions

Vinca alkaloids are very effective against mouse leukemia, Ehrlich ascites cells, and sarcoma cells. They act on the mitotic stage of the cells by causing crystallization of the microtubular protein and interfering with cell division; this results in a metaphase arrest of tumor cells. Although it behaves similarly to other anticancer agents such as colchicine and maytansine, there is no cross resistance between these anticancer agents.

VLB can be converted into VCR and VDS with higher antitumor activity and less toxicity by oxidation and aminolysis, respectively. This leads to an increase of hydrogen-bonding site in the vindoline moiety, decreasing lipophilic activity.

The intracellular uptake of VCR is higher than that of VLB or VDS, especially in nerve cells and rat lymphoma and L-1578 cells. Thus, VCR has a greater relative neurotoxicity and antitumor activity than either VLB and VDS (see Lien and Li).

Small doses of the alkaloids raise thrombocyte count, while large doses decrease it. The herb increases blood pressure by stimulation of the vasomotor center in the medulla oblongata. It is a diuretic and can lower blood-sugar levels.

Toxicity

Vinca alkaloids are bone marrow depressants, especially to the formation of leukocytes and thrombocytes. Patients taking these alkaloids show some severe side effects, including loss of hair, nausea, vomiting, abdominal distention, and constipation. Occasionally, perineuritis occurs. The alkaloids are irritating to the blood vessels, and when administered intravenously, can cause thrombophlebitis.

The LD_{50} of VCR, VLB, and the semisynthetic VDS in mice are 2.1 mg/kg, 10 mg/kg, and 6.3 mg/kg (intravenous), respectively. The alkaloids can damage the nervous system of chickens.

Therapeutic Uses

VLB and VCR are commonly used for anticancer therapy in cases of chronic lymphocytic leukemia and Hodgkin's disease. VCR is less toxic than VLB, and has broader therapeutic applications. It is also used to treat acute lympocytic leukemia in children.

Clinical trials with VDS have shown remarkably promising results. Administration of this semisynthetic alkaloid produces a high remission rate in patients with acute lymphatic and chronic granulocytic leukemia; it also has a higher effective rate in the treatment of breast cancer than either VCR or VLB. The neurotoxicity caused by VDS is slightly less frequent than with either VCR or VLB.

The recommended dose for VDS is 1.0 to 1.3 mg/m^2 surface area, continued for 5 to 7 d. After resting for 3 to 4 weeks, a second course can be started with an increased dose of 3 mg/m^2; on occasion, this dose can be raised to 4 mg/m^2.

Reference

Lien, E. J. and Li, W. Y., in *Folk Medicine,* Steiner, R. P., Ed., American Chemical Society, Washington, D.C., 1986, 153.

MEI DENG MU (美登木) — the fruit, bark, and rhizome of *Maytenus serrata* or *M. buchananii, M. hookeri* Loes., or *M. conterliflories* J. Y. Lo et X. X. Chen.

Chemistry

The herb contains several substances, including maytansine, maytanprine, maytanbutine, maytanvaline, maytanacine, and maytansinol.

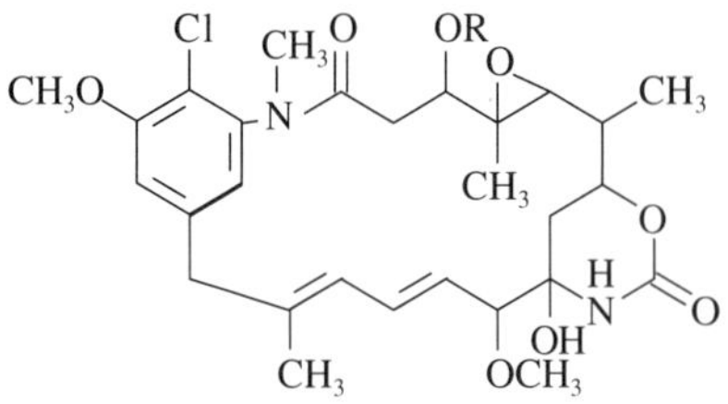

R =

Macrolides-maytanvaline: $-CO-CH(CH_3)-N(CH_3)-C(=O)-CH_2CH(CH_3)_2$

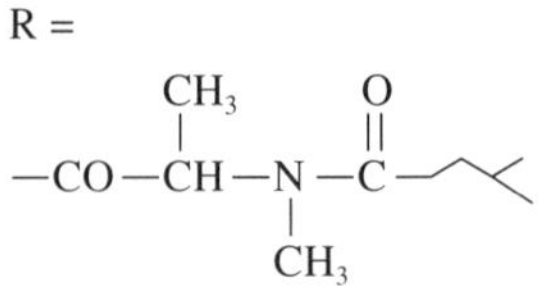

Maytansine: $-COCH(CH_3)-N(CH_3)-C(=O)-CH_3$

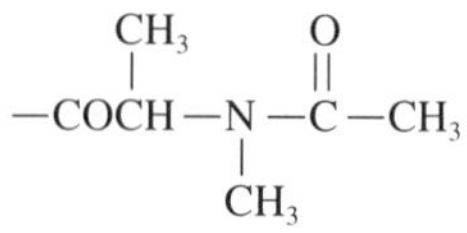

Actions

The anticancer activity of maytansine has been studied. It was found that this alkaloid can inhibit cell mitosis at a concentration of $6 \times 10^{-8}M$. It inhibits DNA, RNA, and protein synthesis in mouse leukemia cells. In cell-culture experiments, maytansine was found to be 20 to 100 times more potent than VCR in inhibiting cell growth. There was a lack of cross resistance between maytansine and the vinca alkaloids.

Toxicity and adverse effects of this agent occur mainly on the gastrointestinal tract. Nausea, vomiting, and diarrhea are common side effects; they are dose dependent. Clinically, some patients may develop liver toxicity. Thrombophlebitis may also be observed if the drug is given by parenteral administration.

Therapeutic Uses

This herb is used in the treatment of lung cancer, breast and ovary cancer, acute lympocytic leukemia, colon carcinoma, and kidney carcinoma. The maximal tolerated daily dose is 0.25 to 0.6 mg/m^2; the maximal weekly dose is 0.75 to 1.25 mg/m^2. A full course extending 3 to 4 weeks should not exceed 2 mg/m^2.

BRUCEANTIN (鸦胆宁) a quassinoid derivative, isolated from *Brucea* or *Ya Dan Zi* (鸦胆子) (see Antiamebial Herbs)

Chemistry

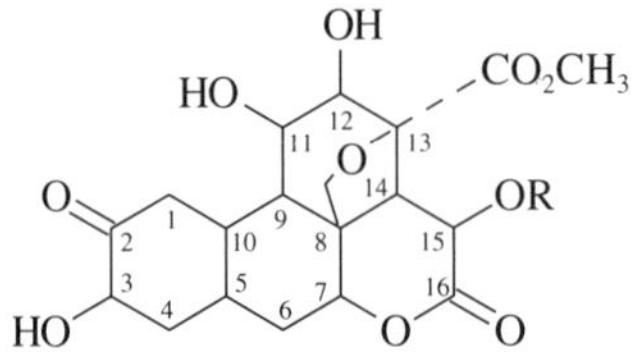

Name	Structure	Source
Bruceantin	$R = COCH{=}C(CH_3)-CH(CH_3)_2$	*Brucea antidysenteria*
Bruceantinol	$R = COCH{=}C(CH_3)-C(OH)(CH_3)_2$	*Brucea antidysenteria*
Bruceantarin	$R = CO-C_6H_5$ 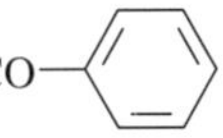	*Brucea antidysenteria*
Dihydrobruceantin	$R = COCH_2CHCH(CH_3)_2$	*Brucea antidysenteria*
Brucein A	$R = COCH_2CH(CH_3)_2$	*Brucea javanica*
Brusatol	$R = COCH{=}CH(CH_3)_2$	*Brucea javanica*
Bruceolide	$R = H$	*Brucea javanica*
Brucein B	$R = -COCH_3$	*Brucea javanica*

(From Lien and Li.[5])

Actions

Bruceantin is a very potent anticancer agent against mouse P-388 lymphocytic leukemia with cytotoxicity in the order of 10^{-2} μg/ml. This agent also significantly inhibits the growth of L-1210 lymphoid leukemia cells, Lewis lung carcinoma, and B-16 melanocarcinoma.

The primary mechanism of action is inhibition of protein synthesis. Bruceantin is a potent inhibitor of polyphenylalanine synthesis, with a secondary effect on DNA synthesis; it has little effect on RNA synthesis.

General References

1. **Chou, C. H., Ed.,** *Chinese Herbs Pharmacology,* Shanghai Science Technology Publisher, 1986, 302.
2. **Hsu, B.,** The use of herbs as anticancer agents, *Am. J. Chin. Med.,* 8(4), 302–306, 1980.
3. **Lee, K. H.,** Chinese plant antitumor agents, in *Advances in Chinese Medical Materials Research,* Chang, H. M. et al., Eds., World Scientific Publishing Co., Singapore, 353–367, 1985.
4. **Li, X. M. et al.,** *Acta Pharmacol. Sin.,* 8, 253–255, 1987.
5. **Lien, E. J. and Li, W. Y.,** Anticancer chinese drugs, in *Advances in Chinese Medical Materials Research,* Chang, H. M. et al., Ed., 1984, 433–452.
6. **Lien, E. J. and Li, W. Y.,** Anticancer chinese drugs, in *Folk Medicine,* Steiner, R. P., Ed., American Chemical Society, Washington, D.C., 1986, 153–168.
7. **Wu, P. C., Ed.,** *Pharmacology of Chinese Herbs* [in Chinese], People's Health Publisher, Beijing, 1982, 282–292.

TABLE 1
Other Herbs Which Possess Anticancer Activity

Chinese name	Source	Active principle	Actions
Yuan Hua (芫花)	*Daphne genkwa* Sieb.	Guidilatichin	Antileukemia
Liang Mian Chen (两面針)	*Zanthoxylum nitidum*	Nitidine Cl	Against mouse leukemia, L-1210 and P-388 lung carcinoma
Han Fang Ji (汉防己)	*Stephania tetranda*	Tetrandrine, fangchinoline	Walker-256 carcinoma
Yuan Xi Huang San (元喜黄山)	*Allamanda cathartica*	Allamandin	P-388 leukemia
Bai Hua She-She Cao (白花蛇舌草)	*Oldenlandia diffusa* Roxb.	Asperuloside, palderoside, desacetylasperuloside, oldenlandoside	Treats malignant tumors, stimulates reticuloendothelial system
Tan Seng (丹参)	*Salvia przewalskii*	Przewaquinone A, B	Against mouse tumors
Ma Lian Zi (马兰子)	*Iris pollasii* var. *chinesis*	Irisquinone	Cervix carcinoma, mouse U-14 cancer, hepatoma, lymphatic sarcoma, Ehrlich carcinoma
Ku Don Zi (古冬子)	*Sophora alopecurosides*	Sophocarpine, sophoridine	U-14, S-37, S-180 cancer cells
Pa Jiao Lian (八角蓮)	*Dysosma pleiantha*	Podophyllotoxin, deoxypodophyllotoxin, isopicropodophyllodine	TLX-5 cells, carcinoma of nasopharynx
Zhu Ling (豬苓)	*Polyporus umbellatus*	Zhu-ling polysaccharides	S-180 cells, has immunostimulating effect
Ling Zhi Cao (灵芝草)	*Ganoderma lucidum*	Polysaccharides Gl-1, Gl-2, and Gl-3	S-180 cells
Di Dan Tou (地胆头)	*Elephantopus elatus, E. scaber*, or *E. mollis*	Elephantopin, molephantinin, deoxoelphantopin	Against P-388 lymphocytic tumor, ascites cells

TABLE 1 (continued)
Other Herbs Which Possess Anticancer Activity

Chinese name	Source	Active principle	Actions
Ma Bo (马勃)	*Lasiosphaera nipponia*	Cavacin, uric acid	Treats larynx carcinoma and lung cancer
Feng Wei Cao (凤尾草)	Whole plant of *Pieris multifida*	Flavonoids	Treats uterus and bladder cancer
Long Kui (龙葵)	*Solanum nigrum*	Alkaloids, saponins	Treats stomach cancer, hepatoma, ascites cancer
Shei Yiang Mei Gen (水杨梅根)	Root of *Adina rubella*	β-Sitosterol, salicylic acid	Cervical cancer, lymphoma, and gastrointestinal cancer
Tian Kui Zi (天葵子)	*Semiaquilegia adoxoides*	Alkaloids	Malignant lymphoma; prostate, lung and bladder cancer
Shi Shang Bai (石上柏)	Leaf of *Selaginella doederleinii*	Alkaloids, sterol saponins, skikimic acid	Malignant hydatid mole, chorionic epithelioma
Ban Zhi Lian (半枝莲)	Whole plant of *Scutellaria barbata*	Alkaloids, flavonoids	Mouse S-180, Ehrlich ascites cells
Bai Ying (白英)	*Solanum lyratum*	Alkaloids	Treats uterus cancer, hepatoma
Fu Rong Yie (芙蓉叶)	Leaf of *Hibiscus mutabilis*	Flavonoids	Cancer of stomach, breast, and lung
Gou Shi Cao (狗舌草)	*Senecio campestris*	Alkaloids	Mouse L-1210 leukemia
Cao He Che (草河车)	Rhozome of *Polyganum bistoria*	β-Sitosterol, tannin	Inhibits mouse S-180
Zhu Yin Yin (猪殃殃)	*Galium spurium* L. var. *echinospermen*	Saponin	Leukemia, breast cancer
She Mei (蛇莓)	*Duchesnea indica*		Inhibits mouse S-180 and Ehrlich ascites cells, treats thyroid cancer and hepatoma
Mu Tom Hui (墓头回)	Root or whole plant of *Patrinia heterophylla* or *P. scabra*	Essential oil	Leukemia, cervix carcinoma
Chuen Gen Bi (椿根皮)	*Ailanthus altissima*	Quassin, saponin	Mouse S-180, cervix carcinoma and intestinal cancer
Lour Lu (漏芦)	*Rhaponticum uniflorum*	Essential oils	Hepatoma, stomach and breast cancer
Teng Li Gen (藤梨根)	Root and fruit of *Actinidia chinensis*	Alkaloids and vitamin C	Mouse S-180, stomach and esophagus cancer
Shan Dou Gen (山豆根)	*Sophora subprostrata* Chum.	Matrine, dauricine, anagyrine	Mouse S-180, leukemia, hepatoma, antimicrobial
Tian Long (天龙)	Whole body of *Gekko chinensis*	Toxin	Esophagus, stomach, and lung cancer
Shui Zhi (水蛭)	*Hirudo nipponica*	Anticoagulant substances	Cancer of ovary, cervix, stomach, and esophagus
Shui Hong Hua (水红花)	Fruit of *Polyganum orientale*	β-Sitosterol glucosides	Cancer of stomach, intestine, and liver
Shi Jian Chuan (石見穿)	Whole plant of *Salvia chinensis*	Sitosterol, and amino acids	Mouse S-180
Yang Ti Gen (羊蹄根)	Root of *Rumex japonica* or *R. crispus*	Emodin	Leukemia, malignant lymphoma
Liu Xin Zi (留行子)	Seeds of *Vaccaria pyramidata*	Saponins, alkaloids	Ehrlich ascites cells, breast cancer, hepatoma
Feng Xian Hua (凤仙花)	*Impatiens basamina*	Saponins, fatty acid	Esophagus and stomach cancer

TABLE 1 (continued)
Other Herbs Which Possess Anticancer Activity

Chinese name	Source	Active principle	Actions
Tie Shu Ye (铁树叶)	Leaf of *Cycas revoluta*	Glucosides, choline	Cancer of stomach, lung, liver, uterus, and nose
Kui Shu Zi (葵树子)	Seed of *Livistona chinensis*	Tannin, phenols	Chorionic epithelioma, esophageal cancer
Jia Mu (楤木)	Bark of *Aralia chinensis*	Saponins, essential oils	Cancer of the alimentary tract and gall bladder
Tian Dong (天冬)	Root of *Asparagus cochinensis*	β-Sitosterol	Mouse S-180 leukemia, lung cancer
Bu Gu Zhi (补骨脂)	Seed of *Psoralea corylifolia* L.	Furocumarin, psoralen, corylifolinin, bavachinin, etc.	Mouse S-180, Ehrlich ascites cells, bone sarcoma, lung cancer; treats neurosis, impotence
Gui Ban (龟板)	The abdominal plate of the water tortoise, *Chinemys (Geoclemys) reevesii*	Gelatin, Ca^{2+}, and phosphorus	Lymphoma, hepatoma; increases immunity to anticancer activity
He Tao Shu Zhi (核桃树枝)	Branch and unripe fruit skin of *Juglans regia*	Tannin, glucosides	Mouse S-37 cells and solid tumors
Bi Li Guo (薜荔果)	Fruit of *Ficus pumila*	β-Sitosterol, glucosides	Cervix, breast, prostate, testes, and colon cancer
Mian Hua Gen (棉花根)	Root of *Gossypium hirsutum, G. herbaceum*, or *G. arboreum*	Cottonphenol, $MgSO_4$	Mouse S-180, Walker sarcoma; cancer of the lung, liver, stomach, esophagus, and larynx
Yi Yi Ren (薏苡仁)	Seed of *Coix lachryma*	α, β-Sitosterol, fat, and amino acids	Mouse S-180, Yoshida sarcoma, Ehrlich ascites cells; lung, cervix cancer; chorionic epithelioma
Shan Ci Gu (山慈姑)	Tuber rhizome of *Iphigenia indica* Kunth.	Colchicine	Mouse S-180, Walker sarcoma; cancer of the breast, thyroid, and esophagus
Xia Ku Cao (夏枯草)	Flower petal of *Prunella vulgaris*	Saponins, alkaloids	Mouse S-180, cervix cancer-14, thyroid and breast cancer, hepatoma
Mao Zhua Cao (猫爪草)	Root of *Ranunculus ternatas*	Amino acids, organic acids	Mouse S-180, Sarcoma-37, Ehrlich ascites cells, lymphoma and thyroid cancer
Zao Ci (皂刺)	Needle of *Gleditsia sinensis*	Flavonoids	Mouse S-180, cancer of alimentary tract, breast, and cervix
Nan Xing (南星)	Rhizome of *Arisaema consanguineum, A. heterophyllum*, and *A. amurense*	Saponins, β-Sitosterol	Mouse S-180, cervix and esophageal cancer
Lu Sha (硇砂)	*Sal ammoniac*	Na^+, Mg^{2+} salt of ammonium,	Esophageal cancer
Shi Da Chuan (石打穿)	Whole plant of *Oldenlandia chrysotricha*	α, β-Sitosterol	Mouse U-14 cervix cancer
Ban Bian Lian (半边莲)	*Lobelia chinensis*	Alkaloids, glucosides, saponins	Mouse S-180, hepatoma, stomach and intestine cancer

TABLE 1 (continued)
Other Herbs Which Possess Anticancer Activity

Chinese name	Source	Active principle	Actions
Gang Ban Gui (杠板归)	Whole plant of *Polygonum perfoliatum*	Cardiac glucosides	Esophagus, gastrointestinal, and prostate cancer
Yie Pu Tao Teng (野葡萄藤)	Root and branch of *Ampelopsis brevipedunculata*	Flavonoids, phenols	Mouse S-180, cancer of alimentary tract and urinary tract, malignant lymphoma

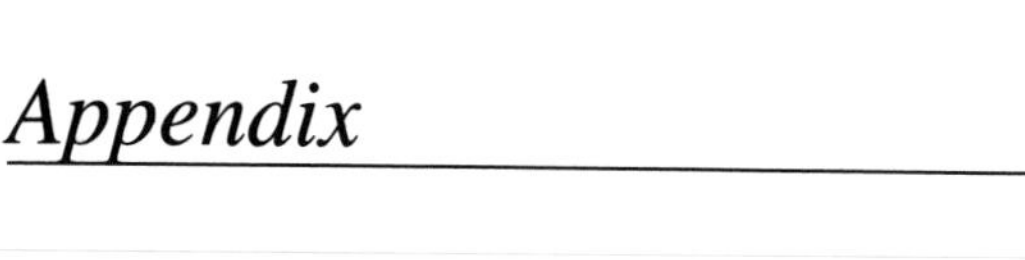

Appendix

APPENDIX

CHINESE EQUIVALENTS OF NAMES OF "HERBS"

A Wei	*Ferula assafoetida*
Ai Di Cha	*Ardisia japonica*
Ai Ye Yu	*Artemisia argyi*
An Xi Xiang	*Styrax tonkinensis*
Ba Dou	*Croton tiglium*
Ba Ji Tian	*Morinda officinalis*
Ba Jiao Feng	*Alangium chinense*
Ba Jiao Hui Xiang	*Illicium verum*
Ba Li Ma	*Rhododendron molle*
Bai Bu	*Stemona japonica*
Bai Guo	*Ginkgo biloba*
Bai He	*Lilium lancifolium*
Bai Hua She	*Agkistrodon acutus*
Bai Hua She She Cao	*Hedyotis diffusa, Oldenlandia diffusa*
Bai Ji	*Bletilla striata*
Bai Jiang Cao	*Patrinia scabiosifolia*
Bai Jie Zi	*Brassica alba*
Bai Lian	*Ampelopsis japonica*
Bai Mao Gen	*Imperata cylindrica*
Bai Qu Cai	*Chelidonium majus*
Bai Shao	*Paeonia lactiflora*
Bai Tou Weng	*Pulsatilla chinensis*
Bai Xian Pi	*Dictamnus dasycarpus*
Bai Yao Zi	*Stephania cepharantha*
Bai Ying	*Solanum lyratum*
Bai Zhi	*Angelica daphurica*
Ban Bian Lian	*Lobelia chinensis*
Ban Lan Gen	*Isatis tinctoria, Baphicacanthus cusia*
Ban Sao	*Mylabris phalerata*
Ban Xia	*Pinellia ternata*
Ban Zhi Lian	*Scutellaria barbata*
Bei Fan	*Alum*
Bei Mu	*Fritillaria verticillata*
Bi Ba	*Piper longum*
Bi Li Guo	*Ficus pumila*
Bi Ma Zi	*Ricinus communis*
Bian Xu	*Polygonum aviculare*
Bing Lang	*Areca catechu*
Bing Lian Lua	*Adonis amurensis*
Bo He	*Mentha haplocalyx*
Bo Lor Hui	*Macleaya cordata*
Bu Gu Zhi	*Psoralea corylifolia*
Can Cao	*Glycyrrhiza uralensis*
Cang Er Zi	*Xanthium sibiricum*
Cang Zhu	*Atractylodes lancea*
Cao Guo	*Amomum tsao-ko*
Cao He Che	*Polygonum bistorta*
Cao Wu	*Aconitum kusnezoffii*
Ce Bai Ye	*Biota orientale*
Cha	*Camellia sinensis*
Chai Hu	*Bupleureum chinense*
Chan Su	*Bufo bufo gargarizans*
Chang Bai Ruei Xiang	*Daphne koreane*
Chang Chu Hua	*Catharanthus roseus*
Chang Nao	*Cinnamomum camphora*
Chang Shan	*Dichroa febrifuga*
Chen Pi	*Citrus reticulata*
Chen Xiang	*Aquilaria sinensis*
Chi Shao	*Paeonia veitchii*
Chi Shi Zhi	*Halloysitum rubrum*
Chong	*Allium fistulosum*
Chou Wu Tong	*Clerodendron trichotomum*
Chu Mao Cai	*Salsola collina*
Chu Pen Cao	*Sedum sarmentosum*
Chuan Duan Chang Cao	*Corydalis incisa*
Chuan Lian Pi	*Melia toosendan, Melia azedarach*
Chuan Shan Long	*Dioscorea niponica*
Chuan Xiang	*Ligusticum chuanxiang*
Chuan Xin Lian	*Andrographis paniculata*
Chuen Gen Pi	*Ailanthus altissima*
Chun Jin Pi	*Hibiscus syriacus*
Chun Pi	*Ailanthus altissima*
Ci Ji Li	*Tribulus terrestris*
Ci Wu Jia	*Acanthopanax senticosus*
Da Feng Zi	*Hydnocarpus anthelmintica*
Da Ji	*Euphorbia pekinensis*
Da Qing Ye	*Isatis indigotica, Polygonum tinctorium*
Da Suan	*Allium sativum*
Da Ye An	*Eucalyptus robusta*
Dan Zhi	Bile concentrate
Dan Zhu Ye	*Lophatherum gracile*
Dang Gui	*Angelica sinensis*
Dao Dou	*Canavalia gladiata*
Deng Tai Ye	*Alstonia scholaris*
Di Dan Tou	*Elephantopus elatus*
Di Er Cao	*Hypericum japonicum*
Di Gu Pi	*Lycium chinense*
Di Huang	*Rehmannia glutinosa*
Di Jin Cao	*Euphorbia humifusa*
Di Long	*Pheretima aspergillam*
Ding Xiang	*Eugenia caryophyllata*
Dong Chong Xia Cao	*Cordyceps sinensis*
Dong Ling Cao	*Rabdosia rubescens*
Dong Yao	*Swertia pseudochinensis*
Dou Kou	*Alpinia katsumadai*
Du Huo	*Angelica pubescens*
Du Zhong	*Eucommia ulmoides*

CHINESE EQUIVALENTS OF NAMES OF "HERBS" (continued)

Duan Xue Liu	*Clinopodium polycephalum*
E Jiao	*Equus asinus*
E Zhu	*Curcuma zedoaria*
Er Cha	*Acacia catechu*
Fan Shi Liu	*Psidium guajava*
Fan Xie Ye	*Cassia angustifolia*
Fang Feng	*Ledebouriella divaricata*
Fang Ji	*Stephania tetrandra*
Feng Wei Cao	*Pteris multifida*
Feng Xian Hua	*Impatiens balsamina*
Fu Ling	*Portia cocos*
Fu Ping	*Spirodela polyrhiza*
Fu Rong Yie	*Hibiscus mutabilis*
Fu Shou Cao	*Adonis amurensis, Adonis chrysocyathus*
Fu Zi	*Radix aconiti praeparata*
Gan Shui	*Euphorbia kansui*
Gang Ban Gui	*Polygonum perfoliatum*
Gao Ben	*Ligustrum sinense*
Gao Liang Jiang	*Alpinia officinarum*
Ge Gen	*Pueraria lobata*
Giang Huo	*Notopterygium incisium*
Ginseng	*Panax ginseng*
Gou Gi	*Poncirus trifoliata*
Gou Gi Guan Zhong	*Woodwardia japonica*
Gou Mei	*Myrica rubra*
Gou Min	*Gelsemium sempervirens*
Gou Qi Zi	*Lycium barbarum*
Gou Shi Cao	*Senecio campestris*
Gou Teng	*Uncaria rhynchophylla*
Gu Ko Yi	*Erythroxylum coca*
Gua Di	*Cucumis melo*
Gua Lou	*Trichosanthes kirilowii*
Guan Zhong	*Dryopteris crassirhizoma*
Guang Dou Gen	*Sophora subprostrata*
Gui Ban	*Chinemys (Geoclemys) reevesii*
Gui Hua	*Osmanthus fragrans*
Gui Jian Yu	*Euonymus alatus*
Gui Zi	*Cinnamomum cassia*
Guo Gang Long	*Entada phaseoloides*
Hai Dai	*Laminaria japonica*
Hai Jin Sha Teng	*Lygodium japonicum*
Hai Piao Xiao	*Sepia esculenta*
Hai Tong Pi	*Erythrina variegata*
Hai Zao	*Sargassum pallidum*
Han Cai	*Rorippa indica*
Hang Fang Ji	*Stephania tetrandra*
Han Xiou Cao	*Mimosa pudica*
He Shi	*Carpesium abrotamoides*
He Shou Wu	*Polygonum multiflorum*
He Tao	*Juglans regia*
He Tao Shu Zhi	*Juglans regia*
He Ye	*Nelumbo nucifera*
He Zi	*Terminalia chebula*
Hong Hua	*Carthamus tinctorius*
Hong Teng	*Sargentodoxa cuneata*
Hong Tuan Yao	*Aster ageratoides*
Hou Po	*Magnolia officinalis*
Hu Chang	*Polygonum cuspidatum*
Hu Gu	Tigridis
Hu Ji Shang	*Viscum coloratum*
Hu Jiao	*Piper nigrum*
Hu Tao Ren	*Juglans regia*
Hu Tin Ye	*Elaeagnus pungens*
Hu Zhang	*Polygonum cuspidatum*
Hua Jiao	*Zanthoxylum schinifolium*
Hua Qian Jin Teng	*Stephania sinica*
Hua Shan Seng	*Physochlaina infundibularis*
Huai Hua	*Sophora japonica*
Huang Bai	*Phellodendron chinense*
Huang Hua Guo	*Artemesia annua*
Huang Hua Jia Zhu Tao	*Thevetia peruviana*
Huang Lian	*Coptis chinensis*
Huang Lu	*Cotinus coggygria*
Huang Qin	*Scutellaria baicalensis*
Huang Teng	*Fibraurea recisa*
Huang Yao Zi	*Dioscorea bulbifera*
Huo Ma Ren	*Cannabis sativa*
Huo Xiang	*Pogostemon cablin*
Huo Yu Jin	*Euphorbia antiquorum*
Japan Hang Fang Ji	*Sinomenium acutum*
Japan Mu Fang Ji	*Cocculus trilobus*
Je Koo Cai	*Calloglossa epieurii*
Ji Mu	*Loropetalum chinense*
Ji Nei Jin	*Gallus gallus*
Ji Xue Cao	*Centella asiatica*
Ji Xue Teng	*Spatholobus suberectus*
Jia Mu	*Aralia chinensis*
Jia Zhu Tao	*Nerium indicum*
Jian Can	*Bombyx mori*
Jian Sui Fuan Hou	*Antiaris toxicaris*
Jie Cai	*Capsella bursa-pastoris*
Jie Geng	*Platycodon grandiflorum*
Jin Deng Long	*Physalis alkekengi*
Jin Gi Er	*Caragana sinica*
Jin Gu Cao	*Ajuga decumbens*
Jin Quian Cao	*Lysimachia christinae*
Jin Xian Diao Wu Gui	*Stephania cepharantha*

CHINESE EQUIVALENTS OF NAMES OF "HERBS" (continued)

Jin Yin Hua	*Lonicera japonica*
Jin Ying Zi	*Rosa laevigata*
Jing Jie	*Schizonepeta tenuifolia*
Jing Tin San Qi	*Sedum aizoon*
Jiu Li Xiang	*Murraya paniculata*
Ju Hong	*Citrus reticulata*
Ju Hua	*Chrysanthemum morifolium, Chrysanthemum boreale*
Juan Bai	*Selaginella tamarisina*
Jue Ming Zi	*Cassia obtusifolia, Cassia tora*
Kong Xin Lian Zi Cao	*Alternanthera philoxeroides*
Ku Don Zi	*Sophora alopecurosides*
Ku Lian Pi	*Melia toosendan, Melia azedarach*
Ku Seng	*Sophora flavescens*
Kui Shu Zi	*Livistona chinensis*
Kun Bu	*Laminaria japonica*
La Liao	*Polygonum hydropiper*
Lao Jiun Xiu	*Usnea diffracta*
Lei Gong Teng	*Tripterygium wilfordii*
Lei Wan	*Omphalia lapidescens*
Li Lu	*Veratrum nigrum*
Lian Mian Zhen	*Zanthoxylum nitidum*
Lian Qiao	*Forsythia suspensa*
Lian Zi Xin	*Nelumbo nucifera*
Liao Ge Wang	*Wikstroemia indica*
Lie Xiang Du Juan	*Rhododendron anthopogonoides*
Ling Lan	*Convallaria keiskei*
Ling Yang Jiao	*Saiga tatarica*
Ling Zhi	*Ganoderma lucidum*
Liu Xin Zi	*Vaccaria pyramidata*
Liu Ye	*Salix babylonica*
Long Dan	*Gentiana manshurica*
Long Kui	*Solanum nigrum*
Lou Lu	*Rhaponticum uniflorum*
Lu Cao	*Humulus scandens*
Lu Rong	*Cervus nippon*
Lu Sha	Sal ammoniac
Lu Xian Cao	*Pyrola decorata, Pyrola rotundifolia chinensis*
Luo Bu Ma	*Apocynum venetum*
Luo De Da	*Centella asiatica*
Luo Fu Mu	*Rauwolfia verticillata*
Luo Han Guo	*Momordica grosvenori*
Huo Hua Xi Zhu	*Callicarpa nudiflora*
Luo Shi Teng	*Trachelospermum jasminoides*
Ma Bian Cao	*Verbena officinalis*
Ma Bo	*Lasiosphaera nipponica*
Ma Chi Xian	*Portulaca oleracea*
Ma Dou Ling	*Aristolochia debilis*
Ma Han Lian	*Eclipta prostrata*
Ma Huang	*Ephedra sinica*
Ma Lian Zi	*Iris pallasii*
Ma Quian Zi	*Strychnos pieriana*
Ma Wei Lian	*Thalictrum glandulissimum*
Mai Jiao	*Claviceps purpurea*
Mai She Cao	*Calloglossa epieurii*
Mai Ya	*Hordeum vulgare*
Man Shan Hong	*Rhododendron dahuricum*
Man Tao Luo	*Datura stramonium*
Mang Xiao	Sodium sulfate
Mao Dong Qing	*Ilex pubescens*
Mao Shua Cao	*Ranunculus ternatus*
Mar Dong	*Ophiopogon japonicus*
Mei Deng Mu	*Maytenus serrata*
Mi Meng Hua	*Buddleja officinalis*
Mian Hua Gen	*Gossypium hirsutum*
Mo Yao	*Commiphora myrrha*
Mu Dan Pi	*Paeonia suffruticosa*
Mu Fang Ji	*Cocculus thunbergii*
Mu Jing	*Vitex jeguado*
Mu Li	*Ostrea gigas*
Mu Tom Hui	*Patrinia heterophylla*
Mu Tong	*Clematis armandii*
Mu Xiang	*Aucklandia lappa*
Nan Gua Zi	*Cucurbita moschata*
Nan He Shi	*Daucus carota*
Nan Xing	*Arisaema consanguineum*
Nao Yang Hua	*Rhododendron molle*
Niu Bang	*Arctium lappa*
Niu Huang	*Bos taurus*
Nu Zhen Zi	*Ligustrum lucidum*
Pa Chiao Lien	*Dysosma pleiantha*
Pai Chien Cao	*Desmodium pulchellum*
Pei Go Su Ye	*Ginkgo Biloba*
Pei Lan	*Eupatorium odoratum*
Pi Jiu Hua	*Humulus lupulus*
Pin Di Mu	*Ardisia japonica*
Pu Gong Ying	*Taraxacum mongolicum*
Pu Huang	*Typha angustifolia*
Qian Cao	*Rubia cordifolia*
Qian Hu	*Peucedanum praeruptorum*
Qian Jin Zi	*Euphorbia lathyris*
Qian Jin Teng	*Stephania japonica*
Qian Li Guang	*Senecio scandens*
Qian Niu Zi	*Pharbitis nil*
Qian Ri Hong	*Gomphrena globosa*

CHINESE EQUIVALENTS OF NAMES OF "HERBS" (continued)

Qian Shi	*Euryala ferox*
Qin Cai	*Apium graveolens*
Qin Jiu	*Gentiana macrophylla*
Qin Mu Xing	*Aristolochia debilis*
Qin Pi	*Fraxinus rhynchopylla*
Qing Feng Teng	*Sinomenium acutum*
Qing Teng	*Sinomenium acutum*
Qing Ye Dan	*Swertia mileensis*
Qiong Zhi, Agar agar	*Gelidium amansii*
Qu Mai	*Dianthus superbus*
Quan Xie	*Buthus martensi*
Quan Ye Qing Lan	*Dracocephalum integrifolium*
Rou Dau Kou	*Myristica fragrans*
San Hai Ton	*Tripterygium hypoglaucum*
San Ke Zhen	*Berberis julianae*
San Jin Shan	*Cephalotaxus fortunei*
San Long Zhi	*Scopolia tangutica*
San Qi	*Panax zingiberensis*
Sang Zhi	*Morus alba*
Sha Ren	*Amomum villosum*
Sha Seng	*Adenophora tetraphylla*
Sha Yuan Zi	*Astragalus complanatus*
Shan Ci Gu	*Iphigenia indica*
Shan Cong Zi Oil	*Litsea cubeba*
Shan Dou Gen	*Menispermum dauricum*
Shan Dou Gen	*Sophora subprostrata*
Shan Zha	*Crataegus pinnatifida*
Shan Zhu Yu	*Cornus officinalis*
Shang Lu	*Phytolacca acinosa*
She Can	*Belamcanda chinensis*
She Chuang Zi	*Cnidium monnieri*
She Mei	*Duchesnea indica*
She Xiang	*Moschus sifanicus*
Shei Yiang Mei Gen	*Adina rubella*
Shen Jin Cao	*Lycopodium clavatum*
Sheng Jiang	*Zingiber officinale*
Sheng Ma	*Cimicifuga heracleifolia*
Shi Cao	*Achillea alpina*
Shi Da Chuan	*Oldenlandia chrysotricha*
Shi Diao Lan	*Lysionotus pauciflorus*
Shi Gao	*Gypsum fibrosum*
Shi Jian Chuan	*Salvia chinensis*
Shi Jiun Zi	*Quisqualis indica*
Shi Liu Pi	*Punica granatum*
Shi Shang Bai	*Selaginella doederleinii*
Shi Suan	*Lycoris radiata*
Shu Liang	*Dioscorea cirrhosa*
Shu Qu Cao	*Gnaphalium affine*
Shui Fei Ji	*Silybum marianum*
Shui Hong Cao	*Polygonum orientale*
Shui Hong Hua	*Polygonum orientale*
Si Ji Qing	*Ilex chinensis*
Song Ji Shang	*Loranthus parasiticus*
Song Ta	*Pinus bungeana*
Su Mu	*Caesalpinia sappan*
Suan Zao Ren	*Ziziphus spinosa*
Suo Luo Zi	*Aesculus chinensis*
Suo Yang	*Cynomorium songaricum*
Tai Huang	*Rheum palmatum*
Taiwan Pei Lan	*Eupatorium formosanum*
Tan Seng	*Salvia miltiorrhiza*
Tan Xiang	*Santalium album*
Tang Jie	*Erysimum cheiranthoides*
Teng Li Gen	*Actinidia chinensis*
Tian Dong	*Asparagus cochinensis*
Tian Hua Fen	*Trichosanthes kirilowii*
Tian Kui Zi	*Semiaquilegia adoxoides*
Tian Long	*Gekko chinensis*
Tian Ma	*Gastrodia elata*
Tie Shu	*Cycas revoluta*
Tie Xian Cai	*Acalypha australis*
Tong Guan Teng	*Marsdenia tenacissima*
Tou Gu Cao	*Impatiens balsamina*
Tu Si Zi	*Cuscuta chinensis*
Tu Tai Huang	*Rumex patientia*
Wa Leng Zi	*Arca subcrenata*
Wang Bu Liu Xing	*Vaccaria segetalis*
Wang Jiang Nan	*Cassia occidentalis*
Wei Ling Xian	*Clematis chinensis*
Wei Mao	*Euonymus alatus*
Won Nian Qing	*Rhodea japonica*
Wu Bei Zi	*Rhus chinensis*
Wu Gong	*Scolopendra subspinipes*
Wu Jia Pi	*Acanthopanax gracilistylus*
Wu Jiu	*Sapium sebiferum*
Wu Mei	*Prunus mume*
Wu Tong	*Firmiana simplex*
Wu Wei Zi	*Schisandra chinensis*
Wu Tao	*Radix aconiti praeparata*
Wu Yao	*Lindera strychnifolia*
Wu Zhu Yu	*Evodia rutaecarpa*
Xao Yie Pi Pa	*Rhododendron anthopogonoides*
Xi Sheng Teng	*Cissampelos pareira*
Xi Xin	*Asarum heterotropoides*
Xi Xian Cao	*Siegesbeckia orientalis*
Xia Ku Cao	*Prunella vulgaris*
Xia Tian Wu	*Corydalis decumbens*

CHINESE EQUIVALENTS OF NAMES OF "HERBS" (continued)

Xian He Cao	*Agrimonia pilosa*
Xian Mao	*Curculigo orchiodes*
Xiang Fu	*Cyperus rotundus*
Xiang Jia Pi	*Periploca sepium*
Xian Si Zi	*Abrus precatorius*
Xiao Hui Xiang	*Foeniculum vulgare*
Xiao Ji	*Cephalanoplos segetum*
Xin Yi Hua	*Magnolia liliflora*
Xing Ren	*Prunus armeniaca*
Xu Chang Qing	*Cynanchum paniculatum*
Xu Duan	*Dipsacus asper*
Xuan Seng	*Scrophularia mingpoensis*
Xue Shang Yi Zhi Gao	*Aconitum brachypodum*
Ya Dan Zi	*Brucea javenica*
Ya Zhi Cao	*Commelina communis*
Yan Hu Suo	*Corydalis turischaninovii*
Yang Guo Nau	*Strophanthus divaricatus*
Yang Ti Gen	*Rumex japonica*
Yao Jiu Hua	*Chrysanthemum indicum*
Ye Bai He	*Crotalaria sessiliflora*
Ye Dou Gen	*Menispermum dauricum*
Ye Huang Hua	*Patrinia scabiosaefolia*
Ye Kuan Mun	*Lespedeza cuneata*
Ye Xia Zhu	*Phyllanthus urinaria*
Yen Xing Leaf	*Ginkgo biloba*
Yi Mu Cao	*Leonurus heterophyllus*
Yi Ye Chau	*Securinega suffruticosa*
Yi Yi Ren	*Coix lachryma*
Yi Zhi Jian	*Ophioglossum vulgatum*
Yiang Jin Hua	*Datura tatula*
Yie Pu Tao Teng	*Ampelopsis brevipedunculata*
Yin Chen	*Artemisia capillaris*
Yin Yang Huo	*Epimedium brevicorum*
Yu Gan Zi	*Phyllanthus emblica*
Yu Jin	*Curcuma aromatica*
Yu Mi Xiu	*Zea mays*
Yu Xing Cao	*Houttuynia cordata*
Yuan Hua	*Daphne genkwa*
Yuan Zhi	*Polygala tenuifolia*
Yuan Xi Huang San	*Allamanda cathartica*
Yue Li Ren	Bush chery
Yun Xiang Cao	*Cymbopogon distans*
Ze Qi	*Euphorbia helioscopia*
Ze Xie	*Alisma orientalis, Alisma plantago-aquatica*
Zhang Zhu	*Atractylis orata*
Zhi Mu	*Anemarrhena asphodeloides*
Zhi Qiao	*Citrus aurantium*
Zhi Shi	*Citrus aurantium*
Zhi Zi	*Gardenia jasminoides*
Zhu Je Ginseng	*Panax japonicum*
Zhong Jie Feng	*Sarcandra glabra*
Zhu Ling	*Polyporus umbellatus*
Zhu Zi Tou	*Crotalaria mucronata*
Zi Cao	*Arnebia euchroma*
Zi Kee Guan Zhong	*Osmunda japonica*
Zi Su	*Perilla frutescens*
Zi Yu	*Sanguisorba officinalis*
Zi Zhu Cao	*Callicarpa macrophylla*
Zu Si Ma	*Daphne giraldii*

Indexes

INDEX TO HERBS

A

B

Q

R

S

T

U

V

W

X

Z

O

P

Q

R

S

T

U

V

W

X

Y

Z